STATISTICAL THEORY OF SAMPLE SURVEY DESIGN AND ANALYSIS

STATISTICAL THEORY OF SAMPLE SURVEY DESIGN AND ANALYSIS

BY

H. S. KONIJN

Tel-Aviv University, Israel

1973

NORTH-HOLLAND PUBLISHING COMPANY – AMSTERDAM – LONDON

AMERICAN ELSEVIER PUBLISHING COMPANY, INC. – NEW YORK

Library of Congress Catalog Card Number: 73-146183
North-Holland ISBN: 0 7204 2364 3
American Elsevier ISBN: 0 444 10486 0

Publishers:

NORTH-HOLLAND PUBLISHING COMPANY – AMSTERDAM
NORTH-HOLLAND PUBLISHING COMPANY, LTD. – LONDON

Sole distributors for the U.S.A. and Canada:
AMERICAN ELSEVIER PUBLISHING COMPANY, INC.
52 VANDERBILT AVENUE
NEW YORK, N.Y. 10017

PRINTED IN ENGLAND

CONTENTS

Foreword

This treatise is directed both to professionals and to students. *Professionals* may, of course, skip many of its pages, although at the risk of missing some scattered points that may be of value to them. Incidentally, they will notice that some matters commonly presented have been played down or omitted; in many cases the reason is that I do not consider that they have (or continue to have) much value. In my judgement it is also not yet justified to devote, in a work like the present one, space to the few recent attempts at formulating or utilizing ideas of optimality appropriate to the field of sample surveys. As a rule, bibliographical references have been given only in case significantly more material on the subject under discussion may be found there.

I shall now set out my *pedagogical* aims. The students I have in mind are those who have had a previous elementary introduction to probability and statistics. Not that I have relied much on such previous knowledge. On the contrary, convinced that very few students fully grasp the basics of that subject at their first exposure, I have carefully set out nearly all fundamental matters contained in the book (even though, sometimes in a rather condensed form). But more than that – I have aimed at presenting the theory of sampling surveys as a *bridge* between the most elementary introduction to probability and statistics on the one hand, and the more sophisticated and abstract uses of probability in statistical inference on the other.

In survey theory most of the probabilities are not highly abstract entities imputed by some theory to "nature"; they are there by the *design* of the surveyor, and are therefore reducible to the use of a table of random numbers (or equivalently to the use of some simple chance mechanism such as was familiar to the founders of the theory of probability). Careful study of survey theory should therefore enable the student to grasp very thoroughly the meaning of sampling distributions of statistics and related matters – before he has to rise very high into the more rarefied and sometimes treacherous atmosphere of "statistical inference". I have, however, also attempted to aid the student in beginning this ascent. For one thing, no other aids seem to be available to him that make clear how to make the transition. For another, the modern theory of surveys also requires the use of these more abstract notions.

Other matters which may be easier to grasp in connection with surveys than with experiments are: the safe use of not fully established a priori data without sacrificing objectivity, and the inevitable presence in statistical investigations of considerations of expenditure of money, effort, time or resources.

The *mathematical knowledge requisite* for reading of the present work rarely goes beyond what is learned in general high school courses. Only occasionally is it necessary or even helpful to use elementary calculus. It is only fair to say, however, that the argument can sometimes get rather involved, and may be beyond the capacities of most of those who have had only such a high school course. Nevertheless, it is hoped that the absence of "advanced" mathematics in virtually all of the book may enable the student to focus on essentials. Notations have been kept rather explicit and consistent, even at the expense of losing typographical simplicity.

A guide to a number of possible uses of the book in different *courses* of study follow this foreword. Some explanations occur in more than one place in order to make it possible for some of the chapters to be read independently of others. In the earlier chapters I have exemplified fairly many results by simple numerical applications, but I have assumed that these will be much less necessary after the student has mastered these chapters.

I have taught this subject off and on over a period of years, and have always found it hard to avoid overwhelming the students with details. On the other hand, detailed proofs are quite necessary for him, since he must be prepared to derive any of the almost unlimited number of variants and combinations of methods that commonly arise in practice. With a detailed treatment available, the teacher can spend more class time emphasizing main points. He can also devote more of the lecture time to examples that are live *to him.* Anyway, many examples suitable for use in class tend to "fade" on the printed page.

Most of the material in the book has been used during the last few years in the way referred to above and much of the remainder in advanced courses and seminars. An early draft and subsequently two revised editions appeared in Hebrew. They include results in the literature that came to hand up to 1967. Regretfully, it has not been feasible to include in the present work most of the contributions published since then; however, in a few cases these appear to have been anticipated by the present author.

I am obliged to students, assistants and colleagues for provoking clarifications, among these particularly D. Bradu of the Hebrew University and G. Natan, Chief of the Methods Section of the Central Bureau of Statistics of the State of Israel.

H. S. KONIJN

Uses of the Book for Various Courses of Study

1. The following sections constitute a short introductory and fairly elementary course on the subject:

Chapter I.	Sections 1–6, 8–10, possibly omitting 5.5.
Chapter II.	Sections 1–10, 11.1–11.5, 12–14, 15.1 (omitting small print), 15.2, 15.3, 15.6, 15.7.
Chapter III.	Sections 1–3, 4 (except for Remark), 5, 6.1 (except for the discussion of bias and the Note at the end), 6.2 (except for the calculation of bias), 8, 11.1, 12.1, 12.4, 13.
Chapter X.	Sections 1, 2.1, 2.2, 2.5.
Chapter XI.	Section 1.

2. A longer course at the same level might include in addition:

Chapter II.	Sections 15.8, 15.10, 15.11.
Chapter IIA.	Sections 1–3, 6, 8, 10, 13.1, 14.
Chapter III.	Sections 6.1 (discussion of bias, not the Note), 6.2 (bias), 6.3.
Chapter V.	Sections 1, 2, 4.1–4.4.
Chapter VII	Sections 1, 2, 3.1.
Chapter VIII.	Sections 1–5, 6.1, 6.2, 7, 8 (without small print).
Chapter IX.	Sections 1, 2, 3.1, 4, 5, 8, 11.
Chapter X.	Section 2.3.
Chapter XII.	Sections 1, 2, 3 (first two paragraphs only), 4 (first four paragraphs only).

3. If time permits one might add to that:

Chapter II.	Section 16 (up to (ii)).
Chapter IIA.	Sections 12, 13.2.
Chapter III.	Sections 7 (without Remark), 15, 16.
Chapter XI.	Section 2.
Chapter XII.	Section 3 (third paragraph).

CHAPTER I

INTRODUCTION

1. Fields of application of sample survey theory and techniques; populations, some of their overall properties; samples

Sample surveys are or should be widely used as a means of gathering information in government, industry and trade, the investigation of social, psychological, educational and health problems, physical and life sciences and technology, and nowadays even in humanistic studies (history and archaeology, study of languages, literature, religion and the arts). Moreover, statistical techniques and theory for their design and analysis have *at least partial* applicability to investigations which are not usually referred to as surveys, such as quality control and inspection, auditing, experimentation and computer simulation. It is not possible to include in this book a full range of applications, but it is believed that the student who has acquired a firm grasp of the *theory* here presented – and who is familiar with a field of application – will be well equipped to use or develop techniques suitable to it.

We shall study problems of the following kind: We are faced with a (usually large) collection $\mathscr{P}$ (called *population*),

$$\mathscr{P} = \{u_1, . . ., u_N\}$$

of objects $u_1, u_2, . . ., u_N$ (called *units* or *elements*), of which we wish to know some *property* $\mathscr{Y}$, defined for each u_i.† However, we are not interested in knowing $\mathscr{Y}$ for each or certain given u_i, but are interested in some *overall property* of the numbers‡

$$y(u_1), . . ., y(u_N),$$

† Often but not always the ordering in $\mathscr{P}$ is arbitrary, in the sense that there is no property of interest or auxiliary property whose value for any u_i is a given function of i.

‡ Usually N is known; we shall often make use of this fact. However, there are cases in which N is unknown. For most of the methods for estimating N the sample size is not determined in advance; such methods are not discussed here. See, e.g., SAMUEL, Ann. Math. Statist., **39** (1968) 1057–1068, together with references quoted there, and its sequel in Biometrics, **25** (1969) 517–528; KINDAHL, J. Amer. Statist. Ass., **57** (1962) 61–91. However, some fixed sample-size methods are given in the footnote of section 13.3 of Chapter II, in the last paragraphs of section 10 of Chapter IIA and in the first footnote of Chapter IV.

usually a property which is independent of the numbering of the elements of $\mathscr{P}$ (a symmetric function of the u's).

Much of the time interest will center on the following overall properties:†

the *total* of the values: $Y = y(u_1) + \ldots + y(u_N)$,
the *average* of the values: $\bar{Y} = Y/N$,

a measure of *spread* (*dispersion*) of the values:

$$\sigma^2(\mathscr{Y}) = \sum_{i=1}^{N} [y(u_i) - \bar{Y}]^2/N,$$

a measure of their *relative spread* (relative to $|\bar{Y}|$, assumed $\neq 0$): $\sigma^2(\mathscr{Y})/\bar{Y}^2$.

$\mathscr{Y}$ will always be a numerical property. Sometimes we are interested in several properties: $\mathscr{Y}$, $\mathscr{Z}$, etc. (Some properties may be of interest only as *auxiliaries* in the estimation of others.) A measure of *joint dispersion* between $\mathscr{Y}$ and $\mathscr{Z}$ of the units in $\mathscr{P}$ is

$$\sigma(\mathscr{Y}\mathscr{Z}) = \sum_{i=1}^{N} \{y(u_i) - \bar{Y}\}\{z(u_i) - \bar{Z}\}/N;$$

and $\rho(\mathscr{Y}\mathscr{Z})$, the *correlation coefficient* is defined, when $\sigma(\mathscr{Y})$ and $\sigma(\mathscr{Z})$ do not vanish, as

$$\rho(\mathscr{Y}\mathscr{Z}) = \sigma(\mathscr{Y}\mathscr{Z})/\{\sigma(\mathscr{Y})\sigma(\mathscr{Z})\}.$$

When $X \neq 0$ we are often also interested in the *ratio*

$$R = Y/X.$$

We assume that it is either not feasible (e.g., if measurements of the elements destroy the elements), or that it is too time consuming or too costly to measure $\mathscr{Y}$, etc., on each‡ u_i; or finally that, if performed on all elements, measurements would be too inexact.§ We therefore examine $\mathscr{Y}$, etc., on only some of the elements (a *sample* of them). We shall have to discuss how to *obtain* such a sample, and how to *estimate* from a sample quantities like Y or $\bar{Y}$.

2. Descriptive surveys

Let us give some examples of surveys conducted in order to estimate some overall properties of given populations.

† Some other properties are discussed, e.g. in sections 12 and 16 of Chapter II.

‡ Even when we observe *some* property on *all* elements of $\mathscr{P}$. For example, often part of the information collected during a Census is obtained on a sampling basis.

§ Among the costs of and lack of exactness in measuring we include here those of gathering and identifying the units, of recording the results, and of obtaining the estimates.

EXAMPLE 1.

$\mathscr{P}$ = the set of students enrolled in a certain university during the current term.

$\mathscr{Y}$ = stature as measured to within $\frac{1}{2}$ cm (without shoes).

We want to know $\bar{Y}$ and $\sigma^2(\mathscr{Y})$ to within a certain accuracy, and time does not allow measuring all the students.

EXAMPLE 2. Let $\mathscr{P}$ be as above. Let $\mathscr{Z}$ be the property of emotional balance, and let us distinguish two classes: $z^{(1)}$ – well-balanced, and $z^{(2)}$ – not well-balanced. Let $\mathscr{Y}$ be defined as follows:

$$y(u_i) = 1 \quad \text{if } u_i \text{ has property } z^{(1)} \text{ [i.e. if } z(u_i) = z^{(1)}\text{]},$$
$$y(u_i) = 0 \quad \text{otherwise.}$$

Then

$$Y = \sum_{i=1}^{N} y(u_i)$$

is the number of emotionally well-balanced students registered in this university during the current term, and $\bar{Y}$ the proportion of such students. In this case $\sigma^2(\mathscr{Y}) = \bar{Y}(1 - \bar{Y})$, since†

$$\sigma^2(\mathscr{Y}) = \frac{1}{N}\sum_{i=1}^{N} y^2(u_i) - \bar{Y}^2$$
$$= \frac{1}{N}\sum_{i=1}^{N} y(u_i) - \bar{Y}^2 = \bar{Y} - \bar{Y}^2 = \bar{Y}(1 - \bar{Y}).$$

Suppose the determination of $\mathscr{Y}$ by the method to be adopted takes on the average a given amount of time, too large to make it possible to examine all students. We wish to assess $\bar{Y}$ and $\sigma^2(\mathscr{Y})$ with certain accuracies.

EXAMPLE 3. Same as Example 2, except that there are different methods of assessing $\mathscr{Z}$ which are not equally precise and take different amounts of time. One crude method allows examining all students.

The latter is perhaps an extreme example of varying degrees of exactness or costs. But in very many cases there is a range of available methods, leading to different costs and degrees of exactness; a complete census is by no means always the best thing to do, even if feasible.

EXAMPLE 4. Same as Example 2, but in addition we administer an intelligence test. Let $x(u_i)$ be the intelligence quotient of student i as measured by the test we administer. It may be that it is quite feasible and economical to measure $\mathscr{X}$ for all students, but not $\mathscr{Z}$.

† As $y^2(u_i)$ equals 1 if $y(u_i) = 1$ and 0 if $y(u_i) = 0$, $\mathscr{Y}^2 = \mathscr{Y}$.

EXAMPLE 5. Same as Example 1, but in addition we record $x(u_i) = 1$ if u_i is male and 0 otherwise. We may be interested in

$$\bar{Y}_{\mathrm{m}} = \frac{\Sigma y(u_i)\, x(u_i)}{\Sigma x(u_i)} \quad \text{and} \quad \bar{Y}_{\mathrm{f}} = \frac{\Sigma y(u_i)\{1 - x(u_i)\}}{\Sigma\{1 - x(u_i)\}},$$

the average stature of male and of female students, respectively.

EXAMPLE 6. Same as Example 2, but we adopt a trichotomous scale for $\mathscr{Z}$:

$z^{(1)}$ (well-balanced),
$z^{(2)}$ (moderately balanced),
$z^{(3)}$ (ill-balanced).

Let

$$y(u_i) = \begin{cases} 1 & \text{if } z(u_i) = z^{(1)}, \\ 0 & \text{if } z(u_i) \neq z^{(1)}; \end{cases}$$

$$x(u_i) = \begin{cases} 1 & \text{if } z(u_i) = z^{(2)}, \\ 0 & \text{if } z(u_i) \neq z^{(2)}. \end{cases}$$

We may wish to know $\bar{Y}$ and $\bar{X}$; or $Y/(Y + X)$, the proportion of well-balanced among well or moderately balanced.

3. Analytic surveys

We may be interested in *relations between various properties*, $\mathscr{Y}, \mathscr{X}, \ldots$, *of the units*. A survey conducted primarily with a view to such relations is called an "analytic survey", in contrast to a "descriptive survey". The term "analytic" refers to the fact that the interest in relations usually arises out of a theoretical framework. On the other hand, the interest in totals, averages, measures of dispersion, and even ratios, often arises out of the need for direct† knowledge of such quantities (e.g., if the purpose of the survey is to answer the question: "How many children are of such an age that they have to enter elementary school next September?"). There are, however, many exceptions.

EXAMPLE 7. Let $\mathscr{P}$ be as before. $\mathscr{X}$ is annual income (including income received in terms of free lodging, food etc.) of a student, and $\mathscr{Y}$ annual expenditure on books of a student. If $\mathscr{Z}$ is defined by

$$z(u_i) = y(u_i)/x(u_i),$$

† Note, however, that it may well be more efficient to estimate averages, sums, or measures of dispersion via an assessment of such relations than to estimate them by a direct method. See section 10, and Chapter VIII, especially section 4.

we may be interested† in $\bar{Z}$, the average ratio. (In this case we would also want to characterize the amount of deviation in $\mathscr{P}$ from this average relation, e.g. by $\Sigma[z(u_i) - \bar{Z}]^2/N$.) Note the difference between $\bar{Z}$ and the relation $R = \bar{Y}/\bar{X}$ between the averages for the population. Actually R also equals the average of the ratios *weighted* by incomes, but incomes need not, of course, constitute the most appropriate weights in all contexts that call for weighting.

EXAMPLE 8. Same as Example 7, but we now postulate two classes of expenditure on books, one which does not depend on income (fixed outlays) and one which is proportional to income (variable outlays). This may be expressed in the form

$$y(u_i) = z_0(u_i) + z(u_i)\, x(u_i),$$

and we may wish to know $\bar{Z}$ (and perhaps secondarily $\bar{Z}_0$). In particular, if we write

$$w(u_i) = \bar{Z}_0 + \bar{Z}\, x(u_i),$$

we may expect that

$$\tau^2 = \Sigma[y(u_i) - w(u_i)]^2/N$$

is small.

EXAMPLE 9. Same as Example 5, but we are interested not so much in assessing $\bar{Y}_m$ and $\bar{Y}_f$ with a certain precision, as in assessing $\bar{Y}_m - \bar{Y}_f$ with a certain precision.

$\bar{Y}_m - \bar{Y}_f$ is a way of expressing a relation between $\mathscr{Y}$ (stature) and $\mathscr{X}$ (sex). Indeed (applying the notations w, $\bar{Z}_0$ and $\bar{Z}$ of the previous example to $\mathscr{Y}$ = stature and $\mathscr{X}$ = sex), if we average the $y(u_i)$ and the $w(u_i)$ over all u_i for which $x(u_i) = 0$, we get $\bar{Y}_f$ and $\bar{Z}_0$ respectively; whereas, if we average the $y(u_i)$ and the $w(u_i)$ over all u_i for which $x(u_i) = 1$, we get $\bar{Y}_m$ and $\bar{Z}_0 + \bar{Z}$ respectively. For small τ^2 we may approximately identify the corresponding averages, so

$$\bar{Z} \approx \bar{Y}_m - \bar{Y}_f, \quad \bar{Z}_0 \approx \bar{Y}_f$$

with interest centred on $\bar{Z}$.

Of course, surveys may be both descriptive and analytic. Question: Why does a survey designed to estimate $Y/(Y + X)$ of Example 6 not fall under the present heading?

† Often we are interested in $\bar{Z}$ for $\mathscr{X}$ defined as income above some minimum sum.

4. Repeated surveys

EXAMPLE 10. Let $\mathscr{P}$ be as before, and $\mathscr{Y}$ as in Example 2 for $\mathscr{Z}$ measured at the beginning of the year. Let $\mathscr{X}$ be $\mathscr{Z}$ measured in the same manner at the end of the year. (For simplicity assume no drop-outs.) We are interested in whether there is a change in emotional balance. More precisely we may be simply interested in whether $\bar{X}$ is larger or smaller than $\bar{Y}$, or we may be interested in

$$\bar{A} = \Sigma\{y(u_i) - x(u_i)\}/N,$$

summed over those u_i for which $y(u_i) > 0$; or in

$$\bar{B} = \Sigma\{x(u_i) - y(u_i)\}/N,$$

summed over those u_i for which $y(u_i) = 0$.

$N\bar{A}$ is the number of students that became unbalanced, whereas $N\bar{B}$ is the number that became balanced during the year. (There is no implication that the sole cause was attendance at the University! If this is the object of study, further comparisons with comparable non-students have to be made.) Hopefully, $\bar{B}$ is large and $\bar{A}$ is small. [Question: Is there a relation between $\bar{X} - \bar{Y}$ and $\bar{B} - \bar{A}$?] We may, of course, be interested in $\bar{X}$, $\bar{Y}$ *and* $\bar{X} - \bar{Y}$ (or $\bar{A}$ *and* $\bar{B}$).

If we cannot examine all students at both times, one of the special questions arising is: Should we examine the same students at the beginning and the end of the year, or not? If the purpose is analytic we must study identical students, but for descriptive purposes this may not be best.

One of the special features of repeated surveys is that the knowledge gained in one survey can be used to make a later survey more efficient than it could be if no previous survey had been carried out.

5. Sampling theory†

5.1. SAMPLE SPACE, PROBABILITY FUNCTION, DESIGN

Sampling theory is firstly concerned with ways of obtaining *samples*, i.e., subsets from $\{u_1, \ldots, u_N\}$, or sequences of elements taken from $\{u_1, \ldots, u_N\}$, in order to estimate functions like Y, $\bar{Y}$, $\sigma^2(\mathscr{Y})$, Y/X. A useful theory can be obtained on the basis of the theory of probability. In this theory we conceive of listing all possible different samples consistent with the method of sampling adopted, calling it the *sample space* $\mathscr{S}$. The possible different samples are

† It is assumed that students have already some familiarity with most of the concepts referred to in this section.

called *sample points* (to continue the geometrical metaphor). The method of sampling must also define the probability $P(\mathscr{s})$ that we will draw any particular sample $\mathscr{s}$, and

$$\sum_{\mathscr{s} \in \mathscr{S}} P(\mathscr{s}) = 1.$$

($\mathscr{s} \in \mathscr{S}$ stands for: $\mathscr{s}$ belongs to $\mathscr{S}$.) P is called the *probability function.*†

Actually, when specifying a method of sampling, it is rare that one gives an explicit list of all possible different samples $\mathscr{s}$ together with the corresponding values of $P(\mathscr{s})$; instead one states a way of generating $\mathscr{S}$ and $P(\mathscr{s})$ from a smaller number of specifications. For example, one may state that n observations are to be drawn by taking one observation u_i from $\mathscr{P}$ with probability $1/N$; taking, independently of the first draw, a second one from $\mathscr{P} - \{u_i\}$ with probability $1/(N-1)$; etc. In this case the sample space consists of all different possible‡ ordered subsets of n distinct u's from $\mathscr{P}$ and

$$P(\mathscr{s}) = \frac{1}{N(N-1)\ldots(N-n+1)} = \frac{1}{N^{(n)}}$$

for all $\mathscr{s}$ in $\mathscr{S}$. This is called *simple random sampling* from $\mathscr{P}$ (of samples of n) without replacement. "Simple" refers to the fact that $P(\mathscr{s})$ is a constant. "Without replacement" stands in contrast with the method in which one takes an element u_i from $\mathscr{P}$ with a given positive probability, replaces it, takes again an element from $\mathscr{P}$ with a prescribed probability (usually independently of the outcome of the first draw), replaces it, etc., doing his operation n times; this is called "sampling of n with replacement". How to carry out such processes to obtain the specified probabilities will be discussed in the next section.§ The choice of sampling method (*design*) will be discussed in most of the chapters following Chapter IIA, but especially in Chapters III, VII and IX.

† It is sometimes useful to consider more than one probability function over the same sample space. In this case $\mathscr{S}$ may contain points to which some of those functions assign zero probability.

‡ Methods of random sampling in which some of these subsets (or sequences) cannot be obtained are called *restricted*, in contrast to the unrestricted random sampling methods described above. Restricted methods are discussed in the chapters on stratified sampling, cluster sampling and systematic sampling.

§ These prescriptions may also be looked upon as furnishing an *operational* definition of the probability of certain kinds of events, from which probabilities of other kinds of events may be obtained by the well-known rules for combining probabilities. (Eventually the need for generalization arises; see Chapter VIII.) Of course, the corresponding *abstract*, mathematical notion of probability (sufficiently for our purpose: the notion of *equiprobability* of all points in the sample space), like all basic notions of mathematics (such as that of "point" in geometry), *cannot* be defined in terms of other mathematical notions, unless one introduces still more basic notions (which then will have to remain undefined).

In practice, the *definitions* of $\mathscr{P}$, of units, and of the characteristics (including the definition of the method of determination of any given unit's characteristic) may involve many difficulties. Since these vary for different fields of application, we shall not be able to give them much attention, but they vitally affect the choice of the sampling method.

5.2 ESTIMATORS AND THEIR SAMPLING DISTRIBUTIONS, CONFIDENCE INTERVALS

Next, sampling theory concerns itself with methods of *estimating* overall properties of the characteristics of the units over $\mathscr{P}$. A prescription (formula) for estimating is called an *estimator*. The value the estimator takes on in any particular case (sample) is then its *estimate*.

We call any quantity which is a function of the sample (and possibly of other variables as well) a *random variable*; such a function has a distribution over $\mathscr{S}$, called its *sampling distribution*. It follows that an estimator is a random variable and has a sampling distribution.† For example,

$$\bar{y} = \sum_{u_i \in \sigma} y(u_i)/n$$

is an estimator of $\bar{Y}$ often used if the method of sampling is one of those discussed above, and in some simple cases we may, from given numbers $y_1, \ldots, y_N$, defined by

$$y_1 = y(u_1), \ldots, y_N = y(u_N),$$

be able to obtain explicitly

$$F(z^{(k)}) = \mathrm{P}\{\bar{y} \leq z^{(k)}\} \quad (k = 1, \ldots, K).$$

This latter function is called the *distribution function* of $\bar{y}$; $z^{(1)}, \ldots, z^{(K)}$ are the possible values of $\bar{y}$. The *frequency function*‡ of $\bar{y}$ is defined by

$$f(z^{(k)}) = \mathrm{P}\{\bar{y} = z^{(k)}\} \quad (k = 1, \ldots, K);$$

therefore

$$F(z^{(k)}) = \sum_{t \leq z^{(k)}} f(t).$$

However, in most cases it is not practicable to derive explicitly the distribution function of the estimator, and we only obtain some numbers which *characterize* this function, such as the *first moment* of the distribution of $\bar{y}$

$$\sum_{k=1}^{K} z^{(k)} f(z^{(k)})$$

† In fact, an estimator is generally a *statistic*, i.e., a function of the sample *only*.

‡ Some call f a "distribution function" or use this word in a general sense which includes both f and F. Then they call F by the name *cumulative* distribution function. One even encounters for f the name probability function, which we have reserved for P.

and its *second central moment*

$$\sum_{k=1}^{K} \{z^{*(k)}\}^2 f(z^{(k)}),$$

with

$$z^{*(k)} = z^{(k)} - \sum_{h=1}^{K} z^{(h)} f(z^{(h)}).$$

(These are, respectively, equal to the *expected value* of $\bar{y}$, $\mathscr{E}\bar{y}$, and the *variance* of $\bar{y}$, $V(\bar{y})$ – see section 5.4 below.)

Moreover, the characteristics, such as the first and second moments of the sampling distributions of an estimator, are often themselves functions of only a small number of simple characterizing functions† of $y_1, \ldots, y_N$. Thus we prove in the next chapter that in the case of simple random sampling $\mathscr{E}\bar{y}$ is a function of $\bar{Y}$; and in particular that $\bar{y}$ is an *unbiased* estimator of $\bar{Y}$, meaning that $\mathscr{E}\bar{y} = \bar{Y}$. We also show that the variance of $\bar{y}$ is a certain function of $\sigma^2(\mathscr{Y})$. [The student should carefully note that the proofs of these statements are but more or less ingenious condensations of the enumeration of all possible different cases.]

Of course, in practice we cannot compute the moments from such formulae, since we know neither $y_1, \ldots, y_N$, nor the values of its characterizing functions $\bar{Y}$, $\sigma^2(\mathscr{Y})$, etc. This is simply the *distinction between probability and statistics:* in probability we derive, from an assumed knowledge of a population or its characteristics, something about functions whose values are generated by sampling from this population; in statistics we try to infer properties of the population from samples (or to reach practical decisions which may be regarded as implicitly based on such inferences). The following are two illustrations of the kind of statements we *can* make about $\mathscr{P}$: With the above sampling method, $\bar{y}$, as an estimator of $\bar{Y}$, is *unbiased.*‡ Moreover, if $\bar{y}$ has approximately the well-known "normal" distribution, we can, using $\bar{y}$, give a *confidence interval,*§ $I_{\bar{y}}$ for $\bar{Y}$ with given *confidence coefficient;* and we know

† Any function or set of functions of $y_1, \ldots, y_N$, including the set of numbers $y_1, \ldots, y_N$ themselves, are called *parameters of the property* $\mathscr{Y}$ *over* $\mathscr{P}$. On the other hand, $\mathscr{E}\bar{y}$ and $V(\bar{y})$, f and F, mentioned above, are *parameters of the sampling distribution of* $\bar{y}$, i.e., of the distribution of $\bar{y}$ over $\mathscr{S}$.

‡ An advantage of *approximate* unbiasedness of an estimator is pointed out below.

§ We say that an interval I_x, which depends only on the statistic x, is a confidence interval for a parameter, of confidence coefficient β, if the probability of I_x covering the parameter is (at least) β, no matter what the value of all the parameters. Since I_x depends on x, the interval is random. The concept is *illustrated in detail* in Chapter II, section 7.

that, if n is sufficiently large, the distribution of $\bar{y}$ is, indeed, close to normal (for further discussion see Chapter II, section 15).†

5.3 LOSSES FROM ERRORS OF ESTIMATION, ESTIMATORS OF RELATIVELY SMALL BIAS

At any rate, the *mean square error*‡ of an estimator x of a parameter μ is of some value as a measure of the expected value of the loss due to inexact estimation. For, if in any one case we measure the loss due to estimating μ by x in terms proportional to $(x - \mu)^2$, i.e., to the squared§ deviation of the estimator from what it estimates, then MSE $= \mathscr{E}(x - \mu)^2$ is proportional to the expected value of the loss due to estimation. And if $B = \mathscr{E}x - \mu$, the *bias* of x for estimating μ, is numerically relatively small, then MSE differs little from

$$V(x) = \mathscr{E}(x - \mathscr{E}x)^2,$$

for

$$\mathscr{E}(x - \mu)^2 = \mathscr{E}(x - \mathscr{E}x)^2 + B^2.$$

In addition it may be desirable for B^2 (and, of course, MSE) to be small relative to μ^2.

In this book we shall largely confine ourselves to *estimators with relatively small numerical bias*.§§ One reason is that for other estimators we cannot usually guarantee that, no matter what be the values of the characteristics

† The reasoning for $\bar{y}$ also holds for any other approximately normal statistic. In particular, if x is a normally distributed statistic with mean μ and known variance σ_n^2, we have, e.g., that the probability that the *random interval* I_x: $x - 1.96\sigma_n$ to $x + 1.96\sigma_n$ covers the unknown μ is 0.95, since μ belongs to I_x if and only if the standard normal variable belongs to the interval from -1.96 to 1.96, which has probability 0.95. We shall also discuss estimators of Y, $\bar{Y}$ and $\sigma^2(\mathscr{Y})$ for nonsimple sampling. We shall see that it is not very easy to make helpful statements about how good confidence intervals for $\bar{Y}$ based on these estimators are.

‡ The mean square error (MSE) of an estimator t of a parameter θ is defined as $\mathscr{E}(t - \theta)^2$.

§ The *square* rather than some other positive power of $|x - \mu|$ is chosen because it gives results which usually can be handled most easily mathematically. (Some of the relevant mathematical properties are recalled in 5.4 below.) Anyway, one may raise the question as to why one should want to consider the *expected value* of the loss. An answer may be given in terms of the theory of measurement of the utility of "prospects", which we sketch very briefly in section 5.5 below.

Instead of raising questions like those of the present footnote, one might ask oneself whether one would be prepared to choose a certain estimator if under the same circumstances one with a substantially smaller MSE is available.

§§ In many cases it is difficult to find such a class of estimators. Therefore we often have little choice but to confine ourselves to unbiased estimators (if such exist), even though the possibility is not excluded that, among estimators of small bias, there exists one with MSE much smaller than that of any unbiased estimator.

of the units of $\mathscr{P}$, the MSE will be less than an amount which is both acceptable and attainable with an estimator of the kind we consider.† Within this class of estimators we may then look for one which makes $V(x)$ as small as possible, given the resources at our disposal; or which minimizes expenditure of resources, given an acceptable level of $V(x)$.

(One should note, however, that occasionally there may be reason to measure loss quite differently. Thus it may be that a negative deviation of x from μ causes a much more serious loss than a positive one, or that it is highly undesirable to estimate a parameter by an estimate which is not a possible value of the parameter – e.g., to estimate a frequency by a number which is not an integer, or a variance by a negative number.)

To the extent that we can limit ourselves to estimators on which we can base confidence intervals of small expected length whenever they contain the parameter estimated, and whose confidence coefficient is close to that based on the assumption of normality,‡ the above considerations need not enter. But evidently, such estimators can only have a bias whose square is small relative to both μ^2 and the variance of the estimator.§

5.4. EXPECTATIONS, VARIANCES AND COVARIANCES OF RANDOM VARIABLES

If $\mathscr{X}$ is a random variable, which on $\mathscr{s}$ takes on the value $x(\mathscr{s})$, we call

$$\mathscr{E}\mathscr{X} = \sum_{\mathscr{s}\in\mathscr{S}} x(\mathscr{s})\, P(\mathscr{s})$$

† The effect of bias on confidence intervals is discussed, for the normal case, in sections 15.8 and 15.11 of Chapter II. Sometimes absence or near absence of bias has independent merit – see section 5.6 below.

‡ These estimators play a leading role in our subject, in which, as compared with other subjects, sample sizes are frequently *large* and in which it is often desirable to make only a *minimal* amount of *assumptions* about the $\mathscr{Y}$-values of $\mathscr{P}$ (see section 10).

§ Confidence intervals for such estimators are usually of the form $I'_x = [x - t\hat{\sigma}_n, x + t\hat{\sigma}_n]$, where $\hat{\sigma}_n^2$ is an estimator of the variance of x and t is read from a standard normal table corresponding to the desired confidence coefficient β_0. One of the assumptions of the text states that β, the probability that the interval covers μ, is close to β_0. Moreover, under the circumstances mentioned, I'_x is not much affected by replacing it by I_x in which the estimated variance $\hat{\sigma}_n^2$ is replaced by the actual variance σ_n^2. Now

$$\begin{aligned} 1 - \beta &\approx \mathrm{P}\{x - \mu < -t\sigma_n\} + \mathrm{P}\{x - \mu > t\sigma_n\} \\ &= \mathrm{P}\{x - \mathscr{E}x < -t\sigma_n - B\} + \mathrm{P}\{x - \mathscr{E}x > t\sigma_n - B\} \\ &= \Phi_n\{-t - (B/\sigma_n)\} + 1 - \Phi_n\{t - (B/\sigma_n)\} \end{aligned}$$

where Φ_n is the distribution of $u = (x - \mathscr{E}x)/\sigma_n$; and by assumption this is close to $\Phi(-t) + 1 - \Phi(t) = 1 - \beta_0$, where Φ is the standard normal distribution. Reference to standard normal tables shows that this can only be so for the β_0-values in common use if $|B|/\sigma_n$ is small. For these β_0-values t is between 1.6 and 2.6. The assumption of the text can also be shown (see the first footnote in section 15.10 of Chapter II) to imply that $2t\sigma_n/|\mu|$ is small (by smallness of the expected length of a confidence interval for μ we of course mean smallness relative to $|\mu|$). But this implies that $|B|/|\mu|$ is very small.

its *expectation* (or *expected value*, or *mean value*, or *mean*), and

$$V(\mathcal{X}) = \sum_{\mathscr{s}\in\mathscr{S}} x^*(\mathscr{s})^2 P(\mathscr{s}) \quad (x^*(\mathscr{s}) = x(\mathscr{s}) - \mathscr{E}\mathcal{X})$$

its *variance* or *sampling variance*† (the square root of $V(\mathcal{X})$ is called the *standard deviation* of $\mathcal{X}$ or its *standard error*). Actually $\mathscr{E}\mathcal{X}$ is *equal* to the first moment and $V(\mathcal{X})$ to the second central moment of the distribution of $\mathcal{X}$. Thus, if the possible different values of $x(\mathscr{s})$ are $x^{(1)}, \ldots, x^{(K)}$ and, $S(x_0)$ is the set of $\mathscr{s}$ in $\mathscr{S}$ for which $\mathcal{X}$ takes on the value x_0,

$$\sum_{\mathscr{s}\in\mathscr{S}} x(\mathscr{s})\, P(\mathscr{s}) = \sum_{k=1}^{K} x^{(k)} \sum_{\mathscr{s}\in S(x^{(k)})} P(\mathscr{s})$$
$$= \sum_{k=1}^{K} x^{(k)} f(x^{(k)}).$$

(Sometimes the left-hand side is easier to compute than the right hand side, and sometimes the reverse holds.) As an example, see the computation (with $\mathcal{X} = \bar{y}$) in section 6 of the next chapter. [The student should study the example very cosely.] The proof of equality of $V(\mathcal{X})$ and the second central moment is similar.

For two random variables $\mathcal{X}$ and $\mathcal{Y}$ over $\mathscr{S}$ we define also the *covariance* of $\mathcal{X}$ and $\mathcal{Y}$:

$$\text{Cov}\,(\mathcal{X}, \mathcal{Y}) = \sum_{\mathscr{s}\in\mathscr{S}} x^*(\mathscr{s})\, y^*(\mathscr{s})\, P(\mathscr{s}),$$

which is equal to the central *cross-moment* of the *joint distribution* of $\mathcal{X}$ and $\mathcal{Y}$, defined by

$$\sum_{l=1}^{L} \sum_{k=1}^{K} x^{*(k)} y^{*(l)} f(x^{(k)}, y^{(l)});$$

here f, defined by

$$f(x^{(k)}, y^{(l)}) = \text{P}\{\mathcal{X} = x^{(k)}, \mathcal{Y} = y^{(l)}\},$$

is the joint frequency function of $\mathcal{X}$ and $\mathcal{Y}$. In particular,

$$\text{Cov}\,(\mathcal{X}, \mathcal{X}) = V(\mathcal{X}).$$

A few often used properties of $\mathscr{E}$, V and Cov are:

(i) If α is not random:

$$\mathscr{E}\alpha\mathcal{X} = \alpha\mathscr{E}\mathcal{X}, \quad V(\alpha\mathcal{X}) = \alpha^2 V(\mathcal{X}).$$

† The terms sampling variance and standard error were introduced in order to emphasize that reference is made to the sampling distributions – in this case to the sampling distribution of $\mathcal{X}$.

(ii) If α and β are not random

$$\mathrm{Cov}\,(\alpha\mathscr{X}, \beta\mathscr{Y}) = \alpha\beta\,\mathrm{Cov}\,(\mathscr{X}, \mathscr{Y}).$$

(iii) $\mathscr{E}(\mathscr{X} + \mathscr{Y}) = \mathscr{E}\mathscr{X} + \mathscr{E}\mathscr{Y}$,

$$V(\mathscr{X} + \mathscr{Y}) = V(\mathscr{X}) + 2\,\mathrm{Cov}\,(\mathscr{X}, \mathscr{Y}) + V(\mathscr{Y}).$$

(iv) $\mathrm{Cov}\,(\mathscr{X}, \mathscr{Y}) = \mathscr{E}\mathscr{X}\mathscr{Y} - \mathscr{E}\mathscr{X}\,\mathscr{E}\mathscr{Y}$.

5.5. APPENDIX: UTILITY OF PROSPECTS

Suppose an individual or an organization wishes to make comparisons among a set $\mathscr{A}$ of possible or available actions (decisions). Let action A be known to lead to a certain situation T^A. We may (by way of definition) say that the individual or organization chooses that action A which leads to the greatest *utility* among the actions of $\mathscr{A}$, if we can define a function u^* over $\mathscr{A}$ satisfying $u^*(A) > u^*(A')$ if and only if the situation resulting from A is preferred to that resulting from A'. We shall then also define the function u on the set of situations: $u(T^A) = u^*(A)$. It is clear that such a function is determinate only up to an increasing transformation, and so is not particularly useful.

The situation is different when there is no certainty about the situation to which an action will lead. In particular, we shall suppose that A leads to $\mathscr{T}^A$ which may have the "values" T_i^A with probabilities p_i^A $(i = 1, \ldots, k_A)$. Situations like $\mathscr{T}^A$ are called *prospects*. Under certain, rather plausible assumptions on the nature of preferences of an individual, it has been shown† that there exists a function u on the set of situations, such that the individual will choose A rather than A' if and only if $u^*(A) > u^*(A')$, and u^* is defined by $u^*(A) = \Sigma u(T_i^A)p_i^A$, that is, by $u^*(A) = \mathscr{E}u(\mathscr{T}^A)$. Moreover, after arbitrarily fixing a 0 point and a unit of measurement, these functions are now unique.

To bring out the significance of the result, we discuss two examples of its application.

(i) A man shoots at a target. Suppose that, if he aims at a point which is at a distance, A, to the left of the target, he hits a point which is $\mathscr{T}^A$ to the left of the target, where $\mathscr{T}^A = A - z$, where z is a random variable (which reflects faults in the gun, effect of winds, etc.) with a known probability dis-

† The reader may find an elementary discussion of such assumptions and a proof in CHERNOFF and MOSES, Elementary Decision Theory (Wiley, 1959). It is not difficult to think of situations in which the assumptions are not likely to be fulfilled; nonetheless, if assumptions are sought which apply to a wide range of applications, assumptions such as the ones referred to are reasonable. The reader may also want to consider their applicability to a group situation.

tribution. Suppose, moreover, that if he hits T to the right or to the left of the target, he gains

$$u(T) = -kT^2,$$

where k is a positive constant (which may well depend on the location of the target). How should he select A?

$$\begin{aligned} u^*(A) &= \mathscr{E}u(\mathscr{T}^A) = -k\mathscr{E}(\mathscr{T}^A)^2 = -k\mathscr{E}(z-A)^2 \\ &= -k\{\mathscr{E}(z-\mathscr{E}z)^2 + (\mathscr{E}z - A)^2 + 2\mathscr{E}(z-\mathscr{E}z)(\mathscr{E}z - A)\} \\ &= -k\sigma_z^2 - k(\mathscr{E}z - A)^2, \end{aligned}$$

which is maximized when we take A equal to $\mathscr{E}z$. On the other hand, if $u(T) = -k|T|$, it can be shown that $u^*(A)$ is maximized when we take A equal to a median of z, that is, a number which is just as likely to exceed as to fall short of z. If the distribution of z is not highly asymmetric, $-k\mathscr{E}|\mathscr{T}^A|$ for $A = \mathscr{E}z$ may not differ much from max $\{-k\mathscr{E}|\mathscr{T}^A|\}$. Still another case of interest would be one in which, instead of one constant k, we have two, much different, constants, one for positive and one for negative T.

(ii) Consider the case in which it is desired to estimate a certain parameter θ by an estimator A which leads to prospect $\mathscr{T}^A$ with a distribution known for each A in $\mathscr{A}$, and

$$u(T) = -k(T-\theta)^2.$$

(The T_i^A are the possible results of using the estimator A; as a special case they may be the possible values that this estimator can take on.)

$$u^*(A) = -k\mathscr{E}(\mathscr{T}^A - \theta)^2 = -k \text{ MSE } (\mathscr{T}^A) = -kV(\mathscr{T}^A) - k(B^A)^2,$$

where $B^A = \mathscr{E}\mathscr{T}^A - \theta$. This case is somewhat like the previous one, but here both terms on the right-hand side depend on A. As explained earlier in this section, we shall confine ourselves to estimators A for which $(B^A)^2$ is small compared with $V(\mathscr{T}^A)$ and choose among these estimators one with small $V(\mathscr{T}^A)$. Again, one may consider other functions u.

5.6. APPENDIX: SOME REMARKS ON UNBIASED ESTIMATORS

In certain cases one wishes to estimate a parameter μ not only from the survey as a whole, but also from the different parts of that survey (see, e.g., the last section of Chapter IX). If we restrict ourselves to unbiased estimators, an estimate based on the entire survey may be obtained by combining the estimates that are based on the parts, without referring back to the original data, viz., as a (weighted or unweighted) average of the latter. This follows from properties (i) and (iii) recalled in section 5.4. (The remaining properties may be used to obtain suitable weights, see, e.g., Chapter XI.)

More frequently one wishes to estimate from a survey, or from a set of surveys, a number of parameters, say, $\mu_1, \ldots, \mu_k$, and also one or more functions of the form $c_{l1}\mu_1 + \ldots + c_{lk}\mu_k$ $(l = 1, \ldots, m)$. For example, $\mu_1, \ldots, \mu_k$ are subtotals for different parts of the population, and one may wish to estimate the grand total as well as its component parts (then $m = 1$ and $c_{11} = \ldots = c_{1k} = 1$), or the overall average as well as the averages for the several parts or differences between averages. Especially if different users of survey may be interested in different functions of the above form, it is advantageous if these can be estimated with small bias solely on the basis of the separate estimates of $\mu_1, \ldots, \mu_k$. Within the class of unbiased estimators this is always possible; within the class of somewhat biased estimators these properties do not necessarily hold, and have to be examined in each particular case.

Finally there are cases which involve a long sequence of estimators, and in which it is of prime importance that their average is close to what is estimated. Thus when farmers are paid according to the estimated mean sugar content of each batch of beets they deliver to the factory, moderate inaccuracies in the separate determinations are unimportant, provided they nearly cancel each other.

6. Obtaining a random sample by means of a table of random digits†

6.1. OBTAINING THE TABLE FROM A RANDOM PHYSICAL PROCESS

The word "simple random sampling" refers to any procedure that gives every possible sample the same probability of being the one obtained. A convenient way to achieve this is to use a table of random digits. Such a table is a finite sequence of digits 0, 1, . . ., 9, obtained by means of a process (like tossing a ten-faced "die") that is as close as possible to a *random physical process*, i.e., a process satisfying the conditions:

(1) at each stage there is complete physical symmetry‡ in the process for the ten possible outcomes 0, 1, . . ., 9 (i.e., the die is balanced and symmetrically shaped, and the throwing process is symmetric);

(2) the outcome at each stage is unaffected by the results obtained at previous stages.

The table is the result of a past experiment and may be used instead of conducting an actual experiment.

† Sometimes the sample is obtained *directly* by an uncontrolled or partially controlled physical process which one takes to be of an approximately random nature. Thus sampling of wild life on a plain, insects in a field, or fish in a lake is often conducted by placing traps at randomly spaced locations; whereas, in sampling from a container of fluids or small particles, the material is usually thoroughly mixed prior to withdrawing a portion.

‡ Leading to complete unpredictability of the outcome at that stage.

6.2. OBTAINING RANDOM SEQUENCES FROM THE TABLE

Any sequence of n successive digits in the table, taken from a starting point chosen without regard to the value of the digit appearing at that place, will be considered to constitute a simple random sample of size n. It is not advisable to repeatedly use one and the same sequence of n digits. To avoid repeated use of the same digits, one may either use bigger tables, or introduce some more random elements – like random determination of the starting point and of the direction of reading the table.

We shall use a small table of 1 000 random digits arranged in 20 lines (numbered 0–19) and 50 columns (numbered 1–50 and grouped in groups of 5).

The first thousand digits from the book: The Rand Corporation: A Million Random Digits with 100 000 Normal Deviates (Free Press, 1955) are:

Line Number	Column Number 1–5	6–10	11–15	16–20	21–25	26–30	31–35	36–40	41–45	46–50
0	10097	32533	76520	13586	34673	54876	80959	09117	39292	74945
1	37542	04805	64894	74296	24805	24037	20636	10402	00822	91665
2	08422	68953	19645	09303	23209	02560	15953	34764	35080	33606
3	99019	02529	09376	70715	38311	31165	88676	74397	04436	27659
4	12807	99970	80157	36147	64032	36653	98951	16877	12171	76833
5	66065	74717	34072	76850	36697	36170	65813	39885	11199	29170
6	31060	10805	45571	82406	35303	42614	86799	07439	23403	09732
7	85269	77602	02051	65692	68665	74818	73053	85247	18623	88579
8	63573	32135	05325	47048	90553	57548	28468	28709	83491	25624
9	73796	45753	03529	64778	35808	34282	60935	20344	35273	88435
10	98520	17767	14905	68607	22109	40558	60970	93433	50500	73998
11	11805	05431	39808	27732	50725	68248	29405	24201	52775	67851
12	83452	99634	06288	98083	13746	70078	18475	40610	68711	77817
13	88685	40200	86507	58401	36766	67951	90364	76493	29609	11062
14	99594	67348	87517	64969	91826	08928	93785	61368	23478	34113
15	65481	17674	17468	50950	58047	76974	73039	57186	40218	16544
16	80124	35635	17727	08015	45318	22374	21115	78253	14385	53763
17	74350	99817	77402	77214	43236	00210	45521	64237	96286	02655
18	69916	26803	66252	29148	36936	87203	76621	13990	94400	56418
19	09893	20505	14225	68514	46427	56788	96297	78822	54382	14598

To see how we proceed, suppose we wish to obtain a sequence of 30 random digits (with replacement). We pick a five digit number, K, from the table (e.g., by marking blindly with a pencil). We denote u, v, w as follows:

$u =$ the remainder of the first two digits of K after division by the number of lines in the table (here 20);

$v =$ the remainder of the next two digits of K after division by the number of columns in the table (here 50);

$w =$ the last digit† of K.

Let us take u and $v + 1$ as the line and column of the starting point, and determine the direction of reading according to w: If $w = 0$ read to the left; if $w = 1$ read diagonally up and left (north-west); if $w = 2$ read upwards: if $w = 3$ read diagonally up and right (north-east); if $w = 4$ read to the right; if $w = 5$ read diagonally down and right (south-east); if $w = 6$ read downwards; if $w = 7$ read diagonally down and left (south-west).

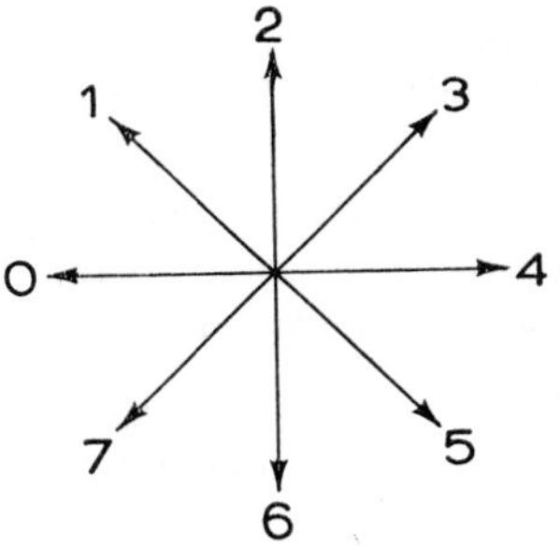

Every time we reach the margin of the table, we have to begin anew.

Suppose we marked the 6 in line 7, column 16, and got $K = 65692$. Then $u = 5$, $v = 19$, $w = 2$. We start from line 5, column 20, upwards, and obtain 0, 7, 5, 3, 6, 6. Now we have to pick another five-digit number. Suppose we marked the 7 in line 17, column 40. Then $K = 79628$, $u = 19$, $v = 12$. Since the last digit is an 8, we have to draw the next digit, which is 6; so $w = 6$. We retain only 2 in line 19, column 13, as we are unable to go downwards. We again mark some number blindly, say 0, in line 13, column 14. Then $K = 07584$, $u = 7$, $v = 8$, $w = 4$. We start from line 7, column 9, and read to the right 23 digits: 0, 2, 0, ... To sum up, the sequence of 30 digits we obtained is

075366202020516569268665748187.

We obtained it by a process which may well be considered to give to any sequence of 30 decimal digits the same probability of being selected.

Should we wish to obtain, instead, 15 two-digit numbers, we need only group each of two successive digits of the sequence as previously obtained. In the above example this would yield the sequence

07, 53, 66, 20, 20, 20, 51, 65, 69, 26, 86, 65, 74, 81, 87

† If this digit is 8 or 9, we continue taking further digits, until we get a digit different from 8 or 9.

of 15 two-digit numbers; the process assures every sequence of 15 two-digit numbers from 00 up to 99 the same probability of being selected. Or, grouping by threes, we have a random sequence of 10 three-digit numbers,

075, 366, 202, 020, 516, 569, 268, 665, 748, 187;

and so on.

6.3. OBTAINING VARIOUS KINDS OF SAMPLES FROM THESE SEQUENCES

By means of examples, it will be shown how to use such sequences in order to form random samples of different types.

(i) *Sampling with replacement – equal probabilities*. Suppose we wish to draw a simple random sample of 15, with replacement, out of a population of 475. We assign a number from 1 to 475 to each member and call this population $u_1, u_2, \ldots, u_{475}$. We form a random sequence of three-digit numbers – rejecting at each stage the number 000 and any numbers exceeding 475 – until we have a sequence of 15 integers in the interval from 001 to 475 (inclusive). Suppose we found successively

541, 784, 561, 418, 099, 337, 143, 053, 351, 296, 986, 127,
192, 553, 604, 090, 324, 116, 644, 988, 337, 079, 848, 275.

The simple random sample drawn will be

$$u_{418};\ u_{99};\ u_{337};\ u_{143};\ u_{53};\ u_{351};\ u_{296};\ u_{127};$$
$$u_{192};\ u_{90};\ u_{324};\ u_{116};\ u_{337};\ u_{79};\ u_{275}.$$

24 drawings were necessary in order to obtain 15 numbers between 1 and 475.

REMARK. The procedure may be made more efficient if to each member of the population are assigned two numbers:

to u_1: 1 and 501,
to u_2: 2 and 502,
. . . .
to u_{475}: 475 and 975.

Only 000, 476–500 and 976–999 will be excluded. Now, only 16 steps are enough:

541, 784, 561, 418, 099, 337, 143, 053,
351, 296, 986, 127, 192, 553, 604, 090;

and the sample obtained is now

$$u_{41},\ u_{284},\ u_{61},\ u_{418},\ u_{99},\ u_{337},\ u_{143},\ u_{53},$$
$$u_{351},\ u_{296},\ u_{127},\ u_{192},\ u_{53},\ u_{104},\ u_{90}.$$

(ii) *Sampling with replacement – unequal probabilities.* Suppose we wish to take a sample of size 11 with replacement out of the population

$$\mathscr{P} = \{u_1, u_2, u_3, u_4\},$$

such that the probability of obtaining u_1 is 0.15, of obtaining u_2 is 0.42, of obtaining u_3 is 0.08, and of obtaining u_4 is 0.35. We assign

to u_1 the 15 numbers 00–14,
to u_2 the 42 numbers 15–56,
to u_3 the 8 numbers 57–64,
to u_4 the 35 numbers 65–99,

and form a sequence of 11 two-digit random numbers. Let it be

61, 19, 69, 04, 46, 26, 45, 74, 77, 74, 51.

The sample is then

$$u_3, u_2, u_4, u_1, u_2, u_2, u_2, u_4, u_4, u_4, u_2.$$

(iii) *Sampling without replacement.* For simple random sampling the only difference is that here we also have to discard any random numbers that have turned up previously.

For a discussion of how nonsimple random sampling without replacement may be defined and carried out see section 4 of Chapter VI.

6.4. JUSTIFICATION OF THE PROCEDURES DESCRIBED

The procedures described above have to be justified theoretically. Thus, it has to be proved, e.g., that in the case of simple random sampling all possible samples are equiprobable.

We shall outline the idea of the proof only for simple random sampling.

Suppose we wish to obtain a sample of size n out of a population

$$\mathscr{P} = \{u_1, . . ., u_N\},$$

where N is an integer less than 1 000. By the procedure described above, we obtain, successively three-digit random numbers. A draw will be called "successful" if it leads to a three-digit number between 1 and N; and, for sampling without replacement, if in addition it does not repeat a three-digit number drawn previously.

Let us compute the probability that n successive successful draws from a specified sample sequence $u_{i_1}, . . ., u_{i_n}$ (for sampling without replacement the u's are supposed different units), where each of $i_1, . . ., i_n$ is a number between 1 and N.

In the 1st successful draw, the probability of obtaining i_1 is $1/N$.

In the 2nd successful draw, with replacement, the probability of obtaining

i_2 is $1/N$; and without replacement, the conditional probability of obtaining i_2 given some other number was drawn on the first successful draw, is $1/(N-1)$.

Continuing this way, we find that the probability of obtaining a given sample $u_{i_1}, \ldots, u_{i_n}$, is

$1/N^n$ for sampling with replacement,
$1/N^{(n)}$ for sampling without replacement.

6.5. EXAMPLE

Suppose we wish to select 120 addresses from a directory. The book has 203 pages, none containing more than 150 addresses per page. One procedure is to select 120 six-digit random numbers, the first half representing the page number and the last half the location of the address on the page, counting from the top.† To avoid requiring an upper bound for the number of addresses on a page, it might be suggested that one could proceed as follows: count the number of addresses on the selected page only, and if there are, e.g., 50, use a two-digit number for that page, while, if on another page there are 130, use a three-digit number for that page. The student should prove that this procedure does not give each address an equal probability and so certainly does not lead to a simple random sample. (Actually, an upper bound is not required for all the pages of the directory, only for the selected pages.) Similar problems arise in obtaining a sample of households by using maps showing city blocks.

6.6. USE OF COMPUTERS

A computer is often used to generate a sequence of numbers to be used as random numbers. We shall not discuss the generating methods employed; the reader may consult any up-to-date book on numerical methods. At this point we merely wish to remark that, given any such method and assuming its validity, one can use the computer to draw random samples, restricted or unrestricted ones. Some of these methods are described in FAN, MULLER and REZUCHA, J. Amer. Statist. Ass., **57** (1962) 387–402.

7. Purposive selection and random sampling‡

Suppose we wish to estimate quickly $\bar{Y}$, the average weight of male undergraduate university students in a class of 600, weighed without clothes, just before lunch time, correct within 1 kg.

† Of course, on any one page, the entire 6-digit number must be rejected whenever the number defined by the last 3 digits exceeds the number of addresses on that page.

‡ Beginners should postpone studying this section in detail at least until they have studied Chapter II.

A somewhat experienced person could easily pick a "representative sample" of 20 of the students and then weigh them. He could primarily look for apparent height, girth and type of build, since jointly these are very highly correlated with weight. From published statistics, or from his general knowledge, he may know approximately the functional form of the (multivariate!) distribution† of these characteristics (among men of this age group and standard of living). Contrast this with his taking a simple random sample of 20 students from the class.

The following may well be the true situation:

Mean weight of the students: 70 kg;
Standard deviation of the weights of the students: 10 kg;
Approximate distribution of weights: normal.

In that case the fraction of the students that weigh more than 90 kg or less than 50 kg may be expected to be about 0.05 (why?). For a simple random sample of 20 from such a population, what is the chance that its sample mean weight will deviate from the correct one by more than 1 kg?

The variance of the sample mean of 20 from such a population is approximately $100/20 = 5$, giving a standard error of about 2.24. Now

$$\mathrm{P}\{|\bar{y} - 70| > 1\} = \mathrm{P}\left\{\left|\frac{\bar{y} - 70}{2.24}\right| > \frac{1}{2.24}\right\}$$

and, as $(\bar{y} - 70)/2.24$ is approximately a standard normal variate, this probability is 0.65. So it is not at all so unlikely that a *simple random sample* may give a result off by as much as 1 kg. A *judgement sample* may well do better if the conditions outlined are fulfilled and the multiple correlation of the weights with the characteristics according to which we select the students is very high. Suppose, for simplicity, that only height (and not girth or build) is taken into account, and that the following is the true situation: the joint distribution of heights ($\mathscr{X}$) and of weights ($\mathscr{Y}$) is approximately normal with correlation ρ, and the mean height of the students is 170 cm, the standard deviation of their heights is 10 cm. It is known that, in the case of bivariate normality, the following holds:

$$\mathscr{Y} = \mu(\mathscr{Y}) + \rho\,\frac{\sigma(\mathscr{Y})}{\sigma(\mathscr{X})}\,(\mathscr{X} - \mu(\mathscr{X})) + \mathscr{U} = 70 + \rho(\mathscr{X} - 170) + \mathscr{U}$$

with the conditional mean and variance of $\mathscr{U}$ equal to 0 and

$$\sigma^2(\mathscr{Y})(1 - \rho^2) = 100(1 - \rho^2),$$

respectively, and with $\mathscr{U}$ and $\mathscr{X}$ uncorrelated.

† Distribution in $\mathscr{P}$ as defined in section 12 of Chapter II.

Since we try to select $x_1, \ldots, x_n$ in a "representative" way, it follows that the quantities δ and Δ in the equations

$$\bar{x} = \mu(\mathscr{X})(1 + \delta) = 170(1 + \delta)$$

and

$$\frac{1}{n} \sum_{i=1}^{n} (x_i - \bar{x})^2 = \sigma^2(\mathscr{X})(1 + \Delta) = 100(1 + \Delta)$$

are small fractions if the distribution of $\mathscr{X}$ is reasonably well known.

In judgement sampling the choice of students with given representative heights is not random, even if efforts are made to avoid biases. But without further assumptions it is impossible to make a comparison of simple random sampling, which is based on a probability model, with a kind of sampling which is not based on a probability model.

Suppose, therefore,† that for each i the choice of students from the (conditional) population for given x_i (where the scatter is small when ρ is large!) may be considered (approximately) equivalent to random sampling. Then the y_i have a distribution, with (conditional) mean

$$\mu(\mathscr{Y}) + \rho \frac{\sigma(\mathscr{Y})}{\sigma(\mathscr{X})} (x_i - \mu(\mathscr{X}))$$

and (conditional) variance

$$\sigma^2(\mathscr{Y})(1 - \rho^2),$$

and so

$$\bar{y} = \mu(\mathscr{Y}) + \rho \frac{\sigma(\mathscr{Y})}{\sigma(\mathscr{X})} \delta\mu(\mathscr{X}) + \bar{u},$$

$$V(\bar{y}) = V(\bar{u}) = \sigma^2(\mathscr{Y}) \frac{1 - \rho^2}{n} = \frac{100}{20} (1 - \rho^2) = 5(1 - \rho^2).$$

Also

$$s^2(\mathscr{Y}) = \frac{1}{n-1} \sum_{i=1}^{n} (y_i - \bar{y})^2$$

$$= \left(\rho \frac{\sigma(\mathscr{Y})}{\sigma(\mathscr{X})}\right)^2 \frac{n}{n-1} \sigma^2(\mathscr{X})(1 + \Delta) + \frac{1}{n-1} \sum_{i=1}^{n} (u_i - \bar{u})^2$$

$$+ 2\rho \frac{\sigma(\mathscr{Y})}{\sigma(\mathscr{X})} \frac{1}{n-1} \sum_{i=1}^{n} (u_i - \bar{u})(x_i - \bar{x}).$$

So the biases of $\bar{y}$ and $s^2(\mathscr{Y})/n$ as estimators of $\mu(\mathscr{Y})$ and $V(\bar{y})$ are, respectively,

$$\mathscr{E}\bar{y} - \mu(\mathscr{Y}) = \rho\delta \frac{\sigma(\mathscr{Y})}{\sigma(\mathscr{X})} \mu(\mathscr{X}) = 170\rho\delta,$$

† The use of such assumptions is examined in the chapter on the use of models. On "representativeness" of a sample, see section 12 of Chapter III.

and

$$\mathscr{E}s^2(\mathscr{Y})/n - V(\bar{y}) = \rho^2\sigma^2(\mathscr{Y})\frac{1+\Delta}{n-1} = \frac{100}{19}\rho^2(1+\Delta).$$

Thus, if $\rho = 0.8$ and $\delta = 0.002$, the bias of $\bar{y}$ is $0.8 \times 0.002 \times 170 = 0.272$ and the standard error of $\bar{y}$: $\{5 \times 0.36\}^{\frac{1}{2}} = 1.341\,6$. The probability that $\bar{y}$ falls outside the interval from 69 to 71 kg for this sample is computed as follows:

$$\begin{aligned} 1 - \mathrm{P}\{69 \leqslant \bar{y} \leqslant 71\} &= 1 - \mathrm{P}\{-1.272 \leqslant \bar{y} - 70.272 \leqslant 0.728\} \\ &= 1 - \mathrm{P}\left\{\frac{-1.272}{1.341\,6} \leqslant \frac{\bar{y} - 70.272}{1.341\,6} \leqslant \frac{0.728}{1.341\,6}\right\} \\ &= 1 - \mathrm{P}\left\{-0.948 \leqslant \frac{\bar{y} - 70.272}{1.341\,6} \leqslant 0.543\right\}, \end{aligned}$$

which, since $(\bar{y} - 70.272)/1.341\,6$ is approximately standard normal, equals about 0.466.

With multiple regression we may get (multiple) ρ much closer to 1, perhaps $\rho = 0.99$. With $\delta = 0.002$, this gives a bias of $0.99 \times 0.002 \times 170 = 0.337$, and standard error of $\bar{y}$ of $\{5 \times 0.019\,9\}^{\frac{1}{2}} = 0.315\,4$; and the required probability becomes

$$1 - \mathrm{P}\left\{\frac{-1.337}{0.315\,4} \leqslant \frac{\bar{y} - 70.337}{0.315\,4} \leqslant \frac{0.663}{0.315\,4}\right\} = 1 - \mathrm{P}\left\{-4.329 \leqslant \frac{\bar{y} - 70.337}{0.315\,4} \leqslant 2.102\right\},$$

which is about 0.018. That is much better than in the case of simple random sampling.

The relative bias in the estimator $s^2(\mathscr{Y})/n$ of

$$V(\bar{y}) = 5\{1 - 0.99^2\} = 0.099\,5$$

is $5.16(1 + \Delta)/0.099\,5$ – enormous! (Much better estimates of the variance of $\bar{y}$ can be obtained by estimating the residuals of the regression of $\mathscr{Y}$ on height, etc.)

In practice, as we shall see later (Chapter IIA), it may be possible to obtain, in the class of simple random sampling, a better estimator than the sample mean of the $\mathscr{Y}$-values by employing a formula which involves actual measurements, or even eye estimates, of heights, girths and builds.† Another approach is to use such information to *stratify* the sample by these characteristics (Chapter III); or, more generally, to sample students with *unequal probabilities*, where we make the probabilities depend in a certain way on this information (Chapters VI and VII). It is probably almost always advisable and possible to introduce *some* element of randomness in the *design* of the survey in order to guard against unforeseen influences and in order to enable the data themselves to provide a usable estimate of the variability of the estimator. Most of the following chapters will deal with ways of doing this.

Nonetheless, there are situations in which we can only take so small a number of observations, that the chance of getting a very unrepresentative

† Under the assumptions given, and even under weaker ones, it can be shown that a "regression estimator" mentioned in Chapter IIA, section 7, has a variance of the same order of magnitude as that of $\bar{y}$ in the judgement sample, and other estimators discussed in that chapter may have a not much larger variance. In many cases, however, the cost of judgement sampling is very much less than the cost of probabilistic sampling.

sample is unacceptably large, even with the most efficient scheme of stratification, etc. Suppose, for example, that we can afford to study only 3 cases, because each case takes a year's investigation. (Even if the situation allows us to form 3 groups (strata) from the population in which the characteristics under investigation have quite different distributions, randomness would require us taking a sample of just one from each by some random method.) An "expert" may be confident that he can obtain by purposive selection a sample which is more representative than a random sample (although, were I a user of this report, I would then want to consider the "risk" that the "expert" is not as "expert" as he thinks he is!).

The conclusion may well be that this – very common – type of investigation *cannot* yield "reliable" information,† only *suggestive* information (by saying this I do not mean to deprecate suggestive information!), and that, once he knows this, the investigator should reach a decision on how to proceed in the light of this impossibility. One such decision may be not to attempt to cover as broad a population but only one of the strata.

8. Conditional mean and variance

We consider two further expressions connected with the sampling distribution of properties of $\mathscr{P}$. Let $\mathscr{T}$ be a statistic, that is a function defined over $\mathscr{S}$, or more generally an (ordered) set of functions over $\mathscr{S}$. In the latter case, it is convenient to continue speaking of "a value t_0 of $\mathscr{T}$". Let $S(t_0)$ be the set of s in $\mathscr{S}$ for which $\mathscr{T}$ takes that value t_0; define

$$P(t_0) = \sum_{s \in S(t_0)} P(s).$$

Let $t^{(1)}, \ldots, t^{(L)}$, be the possible values of $\mathscr{T}$ on $\mathscr{S}$, so that $P(t^{(l)}) > 0$ for $l = 1, \ldots, L$, and

$$\sum_{l=1}^{L} P(t^{(l)}) = 1.$$

Let $\mathscr{Y}$ be any (numerical-valued) function defined over $\mathscr{S}$ such that, for each $s \in \mathscr{S}$, $y(s)$ is a well defined number.‡ We define for each $l = 1, \ldots, L$

$$g(t^{(l)}) = \sum_{s \in S(t^{(l)})} \frac{y(s)P(s)}{P(t^{(l)})}.$$

† It must not be concluded that the same (high) standard of "reliability" should be set for all fields of investigation, even in pure science. One consequence of that would be that some very important features of reality would become out-of-bounds for the scientific community, which would not, of course prevent their being examined by others at standards far below those that are feasible.

‡ E.g., $y(s)$ may be the average of a certain characteristic for the sample s. However, $\mathscr{Y}$ may depend also on $\mathscr{P}$, that is, it need not be a statistic. E.g., $\mathscr{Y}'$ defined below depends on $\mathscr{X}$, with the latter depending on $\mathscr{S}$ and parameters of $\mathscr{P}$.

EXAMPLE. $L = 4$ and

$$P(t^{(1)}) = P(t^{(2)}) = P(t^{(3)}) = P(t^{(4)}) = \tfrac{1}{4}.$$

$\mathscr{S}$ has 16 points, each† has probability $\frac{1}{16}$.

Values‡ of $\mathscr{Y}$ on $S(t^{(1)})$: 6, $6\frac{1}{2}$, 9, 10.
Values of $\mathscr{Y}$ on $S(t^{(2)})$: 4, 5, 7, 11.
Values of $\mathscr{Y}$ on $S(t^{(3)})$: 13, $13\frac{1}{2}$, 14, $14\frac{1}{2}$.
Values of $\mathscr{Y}$ on $S(t^{(4)})$: 1, 2, $12\frac{1}{2}$, 15.

Therefore,

$$\begin{aligned} g(t^{(1)}) &= \{6 \times \tfrac{1}{16} + 6\tfrac{1}{2} \times \tfrac{1}{16} + 9 \times \tfrac{1}{16} + 10 \times \tfrac{1}{16}\}/\tfrac{1}{4} = 7\tfrac{7}{8}, \\ g(t^{(2)}) &= 6\tfrac{3}{4}, \\ g(t^{(3)}) &= 13\tfrac{3}{4}, \\ g(t^{(4)}) &= 7\tfrac{5}{8}. \end{aligned}$$

We then define the function $\mathscr{X}$ on $\mathscr{S}$ (and $\mathscr{P}$) as follows

$$x(s) = \begin{cases} g(t^{(1)}) \text{ for all } s \text{ for which } \mathscr{T} \text{ takes the value } t^{(1)}, \\ \cdot\ \cdot\ \cdot\ \cdot\ \cdot\ \cdot\ \cdot\ \cdot\ \cdot\ \cdot\ \cdot\ \cdot \\ g(t^{(L)}) \text{ for all } s \text{ for which } \mathscr{T} \text{ takes the value of } t^{(L)}; \end{cases}$$

and call it the *conditional expectation* (*conditional mean*) *of* $\mathscr{Y}$ *given* $\mathscr{T}$;§ we also write it $\mathscr{E}\{\mathscr{Y}\,|\,\mathscr{T}\}$.

In our example $\mathscr{X} = \mathscr{E}\{\mathscr{Y}|\mathscr{T}\}$ takes on the values $7\frac{7}{8}$, $6\frac{3}{4}$, $13\frac{3}{4}$ and $7\frac{5}{8}$ when $\mathscr{T}$ takes on the values $t^{(1)}, t^{(2)}, t^{(3)}$ and $t^{(4)}$, respectively. Note that $x(s)$ cannot be computed from the knowledge of ($\mathscr{Y}$ on) s alone, so that $\mathscr{X}$ is not a statistic. Thus, for s consisting of the unit on which $y(s) = 6$, the computation of $x(s)$ requires the knowledge of $\mathscr{Y}$ on 3 other elements of $\mathscr{S}$ as well.

Since $\mathscr{X}$ is a function on $\mathscr{S}$, it has an expected value and a variance.

For example in our problem

$$\begin{aligned} \mathscr{E}\mathscr{X} &= \mathscr{E}\mathscr{E}\{\mathscr{Y}|\mathscr{T}\} = 7\tfrac{7}{8} \times \tfrac{1}{4} + 6\tfrac{3}{4} \times \tfrac{1}{4} + 13\tfrac{3}{4} \times \tfrac{1}{4} + 7\tfrac{5}{8} \times \tfrac{1}{4} = 9, \\ V(\mathscr{X}) &= V(\mathscr{E}\{\mathscr{Y}|\mathscr{T}\}) = \mathscr{E}\mathscr{X}^2 - (\mathscr{E}\mathscr{X})^2 \\ &= (7\tfrac{7}{8})^2 \times \tfrac{1}{4} + (6\tfrac{3}{4})^2 \times \tfrac{1}{4} + (13\tfrac{3}{4})^2 \times \tfrac{1}{4} + (7\tfrac{5}{8})^2 \times \tfrac{1}{4} - 9^2 = 7\tfrac{89}{128}. \end{aligned}$$

We also define for all $l = 1, \ldots, L$

$$h(t^{(l)}) = \sum_{s \in S(t^{(l)})} y'(s)^2 P(s)/P(t^{(l)})$$

with $y'(s) = y(s) - x(s)$.

† In general the different sample points do not need to have equal probabilities, nor need all values of $\mathscr{T}$ have equal probability.

‡ In our example the values of $\mathscr{Y}$ on the different sample points are all different, which makes for easy calculation and graphical presentation. However, this is not always the case.

§ Actually, all that is used about $\mathscr{T}$ is that it defines a subdivision of $\mathscr{S}$ into the collection $\{S(t^{(1)}), \ldots, S(t^{(L)})\}$ of L nonoverlapping sets, called a *partition* of $\mathscr{S}$. Therefore one also speaks of the conditional mean of $\mathscr{Y}$ given a certain partition of $\mathscr{S}$.

In our example

$$h(t^{(1)}) = \{(6 - 7\tfrac{7}{8})^2 \times \tfrac{1}{16} + (6\tfrac{1}{2} - 7\tfrac{7}{8})^2 \times \tfrac{1}{16} + (9 - 7\tfrac{7}{8})^2 \times \tfrac{1}{16} + (10 - 7\tfrac{7}{8})^2 \times \tfrac{1}{16}\}/\tfrac{1}{4}$$
$$= \tfrac{1}{4}[\{6^2 + (6\tfrac{1}{2})^2 + 9^2 + 10^2\} - (6 + 6\tfrac{1}{2} + 9 + 10)^2/4]$$
$$= \tfrac{1}{4}[259\tfrac{1}{4} - (31\tfrac{1}{2})^2/4] = 2\tfrac{51}{64},$$
$$h(t^{(2)}) = \tfrac{1}{4}[211 - 27^2/4] = 7\tfrac{3}{16},$$
$$h(t^{(3)}) = \tfrac{1}{4}[757\tfrac{1}{2} - 55^2/4] = \tfrac{5}{16},$$
$$h(t^{(4)}) = \tfrac{1}{4}[386\tfrac{1}{4} - (30\tfrac{1}{2})^2/4] = 38\tfrac{27}{64}.$$

We then also define the function $\mathscr{Z}$ on $\mathscr{S}$ as follows:

$$z(s) = \begin{cases} h(t^{(1)}) \text{ for all } s \text{ for which } \mathscr{T} \text{ takes the value } t^{(1)}, \\ \cdots\cdots\cdots \\ h(t^{(L)}) \text{ for all } s \text{ for which } \mathscr{T} \text{ takes the value } t^{(L)}; \end{cases}$$

and call it the *conditional variance of* $\mathscr{Y}$ *given* $\mathscr{T}$; we also write it $V\{\mathscr{Y}|\mathscr{T}\}$.

In our example $\mathscr{Z} = V\{\mathscr{Y}|\mathscr{T}\}$ takes on the values $2\tfrac{51}{64}$, $7\tfrac{3}{16}$, $\tfrac{5}{16}$ and $38\tfrac{27}{64}$ as $\mathscr{T}$ takes on the values $t^{(1)}$, $t^{(2)}$, $t^{(3)}$ and $t^{(4)}$, respectively.

Since $V\{\mathscr{Y}|\mathscr{T}\}$ is a function on $\mathscr{S}$, it has a mean.

In our example

$$\mathscr{E}\mathscr{Z} = \mathscr{E}V\{\mathscr{Y}|\mathscr{T}\} = 2\tfrac{51}{64} \times \tfrac{1}{4} + 7\tfrac{3}{16} \times \tfrac{1}{4} + \tfrac{5}{16} \times \tfrac{1}{4} + 38\tfrac{27}{64} \times \tfrac{1}{4} = 12\tfrac{23}{128}.$$

We now also show *three important results* concerning the mean and variance of $\mathscr{X}$ and the mean of $\mathscr{Z}$:

(a) $\mathscr{E}\mathscr{X} = \mathscr{E}\mathscr{Y}$; i.e., $\mathscr{E}\mathscr{E}\{\mathscr{Y}|\mathscr{T}\} = \mathscr{E}\mathscr{Y}$.

(b) $V(\mathscr{X}) = V(\mathscr{Y}) - \mathscr{E}\mathscr{Z}$; i.e., $V(\mathscr{Y}) = \mathscr{E}V\{\mathscr{Y}|\mathscr{T}\} + V(\mathscr{E}\{\mathscr{Y}|\mathscr{T}\})$.

(c) $V(\mathscr{X}) < V(\mathscr{Y})$ if, for some s with $P(s) > 0$, $y(s) - x(s) \neq 0$; and $V(\mathscr{X}) = V(\mathscr{Y})$ otherwise.†

Proofs:

(a)

$$\mathscr{E}\mathscr{X} = \sum_{l=1}^{L} P(t^{(l)})\, g(t^{(l)})$$
$$= \sum_{l=1}^{L} P(t^{(l)}) \sum_{s \in S(t^{(l)})} y(s)P(s)/P(t^{(l)})$$
$$= \sum_{l=1}^{L} \sum_{s \in S(t^{(l)})} y(s)\, P(s)$$
$$= \sum_{s \in \mathscr{S}} y(s)\, P(s) = \mathscr{E}\mathscr{Y}.$$

(b) According to (a)

$$\mathscr{E}\mathscr{Y} = \mathscr{E}[\mathscr{E}\{\mathscr{Y}|\mathscr{T}\}]$$

† This inequality is generally referred to as the Rao-Blackwell inequality.

and similarly

$$\mathscr{E}\mathscr{Y}^2 = \mathscr{E}[\mathscr{E}\{\mathscr{Y}^2|\mathscr{T}\}],$$

and so

$$V(\mathscr{Y}) = \mathscr{E}\mathscr{Y}^2 - \mathscr{E}^2\mathscr{Y} = \mathscr{E}[\mathscr{E}\{\mathscr{Y}^2|\mathscr{T}\}] - \mathscr{E}^2[\mathscr{E}\{\mathscr{Y}|\mathscr{T}\}],$$

or, subtracting and adding back $\mathscr{E}\mathscr{E}^2\{\mathscr{Y}|\mathscr{T}\}$:

$$V(\mathscr{Y}) = \mathscr{E}[\mathscr{E}\{\mathscr{Y}^2|\mathscr{T}\} - \mathscr{E}^2\{\mathscr{Y}|\mathscr{T}\}] + \mathscr{E}\mathscr{E}^2\{\mathscr{Y}|\mathscr{T}\} - \mathscr{E}^2[\mathscr{E}\{\mathscr{Y}|\mathscr{T}\}].$$

This equals

$$\mathscr{E}[V\{\mathscr{Y}|\mathscr{T}\}] \qquad\qquad + V(\mathscr{E}\{\mathscr{Y}|\mathscr{T}\}).$$

(c) This follows from the fact that $\mathscr{E}\mathscr{Z} > 0$ if $y(s) \neq x(s)$ for some s with $P(s) > 0$.

REMARKS. These results will be used again and again in the subsequent chapters. (b) can be expressed as: "The variance equals the expectation of the conditional variance plus the variance of the conditional expectation".

In the above example

$$\begin{aligned}
\mathscr{E}\mathscr{Y} &= \{(6 + 6\tfrac{1}{2} + 9 + 10) + \ldots + (1 + 2 + 12\tfrac{1}{2} + 15)\}/16 \\
&= (31\tfrac{1}{2} + 27 + 55 + 30\tfrac{1}{2})/16 = 9; \\
V(\mathscr{Y}) &= \mathscr{E}\mathscr{Y}^2 - \mathscr{E}^2\mathscr{Y} \\
&= \{6^2 + (6\tfrac{1}{2})^2 + 9^2 + 10^2 + \ldots + 15^2\}/16 - 9^2 \\
&= (259\tfrac{1}{4} + 211 + 757\tfrac{1}{2} + 386\tfrac{1}{4})/16 - 81 = 19\tfrac{7}{8}.
\end{aligned}$$

Therefore,

$$\mathscr{E}\mathscr{X} = 9 = \mathscr{E}\mathscr{Y};$$

$$V(\mathscr{X}) = 7\tfrac{89}{128} = 19\tfrac{7}{8} - 12\tfrac{23}{128} = V(\mathscr{Y}) - \mathscr{E}\mathscr{Z}.$$

In the diagram on the following page we are shown $S(t^{(1)})$, $S(t^{(2)})$, $S(t^{(3)})$, $S(t^{(4)})$, and on the left margin is shown $\mathscr{S}$. The crosses show $g(t^{(1)}), \ldots, g(t^{(4)})$, and the mean of $\mathscr{Y}$. We see that the dispersion about $\mathscr{E}\mathscr{Y}$ in $\mathscr{S}$ is much larger than the dispersion in $S(t^{(l)})$ about $g(t^{(l)})$ for each† l (or than the average of these dispersions), and that this fact is due to the difference between the $g(t^{(l)})$.‡

REMARK. $g(t^{(l)})$ is called the "*conditional expectation of* $\mathscr{Y}$*, given* $\mathscr{T}$ *equals* $t^{(l)}$". $h(t^{(l)})$ is called "*the conditional variance of* $\mathscr{Y}$*, given* $\mathscr{T}$ *equals* $t^{(l)}$". g is

† For $S(t^{(4)})$ the range is exactly that of $\mathscr{S}$, but for measures of dispersion in use which depend on more than just the two extreme values, the dispersion in $S(t^{(4)})$ is less than in $\mathscr{S}$.

‡ This is even easier to see in the case in which the dispersion in each $S(t^{(l)})$ is the same but the $g(t^{(l)})$ are different.

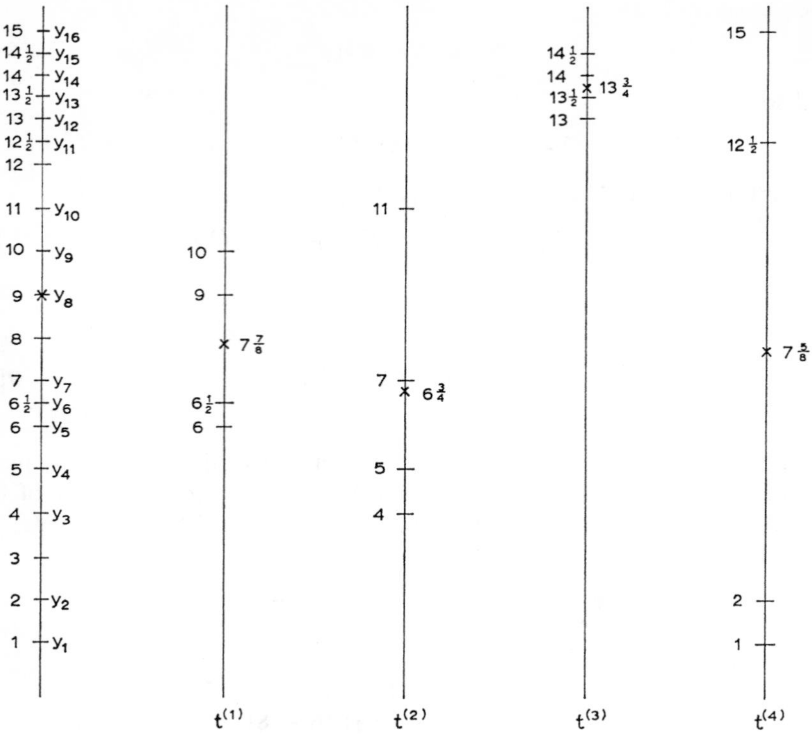

called the *regression function*† *of* $\mathscr{Y}$ *on* $\mathscr{T}$. The numbers $y'(s)$ are called *residuals* from the regression function. If the $h(t^{(l)})$ are equal for all l, we say that the regression is *homoskedastic*, otherwise that it is *heteroskedastic*.

It follows from (c) that if $\mathscr{E}\{\mathscr{Y}|\mathscr{T}\}$ is a statistic, then it is an unbiased estimator of $\mathscr{E}\mathscr{Y}$, of smaller variance than $\mathscr{Y}$, unless it coincides with $\mathscr{Y}$ (i.e., equals it on all samples which have a positive probability). Actually a convenient condition for $\mathscr{E}\{\mathscr{Y}|\mathscr{T}\}$ to be a statistic is known (viz., that $\mathscr{T}$ be socalled "*sufficient* for $\mathscr{Y}$").

EXERCISE. Let $y(s) = 1$ if s has property A and 0 otherwise. Show that $\mathscr{E}\mathscr{Y} = P(A)$, and that $\mathscr{E}\{\mathscr{Y}|\mathscr{T} = t^{(l)}\}$ is the conditional probability of A given $\mathscr{T} = t^{(l)}$: $P(A|t^{(l)})$; it then follows from (a) that $\mathscr{E}P(A|\mathscr{T}) = P(A)$.

† If the $t^{(l)}$ are numbers and, apart from a constant, g is a linear function, g is called a *linear* regression function. Thus, in our example, if $t^{(1)} = 3$, $t^{(2)} = -6$, $t^{(3)} = 50$ and $t^{(4)} = 1$, the regression function of $\mathscr{Y}$ and $\mathscr{T}$ is

$$g(t^{(l)}) = 7\tfrac{1}{2} + \tfrac{1}{8}t^{(l)} \quad (l = 1, \ldots, 4);$$

if in the diagram the $S(t^{(l)})$ were placed at horizontal distances 3, −6, 50 and 1 (centimeters or inches) from some point of origin, the $g(t^{(l)})$ would be *collinear*, lie on a (straight) *line* of slope $\tfrac{1}{8}$ (with height $7\tfrac{1}{2}$ above the point of origin).

9. Preliminaries to a survey

Evidently, for any kind of probability sampling we need to define the units $u_1, \ldots, u_N$ of the population in such a manner that from it the people "in the field" can determine exactly whether or not any given unit belongs to $\mathscr{P}$ or not. In practice such a definition is often hard or even impossible to achieve, but one should strive for it. Also, it often turns out that it is not completely feasible to have the population from which we sample (*the sampled population*) coincide with the one we are really interested in (*the target population*).

For example:

(a) We may wish to survey all those who may possibly be interested in becoming students at a given university during the next four years, should financial aid be available. To be sure not to leave out anyone, we would have to cover not only the local population of high school students, but also potential immigrants, people already out of school, etc. But some of these groups may really be impossible to include in the sampled population.

(b) There is almost always some discrepancy† between the way questions are interpreted by respondents or interviewers and the way they are meant by the one who composed the questionnaire. Let us for instance consider one particular question. We may take as the target population the collection of potential answers to the *intended* meaning of the question. Then the sampled population is the collection of potential answers to the question as *understood* by the potential respondents.

(c) Respondents make factual errors (especially if they have to rely on memory, or have some conscious or unconscious motives for doing so), records contain mistakes, errors are committed in the field and in processing data. These mistakes may be considerable, but the sampled population is the one *with* these errors. Errors such as these are not even uncommon in the physical sciences or technology. (Further information on methods of estimating the dispersion that is caused by such errors is contained in Chapter X.)

(d) Often we may be interested in the state of affairs at a given point of time (for instance, the inventory on May 1st of the current year) or transactions over a well-defined period (for instance, a calendar year), whereas the data available to different respondents may relate to different points or periods of time (for example, respondents use a book year different from the calendar year).

(e) Sometimes it is impossible to obtain information from certain members of the target population, notwithstanding intensive efforts. (We shall discuss this in Chapter X.) The sampled population then excludes these members.

† In order to reduce this discrepancy as much as possible, questionnaires should be pretested by interviewing some members of the population and observing whether they understood the questions in the sense intended by the surveyor.

If the sampled and target populations do not coincide, the final report should clearly point this out.

It is rarely feasible to do the selection by some probabilistic sampling process directly from a human population, a population of farms, apartments, galaxies, accidents, etc. – we cannot toss people, farms, apartments, galaxies, or accidents like dice. Usually a list has to be prepared (or a serial order has to be assigned to all objects in the population in some other way) as a first step towards random sampling by the use of a table of random numbers. This creates what is called the *sampling frame*. In some cases there exists a directory, map† or card file which covers the population, but often these contain inaccuracies, are not up to date, or leave out important parts of the population of interest.

Because many practical difficulties only show up after one puts a plan of survey in operation, it is often advisable first to carry out some of these operations by way of trial. Sometimes even the feasibility of the survey itself needs to be tested. Also some preliminary trials are often useful for training purposes. If the trial itself amounts to a little survey (though usually of a more restricted population) we call it a *pilot survey*. Such pilot surveys are also helpful in getting some preliminary estimates of population characteristics, which will enable us to make the survey design more efficient. Thus it will be seen in later chapters that to determine a reasonably efficient method of sampling, to determine the number of observations needed for getting usable results at nonprohibitive costs by the selected method of sampling, to fix strata if stratification is used, and for many other aspects of the *design* of the survey, we need some advance information about the population. In repeated surveys, such data are already available; in other cases we may get such information from research on populations which we may expect to be similar in respect to the characteristics under discussion (compare the example in section 7).

10. Model-free and other approaches to estimation

There are situations in which, on the basis of previous experience with populations and characteristics which are judged to be similar to the ones under study, one can specify in advance certain mathematical properties of the

† When sampling units for which lists are not available, but which can be located on a map, we generally obtain a simple random sample of coordinates and select the units within which the chosen coordinate points fall. The coordinate net may be a Cartesian one (e.g., if one samples fields); it may constitute equal subdivision of the length of a curve (e.g., if one samples farms in an area where all farms lie on a road – here a curve refers to total extent of the road system). The purpose is to sample with probabilities proportional to the sizes of the fields or the frontages of the farms. This type of sampling is called *area sampling*.

$\mathscr{Y}$-values in $\mathscr{P}$ that may be of great help in the estimation problems previously described. Let us give a simple example.

It has frequently been observed that, if one orders members of a collection of objects by the value of some measure of size from the largest to the smallest: $y_1 \geq y_2 \geq \ldots \geq y_N$, there often is much regularity in the graph of the points (i, y_i). Examples are: firms in a certain industry or trade by turnover, birds by number of species, books by number of words, customers of a firm by amount of purchases, communities of organisms of a certain species by number of individuals, entrepreneurial incomes by magnitude. Typically, in such an ordering, the first few units account for a very substantial part of the aggregate size of the entire collection. For the above examples, a function of the form $y_i = y_1/i$ often gives a rather good fit to the graph.†

Thus, if we examine 1920 Census data for the number of inhabitants of the larger U.S. cities, we find that this function holds approximately. (For later Census years, the approximation gets progressively worse; in part this may be ascribed to the increasing artificiality of the legal boundaries of the cities.) The 1920 population of New York City's five boroughs is reported as 5 620 048. If $y_i = y_1/i$, then

$$Y = y_1 + \ldots + y_{200}$$

is cy_1, where

$$c = \sum_{i=1}^{200} 1/i = 5.878\,03.$$

So, if the formula were exact, we would have $Y = 33\,035\,000$, whereas the Census gave 34 053 000.

Suppose we wish to estimate

$$Z = z_1 + \ldots + z_{200},$$

the total number of inhabitants of the 200 cities in 1921. At this point we cannot discuss the different ways in which this might be done, but it is perhaps intuitively clear that estimation of Z by c times an efficient estimator of z_1 will usually be much cheaper than any other estimator of comparable precision.

In most problems of this kind we do not expect that a simple formula like the above holds exactly, but at best that it leads to accurate estimates of certain overall characteristics. Thus we would not expect to obtain very good estimates of each one of $y_2, y_3, y_4, \ldots$ by division of y_1 by 2, 3, 4, . . ., respectively. Indeed, whereas $y_1/y_2 = 2.1$ and $y_1/y_3 = 3.1$, we have $y_1/y_4 = 5.7$. As will be clarified in Chapter VIII, it often turns out to be helpful to describe a situation like this by saying that $y_1, \ldots, y_N$ are *not constants*, but are *random variables;* that is, that they may be looked upon as themselves

† There are numerous examples in G. K. Zipf, Human Behavior and the Principle of Least Effort (Addison Wesley 1944), who also considers the formula $y_i = y_1/i^{\alpha}$ with α a positive constant.

having been obtained by a (*conceptual*) *sampling process* from some population or set of populations for which certain simple relationships hold,† giving rise to a *sampling distribution for* $y_1, \ldots, y_N$! These matters are further discussed in the chapter on models. In contrast to the probabilities so far discussed, those involved in this sampling distribution are not due to the way we sample from $\mathscr{P}$. In this book, when considering estimation problems, we have *not* supposed that we know this sampling distribution, nor that we know that it belongs to a specific family of such distributions (such as a family for which α defined in the first footnote lies between 0.8 and 1.5 and has some further properties stated in the reference of the second footnote). We indicate this by saying that in this book we largely follow a model-free approach. In cases in which such knowledge is available,‡ it may be grossly inefficient not to utilize it. Frequently, however, our knowledge is far from certain; it may then only be expressible by specifying as known an extremely wide family of distributions, and this is seldom helpful.§

Another point we should already note at this stage is that the y_i yielded by the Census are not the actual sizes of the cities; this has often been shown by critical analysis of Census data and by so called postenumerative surveys. A principal reason for this is the simple fact that some people are overlooked in the enumeration process. No doubt, if instantly afterwards another count would be made, some of these would not be overlooked (although others would). For that reason alone, each y_i should be considered a sample of 1 from a distribution of possible values, distributed (more or less symmetrically) not about the actual size (to wit, the size in the target population), but about some central value ν_i, which is less than the actual size μ_i (compare (c) in the previous section). A way of examining this type of question is to consider the differences $e_i = y_i - \nu_i$ as constituting a sample from a *population* Π *of measurement errors* with zero mean. This is discussed in more detail in section 6.2 of Chapter VIII. In contrast to the case mentioned above, one needs only a minimum of information about Π to be able to draw some helpful conclusions, and such information can often be obtained if the survey is designed with this in view (see Chapter XI). Here we merely remark that it may obviously be desirable to make a confidence interval reflect not only the uncertainty due to sampling from $\mathscr{P}$, but also the uncertainty introduced by the measurement errors.

† Probabilistic mechanisms that may account for results of the above type were first considered by Yule in 1924 and are discussed in detail by SIMON, Biometrika, **42** (1955) 425–440.

‡ To *a very limited extent* the data from the survey itself may serve to test "reasonable assumptions" about the family of distributions involved (methods of doing this fall outside the scope of this book).

§ In questions of *design* our attitude is to allow even for untested reasonable assumptions, when using these will at worst lead to an inefficient design, but not to invalid estimators.

CHAPTER II

SIMPLE RANDOM SAMPLING

(Including Simple Cluster Sampling)

1. With replacement

In sampling "with replacement" we take a sequence of length n from the collection of N different units $\{u_1, \ldots, u_N\}$. The sample space consists of all the N^n different possible sequences. Sampling is simple random sampling if every point (here: every possible sequence of n) has equal probability.†

2. Without replacement

In sampling "without replacement" we take a subset of size n different units from the set (population). The sample space consists of all the different possible subsets of size n. The number of points in the sample space is

$$N^{(n)} = N(N-1)\ldots(N-n+1) \quad \text{or} \quad \binom{N}{n},$$

depending on whether we take all ordered subsets, or all unordered subsets. Often we wish to disregard order, and thus, for example, not to distinguish between the subset which consists of u_1 followed by u_2, and the subset which consists of u_2 followed by u_1. The notation using curly brackets stands for unordered sets. Sampling is simple random sampling if every sample point (here: every possible subset of size n) has the same probability.‡

† It follows from the symmetry of the sampling process that every unit of $\mathscr{P}$ has the same probability to fall in the sample, but the latter property is clearly insufficient to characterize simple sampling with replacement. E.g., if $\mathscr{P} = \{u_1, u_2, u_3\}$ and $P(u_1, u_1) = 3/23$, $P(u_1, u_2) = P(u_2, u_1) = 2/23$, $P(u_1, u_3) = P(u_3, u_1) = 2/23$, $P(u_2, u_2) = 5/23$, $P(u_2, u_3) = P(u_3, u_2) = 1/23$ and $P(u_3, u_3) = 5/23$, then each unit has a probability of falling in the sample of 11/23. Moreover, there are forms of simple random sampling for which the property does not hold. Thus, if $\mathscr{S}$ consists of (u_1, u_2) and (u_1, u_3), and both have the same probability, then the probability for u_i to be in the sample is 1 for $i = 1$, and $\frac{1}{2}$ for $i = 2$ and for $i = 3$.

‡ From the symmetry of the sampling process it follows that every unit of $\mathscr{P}$ has the same probability of falling in the sample. That property does not characterize simple sampling without replacement. Thus, if $\mathscr{P} = \{u_1, u_2, u_3, u_4\}$ and $P\{u_1, u_2\} = 1/12 = P\{u_3, u_4\}$, $P\{u_1, u_3\} = 2/12 = P\{u_2, u_4\}$, and $P\{u_1, u_4\} = 3/12 = P\{u_2, u_3\}$, then each unit has probability $\frac{1}{2}$ of falling in the sample. See also the last part of the previous footnote.

3. Notation

We denote by Y the *total* for the values of the characteristic $\mathscr{Y}$ for all the elements of the population. By $\bar{Y}$ we denote Y/N, the population *average*. By y we denote the *total* of the values of $\mathscr{Y}$ for all the elements in the sample, and by $\bar{y}$ we denote y/n (the sample *average*). We shall usually write

$$y = \sum_{i=1}^{N} y_i t_i$$

for sampling with replacement, where t_i is the number of times that the ith unit comes into the sample, (so that

$$\sum_{i=1}^{N} t_i = n);$$

and

$$y = \sum_{i=1}^{N} y_i a_i,$$

for sampling without replacement, where a_i is 1 if u_i comes into the sample, and zero if it is not in the sample (so that for such sampling

$$\sum_{i=1}^{N} a_i = n).\dagger$$

Most writers denote the sum of the values of the characteristic $\mathscr{Y}$ in the sample by

$$\sum_{i=1}^{n} y_i.$$

NOTE. a_i and t_i are random variables; y_i is *not* a random variable. When the sampling with or without replacement is simple, it follows from the symmetry of the sampling process that $\mathscr{E}a_i$ and $\mathscr{E}t_i$ do not depend on i and so are equal to the expectation of

$$N^{-1}\Sigma a_i = n/N = N^{-1}\Sigma t_i.$$

As mentioned above, in the population we define σ^2 (or $\sigma^2(\mathscr{Y})$), the measure of dispersion of the values of $\mathscr{Y}$, by

$$\sigma^2 = \sum_{i=1}^{N} (y_i - \bar{Y})^2/N.$$

† Note that in sampling without replacement $t_i = a_i$, so that one may use t_i in both cases. Nonetheless, in the former case we shall usually prefer the notation a_i, which may serve to remind us of the fact that there is no replacement.

(This is often also referred to as the *variance* of $\mathscr{Y}$ in $\mathscr{P}$, though it is not a sampling variance as defined in Chapter I, section 5.) It is convenient also to define S^2 (or $S^2(\mathscr{Y})$) by

$$S^2 = \sigma^2 N/(N-1) = \sum_{i=1}^{N} (y_i - \bar{Y})^2/(N-1).$$

We may call S^2 the *adjusted* variance of $\mathscr{Y}$ in $\mathscr{P}$. In an analogous way we define s^2 (or $s^2(\mathscr{Y})$) by:

$$\sum_{i=1}^{N} (y_i - \bar{y})^2 t_i/(n-1)$$

for sampling with replacement,

$$\sum_{i=1}^{N} (y_i - \bar{y})^2 a_i/(n-1)$$

for sampling without replacement.

REMARK. For computing S^2, it is more efficient to use, instead of $\sum (y_i - \bar{Y})^2/(N-1)$,

$$\left\{\sum_{i=1}^{N} y_i^2 - \frac{1}{N}(\Sigma y_i)^2\right\}\Big/(N-1),$$

because in this formula there are fewer stages of computation in which we may lose accuracy due to rounding.

4. The sampling distribution

Let $\mathscr{T}$ be a statistic over the sample space $\mathscr{S}$ (for instance $\mathscr{T}$ may be $\bar{y}$). $\mathscr{T}$ takes on different values – these values constitute the sample space of $\mathscr{T}$ – and any such value has a probability which equals the sum of the probabilities of the points of the sample space $\mathscr{S}$ for which $\mathscr{T}$ has this value. The values of $\mathscr{T}$ and their probabilities constitute the sampling distribution of $\mathscr{T}$. The expected value of $\mathscr{T}$, $\mathscr{E}\mathscr{T}$, is obtained by summing the products of the values of $\mathscr{T}$ by the probabilities of these values. The variance of $\mathscr{T}$, which we denote by $V(\mathscr{T})$, is given by $\mathscr{E}(\mathscr{T}^2) - (\mathscr{E}\mathscr{T})^2$.

This and the following section are illustrated by the example of section 6 below.

5. Four propositions

We shall first state two propositions which allow us to obtain $\mathscr{E}\bar{y}$ and $V(\bar{y})$ without having to write out the sampling distribution of $\bar{y}$:

(a) $\mathscr{E}\bar{y} = \bar{Y}$ for both methods of sampling.
(b) In sampling without replacement

$$V(\bar{y}) = \frac{N - n}{N} S^2/n,$$

in sampling with replacement $V(\bar{y}) = \sigma^2/n$.

The formulas for $V(\bar{y})$ contain the quantity S^2 (or σ^2), which as a rule is unknown to us. However, the following proposition states the unbiasedness of certain estimators:

(c) In sampling without replacement $\mathscr{E}s^2 = S^2$, in sampling with replacement $\mathscr{E}s^2 = \sigma^2$. So if $v(\bar{y})$ denotes the formulas for $V(\bar{y})$ with S^2 replaced by s^2:

(d) $v(\bar{y})$ is an unbiased estimator of $V(\bar{y})$ for both methods of sampling.

Proof of (a):

$$\mathscr{E}\bar{y} = n^{-1}\sum_{i=1}^{N} y_i \mathscr{E}a_i \quad \text{or} \quad n^{-1}\sum_{i=1}^{N} y_i \mathscr{E}t_i,$$

and, by the "Note" in section 3,

$$\mathscr{E}a_i = \mathscr{E}t_i = n/N.$$

Proof of (b):

Let $a_{ij} = 1$ if u_i is drawn in the jth draw, and zero otherwise. Note that, for any given set of $j \leq n$, the a_{ij} are defined independently of n. Let

$$g_j = \sum_{i=1}^{N} a_{ij} y_i.$$

It is $\bar{y}$ in a simple random sample of size 1, and so g_1 has variance

$$\Sigma(y_i - \bar{Y})^2/N = \sigma^2.$$

By the symmetry of simple random sampling, $V(g_j)$ is the same for each $j = 1, \ldots, n$, and

$$\gamma = \operatorname{Cov}(g_j, g_k)$$

is the same for each pair (j, k) of different drawings.

Now

$$\bar{y} = \sum_{j=1}^{n} g_j/n,$$

so

$$V(\bar{y}) = \left\{\sum_{j=1}^{n} \sigma^2 + \sum_{j \neq k}\sum \gamma\right\}\Big/ n^2 = \{\sigma^2 + (n-1)\gamma\}/n.$$

In the case of sampling with replacement, g_j and g_k are independent, hence $\gamma = 0$, and $V(\bar{y}) = \sigma^2/n$.

For the case of sampling without replacement, we examine first the special case $n = N$, in which $\bar{y}$ is a constant, namely $\bar{Y}$, and so has variance 0:

$$0 = \{\sigma^2 + (N-1)\gamma\}/n.$$

So in this special case

$$\gamma = -\frac{1}{N-1}\,\sigma^2 = -\frac{1}{N}\,S^2.$$

But, as g_j is independent of n, so is γ. Therefore,

$$\gamma = -\frac{1}{N}\,S^2$$

for *any* n, and

$$V(\bar{y}) = \left\{\frac{N-1}{N}\,S^2 - \frac{n-1}{N}\,S^2\right\}\Big/ n = \frac{N-n}{N}\,S^2/n.$$

Proof of (c):

Denoting by $u(1), \ldots, u(n)$ the units successively drawn,

$$\begin{aligned}\mathscr{E}(n-1)s^2 &= \mathscr{E}\Sigma[(y(\alpha) - \bar{Y}) - (\bar{y} - \bar{Y})]^2 \\ &= \Sigma\mathscr{E}(y(\alpha) - \bar{Y})^2 - n\mathscr{E}(\bar{y} - \bar{Y})^2 \\ &= n\sigma^2 - nV(\bar{y}).\end{aligned}$$

Substituting (b) this gives for sampling with replacement

$$\mathscr{E}s^2 = \frac{n}{n-1}\left\{\sigma^2 - \frac{1}{n}\,\sigma^2\right\} = \sigma^2,$$

and for sampling without replacement

$$\mathscr{E}s^2 = \frac{n}{n-1}\left\{\frac{N-1}{N}\,S^2 - \frac{N-n}{N}\frac{1}{n}\,S^2\right\} = S^2.$$

For alternative proofs see Remarks (iv) and (v) below.

REMARKS. (i) The layman often holds the opinion that the precision† of an estimator is an increasing function of the sampling fraction n/N only.

† *Precision* is defined as the reciprocal of variance.

Proposition (b) shows that, in simple random sampling with replacement, the precision of $\bar{y}$ is an increasing function of n (rather than of n/N), and that this still approximately characterizes the situation of simple† random sampling without replacement, provided n/N is small.

(ii) We see that, in estimating $\bar{Y}$ by $\bar{y}$, sampling of $n > 1$ without replacement leads to a smaller standard error than sampling of n with replacement, as $(N - n)/(N - 1) < 1$.‡

(iii) If $\hat{Y} = N\bar{y}$, then $\mathscr{E}\hat{Y} = Y$ and $V(\hat{Y}) = N^2 V(\bar{y})$.

(iv) In Chapter I, section 5.4 it is noted sub (iv) that

$$V(\bar{y}) = \mathscr{E}\bar{y}^2 - \bar{Y}^2,$$

so that, if $v(\bar{y})$ is any unbiased estimator of $V(\bar{y})$,

$$\bar{Y}^2 = \mathscr{E}\{\bar{y}^2 - v(\bar{y})\}.$$

Now

$$\frac{1}{n}\,\Sigma y(\alpha)^2 - \{\bar{y}^2 - v(\bar{y})\} = (n - 1)s^2/n + v(\bar{y}), \qquad (*)$$

and, since by (a) the mean of a sample average equals the population mean,

$$\mathscr{E}\,\frac{1}{n}\,\Sigma y(\alpha)^2 = \frac{1}{N}\,\Sigma y_i^2.$$

Therefore the former expression (*) is an unbiased estimator of

$$\sigma^2 = \frac{1}{N}\,\Sigma y_i^2 - \bar{Y}^2,$$

and

$$\mathscr{E}(n - 1)s^2 = n\{\sigma^2 - V(\bar{y})\},$$

as proved differently above. This indirect approach is often useful in more complicated situations.

(v) Another interesting proof of (c) is based on the observation that $2S^2$ equals the average of $(y_i - y_j)^2$ over all different pairs of $\mathscr{Y}$-values:

$$2S^2 = \binom{N}{2}^{-1} \sum_{i>j}\sum (y_i - y_j)^2,$$

† In fact, in most nonsimple random sampling situations that characteristic continues to hold.

‡ This result has been generalized to a wider class of loss functions than the one implied here (cf. the end of section 5 of Chapter I) by Hoeffding and Rosén; see Rosén, Ann. Math. Statist., **38** (1967) 382–392. Rosén considers also methods of sampling with equal probabilities other than the two methods we consider. Karlin has since then obtained still more general results. In Chapter IV we study sampling with replacement and the estimator $\Sigma a_i y_i / \Sigma a_i$ and find that it may sometimes be preferable to sampling without replacement using $\bar{y}$.

and also (see footnote in section 3)

$$2s^2 = \binom{n}{2}^{-1} \sum_{i>j}\sum t_i t_j (y_i - y_j)^2.$$

We need only substitute (for $i \neq j$):

for sampling with replacement: $\mathscr{E} t_i t_j = n(n-1)/N^2$,
for sampling without replacement: $\mathscr{E} t_i t_j = n(n-1)/N(N-1)$,

which relations follow from the fact that u_i may be obtained in one of the n drawings and u_j in one of the $n-1$ remaining ones; and that in the first drawing all N units and in the second all remaining units (which in case of replacement means: all N units) have an equal chance of being drawn.†

(vi) A direct proof of (b) uses $\mathscr{E} t_i t_j$ given in (v), and

$$V(t_i) = n\frac{1}{N}\left(1 - \frac{1}{N}\right)$$

for sampling with replacement (binominal case of n independent trials each with probability N^{-1}) and

$$V(t_i) = \mathscr{E} t_i^2 - (\mathscr{E} t_i)^2 = \mathscr{E} t_i - (\mathscr{E} t_i)^2 = \frac{n}{N}\left(1 - \frac{n}{N}\right)$$

for sampling without replacement. Accordingly

$$n^2 V(\bar{y}) = \Sigma y_i^2 V(t_i) + \sum_{i \neq j}\sum y_i y_j \operatorname{Cov}(t_i, t_j)$$

equals for sampling without replacement

$$\frac{N-n}{N}\frac{n}{N}\left(\Sigma y_i^2 - \frac{1}{N-1}\sum_{i \neq j}\sum y_i y_j\right) = \frac{N-n}{N}\frac{n}{N}\left(\frac{N}{N-1}\Sigma y_i^2 - \frac{1}{N-1}Y^2\right),$$

and for sampling with replacement

$$\frac{n}{N}\left\{\Sigma y_i^2\left(1 - \frac{1}{N}\right) - \sum_{i \neq j}\sum y_i y_j \frac{1}{N}\right\} = \frac{n}{N}\left(\Sigma y_i^2 - \frac{1}{N}Y^2\right),$$

where in both cases we used

$$Y^2 = \Sigma y_i^2 + \sum_{i \neq j}\sum y_i y_j.$$

† In terms of a_{ij} (defined as 1 when the jth draw yields u_i and 0 otherwise) we get, when $i \neq j$,

$$\mathscr{E} t_i t_j = \sum_{k \neq l}\sum \mathscr{E} a_{ik} a_{jl} = n(n-1)\mathscr{E} a_{i1} a_{j2},$$

which equals $n(n-1)/N^2$ and $n(n-1)/N(N-1)$, respectively.

(vii) (b) may also be proved quite simply using (c). As the sample sum of the y_i^2 is $(n-1)s^2 + n\bar{y}^2$,

$$\bar{y}^2 = \text{sample av } (y_i^2) - \frac{n-1}{n} s^2,$$

which by (a) and (c) has expected value

$$\Sigma y_i^2/N - \frac{n-1}{n}\sigma^2 \quad \text{or} \quad \Sigma y_i^2/N - \frac{n-1}{n} S^2,$$

respectively in the two cases. Therefore

$$V(\bar{y}) = \mathscr{E}\bar{y}^2 - \bar{Y}^2$$

is

$$\sigma^2 - \frac{n-1}{n}\sigma^2 = \frac{1}{n}\sigma^2$$

or

$$\sigma^2 - \frac{n-1}{n} S^2 = \left(\frac{N-1}{N} - \frac{n-1}{n}\right) S^2 = \frac{N-n}{Nn} S^2,$$

respectively.

A proof along quite different lines is given in section 6.3 of Chapter VIII.

6. Example

The population is

$$\mathscr{P} = \{u_1, u_2, u_3, u_4\},$$

and $\mathscr{Y}$ is a property which for these 4 units has the values

$$y_1 = 3, \quad y_2 = 4, \quad y_3 = 3, \quad y_4 = 5.$$

Computing according to our formulae, we obtain

$$\bar{Y} = 3\tfrac{3}{4}, \quad \sigma^2 = \tfrac{11}{16}, \quad S^2 = \tfrac{11}{12}.$$

Let us take, with and without replacement, all the possible samples of size 2, and let us get the distributions of $\bar{y}$, s^2 and $v(\bar{y})$. From the above propositions we expect to obtain $\mathscr{E}\bar{y} = \bar{Y} = 3\frac{3}{4}$ in both cases, $\mathscr{E}(s^2) = S^2 = \frac{11}{12}$ for sampling without replacement, and $\mathscr{E}(s^2) = \sigma^2 = \frac{11}{16}$ for sampling with replacement. Also in both cases we expect to come out with $\mathscr{E}v(\bar{y}) = V(\bar{y})$; i.e., $\frac{1}{4}(4-2) \times \frac{1}{2} \times \frac{11}{12} = \frac{11}{48}$ for sampling without replacement, and $\frac{1}{2} \times \frac{11}{16} = \frac{11}{32}$ for sampling with replacement, according to the formulae.

Possible samples of size 2 without replacement

Sample	Values of $\mathscr{Y}$	$\bar{y}$	s^2	
$\{u_1, u_2\}$	3,4	$3\frac{1}{2}$	$\frac{1}{2}$	
$\{u_1, u_3\}$	3,3	3	0	
$\{u_1, u_4\}$	3,5	4	2	
$\{u_2, u_3\}$	4,3	$3\frac{1}{2}$	$\frac{1}{2}$	
$\{u_2, u_4\}$	4,5	$4\frac{1}{2}$	$\frac{1}{2}$	
$\{u_3, u_4\}$	3,5	4	2	
		$22\frac{1}{2}$	$5\frac{1}{2}$	Total
		$3\frac{3}{4}$	$\frac{11}{12}$	Average
		$\frac{11}{48}$	$\frac{89}{144}$	MSD†

† MSD = mean square deviation, see below.

Distribution of $\bar{y}$

First Moment			Second Central Moment			
Values	f‡	Product	$(y - 3\frac{3}{4})^2$	f	Product	
3	$\frac{1}{6}$	$\frac{1}{2}$	$\frac{9}{16}$	$\frac{1}{6}$	$\frac{3}{32}$	
$3\frac{1}{2}$	$\frac{2}{6}$	$\frac{7}{6}$	$\frac{1}{16}$	$\frac{2}{6}$	$\frac{1}{48}$	
4	$\frac{2}{6}$	$\frac{4}{3}$	$\frac{1}{16}$	$\frac{2}{6}$	$\frac{1}{48}$	
$4\frac{1}{2}$	$\frac{1}{6}$	$\frac{3}{4}$	$\frac{9}{16}$	$\frac{1}{6}$	$\frac{3}{32}$	
	1	$3\frac{3}{4}$		1	$\frac{11}{48}$	Total

‡ f is the relative frequency.

Distribution of s^2			Distribution of $v(\bar{y})$†		
Values	f	Product	Values	f	Product
0	$\frac{1}{6}$	0	0	$\frac{1}{6}$	0
$\frac{1}{2}$	$\frac{3}{6}$	$\frac{1}{4}$	$\frac{1}{8}$	$\frac{3}{6}$	$\frac{1}{16}$
2	$\frac{2}{6}$	$\frac{2}{3}$	$\frac{1}{2}$	$\frac{2}{6}$	$\frac{1}{6}$
First moment		$\frac{11}{12}$ Total	First moment		$\frac{11}{48}$ Total

† $v(\bar{y}) = \frac{1}{4}(4 - 2) \times \frac{1}{2}s^2 = \frac{1}{4}s^2$.

Possible samples of size 2 with replacement

Sample	Values of $\mathscr{Y}$	Values of $\bar{y}$	Values of s^2	
(u_1, u_1)	3,3	3	0	
(u_1, u_2)	3,4	$3\frac{1}{2}$	$\frac{1}{2}$	
(u_1, u_3)	3,3	3	0	
(u_1, u_4)	3,5	4	2	
(u_2, u_1)	4,3	$3\frac{1}{2}$	$\frac{1}{2}$	
(u_2, u_2)	4,4	4	0	
(u_2, u_3)	4,3	$3\frac{1}{2}$	$\frac{1}{2}$	
(u_2, u_4)	4,5	$4\frac{1}{2}$	$\frac{1}{2}$	
(u_3, u_1)	3,3	3	0	
(u_3, u_2)	3,4	$3\frac{1}{2}$	$\frac{1}{2}$	
(u_3, u_3)	3,3	3	0	
(u_3, u_4)	3,5	4	2	
(u_4, u_1)	5,3	4	2	
(u_4, u_2)	5,4	$4\frac{1}{2}$	$\frac{1}{2}$	
(u_4, u_3)	5,3	4	2	
(u_4, u_4)	5,5	5	0	
		60	11	Total
		$3\frac{3}{4}$	$\frac{11}{16}$	Average
		$\frac{11}{32}$	$\frac{159}{256}$	MSD‡

‡ MSD = mean square deviation, see below.

Distribution of $\bar{y}$

First Moment			Second Central Moment		
Values	f†	Product	$(\bar{y} - 3\frac{3}{4})^2$	f	Product
3	$\frac{4}{16}$	$\frac{12}{16}$	$\frac{9}{16}$	$\frac{4}{16}$	$36/(16)^2$
$3\frac{1}{2}$	$\frac{4}{16}$	$\frac{14}{16}$	$\frac{1}{16}$	$\frac{4}{16}$	$4/(16)^2$
4	$\frac{5}{16}$	$\frac{20}{16}$	$\frac{1}{16}$	$\frac{5}{16}$	$5/(16)^2$
$4\frac{1}{2}$	$\frac{2}{16}$	$\frac{9}{16}$	$\frac{9}{16}$	$\frac{2}{16}$	$18/(16)^2$
5	$\frac{1}{16}$	$\frac{5}{16}$	$\frac{25}{16}$	$\frac{1}{16}$	$25/(16)^2$
	1	$3\frac{3}{4}$		1	$\frac{11}{32}$ Total

† f is the relative frequency.

Distribution of s^2			Distribution of $v(\bar{y})$‡		
Values	f	Product	Values	f	Product
0	$\frac{6}{16}$	0	0	$\frac{6}{16}$	0
$\frac{1}{2}$	$\frac{6}{16}$	$\frac{3}{16}$	$\frac{1}{4}$	$\frac{6}{16}$	$\frac{3}{32}$
2	$\frac{4}{16}$	$\frac{8}{16}$	1	$\frac{4}{16}$	$\frac{8}{32}$
First moment		$\frac{11}{16}$ Total	First moment		$\frac{11}{32}$ Total

‡ $v(\bar{y}) = \frac{1}{2}s^2$.

The "averages" shown above are the expectations, since each sample point has equal probability. Similarly, the mean square deviation (=average of the squared deviations from the average) equals the variance.

Also shown above is the computation of the first moments of the distributions of $\bar{y}$ and s^2, and of the second moment of the distribution of $\bar{y}$. These computations verify that, as mentioned in Chapter I, section 5, these moments are equal to the mean and variance of these respective statistics.

7. Confidence intervals and bounds for $\bar{Y}$ or Y§

As mentioned in section 5 of Chapter I, and as will be expanded upon in section 15, if n is not very small, $\bar{y}$ is often close to normally distributed; so

§ A discussion of confidence bounds for $\sigma^2(\mathscr{Y})$ and $\rho(\mathscr{X}\mathscr{Y})$ falls outside the scope of this book; confidence intervals for Y/X are discussed in section 4 of Chapter IIA.

that we can construct a confidence interval for $\bar{Y}$ and Y, with a confidence coefficient approximately equal to some desired value β, on the basis of $\bar{y}$ if $\sigma^2(\mathscr{Y})$ is known, or on the basis of $\bar{y}$ and $s^2(\mathscr{Y})$ otherwise. The interval generally used in that case is of the form

$$[\bar{y} - t\, V(\bar{y})^{\frac{1}{2}}, \bar{y} + t\, V(\bar{y})^{\frac{1}{2}}],$$

or

$$[\bar{y} - t\, v(\bar{y})^{\frac{1}{2}}, \bar{y} + t\, v(\bar{y})^{\frac{1}{2}}],$$

respectively, where t depends on the choice of β.

If $\bar{y}$ is exactly normally distributed, we can attain exactly any desired probability β that the interval contains $\bar{Y}$. Thus, the first interval with $t = 1.96$ has $\beta = 0.95$. The constant t appearing in the second interval depends also on n; the reason for this is that $(\bar{y} - \bar{Y})v(\bar{y})^{-\frac{1}{2}}$ is not normally distributed but has the so-called *Student or t distribution*. For this case t always exceeds the corresponding normal value (i.e., the value found in a table of the standard normal distribution), but for given β, as n increases, the t-values come reasonably close to the normal value. For example, with $\beta = 0.95$, and 6 observations, the t-value is 2.57 for the Student distribution (as against 1.96 for the normal distribution), but for 12 observations it is already 2.20, for 30 observations 2.04. Since, in most survey applications, approximate normality will not usually be appealed to unless we have several tens of observations (and usually a good deal more), we shall generally just use the normal values.†

Confidence intervals for Y are obtained by multiplication by N of those for $\bar{Y}$. Upper (lower) confidence bounds‡ for $\bar{Y}$ of confidence coefficient $\beta = 1 - \delta$ are obtained by taking the upper (lower) limit of the confidence interval for $\bar{Y}$ of confidence coefficient $1 - 2\delta$. For example the normal value for $\beta = 1 - 2\delta = 0.90$ is 1.64; therefore, the normal upper confidence bound for $\bar{Y}$ with confidence coefficient $0.95 = 1 - 0.05 = 1 - \delta$ is $\bar{y} + 1.64 V(\bar{y})^{\frac{1}{2}}$.

To illustrate the meaning of a confidence coefficient, we shall present an example. For this purpose we take again a small population and a small sample, so that we may exhibit the entire sample space. For each point in the sample space we compute the "value" of the interval computed by the above formula (which assumes that $\bar{y}$ is normal – not a bad approximation in this

† In the beginning of section 14.2 of Chapter III we shall mention a method which occasionally may be useful in simple random sampling and involves the use of Student's distribution.

‡ U_z is an *upper confidence bound*, with confidence coefficient β, for a certain parameter if the interval from $-\infty$ (or the lowest possible value of the parameter) to U_z is a confidence interval for this parameter with confidence coefficient β. If $\beta = 0.95$ and x is normally distributed with mean μ and variance σ_n^2, $U_z = x + 1.645\sigma_n$, and $\mu \leqslant U_z$ is equivalent with $(x - \mu)/\sigma_n \geq -1.645$.

example) and then find *the fraction of the sample points for which the interval so computed contains* $\bar{Y}$. In particular, there are 28 sample points, and for 26 of them the confidence interval computed from the formula

$$[\bar{y} - 2.57v(\bar{y})^{\frac{1}{2}}, \bar{y} + 2.57v(\bar{y})^{\frac{1}{2}}]$$

contains the correct parameter value $\bar{Y} = 7$, i.e., the fraction is $\frac{26}{28}$. This should be compared with 0.95, the theoretical value of β for $\bar{y}$ exactly normal. (In that case, since the distribution is continuous, the "fraction" is not limited to multiples of $\frac{1}{28}$; so the approximation is as close as is possible in our example.)

EXAMPLE

$$\mathscr{P} = \{u_1, \ldots, u_8\};$$
$$y(u_1) = 4, \quad y(u_2) = y(u_3) = 6, \quad y(u_4) = 7,$$
$$y(u_5) = y(u_6) = y(u_7) = 8, \quad y(u_8) = 9.$$

So

$$\bar{Y} = 7, \quad S^2 = 2.571\,428\,6.$$

We take samples of 6 by simple random sampling without replacement. Therefore

$$V(\bar{y}) = \frac{8-6}{8}\,\frac{1}{6}\,S^2 = 0.107\,142\,86, \quad v(\bar{y}) = \frac{8-6}{8}\,\frac{1}{6}\,s^2.$$

The adjoined table shows the sample points in so-called "dictionary" order (the "alphabet" being 1, 2, . . ., 8) with the values for $\bar{y}$, s^2, $v(\bar{y})^{\frac{1}{2}}$ and for the confidence interval.

If S^2 were known, the confidence interval with confidence coefficient 0.95 would be

$$[\bar{y} - (1.96)(0.327\,1), \quad \bar{y} + (1.96)(0.327\,1)],$$

and from the $\bar{y}$-values in the table one finds that for a fraction $\frac{26}{28}$ of the sample points the interval includes 7.

We can also get from the table the fraction of the sample points for which the upper confidence bound is not less than 7 (which theoretically is 0.975) namely $\frac{28}{28}$. (The fact that the approximation gives worse results for finding upper or lower confidence bounds than for finding confidence intervals, is of common occurrence for values of n for which we do not get an almost perfect approximation for the latter.)

EXERCISE. Find the normal upper confidence bounds of confidence coefficient 0.95, and the fraction of the points in the sample space for which its value is greater than or equal to 7 (S unknown).

Now consider the class of all problems for which someone will (at one time or other) compute confidence intervals of confidence coefficient β. In each particular case, the parameter in which he will be interested may fall in the interval to be computed, or it may not. If the class of such problems is very large, the fraction of cases in which the parameter of interest will fall in the

Explicit computation of the distribution of the confidence limits from samples of six from $\mathscr{P}$

Indices of units in the samples	Values of $\mathscr{Y}$	Values of $\bar{y}$	Values of s^2	Values of $v(\bar{y})^{1/2}$	Values of $\bar{y} - 2.57v(\bar{y})^{1/2}$	Values of $\bar{y} + 2.57v(\bar{y})^{1/2}$
123456	466788	6.500	2.30	0.3096	5.70	7.30
12345 7	46678 8	6.500	2.30	0.3096	5.70	7.30
12345 8	46678 9	6.667	3.07	0.3564	5.75	7.58
1234 67	4667 88	6.500	2.30	0.3096	5.70	7.30
1234 6 8	4667 8 9	6.667	3.07	0.3564	5.75	7.58
1234 78	4667 89	6.667	3.07	0.3564	5.75	7.58
123 567	466 888	6.667	2.67	0.3332	5.81	7.52
123 56 8	466 88 9	6.833	3.07	0.3742	5.87	7.80
123 5 78	466 8 89	6.833	3.07	0.3742	5.87	7.80
123 678	466 889	6.833	3.07	0.3742	5.87	7.80
12 4567	46 7888	6.833	2.57	0.3271	5.99	7.67
12 456 8	46 788 9	7.000	3.20	0.3606	6.07	7.93
12 45 78	46 78 89	7.000	3.20	0.3606	6.07	7.93
12 4 678	46 7 889	7.000	3.20	0.3606	6.07	7.93
12 5678	46 8889	7.167	3.37	0.3742	6.20	8.13
1 34567	4 67888	6.833	2.57	0.3271	5.99	7.67
1 3456 8	4 6788 9	7.000	3.20	0.3606	6.07	7.93
1 345 78	4 678 89	7.000	3.20	0.3606	6.07	7.93
1 34 678	4 67 889	7.000	3.20	0.3606	6.07	7.93
1 3 5678	4 6 8889	7.167	3.37	0.3742	6.20	8.13
1 45678	4 78889	7.333	3.07	0.3564	6.42	8.25
234567	667888	7.167	0.97	0.2069	6.63	7.70
23456 8	66788 9	7.333	1.43	0.2443	6.71	7.96
2345 78	6678 89	7.333	1.43	0.2443	6.71	7.96
234 678	667 889	7.333	1.43	0.2443	6.71	7.96
23 5678	66 8889	7.500	1.50	0.2500	6.86	8.14
2 45678	6 78889	7.667	1.07	0.2000	7.15	8.18
345678	678889	7.667	1.07	0.2000	7.15	8.18

interval is (in the case of independence) at most a number rather close to β with probability extremely close to 1. E.g., for $\beta = 0.95$ and with 1 000 problems in the class, the probability that the fraction is less than 0.93 is found from the tables to be 0.002.† If the person in question *always* takes the para-

† $\sum_{k=0}^{929} \binom{1\,000}{k} (0.95)^k (0.05)^{1000-k}$.

meter to lie in the interval he computes, he *may be expected to be correct* in at least 930 of these 1 000 problems. This is *a possible interpretation* of the statement that the procedure for determining the interval from the sample leads to a probability β of covering the parameter.

8. Ratio estimator: definition and examples

Let $\mathscr{X}$, $\mathscr{Y}$ be properties which we can measure for each member of the population. Often, if $X \neq 0$, we are interested in the value of $R = Y/X$. For example, y_i is the monthly expenditure of food of family number i in the population, and x_i its monthly total expenditure. Then $100R$ is the percentage of expenditures devoted to food by the entire population. It may be important for us to know R: if we propose to levy an expenditure tax, we should know which portion of the tax we would lose by exemption of food expenditure.

In certain cases we can assume that X or $\bar{X}$ is known and that we want to obtain an estimate $\hat{R}$ of R in order to obtain from it the estimate $\hat{Y}_R = \hat{R}X$ for $Y(=RX)$, or $\hat{\bar{Y}}_R = \hat{R}\bar{X}$ for $\bar{Y}(=R\bar{X})$. For instance we may obtain from the national accounts very good estimates for total expenditures X. With the help of the sample we can obtain the value of an estimator $\hat{R}$ for R, the proportion of the total expenditure of the population devoted to food. From this we obtain an estimate $\hat{Y}_R$ for the food expenditures of the population. This method is used in cases in which we can obtain very good data for X but it is difficult or too expensive to obtain such good data for Y.†

In simple random sampling we use as estimator for R the expression

$$\hat{R} = \bar{y}/\bar{x} = y/x,$$

(where x and y are defined as before), provided circumstances guarantee that $x \neq 0$.

REMARK. At first sight it may seem strange that, if $\bar{X}$ is known, it may be better to estimate $\bar{Y}/\bar{X}$ by $\bar{y}/\bar{x}$ than by $\bar{y}/\bar{X}$.

9. Ratio estimator: bias and variance

If n is large and $\bar{x}$ and $\bar{X}$ are positive, the bias of $\hat{R}$ is small. This may be seen as follows (a more exact evaluation is given in Chapter IIA): Let

$$d_i = y_i - Rx_i,$$

† A sampling plan in which first a large sample is drawn to measure $\mathscr{X}$ and then a smaller one to measure $\mathscr{Y}$ is called a *two-phase* or *double sampling* plan. Such plans are discussed in sections 12 and 13.2 of Chapter IIA, section 9 of Chapter VI, section 12 of Chapter VII, section 4 of Chapter X, and (for $\mathscr{X}$ the stratifying variable) in section 9 of Chapter III.

then

$$\hat{R} - R = \bar{d}/\bar{x} = (\bar{d}/\bar{X})(\bar{X}/\bar{x})$$

may be expected to be close to $\bar{d}/\bar{X}$ with probability close to 1. Therefore, $\mathscr{E}(\hat{R} - R)$ will be close to

$$\mathscr{E}\,\frac{\bar{d}}{\bar{X}} = \frac{\bar{Y} - R\bar{X}}{\bar{X}} = \frac{\bar{D}}{\bar{X}} = 0.$$

We shall only wish to use $\hat{R}$ as an estimator of R if we can be confident that its bias is negligible. In this case

$$V(\hat{R}) = \mathscr{E}(\hat{R} - R)^2 - (\mathscr{E}\hat{R} - R)^2$$

is nearly equal to $\mathscr{E}(\hat{R} - R)^2 = \text{MSE}$. However, the latter is very close to

$$\mathscr{E}\,\frac{\bar{d}^2}{\bar{X}^2} = \frac{1}{\bar{X}^2}\,V(\bar{d})$$

(the last step follows from $\mathscr{E}\bar{d} = \bar{D} = 0$). If we substitute for $V(\bar{d})$ its expression† in terms of $S^2(\mathscr{D})$ (or $\sigma^2(\mathscr{D})$ respectively), we obtain

$$V(\hat{Y}_R) \approx \frac{N-n}{N}\,\frac{1}{n}\,S^2(\mathscr{D}), \quad V(\hat{R}) \approx \frac{N-n}{N}\,\frac{1}{n}\,\frac{1}{\bar{X}^2}\,S^2(\mathscr{D})$$

for sampling without replacement, and

$$V(\hat{Y}_R) \approx \frac{1}{n}\,\sigma^2(\mathscr{D}), \quad V(\hat{R}) \approx \frac{1}{n}\,\frac{1}{\bar{X}^2}\,\sigma^2(\mathscr{D})$$

for sampling with replacement.

Let us now define $\hat{\mathscr{D}}$ by $\hat{d}_i = y_i - \hat{R}x_i$. We cannot use as estimator of $V(\hat{R})$ the expression we obtain by substituting $s^2(\mathscr{D})$ for $S^2(\mathscr{D})$, since it is impossible to obtain $s^2(\mathscr{D})$ from the sample. We therefore substitute instead $s^2(\hat{\mathscr{D}})$. ($s^2(\hat{\mathscr{D}})$ is not an unbiased estimator but its bias is small if n is large.) Note that the average of the $\hat{d}_i$ for the whole sample is zero; therefore

$$s^2(\hat{\mathscr{D}}) = \sum_{i=1}^{N} a_i \hat{d}_i^2/(n-1) \quad \left(\sum_{i=1}^{N} t_i \hat{d}_i^2/(n-1) \text{ with replacement}\right).$$

REMARK. If $\bar{X}$ is not known, we may estimate $V(\hat{R})$ by

$$\frac{N-n}{N}\,\frac{1}{n}\,\frac{1}{\bar{x}^2}\,s^2(\hat{\mathscr{D}}) \quad \left(\text{instead of by } \frac{N-n}{N}\,\frac{1}{n}\,\frac{1}{\bar{X}^2}\,s^2(\hat{\mathscr{D}})\right)$$

in sampling without replacement; in sampling with replacement the fraction $(N-n)/N$ drops out. There is some evidence that this estimator is even preferable if $\bar{X}$ is known (see Chapter VIII, end of section 8).

† As mentioned in section 5, for every characteristic $\mathscr{Z}$, $V(\bar{z}) = ((N-n)/N)(1/n)S^2(\mathscr{Z})$ (for sampling without replacement), $V(\bar{z}) = (1/n)\sigma^2(\mathscr{Z})$ (for sampling with replacement).

We shall learn more about advantages and disadvantages of ratio estimators in subsequent chapters. One thing is, however, clear at once. If, for each u_i, y_i is a *constant multiple* of x_i, $y_i = \beta x_i$, then $\hat{R} = \beta = R$, so that bias and variance of $\hat{R}$ and $\hat{Y}_R$ are both zero. This indicates that, if there is reason to believe that the $y_i - \beta x_i$ are relatively very small, we may expect ratio estimators to be very good estimators.

10. Example of ratio estimation

The following table gives 16 observations on $\mathscr{X}$ and $\mathscr{Y}$ for the example of total expenditures† and food expenditures mentioned above (in Israeli pounds). The population consists of 80 relatively high income earning families. Let us estimate $\bar{Y}$ and R from the sample.

	x_i	y_i	$\hat{R}x_i$	$\hat{d}_i$
1	580	300	314.129 3	−14.129 3
2	590	302	319.545 4	−17.545 4
3	600	318	324.961 4	−6.961 4
4	620	332	335.793 4	−3.793 4
5	640	356	346.625 5	+9.374 5
6	600	325	324.961 4	+0.038 6
7	580	290	314.129 3	−24.129 3
8	690	383	373.705 6	+9.294 4
9	660	340	357.457 5	−17.457 5
10	670	360	362.873 5	−2.873 5
11	610	330	330.377 4	−0.377 4
12	670	381	362.873 5	+18.126 5
13	680	362	368.289 6	−6.289 6
14	720	412	389.953 7	+22.046 3
15	740	400	400.785 7	−0.785 7
16	710	420	384.537 6	+35.462 4
Σ	10 360	5 611	5 610.999 8	0.000 2

REMARK. The sum of the $\hat{d}_i$ column must be zero but for errors of rounding. In the sum row (the last row) we perform the same operations as in any other row; this serves as a check on the calculations.

† Excluding fixed obligations such as apartment rent or mortgage payments.

$$\Sigma a_i x_i^2 = 6\,749\,000, \quad \Sigma a_i y_i^2 = 1\,992\,151, \quad \Sigma a_i x_i y_i = 3\,663\,300$$

$$\bar{x} = \frac{\Sigma a_i x_i}{n} = \frac{10\,360}{16} = 647.50$$

$$\bar{y} = \frac{\Sigma a_i y_i}{n} = \frac{5\,611}{16} = 350\,687\,5$$

$$\hat{R} = \frac{\bar{y}}{\bar{x}} = \frac{\Sigma a_i y_i}{\Sigma a_i x_i} = \frac{5\,611}{10\,360} = 0.541\,602\,31$$

$$\Sigma a_i \hat{d}_i^2 = 3\,752.341\,457\,08$$

$$s^2(\hat{\mathscr{D}}) = \frac{\sum_{i=1}^{N} a_i \hat{d}_i^2}{n-1} = \frac{3\,752.341\,457\,08}{15} = 250.156\,164$$

$$s(\hat{\mathscr{D}}) = 15.82.$$

(a) Estimators of $\bar{Y}$ and of their variances:

(i) $\bar{y} = 350.687\,5$

$$v(\bar{y}) = \frac{N-n}{N}\frac{\Sigma a_i (y_i - \bar{y})^2}{n(n-1)} = \frac{N-n}{N}\left[\Sigma a_i y_i^2 - \frac{(\Sigma a_i y_i)^2}{n}\right]\Big/ n(n-1)$$

$$= \frac{0.8}{16 \times 15}\left[1\,992\,151 - \frac{5\,611^2}{16}\right]$$

$$= \frac{19\,554.752}{240} = 81.478$$

$$v(\bar{y})^{\frac{1}{2}} = 9.03.$$

(ii) Let us assume that we know that $\bar{X} = 650$, then

$$\hat{\bar{Y}}_R = \hat{R}\bar{X} = 0.541\,602\,31 \times 650 = 352.041\,5$$

$$v(\hat{\bar{Y}}_R) = \frac{N-n}{N}\frac{s^2(\hat{\mathscr{D}})}{n} = 12.507\,6$$

$$v(\hat{\bar{Y}}_R)^{\frac{1}{2}} = 3.54.$$

(b) Estimator of R and of its variance:

$$\hat{R} = 0.541\,602\,31.$$

(i) If $\bar{X}$ is not known we can estimate $V(\hat{R})$ by

$$v(\hat{R}) = \frac{N-n}{N}\frac{s^2(\hat{\mathscr{D}})}{n\bar{x}^2} = \frac{200.122\,4}{16 \times 647.50^2} = 0.000\,029\,832$$

$$v(\hat{R})^{\frac{1}{2}} = 0.005\,46.$$

(ii) If it is known that $\bar{X} = 650$, we may estimate $V(\hat{R})$ by

$$v(\hat{R}) = \frac{N-n}{N}\frac{s^2(\hat{\mathscr{D}})}{n\bar{X}^2} = \frac{200.122\,4}{16 \times 650^2} = 0.000\,029\,603$$

$$v(\hat{R})^{\frac{1}{2}} = 0.005\,44.$$

As we remarked above, (i) may be preferable to (ii), even if $\bar{X}$ is known. In practice one would publish $\hat{R}$ with fewer decimal places, since the estimated standard error does not warrant giving as many as shown above.

11. Domains of study

11.1. DEFINITION, USES, NOTATION

Often we wish to obtain information on averages or totals of properties for subpopulations without sampling separately from these subpopulations. Such subpopulations are called domains of study. E.g., in making a survey of all workers in a country, we may wish to know certain particulars concerning professional workers, without having at our disposal a separate list of professional workers from which to sample.†

Suppose the index 1 denotes the first domain of study, $\mathscr{D}_1$. $\bar{Y}_1$ may, e.g., be the average income of people in $\mathscr{D}_1$. Let $\bar{y}_1$ be the corresponding sample average. Suppose that in $\mathscr{D}_1$ there are N_1 members, and that in the sample there are n_1 from $\mathscr{D}_1$. n_1 is a random variable. Let

$$\delta_{i1} = \begin{cases} 1 & \text{if } u_i \in \mathscr{D}_1, \\ 0 & \text{otherwise.} \end{cases} \qquad (i = 1, \ldots, N)$$

Note that the δ_{i1} are not random variables, but constants. If we use the previous notations, we obtain (for sampling without replacement)

$$\bar{y}_1 = \sum_{i=1}^{N} y_i' a_i / n_1 \quad (\text{if } n_1 > 0),$$

† Sometimes it is worthwhile to first draw a large sample, and then draw a subsample from among those units in that sample belonging to the domain in order to measure their $\mathscr{Y}$-values. Appropriate formulae are derived in section 9.7 of Chapter III.

where

$$n_1 = \sum_{i=1}^{N} \delta_{i1} a_i$$

and

$$y'_i = y_i \delta_{i1}.$$

We shall also speak of the characteristic $\mathscr{Y}'$, namely the characteristic whose value for u_i is y'_i. Its sample total is accordingly denoted by y', its population total by Y' (sometimes by Y_1), its sample average by $\bar{y}' = y'/n$, its population average by $\bar{Y}' = Y'/N$.

11.2. ESTIMATION OF $\bar{Y}_1$

We commonly use the ratio estimator $\bar{y}_1$ to estimate $\bar{Y}_1$. The conditional expectation of $\bar{y}_1$ given n_1 is equal to $\bar{Y}_1$ for all $n_1 \geq 1$. Since $\bar{y}_1$ is not defined for $n_1 = 0$, if per chance we obtain a sample with $n_1 = 0$, we shall discard it and obtain another sample or continue sampling until $n_1 > 0$. (Actually this modification makes n a random variable, but if $\mathrm{P}(n_1 = 0)$ is very small, we can neglect this fact.) Since under any such method of sampling n_1 is at least 1,

$$\sum_{k \geq 1} \mathrm{P}\{n_1 = k\} = 1$$

and by section 8 of Chapter I

$$\mathscr{E}\bar{y}_1 = \sum_{k \geq 1} \mathscr{E}\{\bar{y}_1 | n_1 = k\} \, \mathrm{P}\{n_1 = k\}$$

$$= \bar{Y}_1 \sum_{k \geq 1} \mathrm{P}\{n_1 = k\} = \bar{Y}_1,$$

since $\mathscr{E}\{\bar{y}_1 | n_1 = k\}$ is simply the expected value of the average $\mathscr{Y}$-value for a simple random sample of k units taken from a population of N_1 units. Note that the minimum total sample size required to make $\mathrm{P}(n_1 = 0)$ small depends on N_1/N and is large if the latter is small, i.e., if we wish to estimate a characteristic pertaining to a small part of the population without subsampling it separately. See, however, section 14, and Chapter III, section 9.7.

If we use the previous methods of computing the variance, we obtain for the conditional variance of $\bar{y}_1$ given n_1:†

$$\frac{N_1 - n_1}{N_1} \frac{1}{n_1} S_1^2,$$

† One can obtain an approximation for the expected value of the following expression with the aid of the formulas given in section 10 of Chapter IIA. For another approximation, see Remark (ii) in section 11.5.

where, if $N_1 = 1$, $S_1^2 = 0$, and, if $N_1 \neq 1$,

$$S_1^2 = \frac{\sum_{i=1}^{N_1} (y_i - \bar{Y}_1)^2}{N_1 - 1};$$

and as estimator of the (conditional) variance an analogous function (provided $n_1 > 1$)†:

$$\frac{N_1 - n_1}{N_1} \frac{1}{n_1} s_1^2,$$

where

$$s_1^2 = \sum_{i=1}^{N_1} a_i(y_i - \bar{y})^2/(n_1 - 1).$$

Since $\mathscr{E}(n_1/N_1) = n/N$, we may replace this by

$$\frac{N - n}{N} \frac{1}{n_1} s_1^2$$

if N_1 is unknown.‡

11.3. ESTIMATION OF Y_1 AND RATIOS

As estimator for $Y_1 = Y'$ we may take the unbiased ratio estimator $N_1\bar{y}_1$ if N_1 is known.§ (Note that we cannot use the notation y_1 for the sum over all members of $\mathscr{D}_1$ which are in the sample, since we already denoted by y_1 the value of the property $\mathscr{Y}$ for the first member of the population.) If the population is divided into a number of nonoverlapping domains, this estimating formula does not satisfy the property that the sum of the domain estimators equals the usual estimator $N\bar{y}$ of Y; it only satisfies this property in the mean.

† The expected value of this estimator equals the mean of its conditional expectation given n_1, that is, the mean of the conditional variance of $\bar{y}_1$; and the latter is equal to the unconditional variance, since the conditional mean of $\bar{y}_1$ given n_1 equals the *constant* $\bar{Y}_1$ (see section 8 of Chapter I). – In the expressions given for S_1^2 and s_1^2 we assume for notational convenience that the elements of $\mathscr{D}_1$ are the first N_1 of the population.

‡ We may also use other ratio estimators to estimate $\bar{Y}_1$. For example, if $\bar{X}_1$ is known we may use $\hat{\bar{Y}}_{1,R} = (\bar{y}_1/\bar{x}_1)\bar{X}_1$, $v(\hat{\bar{Y}}_{1,R}) = (n_1^{-1} - N_1^{-1})\Sigma a_i(d_i')^2/(n_1 - 1)$ (cp. †), where, if N_1 is unknown, we replace $(N_1 - n_1)/N_1$ by $(N - n)/N$. Occasionally, when $\bar{X}$ rather than $\bar{X}_1$ is known, and when N_1 is close to N, it will be efficient to use $(\bar{y}_1/\bar{x})\bar{X}$.

§ Sometimes X_1 is known; then we may use $(y'/x')X_1 = (\bar{y}_1/\bar{x}_1)X_1$, with variance N_1^2 times that of $(\bar{y}_1/\bar{x}_1)\bar{X}_1$. We may also be interested in estimating Y'/X'. See the example in section 11.4.

If N_1 is not known, but Y is known, we can use the ratio estimator†

$$(n_1/n)(\bar{y}_1/\bar{y})Y = (y'/y)Y,$$

provided $y \neq 0$. As we learned in section 9, its variance may be estimated by

$$N^2 \frac{N-n}{N} \frac{1}{n} s^2(\hat{\mathscr{D}}),$$

where

$$\begin{aligned}(n-1)s^2(\hat{\mathscr{D}}) &= \Sigma a_i(y_i' - \hat{R}y_i)^2 \\ &= \Sigma a_i y_i'^2 - 2\hat{R}\Sigma a_i y_i' y_i + \hat{R}^2\{\Sigma a_i y_i'^2 + \Sigma a_i(y_i^2 - y_i'^2)\} \\ &= (1-\hat{R})^2 \Sigma \delta_{i1} a_i y_i^2 + \hat{R}^2 \Sigma(1-\delta_{i1}) a_i y_i^2 \qquad (\hat{R} = y'/y).\end{aligned}$$

If all $y_i = 1$ or 0, $y_i^2 = y_i$, and this becomes simply

$$\{1-(y'/y)\}^2 y' + (y'/y)^2(y-y') = y'\{1-(y'/y)\}.$$

Also y'/y, being an estimator of Y'/Y, is sometimes of interest.

If neither N_1 nor Y are known, we can use

$$N\bar{y}' = N(n_1/n)\bar{y}_1,$$

where $\bar{y}'$ is the average of the y_i' in the sample:

$$\bar{y}' = \sum_{i=1}^{N} y_i' a_i / \sum_{i=1}^{N} a_i = n_1 \bar{y}_1 / n.$$

$N\bar{y}'$ is an unbiased estimator of Y_1, since $\bar{y}'$ is an average of the values of the $\mathscr{Y}'$ characteristic in simple random samples from $\mathscr{P}$, and thus has for expected value the population average of the N numbers $y'(u_1), \ldots, y'(u_N)$:

$$\frac{1}{N}\sum_{i=1}^{N} y_i' = \frac{1}{N}\sum_{i=1}^{N} y_i \delta_{i1} = \bar{Y}'.$$

The variance of the estimator $N\bar{y}'$ is

$$N^2 \frac{N-n}{N} \frac{1}{n} S^2(\mathscr{Y}'),$$

where

$$S^2(\mathscr{Y}') = \sum_{i=1}^{N} (y_i' - \bar{Y}')^2/(N-1).$$

† Similarly, if X is known, we may use $(y'/x)X$. From section 2 of Chapter IIA it follows that its bias is (in terms defined there, except for the subscript to C, which refers to the domain)

$$\frac{N-n}{N}\frac{1}{n} Y' \left[C(\mathscr{X}\mathscr{X}) - \frac{N}{N-1}\left\{\frac{N_1-1}{N_1}\frac{\bar{X}_1}{\bar{X}} C_1(\mathscr{X}\mathscr{Y}) + \left(\frac{\bar{X}_1}{\bar{X}} - 1\right)\right\}\right] + O\left(\frac{1}{n^2}\right).$$

We estimate this by

$$s^2(\mathscr{Y}') = \sum_{i=1}^{N} a_i(y_i' - \bar{y}')^2/(n-1).$$

These results follow from the principles we have learned earlier.

11.4. EXAMPLE

Suppose in the example of section 10 the 16 families did not constitute a random sample from 80 above-average income families, but constituted those families of a simple random sample of 40 from a population of 200 families that belong to a certain well-defined subpopulation of 80 families (one of whose characteristics is an above-average income).

In the notation of the present section

$$N = 200, \quad N_1 = 80, \quad n = 40, \quad n_1 = 16,$$

and

$$\sum_{i=1}^{N_1} a_i y_i = 5\,611, \quad \sum_{i=1}^{N_1} a_i y_i^2 = 1\,992\,151,$$

$$\sum_{i=1}^{N_1} a_i x_i = 10\,360, \quad \sum_{i=1}^{N_1} a_i x_i^2 = 6\,749\,000,$$

$$\sum_{i=1}^{N_1} a_i x_i y_i = 3\,663\,300.$$

For the entire sample, let

$$\sum_{i=1}^{N} a_i y_i = 10\,296, \quad \sum_{i=1}^{N} a_i y_i^2 = 2\,911\,210,$$

$$\sum_{i=1}^{N} a_i x_i = 17\,560, \quad \sum_{i=1}^{N} a_i x_i^2 = 8\,917\,300,$$

$$\sum_{i=1}^{N} a_i x_i y_i = 5\,072\,800.$$

(a) Estimates of $\bar{Y}_1$:

The previously given estimates of $\bar{Y}_1$ are

$$\bar{y}_1 = 350.687\,5, \quad \hat{\bar{Y}}_{1,R} = \frac{\bar{y}_1}{\bar{x}_1}\bar{X}_1 = 352.041\,5,$$

where in the latter case we were given $\bar{X}_1 = 650$. The estimator for the variance of $\bar{y}_1$:

$$\frac{N_1 - n_1}{N_1}\frac{1}{n_1}\sum_{i=1}^{N} a_i(y_i - \bar{y}_1)^2\delta_{i1}/(n_1 - 1)$$

was computed as 81.478. It is also a valid estimator here; if N_1 is unknown, we replace $(N_1 - n_1)/N_1$ by $(N - n)/N$, which here happens to have the same value.

The estimator for the variance of $\hat{Y}_{1,R}$ was computed as 12.507 6. It is also valid here; if N_1 is unknown, we again replace $(N_1 - n_1)/N_1$ by $(N - n)/N$, which here coincide.

(b) Estimates of Y_1:

To the estimator $\bar{y}_1$ corresponds $N_1\bar{y}_1$, here equal to 28 055.00, with estimated variance $80^2 \times 81.478 = 521\,461$. If N_1 is unknown, but Y is known to be 52 000, we may use

$$\frac{y'}{y}\,Y = \frac{5\,611}{10\,296}\,52\,000 = 28\,349.40;$$

the estimated variance of $(y'/y)\,Y$ is

$$\frac{N(N-n)}{n(n-1)}\left\{\left(1-\frac{y'}{y}\right)^2 \sum_{i=1}^{N_1} a_i y_i^2 + \left(\frac{y'}{y}\right)^2 \sum_{i=N_1+1}^{N} a_i y_i^2\right\}$$

$$= \frac{200 \times 160}{40 \times 39}\,709\,471.7 = 14\,560\,157.$$

If N_1 as well as Y are unknown, we may use

$$N\bar{y}' = 200\,\frac{5\,611}{40} = 28\,055.00;$$

the estimated variance of $N\bar{y}'$ is

$$\frac{N(N-n)}{n(n-1)}\left\{\sum_{i=1}^{N_1} a_i y_i^2 - \frac{\left(\sum_{i=1}^{N_1} a_i y_i\right)^2}{n}\right\} = \frac{200 \times 160}{40 \times 39}\,1\,205\,068 = 24\,719\,343.$$

If we know $X = 90\,000$, we may use

$$\frac{y'}{x}\,X = 28\,757.97$$

and

$$v\left(\frac{y'}{x}\,X\right) = \frac{N(N-n)}{n(n-1)}\left\{\sum_{i=1}^{N_1} a_i y_i^2 - 2\frac{y'}{x}\sum_{i=1}^{N_1} a_i x_i y_i + \right.$$

$$\left. + \left(\frac{y'}{x}\right)^2\left(\sum_{i=1}^{N_1} a_i x_i^2 + \sum_{i=N_1+1}^{N} a_i x_i^2\right)\right\} = \frac{200 \times 160}{40 \times 39}\,561\,529 = 11\,518\,544.$$

If, as some times happens, we know X_1, we may use

$$\frac{y'}{x'} X_1 = \frac{\bar{y}_1}{\bar{x}_1} X_1 = 28\,163.32$$

for $X_1 = 52\,000$, with estimated variance N_1^2 times that of $(\bar{y}_1/\bar{x}_1)\bar{X}_1$: 80 049.

(c) Estimate of Y_1/X_1:

We previously gave

$$y'/x' = 0.541\,602\,31,$$

and

$$v(y'/x') = \begin{cases} 0.000\,029\,603 & \text{if it is given that } \bar{X}_1 = 650, \\ 0.000\,029\,832 & \text{if we use the estimator for } \bar{X}_1. \end{cases}$$

As mentioned before, the second may be preferable even if $\bar{X}_1$ is known.

If N_1 is unknown, we replace $(N_1 - n_1)/N_1$ by $(N - n)/N$, yielding here the same result.

(d) Estimate of Y_1/Y, the share of the families which belong to the domain of study in the total food expenditure of the population:

$$y'/y = 0.544\,968\,92;$$

the estimated variance of y'/y is

$$\frac{14\,560\,157}{52\,000^2} = 0.005\,388\,352 \quad \text{if } Y \text{ is known to be } 52\,000,$$

$$\frac{14\,560\,157}{51\,480^2} = 0.005\,459\,400 \quad \text{otherwise.}$$

As mentioned before, the second may be preferable even if Y is known.

11.5. REMARKS

(i) We may also write†

$$S^2(\mathscr{Y}') = \frac{N_1 - 1}{N - 1} S_1^2 + \frac{N_1}{N}\frac{N - N_1}{N - 1} \bar{Y}_1^2 = \frac{(N_1 - 1)S_1^2 + (1/N_1 - 1/N) Y_1^2}{N - 1}$$

† In particular, $\Sigma_{i=1}^{N} (\delta_{i1} - (N_1/N))^2 = (1 - (N_1/N))N_1$, as in this case $S_1^2 = 0$ (compare the first formula in the next section). By the same method we use for getting $S^2(\mathscr{Y}')$ we also obtain $\Sigma_{i=1}^{N} (\delta_{i1} - (N_1/N))(y_i' - \bar{Y}') = (1 - (N_1/N)) Y'$.

(and an analogous formula for $s^2(\mathscr{Y}')$), because

$$\begin{aligned}
\sum_{i=1}^{N} (y_i' - \bar{Y}')^2
&= \sum_{i=1}^{N_1} (y_i - \bar{Y}')^2 + (N - N_1)\bar{Y}'^2 \\
&= \sum_{i=1}^{N_1} (y_i - \bar{Y}_1)^2 + N_1(\bar{Y}_1 - \bar{Y}')^2 + (N - N_1)\bar{Y}'^2 \\
&= \sum_{i=1}^{N_1} (y_i - \bar{Y}_1)^2 + N_1 \left\{\bar{Y}_1\left(1 - \frac{N_1}{N}\right)\right\}^2 + (N - N_1)\left\{\frac{N_1 \bar{Y}_1}{N}\right\}^2 \\
&= (N_1 - 1)S_1^2 + \frac{N_1(N - N_1)}{N}\,\bar{Y}_1^2.
\end{aligned}$$

(ii) Since $\bar{y}_1 = y'/n_1$, the variance of $N_1\bar{y}_1$ may also be approximated† by using the result of section 9:

$$\begin{aligned}
V(N_1\bar{y}_1) = N_1^2 V(\bar{y}_1) &\approx N(N - n)\frac{1}{n}\sum_{i=1}^{N} (y_i' - \bar{Y}_1\delta_{i1})^2/(N - 1) \\
&= \frac{N(N - n)}{N - 1}\frac{1}{n}(N_1 - 1)S_1^2,
\end{aligned}$$

which may be compared with

$$V(N\bar{y}') = \frac{N(N - n)}{N - 1}\frac{1}{n}\left\{(N_1 - 1)S_1^2 + \left(\frac{1}{N_1} - \frac{1}{N}\right) Y_1^2\right\}.$$

This shows that (insofar as the above approximation holds), if N_1 is known, $N_1\bar{y}_1$ is a better estimator of Y' than $N\bar{y}'$; and that $N\bar{y}'/N_1$ is a worse estimator for $\bar{Y}_1$ than $\bar{y}_1$.

(iii) The estimator corresponding to the approximation to $V(N_1\bar{y}_1)$ is

$$N(N - n)\frac{1}{n}\sum_{i=1}^{N} a_i\left((y_i' - \frac{y'}{n_1}\delta_{i1}\right)^2/(n - 1) = N(N - n)\frac{1}{n(n - 1)}(n_1 - 1)s_1^2.$$

The corresponding estimator for $V(\bar{y}_1)$ is

$$N(N - n)\frac{1}{n(n - 1)}\frac{1}{N_1^2}(n_1 - 1)s_1^2,$$

† A closer approximation is considered (in connection with another problem) in section 11.1 of Chapter III. Instead of $N_1 - 1$ that approximation has

$$N_1\{1 - n^{-1} + N(nN_1)^{-1}\},$$

which exceeds N_1. Therefore it would appear to be somewhat better to use N_1 rather than $(N_1 - 1)$.

and, if N_1 is unknown,

$$\frac{N-n}{N}\frac{1}{n_1}s_1^2\left(\frac{n}{n-1}\frac{n_1-1}{n_1}\right),$$

which, as remarked in section 9, will often be used even if N_1 is known. This differs from the expression we previously gave for the estimator of $V(\bar{y}_1)$ by the multiplicative term in brackets which is less than 1.†

11.6. SUBDOMAINS

Sometimes we wish to estimate the total for a subdomain, say $\mathscr{D}_{1z}$, of $\mathscr{D}_1$ for which a characteristic $\mathscr{Z}$ has a value z. Calling this total $Y'(z)$ and the corresponding sample total $y'(z)$, the above methods give the estimators $Ny'(z)/n$ and $N'(z)y'(z)/n'(z)$ (where $N'(z)$ is the size of $\mathscr{D}_{1z}$ and $n'(z)$ the corresponding sample size). The latter has a smaller variance than the former. However, often $n'(z)$ will have an appreciable chance of vanishing and the $N'(z)$ may not be readily available. One may then consider the estimator

$$N(z)\, y'(z)/n(z),$$

where $N(z)$ is the size of $\mathscr{D}_z$ and $n(z)$ the corresponding sample size. Its variance is approximately

$$N^2\left(\frac{1}{n}-\frac{1}{N}\right)\frac{1}{N-1}\left[\{N'(z)-1\}S_{1z}^2+\{Y'(z)\}^2\left\{\frac{1}{N'(z)}-\frac{1}{N(z)}\right\}\right]$$

(compare (ii) above), which may be estimated by

$$N^2\left(\frac{1}{n}-\frac{1}{N}\right)\frac{1}{n-1}\left[\{n'(z)-1\}s_{1z}^2+\{y'(z)\}^2\left\{\frac{1}{n'(z)}-\frac{1}{n(z)}\right\}\right].$$

An analogous estimator of $Y'(z)$ is $N_1y'(z)/n_1$; in contrast to the previous estimator, it requires knowledge of N_1 instead of $N(z)$, and the sum over all $\mathscr{Z}$-values of such estimators equals the analogous ratio estimator $N'\bar{y}_1$.

11.7. MULTIWAY ADDITIVITY

In cases in which we examine a number of characteristics $\mathscr{Y}$, $\mathscr{Z}$, . . ., we are often interested not only in Y', Z', . . ., but also in their sum, W' (say). For example, y_i, z_i, . . . may be expenditures on various goods by the ith unit and w_i their total; or y_i, z_i, . . . may be 1 or 0 according to whether or not the ith unit has property $\mathscr{Y}$, $\mathscr{Z}$, . . ., respectively, these properties being such that no unit can have more than one of them.

In such cases it may be that W' is known; then the estimators $(y'/w')W'$, $(z'/w')W'$, . . . have the desirable property that their sum equals W'. However, if Y, Z, . . . are also known, the above estimators do not take that into

† A perfectly analogous result holds for the estimators of $V(\hat{Y}_{1,R})$.

account and therefore may be expected to be inefficient. Moreover, if the population is divided into nonoverlapping domains $\mathscr{D}_1, \mathscr{D}_2, \ldots$, the estimators for $Y', Y'', \ldots$ do not add up to Y.

Denote the ratio of Y to the sum of these estimators by $c(\mathscr{Y})$. A new set of estimators which do add to Y would be

$$c(\mathscr{Y})(y'/w')W', \quad c(\mathscr{Y})(y''/w'')W'', \ldots;$$

but, using obvious notation, the sum of the estimators of $Y', Z', \ldots$:

$$c(\mathscr{Y})(y'/w')W', \quad c(\mathscr{Z})(z'/w')W', \ldots$$

will not generally add to W'. One can then multiply the latter estimators by W' divided by their sum. Continuing likewise, we shall usually obtain convergence to a set of estimators of $Y', Z', \ldots, Y'', Z'', \ldots$ with the specified marginal sums $W', W'', \ldots$ and $Y, Z, \ldots$.

In the literature the case of the second example above has been discussed (here $W = N$). It is assumed that the matrix of domain sums

$$\begin{matrix} y' & z' & \ldots \\ y'' & z'' & \ldots \\ \ldots & \ldots & \ldots \end{matrix}$$

consists of positive numbers only (we already assumed that the method of sampling leads to positive marginals of this matrix), and it is shown that the procedure converges and leads to estimators with desirable properties for n large (so-called BAN estimators). See for these and other estimators with the same properties when n is large (minimum chi-squared, minimum modified chi-squared, maximum likelihood in sampling with replacement – all BAN estimators) WEICHSELBERGER, Metrika, **2** (1959) 100–130, 198–229.

If some entries in the above matrix are zero, but the matrix can be partitioned into submatrices A_{jk} such that all entries of the A_{jj} are positive and all those of the other A_{jk} zero, the above method can be used on each of the A_{jj} separately to obtain the corresponding matrix of estimators, and the estimators corresponding to the elements of the other matrices (A_{jk}) are taken to be 0.

12. Specialization to frequencies of properties

If $y_i = 1$ when u_i has a certain property, and 0 otherwise,

$$S^2(\mathscr{Y}) = NP(1-P)/(N-1), \quad \sigma^2(\mathscr{Y}) = P(1-P), \quad (P = \bar{Y});$$
$$s^2(\mathscr{Y}) = np(1-p)/(n-1) \qquad (p = \bar{y}).$$

Note that, therefore, in sampling with replacement $p(1-p)/n$ is not an unbiased estimator of $V(\bar{y})$, but $p(1-p)/(n-1)$ is.

It may be observed that, as an estimator of Y, $N\bar{y}$ has the defect that not all its possible values are integers (e.g., if N and n have no common factor, $N\bar{y}$ is not an integer if $0 < y < n$). For moderate N, the difference between $N\bar{y}$ and the nearest integer is, however, relatively small.

Sometimes we are not so much interested in the relation that the number of elements that have a certain property bears to the size of the population, as in the relation it bears to the size of a certain part of the population (a part which contains the elements which have that property). Thus we may conduct a survey and wish to know for a certain question the proportion yes/(yes + no) when the possible answers are "yes", "no", and "do not know". In the sub-population of those answering "yes" or "no", there are $N' = N_+ + N_-$ members, where N_+ is the number answering "yes" and N_- the number answering "no". In the sample we get $n' = n_+ + n_-$. Evidently, n', n_+ and n_- are random variables. The problem is identical with the problem of domains of study with $N' = N_1$, $n' = n_1$, $N_+ = Y_1$, $n_+ = n\bar{y}'$.

EXAMPLE. Estimate the proportion of "yes" answers among those that answer "yes" or "no", and the ratio of the number of "yes" to the number of "no" answers, and find the formula for estimating the variances of these estimators. – The estimators are $\bar{y}_1$, and $\bar{y}_1/(1 - \bar{y}_1)$. To estimate their variances, we may apply the beginning of the previous section: $\bar{y}_1$ has estimated variance†

$$\frac{N_1 - n_1}{N_1}\frac{1}{n_1}\sum_{i=1}^{N} a_i\delta_{i1}(y_i - \bar{y}_1)^2/(n_1 - 1) = \frac{N_1 - n_1}{N_1}\frac{1}{n_1 - 1}\bar{y}_1(1 - \bar{y}_1).$$

The conditional variance of $\bar{y}_1/(1 - \bar{y}_1)$ is approximately

$$\frac{N_1 - n_1}{N_1}\frac{1}{n_1}\frac{1}{(1 - \bar{Y}_1)^2}\sum_{i=1}^{N}\delta_{i1}\left\{y_i - \frac{\bar{Y}_1}{1 - \bar{Y}_1}(1 - y_i)\right\}^2/(N_1 - 1)$$
$$= \frac{N_1 - n_1}{N_1 - 1}\frac{1}{n_1}\frac{\bar{Y}_1}{(1 - \bar{Y}_1)^3},$$

which may be estimated by

$$\frac{N_1 - n_1}{N_1}\frac{1}{n_1}\frac{1}{(1 - \bar{y}_1)^2}\sum_{i=1}^{N} a_i\delta_{i1}\left\{y_i - \frac{\bar{y}_1}{1 - \bar{y}_1}(1 - y_i)\right\}^2/(n_1 - 1),$$

† Replace $(N_1 - n_1)/N_1$ by $(N - n)/N$ if N_1 is unknown. Also note that from the first formula in (ii) of section 11.5 it follows that

$$V(\bar{y}_1) \approx \frac{N - n}{N - 1}\frac{1}{n}\frac{N}{N_1}\bar{Y}_1(1 - \bar{Y}_1).$$

which, since

$$\Sigma\left\{y_i - \frac{\bar{y}_1}{1-\bar{y}_1}(1-y_i)\right\}^2 a_i\delta_{i1} = \Sigma y_i^2 a_i\delta_{i1} + \frac{\bar{y}_1^2}{(1-\bar{y}_1)^2}\Sigma(1-y_i)^2 a_i\delta_{i1}$$

$$= y' + \frac{\bar{y}_1^2}{(1-\bar{y}_1)^2}(n_1 - y') = n_1\frac{\bar{y}_1}{1-\bar{y}_1},$$

is equal to†

$$\frac{N_1 - n_1}{N_1}\frac{1}{n_1 - 1}\frac{\bar{y}_1}{(1-\bar{y}_1)^3}.$$

Note that this is $(1-\bar{y}_1)^{-4}$ times the estimated variance of $\bar{y}_1$; similarly the (approximate) conditional variance of $\bar{y}_1/(1-\bar{y}_1)$ is $(1-\bar{Y}_1)^{-4}$ times the (approximate) conditional variance of $\bar{y}_1$.‡

In Chapter I, section 1, we already mentioned some overall properties (parameters) concerning $\mathscr{Y}$ in the population $\mathscr{P}$. For any given characteristic $\mathscr{Y}$ and population $\mathscr{P}$, denote by $Nf_N(t)$ the number of elements u_i in the population for which $y(u_i)$ equals t. Denote by $NF_N(t)$ the numbers of elements u_i in the population for which $y(u_i)$ does not exceed t. Both are defined for all real numbers t (of course, $Nf_N(t)$ is zero for most numbers t), and therefore f_N and F_N define functions, which are called the *(relative) frequency function of $\mathscr{Y}$ in $\mathscr{P}$* and the *(cumulative) distribution function of $\mathscr{Y}$ in $\mathscr{P}$*, respectively.

WARNING. The student should not confuse these functions with the frequency function and distribution function of a *statistic*, defined in Chapter I, section 5. They may, however, be *related* to these functions as follows: F_N and f_N also constitute the sampling distribution and frequency function, respectively, of $\bar{y}$ from a simple random sample *of size 1* from $\mathscr{P}$.

$\bar{Y}$ and $\sigma^2(\mathscr{Y})$ are, respectively, the *first moment* and the *second central moment of the distribution of $\mathscr{Y}$ on $\mathscr{P}$*:

$$\bar{Y} = \sum_{t=-\infty}^{\infty} tf_N(t), \quad \sigma^2(\mathscr{Y}) = \sum_{t=-\infty}^{\infty}(t-\bar{Y})^2 f_N(t),$$

and for that reason are also called the *mean* and *variance* of $\mathscr{Y}$ on $\mathscr{P}$.

† Replace $(N_1 - n_1)/N_1$ by $(N-n)/N$ if N_1 is unknown.

‡ From section 10 of Chapter IIA follows that

$$V\left(\frac{\bar{y}_1}{1-\bar{y}_1}\right) \approx \frac{N-n}{N-1}\frac{1}{n}\frac{N}{N_1}\frac{\bar{Y}_1}{(1-\bar{Y}_1)^3}.$$

From a previous footnote it then follows that the ratio of the unconditional variances is the same as that of the conditional variances (approximately).

If we define $\mathscr{Z}$ such that $z(u_i) = 1$ if $y(u_i) = t$ and 0 otherwise, then $Nf_N(t) = Z$ and $f_N(t) = \bar{Z}$. If we define $\mathscr{Z}$ such that $z(u_i) = 1$ if $y(u_i) \leq t$ and 0 otherwise, then $NF_N(t) = Z$ and $F_N(t) = \bar{Z}$. For $s < t$ if we define $\mathscr{Z}$ such that $z(u_i) = 1$ if $y(u_i)$ exceeds s but not t, then $Z = NF_N(t) - NF_N(s)$.

In general

$$F_N(t) = \sum_{s \leq t} f_N(s).$$

If we divide all the real numbers into $K + 1$ nonoverlapping classes by the numbers $c_1 < \ldots < c_K$, and call $c_0 = -\infty$, then the numbers

$$F_N(c_k) - F_N(c_{k-1}) \quad (k = 1, \ldots, K)$$

constitute the *grouped* frequency function of $\mathscr{Y}$ on $\mathscr{P}$.

To most parameters there are sample analogues, which are obtained by replacing, in the definition, the elements of the population by those in the sample (and so N by n). Thus for a given $\mathscr{Y}$ and a given sample we define $nf_n(t)$ to be the number of elements in the sample for which $y(u_i)$ equals t, and $nF_n(t)$ the number of elements in the sample for which $y(u_i)$ does not exceed t; then the functions f_n and F_n vary with $s \in \mathscr{S}$, and so are statistics (in the more general sense referred to in the beginning of Chapter I, section 8). They are called the *empirical frequency function* and the *empirical distribution function* of $\mathscr{Y}$ (in the sample). Again, if $\mathscr{Z}$ is variously defined as above, z equals $nf_n(t)$, $nF_n(t)$ or $nF_n(c_k) - nF_n(c_{k-1})$, respectively. Therefore, we can apply what we learned concerning estimation of frequencies to the estimation of $f_N(t)$, $F_N(t)$ and $F_N(c_k) - F_N(c_{k-1})$.

EXAMPLE. In the example of section 10, what is the proportion, $\bar{Z}$, of the 80 families with food expenditures between 320 and 380 pounds per month? We may estimate $\bar{Z}$ by $\bar{z}$, which here equals $\frac{7}{16}$. The variance of $\bar{z}$ is

$$\frac{80 - 16}{80} \times \frac{1}{15} \times \frac{7}{16} \times \frac{9}{16} = 0.013\,125.$$

13. Simple cluster sampling

13.1. DEFINITION, NOTATION, EXAMPLES

Till now we assumed that the property $\mathscr{Y}$ has for each element u_i a single value y_i; often it has several (M_i) values $y_{i1}, y_{i2}, \ldots, y_{iM_i}$.

For example, each u_i is a family with u_i having M_i members, and $\mathscr{Y}$ is the property which has the value 1 for a male and 0 for a female,

$$z_i = \sum_{j=1}^{M_i} y_{ij}$$

is the number of males in the family u_i; and we may wish to know

$$\bar{\bar{P}} = \sum_{i=1}^{N} z_i / \sum_{i=1}^{N} M_i,$$

the proportion of males in the population, rather than

$$\bar{Z} = \sum_{i=1}^{N} z_i / N,$$

the average number of males per family.

For each characteristic $\mathcal{Y}$, if u_{ij} is the jth member of the ith family, let y_{ij} be the value of $\mathcal{Y}$ for u_{ij}. If we obtain a simple random sample of size n^* from the population

$$\{u_{11}, \ldots, u_{1M_1}, u_{21}, \ldots, u_{2M_2}, \ldots, u_{N1}, \ldots, u_{NM_N}\},$$

we use the estimator

$$\sum_{i=1}^{N} \sum_{j=1}^{M_i} y_{ij} a_{ij} / n^*$$

(in sampling without replacement) or

$$\sum_{i=1}^{N} \sum_{j=1}^{M_i} y_{ij} t_{ij} / n^*$$

(in sampling with replacement) for $\bar{\bar{P}}$, the average of $\mathcal{Y}$ over the population [$a_{ij} = 1$ if u_{ij} is the sample and 0 otherwise, with

$$n^* = \sum_{i=1}^{N} \sum_{j=1}^{M_i} a_{ij};$$

t_{ij} = the number of times u_{ij} is included in the sample, with

$$n^* = \sum_{i=1}^{N} \sum_{j=1}^{M_i} t_{ij}].$$

But if we obtain a simple random sample of *families* $\{u_1, \ldots, u_N\}$ and then record the values of the characteristic $\mathcal{Y}$ for all members of the families in the sample, we clearly do not have a simple random sample from the $\{u_{ij}(j = 1, \ldots, M_i, i = 1, \ldots, N)\}$, and the theory on which the above-mentioned estimator of $\bar{\bar{P}}$ is based does not apply any more: here, if $a_{i1} = 1$, then also $a_{i2} = \ldots = a_{iM_i} = 1$ and, if $a_{i1} = 0$, then also $a_{i2} = \ldots = a_{iM_i} = 0$.

Another example. An organization which provides comprehensive medical services to subscribers assigns each member upon admittance to one of its panel of general practitioners, in such a way that each practitioner will have about 400 patients, who may call on him whenever they should find a need to do so. We wish to estimate what proportion of the subscribers saw "his" doctor during the past year. For that purpose a simple random sample of 25 general practitioners was taken from the panel of 240, and for each selected doctor the proportion of his patients who consulted him during the preceding year was ascertained from his medical records. This averaged 0.610 1 with an (adjusted) sample variance of 0.000 837 2. The variance of $\bar{\bar{p}}$ may therefore be estimated by

$$\frac{240 - 25}{240} \times \frac{1}{25} \times 0.000\,837\,2 = 0.000\,030.$$

The families in the first example, and the set of subscribers assigned to a particular doctor in the second example, are both cases which are called *clusters*. If we define $q_i = z_i/M_i$, the property $\mathcal{Q}$ is defined over $\{u_1, \ldots, u_N\}$, the population of possible clusters.

Let us now return to the general problem and distinguish two cases.

13.2. CASE IN WHICH ALL THE M_i ARE EQUAL

Let $\bar{M}$ be their common value. Since the M_i are equal, the average

$$\bar{Q} = \sum_{i=1}^{N} q_i/N$$

equals the average of y_{ij} over the population of members of the clusters:

$$\bar{\bar{P}} = \sum_{i=1}^{N} \sum_{j=1}^{M_i} y_{ij}/N\bar{M}.$$

Since we have the values of $\mathcal{Q}$ for a simple random sample of n clusters, we can estimate $\bar{\bar{P}}$ by the average of the q_i in the sample, i.e.

$$\bar{\bar{p}} = \sum_{i=1}^{N} q_i a_i / \sum_{i=1}^{N} a_i$$

for sampling without replacement, and

$$\bar{\bar{p}} = \sum_{i=1}^{N} q_i t_i / \sum_{i=1}^{N} t_i$$

for sampling with replacement.

$$(\sum_{i=1}^{N} a_i = n, \quad \text{and} \quad \sum_{i=1}^{N} t_i = n, \quad \text{respectively.})$$

The variance of $\bar{\bar{p}}$ will be

$$\frac{N - n}{N} \frac{1}{n} S^2(\mathcal{Q})$$

in sampling without replacement

$$\left(S^2(\mathscr{Q}) = \frac{1}{N-1}\sum_{i=1}^{N}(q_i - \bar{\bar{P}})^2\right) \quad \text{and} \quad \frac{1}{n}\,\sigma^2(\mathscr{Q})\left(\sigma^2(\mathscr{Q}) = \frac{N-1}{N}\,S^2(\mathscr{Q})\right)$$

in sampling with replacement. Estimators for the variance of $\bar{\bar{p}}$ will be respectively

$$v(\bar{\bar{p}}) = \frac{N-n}{N}\,\frac{1}{n}\,s^2(\mathscr{Q})$$

(without replacement) or

$$v(\bar{\bar{p}}) = \frac{1}{n}\,s^2(\mathscr{Q})$$

(with replacement), where

$$s^2(\mathscr{Q}) = \frac{1}{n-1}\sum_{i=1}^{N}(q_i - \bar{\bar{p}})^2 a_i$$

(without replacement), and

$$s^2(\mathscr{Q}) = \frac{1}{n-1}\sum_{i=1}^{N}(q_i - \bar{\bar{p}})^2 t_i$$

(with replacement).

In the present case, $\mathscr{Q}$ is related to $\mathscr{Z}$ by $q_i = z_i/\bar{M}$, so

$$V(\bar{\bar{p}}) = \bar{M}^{-2}V(\bar{z}), \quad v(\bar{\bar{p}}) = \bar{M}^{-2}v(\bar{z}).$$

The formula for $v(\bar{\bar{p}})$, if – mistakenly – we should consider $n\bar{M}\bar{\bar{p}}$ as a binomial random variable with mean $n\bar{M}\bar{\bar{P}}$, would be

$$\{(N\bar{M} - n\bar{M})/N\bar{M}\}\bar{\bar{P}}(1 - \bar{\bar{P}})/(n\bar{M} - 1),$$

as noted in section 12. In the above example this equals

$$\frac{96\,000 - 10\,000}{96\,000}\,\frac{0.610\,1 \times 0.389\,9}{9\,999} = 0.000\,021,$$

less than the above estimate 0.000 030. Why? Because we did not obtain a simple random sample of 10 000 members, but a sample of 25 doctors (and information on each of their patients). If we denote A_h^i the event "the hth patient of doctor i consulted him during the past year", then the conditional probabilities (of a year ago!)

$$\mathrm{P}(A_h^i|A_k^i), \quad \mathrm{P}(A_h^i|A_k^iA_l^i), \ldots \qquad (\text{all } k, l, \ldots)$$

(where $h, k, l, \ldots$, are different elements of $\{1, \ldots, 400\}$) are not for each h and i equal to $P(A_h^i)$. A measure of their lack of equality is the socalled *intra-cluster correlation coefficient* r (see Remark at the end of this section), which is positive when on the average the conditioning increases the probabilities. It will be shown in a later chapter that with r positive the binomial formula will tend to underestimate the variance, with r negative – a less frequent occurrence – overestimate it. In the present example we have no clear reasons for expecting in advance that r is positive; in the next example we do.

Why do we often use cluster sampling for estimating $\bar{\bar{P}}$ in cases in which we may possibly (in other cases: certainly) obtain a larger standard error than in simple random sampling? Because of the difference in cost or convenience of obtaining the data. In our example, rather than bother all the physicians assigned to a random sample of 10 000 subscribers drawn from the central financial records, it would be better to bother only 25 doctors and check their records, since subscribers in the first kind of sample would probably "belong" to nearly the entire panel of 240 general practioners.

13.3. CASE IN WHICH NOT ALL THE M_i ARE EQUAL

Here we use

$$\sum_{i=1}^{N} z_i a_i / \sum_{i=1}^{N} M_i a_i \quad \left(\text{or} \quad \sum_{i=1}^{N} z_i t_i / \sum_{i=1}^{N} M_i t_i\right).$$

This is a ratio estimator, since the two characteristics $\mathscr{Z}$ and $\mathscr{M}$ are defined over the population of possible clusters, and receive for the ith possible cluster the values z_i and M_i. What we already know about ratio estimators shows that we can write its variance as approximately

$$\frac{N-n}{N}\frac{1}{n}\frac{1}{\bar{M}^2}S^2(\mathscr{D}) \quad \left(\text{or, respectively,} \quad \frac{1}{n}\frac{1}{\bar{M}^2}\sigma^2(\mathscr{D})\right),$$

where

$$\sigma^2(\mathscr{D}) = \frac{1}{N}\sum_{i=1}^{N}(z_i - \bar{\bar{P}}M_i)^2, \quad S^2(\mathscr{D}) = \frac{1}{N-1}\sum_{i=1}^{N}(z_i - \bar{\bar{P}}M_i)^2,$$

and

$$\bar{M} = \sum_{i=1}^{N} M_i/N.$$

Instead of

$$\frac{1}{\bar{M}^2}\sum_{i=1}^{N}\frac{(z_i - \bar{\bar{P}}M_i)^2}{N-1}$$

we may write

$$\sum_{i=1}^{N} \left(\frac{M_i}{\bar{M}}\right)^2 \frac{(q_i - \bar{\bar{P}})^2}{N-1},$$

and we see that, if $M_i = \bar{M}$, we obtain the formulas of 13.2. In the formula for the estimated variance we shall usually use, in place of $1/\bar{M}^2$, $1/\hat{M}^2$, where $\hat{M} = \Sigma a_i M_i / n$.†

EXAMPLE. A simple random sample of 8 boxes of oranges out of a lot of 1 000 boxes was examined for spoilage with the following results:

Box number	1	2	3	4	5	6	7	8	Total
Number of fruit in the box	50	40	45	55	70	65	35	40	400
Number of spoiled fruit in the box	4	21	6	30	50	4	20	15	150

On the basis of these data we wish to estimate the proportion, $\bar{\bar{P}}$, of spoiled fruit in the entire lot and the variance of the estimator:

$$\bar{\bar{p}} = \Sigma a_i z_i / \Sigma a_i M_i = 150/400 = 0.375;$$

$$v(\bar{\bar{p}}) = \frac{N-n}{N} \frac{1}{n\hat{M}^2} \frac{\Sigma a_i z_i^2 - 2\bar{\bar{p}}\Sigma a_i z_i M_i + \bar{\bar{p}}^2 \Sigma a_i M_i^2}{n-1}$$

$$= \frac{1\,000 - 8}{1\,000} \frac{1}{8 \times 50^2} \frac{4\,534 - 2 \times 0.375 \times 8\,020 + 0.375^2 \times 21\,100}{7}$$

$$= 0.010\,53.$$

On the other hand, the formula for the estimated variance of a proportion in simple random sampling of 400 oranges from about $\frac{1\,000}{8}400 = 50\,000$ oranges is (since the finite population multiplier is negligible)

$$p(1-p)/(400-1).$$

This gives 0.000 59, *smaller* than 0.010 53. That it is smaller is to be expected, since a spoiled orange leads to spoilage of other oranges in the same box, and thus makes for positive r. (Note: The fact that the figures given on the number of oranges in each box are all multiples of 5 raises suspicions about the reliability of the data.)

† Incidentally an estimator of the total number of small units in the population is $N\hat{M}$ (cf. first footnote in Chapter I).

We now revert to the very first example of this section, where $\mathscr{Y}$ was sex. Consider a sample of 30 families from 3 005, with

$$\Sigma a_i z_i = 53,\ \Sigma a_i M_i = 104,\ \Sigma a_i M_i z_i = 206,\ \Sigma a_i z_i^2 = 117,\ \Sigma a_i M_i^2 = 404.$$

We find $\bar{\bar{p}} = 53/104 = 0.509\,6$, and $v(\bar{\bar{p}}) = 0.001\,13$. In this case

$$\frac{\Sigma M_i - 104}{\Sigma M_i}\frac{1}{103}p(1-p) \approx 0.002\,4$$

(since $104/\Sigma M_i$ may be estimated to about $n/N = 0.01$), which *exceeds* $v(\bar{\bar{p}})$. $r < 0$, because if the ith family has $M_i > 1$ members, the probability of the event that "there is a male in the ith family", given that in the same family there is a female, exceeds the unconditional probability; and so the probability of "there is more than one male in family i", given that there is at least one male in the family, is less than the unconditional probability.

13.4. REMARKS

$\mathscr{Y}$ does not have to have only values 0 or 1; for example, y_{ij} may be the number of times subscriber u_{ij} consulted the physician during the past year for a certain class of complaints. Also we need not ascertain the $\mathscr{Y}$-values of all the clusters in the sample (*subsampling*). We shall learn various statistical methods appropriate for all cases in a special chapter on cluster sampling. There we shall also see that, rather than to assign to each cluster an equal probability to be selected (*simple* cluster sampling), it is often more efficient to assign to them *unequal probabilities*. The simplest case is sampling with replacement with probabilities p_i at each draw; an unbiased estimator of $\bar{\bar{P}}$ is then $(\Sigma M_i)^{-1}\Sigma z_i t_i/np_i$, as $\mathscr{E}t_i = np_i$. In particular, if $p_i = M_i/\Sigma M_i$, this simply becomes $\bar{\bar{p}} = \Sigma q_i t_i/n$. The variance of $\bar{\bar{p}}$ is then

$$n^{-1}\Sigma\frac{M_i}{\bar{M}}(q_i - \bar{\bar{P}})^2/N,$$

estimated by $n^{-1}\Sigma t_i(q_i - \bar{\bar{p}})^2/(n-1)$; for proofs see section 3.2 of Chapter VI.

For any characteristic $\mathscr{Y}$, r is the average of $\sigma^{-2}(\mathscr{Y})(y_{ij} - \bar{\bar{Y}})(y_{ij'} - \bar{\bar{Y}})$ over all pairs (j, j') of different elements from $\{1, \ldots, M_i\}$ and all i's from $\{1, \ldots, N\}$. More on this in Chapter V, section 4.

14. Sample size

If we wish $V(\bar{y})$ to be less than a preassigned† number V_0, then, in sampling with replacement, the sample size must be at least n_0, where $n_0 = \sigma^2(\mathscr{Y})/V_0$, since in this case $V(\bar{y}) = \sigma^2(\mathscr{Y})/n \leq V_0$ implies $n \geq \sigma^2(\mathscr{Y})/V_0$. In sampling

† In assigning V_0 one should take into account the nonsampling errors (see Chapter X).

without replacement we may use the same formula if n_0/N is small. Otherwise use n_0' defined by

$$n_0' = \frac{n_0}{(N-1)/N + n_0/N} \approx \frac{n_0}{1 + (n_0/N)},$$

since $V(\bar{y})$ is now

$$\{(N-n)/(N-1)\}(1/n)\sigma^2(\mathscr{Y}),$$

and since from

$$\{(N-n_0')/(N-1)\}/n_0' = 1/n_0$$

follows

$$(N/n_0') - 1 = (N-1)/n_0$$

or

$$n_0' = n_0\{[(N-1)/N] + n_0/N\}^{-1}.$$

In general $\sigma^2(\mathscr{Y})$ is not known. Often it is possible to obtain a rough estimate from an earlier survey or from studies on populations which may be expected to be similar with respect to the variance of $\mathscr{Y}$.†

In other cases we may obtain its order of magnitude from a pilot survey. There are also methods of sampling in which one obtains a part of the sample before it is decided what the total sample size should be (*double or two-phase sampling for determining the sample size*). For this subject we refer the reader to papers by Cox, Biometrika, **39** (1952) 217–227, and Samuel, J. Amer. Statist. Ass., **61** (1966) 220–227, and literature cited there. We merely state that the formulas for means and variances of estimators when n is fixed in advance are not necessarily valid when this is not so.

Suppose we also wish to estimate $\bar{Y}_1$ for the domain $\mathscr{D}_1$, such that the variance of the estimator will not exceed a given number V_1^0. The conditional variance of $\bar{y}_1$, with the latter being defined only for $n_1 > 0$, has expected value

$$\mathscr{E}\left(\frac{1}{n_1} - \frac{1}{N_1}\right)\frac{\sum_{i=1}^{N_1}(y_i - \bar{Y}_1)^2}{N_1 - 1},$$

† The reader should consult in this connection Chapter 14 of Deming, Sample Design in Business Research (Wiley, 1960), especially his figure 16, which is based on a knowledge of the range (= maximum − minimum) of the $\mathscr{Y}$-values of all units in the population. If the range is large, stratification will often lead to much smaller ranges and a much more nearly even distribution of $\mathscr{Y}$-values within the strata than in the population, so that it may be much easier to assess within strata variances than overall variances.

which by Chapter IIA, section 10, equals approximately

$$\frac{N-n}{N-1}\frac{1}{n}\frac{N}{N_1}\left(1+\frac{N-N_1}{nN_1}\right)\frac{\sum_{i=1}^{N_1}(y_i-\bar{Y}_1)^2}{N_1-1}.$$

If $n_1 = 0$, $\bar{y}_1$ is not defined. So we should plan the investigation so that this event has a negligible probability (see the first footnote in Chapter IIA, section 10). A simple sample for which this is so may be of prohibitive size if $\mathscr{D}_1$ represents a very small proportion of the whole population which cannot be sampled separately. To solve such problems special plans may be developed (*sequential*, *inverse binomial* and *inverse hypergeometric* sampling†).

A short computation shows that, if we define

$$n_{00} = \frac{N}{N_1}\frac{\sum_{i=1}^{N_1}(y_i-\bar{Y}_1)^2}{N_1-1}\Big/V_1^0$$

and then compute

$$n'_{00} = n_{00}\left(1-\frac{1}{N}+\frac{n_{00}}{N}\right)^{-1}\left(1-\frac{1}{N_1}+\frac{1}{N}\right)$$
$$\times\left[\frac{1}{2}+\left\{\frac{1}{4}+\left(\frac{1}{N_1}-\frac{1}{N}\right)\left(\frac{N-1}{n_{00}}+1\right)\left(1-\frac{1}{N_1}+\frac{1}{N}\right)^{-2}\right\}^{\frac{1}{2}}\right],$$

we should take for n the larger of n'_{00} and n'_0, rounded up to the nearest integer.

15. Further on confidence intervals and bounds for Y and $\bar{Y}$

These concepts were recalled in Chapter I, section 5, and in section 7 of the present chapter, and applied to the case in which an estimator was approximately normally distributed.

15.1. MEANING OF APPROXIMATE NORMALITY AND SOME EMPIRICAL RESULTS

The first question that arises is: what is meant by "approximately normal"? The term arises from a class of theorems that deal with a sequence of sampling distributions for samples $\mathscr{s}_k, \mathscr{s}_{k+1}, \mathscr{s}_{k+2}, \ldots$ of sizes $k, k+1, k+2, \ldots$.

† The literature on this subject is large; on sequential sampling see WETHERILL, Sequential Methods in Statistics (Methuen, 1966); on the other topics BARTKO, Virginia J. of Science, **12** (1961) 18–37; both sources contain further references.

The simplest of these theorems relates to simple random sampling with replacement† and makes a statement about $\Phi_n(t)$, the probability that

$$\bar{y}(n) = \sum_{u_i \in \mathscr{s}_n} y(u_i)/n$$

satisfies

$$\frac{\bar{y}(n) - \mathscr{E}\bar{y}(n)}{V(\bar{y}(n))^{\frac{1}{2}}} \leq t.$$

In particular it states that for all t the limit of $\Phi_n(t)$ as $n \to \infty$ equals $\Phi(t)$, where $\Phi(t)$ denotes the (tabulated) probability that a standard normal variable does not exceed t. This means, of course, that there exists a (smallest) integer n_0, such that for $n \geq n_0$, $\Phi_n(t)$ differs from $\Phi(t)$ by less than a preassigned positive quantity ε. Here n_0 depends on both t and ε, and also on the function F_N defined in section 12. Also $V(\bar{y}(n))$ may be replaced by $v(\bar{y}(n))$,‡ although this may require n_0 to be much larger.

One would like to use this theorem for approximating $\Phi_n(t)$ by $\Phi(t)$, but for this some knowledge of n_0 is required. If the functions F_N and Φ are already "nearly equal",§ then n_0 is small; this follows from the easily shown fact that the average of normal random variables is itself normal. In most cases we have to obtain our knowledge of n_0 from numerical calculations. In a general way one may say that the more the function F_N resembles the function Φ in certain broad features (which do not necessarily include graphical resemblance), the faster the convergence of $\Phi_n(t)$ to $\Phi(t)$.§§

One of these features is *symmetry*. Thus, if, for each i, $y(u_i) = 0$ or 1, the rapidity of convergence of Φ_n depends very much on the proportion P of units for which $y(u_i)$ has the value 1; it is fastest for P in the neighbourhood of $\frac{1}{2}$. E.g., if we look for 95% confidence intervals and do not mind if they are of confidence coefficient 0.94 or 0.96; or if we look for upper confidence bounds which are exceeded with probability not more than 0.025, but do not mind if they are exceeded with probability 0.035; then, for p equal to 0.5, the normal

† For a generalization to sampling without replacement see HÁJEK, Publ. Math. Inst. Hung. Ac. Sci., **5A** (1960) 361–374, especially for the case in which not only n, but also $N - n$, the number in the population but not in the sample, is large. Clearly, if $N - n$ is small, $V(\bar{y}(n))^{1/2}$ is so small that, in order to obtain a distribution at all, one should divide $\bar{y}(n) - \mathscr{E}\bar{y}(n)$ by a larger quantity. More recent results in this field have been obtained by ROSÉN, Ark. Mat., **5** (1965) 383–432.

‡ This follows from section 20.6 of CRAMÉR, Mathematical Methods of Statistics (Princeton University Press, 1946).

§ More precisely, if $F_N(t)$ is nearly equal to $\Phi((t - a)/b)$ for some a and b, i.e., if F_N is nearly a normal distribution, though not necessarily standard normal.

§§ It should be evident that, in the case of *complete ignorance* of F_N (apart from convergence of Φ_n), *nothing* can be said about n_0. Indeed, if for a given F_N, Φ_n converges to Φ at a stated rate, we can always specify another F_N for which Φ_n converges to Φ more slowly.

approximation can be used for $n = 30$; for p equal to 0.4: $n = 50$; for $p = 0.3$: $n = 80$; but for $p = 0.2$: $n = 200$. For P (or p) near 0 or 1 and n large (e.g. for $p = 0.10$ and $n = 200$, or $p \leqslant 0.05$ and $n = 100$) we can use the Poisson† approximation with $nP = \lambda$:

$$\binom{n}{k} P^k(1 - P)^{n-k} \approx \lambda^k \mathrm{e}^{-\lambda}/k!.$$

If the *number of possible values* of $y(u_i)$ is small, the convergence of Φ_n to Φ is slower than in the case in which there are many possible values (within the same range). This is because in the latter case the frequency function of $\bar{Y}$ is much closer to continuity.‡

Many functions f_N arising in practice have two or more widely *separated main peaks*. Thus among grocery stores one will often find very many having sales in a range typical for small neighborhood stores, and a substantially smaller number having sales ten or twenty times that of the typical neighborhood store. The function f_N for all grocery stores is then a composite of two such (largely) nonoverlapping functions, each one with its own peak or peaks. In such cases convergence of Φ_n to Φ is slow.§

For instance, in HANSEN, HURWITZ and MADOW, Sample Survey Methods and Theory (Wiley, 1953), there is an artificial example of a population with the following values for $Nf_N(t)$.

t	$Nf_N(t)$	t	$Nf_N(t)$
900	1 000 000	2 200	1 000 000
1 300	1 000 000	2 700	1 000 000
1 500	1 000 000	3 100	1 000 000
1 800	1 000 000	3 600	1 000 000
1 900	1 000 000	4 800	1 000 000
2 000	1 000 000	20 000	1 000 000

† It is known that, if x is a Poisson variate, $2\{(x + \frac{3}{8})^{\frac{1}{2}} - \lambda^{\frac{1}{2}}\}$ is fairly well approximated in the tails of its distribution (though not always in the extreme tails) by a standard normal variate when $\lambda \geq 2$; for λ between 1 and 2 it is better to use instead $x^{\frac{1}{2}} + (x + 1)^{\frac{1}{2}} - 2\lambda^{\frac{1}{2}}$. The corresponding expressions for the binomial are $2(n + \frac{1}{2})^{\frac{1}{2}}[\text{arc sin}\,\{(x + \frac{3}{8})/(n + \frac{3}{4})\}^{\frac{1}{2}} - \text{arc sin}\, P^{\frac{1}{2}}]$ and $(n + 1)^{\frac{1}{2}}[\text{arc sin}\,\{x/(n + 1)\}^{\frac{1}{2}} + \text{arc sin}\,\{(x + 1)/(n + 1)\}^{\frac{1}{2}} - 2\,\text{arc sin}\, P^{\frac{1}{2}}]$. More on the Poisson approximation and further references are found in ANDERSON and BURSTEIN, J. Amer. Statist. Ass., **62** (1967) 857–861. References to tables are given in section 15.4 below.

‡ Usually the effect of discontinuity will be larger for larger confidence coefficients.

§ This is, in fact, one of the reasons for using stratified sampling in such cases, if possible (see Chapter III).

$\mu = 3\,817$ and $\sigma = 4\,988$. For $n = 12$ there are still widely separated peaks in the frequency function of $\bar{y}$, but not for $n = 24$. For the confidence interval $[\bar{y} - 2V(\bar{y})^{\frac{1}{2}}, \bar{y} + 2V(\bar{y})^{\frac{1}{2}}]$ ($V(\bar{y})$ known!), the confidence coefficient has, instead of the normal value 0.955, the values 0.953 if $n = 12$, 0.958 if $n = 24$, but 0.926 if $n = 6$; for the confidence interval $[\bar{y} - 3V(\bar{y})^{\frac{1}{2}}, \bar{y} + 3V(\bar{y})^{\frac{1}{2}}]$ the confidence coefficients, instead of 0.997, are 0.990, 0.996 and 0.990, respectively. But almost all the difference between these and 1.000 is located in the right part of Φ_n.

For applications we are interested in the case in which $V(\bar{y})$ is estimated. Therefore, a similar investigation was carried out by the author with $2v(\bar{y})^{\frac{1}{2}}$ replacing $2V(\bar{y})^{\frac{1}{2}}$ for $n = 24$; instead of 0.955, the confidence coefficient came out of 0.88, all outliers being on the right-hand side. When n was increased to 99, the confidence coefficient increased to 0.93 with nearly all outliers on the right-hand side.

In most cases the number of possible values of $y(u_i)$ in a population of 12 million units is very much larger than 12. Therefore the author examined the effect of replacing the given distribution with one in which the number of possible values is 48, 4 of which are near the tail end, and the others having a right skewed bell-shaped frequency function between $t = 200$ and $t = 8\,500$ with highest frequencies between $t = 1\,800$ and $t = 2\,200$. Also μ of the original distribution was preserved. (However, σ changed to 42.) For $n = 24$ the confidence coefficient came out lower, namely 0.78 with virtually all outliers on the right-hand side. For $n = 99$ the two distributions gave the same confidence coefficient.

Investigations for n as large as 1 000 and more brought the confidence coefficient close to the value 0.955, but still resulted in far more of the outliers lying on the right-hand than on the left-hand side.

Often there is no notable peak in f_N for large values of $\mathscr{Y}$, but we still have very many units with small values of $\mathscr{Y}$. This is called *positive skewness*; a measure of skewness is

$$\gamma_1 = N^{-1} \sum_{i=1}^{N} (y_i - \bar{Y})^3/\sigma^3.$$

(In the above distribution $\gamma_1 = 2.8$ and in the modified distribuion 3.3.†) If the skewness of f_N is positive but not very large, the nominal confidence coefficient using normal approximation methods is usually almost equal to the true confidence coefficient (for instance the probability that $\bar{Y}$ falls between $\bar{y} - (1.96)\{v(\bar{y})\}^{\frac{1}{2}}$ and $\bar{y} + (1.96)\{v(\bar{y})\}^{\frac{1}{2}}$ is usually almost equal to 0.95); but the lower nominal confidence bound and the upper nominal confidence bound are usually too low.

In case f_N is thought to be very skew‡ and stratification is not feasible, one sometimes uses *transformed* data $y_i^* = g(y_i)$, selecting for g some increasing

† Two frequently used measures of deviation from normality are γ_1 and

$$\gamma_2 = \{N^{-1} \Sigma(y_i - \bar{Y})^4/\sigma^4\} - 3;$$

both are 0 in the normal case. In the above distribution $\gamma_2 = 6.3$, in the modified distribution 9.9.

‡ Of course, if we wish to characterize the possible magnitudes of $\mathscr{Y}$ by a single number, and if f_N is highly asymmetric, we would often use for that purpose not $\bar{Y}$ but some other parameter, e.g. the median. For estimation of the latter see section 16.

function for which it is expected that f_N^* is much less skew. The estimates, confidence intervals, etc., must then be transformed back with the help of the transformation inverse to g. This procedure should be used with caution: f_N^* may actually behave worse than f_N. Moreover, properties of an estimator such as unbiasedness, or even approximate unbiasedness, are not usually preserved under this procedure. E.g., if g is the natural logarithm and the $(y_i^* - \mu)/\sigma$ have (approximately) the distribution Φ, the e^{y_i} have (approximately) expected value $e^{\mu+\frac{1}{2}\sigma^2}$ instead of e^{μ}. (The objection does not hold for transformations applied to cases in which f_N is not very asymmetric but has much *heavier tails* than the normal distribution; there g may be chosen to diminish the impact of extreme values on either side.)

A class of distributions that roughly fit many *very skew* populations is the class of lognormal distributions. By way of example the author examined the case in which $\log_e \mathscr{Y} - 0.46$ is standard normal. Its frequency function is depicted up to the value 7.2 in CRAMÉR, Mathematical Methods of Statistics (Princeton University Press, 1946), Figure 17. Beyond 7.2 its tail keeps diminishing very slowly. $\bar{Y} = 2.61$, $\gamma_1 = 6.2$, $\gamma_2 = 110.9$, $\sigma = 3.42$. By using a normal approximation to the distribution of $\bar{y}$ with $t = 2$, it was found that a sample size of the order of 2 500 was needed to get near the theoretical confidence coefficient of 0.955; in some 10 000 trials† the average percent of cases in which $\bar{Y}$ fell outside the confidence interval was 4.9. The approach here (as well as in the previous example) is very slow: for half this sample size the percentage was 5.2, for a quarter the sample size the percentage was 5.7.‡ It may well be that, from the point of view of width of confidence intervals, a smaller sample size than the ones mentioned here would have been sufficient; but for the last case above the half-width of the confidence interval was usually between 0.26 and 0.28, which is already about 10% of the value of $\bar{Y}$. (For a sample size of 1 250 the half-width was usually between 0.19 and 0.20, for one of 2 500 between 0.13 and 0.14.) *Thus, even if the population distribution of nonnegative y_i should happen to be as extremely skew as the present one, confidence intervals of half-widths equal to some 10% of $\bar{Y}$ based on the normal approximation may have actual confidence coefficient only slightly smaller than the nominal, one of* 0.95.§

Again in most of the cases in which $\bar{Y}$ fell outside the confidence interval, it fell on the right-hand side: for sample size 2 500 in 71% of the cases, for half that size in 76%, and for

† If the confidence coefficient is 0.955, it takes some 10 000 trials to make the probability at least 0.98 that the percentage of the trials in which $\bar{Y}$ falls outside the confidence interval is between 4.0 and 5.0. This calculation uses the fact that x, the number of trials in which this happens, is virtually a Poisson variate so that, by a previous observation in this section, we can (for the order of n involved) use a standard normal table for $2x^{\frac{1}{2}}$ to obtain the confidence interval of confidence coefficient 0.98: $[2\bar{x}^{\frac{1}{2}} - 2.33n^{-\frac{1}{2}}, 2\bar{x}^{\frac{1}{2}} + 2.33n^{-\frac{1}{2}}]$ for $2(\mathscr{E}\bar{x})^{\frac{1}{2}}$, and so for $\mathscr{E}\bar{x}$ of about $[\bar{x} - 2.33(\bar{x}/n)^{\frac{1}{2}}, \bar{x} + 2.33(\bar{x}/n)^{\frac{1}{2}}]$. Setting $2.33(0.045/n)^{\frac{1}{2}}$ equal to $(0.05 - 0.04)/2$ gives the approximate n required.

‡ In view of the trend shown here and in other cases, it was not thought worthwhile to continue increasing the sample size in order to get still closer to the theoretical percentage. One should also consider in this connection that the fraction of cases obtained on the basis of 10 000 trials is subject to some sampling error (compare the previous footnote).

§ We do not mean to imply that this always occurs, in skew distributions of nonnegative variates. For distributions with smaller $\bar{Y}$, a given half-width constitutes a larger percentage of $\bar{Y}$, but in such distributions a smaller $\bar{Y}$ usually means a much smaller skewness.

a quarter in 81 %. To observe the effect of adding a very slight secondary peak, the author also considered a mixture of 9 parts of the above distribution with 1 part of a normal distribution with mean 6 and variance 1. Then $\bar{Y} = 2.95$, $\gamma_1 = 5.4$, $\gamma_2 = 101.2$, $\sigma = 3.42$. With samples of size 3 600 the percentage of cases in which $\bar{Y}$ fell outside the confidence interval was 4.8 (and with samples half this size it was 5.2), 62 % of these on the right-hand side (for the smaller sample size 68 %). It appears that this case requires somewhat more observations than the unmixed case, but the extremely slow approach to the theoretical percentage makes that conclusion rather uncertain†.

Suppose now that a normal approximation may be used, and that we wish to use a confidence interval for Y with confidence coefficient of about 0.95.

15.2. APPROXIMATE INTERVALS IN THE CASE OF FREQUENCIES OF PROPERTIES USING NORMAL TABLES‡

Let $y(u_i) = 1$ if u_i has a specified property and 0 otherwise, so that $Y = NP$ is the number of elements in $\mathscr{P}$ with this property. Assume that P and n are such that $\Phi_n(t)$, the probability that $\bar{y} - P$ is not more than some number $t\{P(1 - P)/n\}^{\frac{1}{2}}$, is well approximated by $\Phi(t)$ for all t of interest (as discussed above). In particular $\Phi(1.96) - \Phi(-1.96) = 0.95$. Therefore in simple random sampling with replacement the probability is about 0.95 that $\bar{y} - P$ lies between $-t\{P(1 - P)/n\}^{\frac{1}{2}}$ and $t\{P(1 - P)/n\}^{\frac{1}{2}}$, with $t = 1.96$; that is, that

$$(\bar{y} - P)^2 \leq t^2P(1 - P)/n$$

or

$$(n + t^2)P^2 - 2(\bar{y}n + \tfrac{1}{2}t^2)P + \bar{y}^2n \leq 0,$$

which is satisfied for NP in

$$\left[N\frac{\bar{y}n + \frac{1}{2}t^2 - t\{n\bar{y}(1 - \bar{y}) + \frac{1}{4}t^2\}^{\frac{1}{2}}}{n + t^2},\ N\frac{\bar{y}n + \frac{1}{2}t^2 + t\{n\bar{y}(1 - \bar{y}) + \frac{1}{4}t^2\}^{\frac{1}{2}}}{n + t^2}\right].$$

(In this case a sometimes better approximation may be obtained§ by subtracting $1/(2n)$ from $\bar{y}$ wherever $\bar{y}$ appears in the lower limit and adding

† An alternative method of obtaining confidence bounds for $\bar{Y}$ is proposed in section 17. In section 14.1 of Chapter III there is a discussion of stratified sampling from these populations. Ratio estimators for some very skew populations are considered in section 4.2 of Chapter IIA. There we also give results on confidence intervals for $\bar{Y}$ based on $\bar{y}$ in a finite population (of 791 U.S. cities) with $\gamma_1(\mathscr{Y}) = 4.78$. Moreover, for a characteristic $\mathscr{X}$ of these cities with $\gamma_1(\mathscr{X}) = 3.16$, the percentages of 20 000 repetitions in which the confidence interval did not include $\bar{X}$ were 5.5, 4.9, 5.0, 4.8, and 4.5 for samples of 300, 400, 500, 600, and 700, respectively, their average half-widths declining from 10 % to less than 3 % of $\bar{X}$. In 1.1 % to 1.4 % of the repetitions $\bar{X}$ was to the left of the confidence interval.

‡ Graphical methods for binomial data are discussed by MOSTELLER and TUKEY, J. Amer. Statist. Ass., **44** (1949) 174–212.

§ Instead, some subtract $N/(2n)$ from the lower limit and add $N/(2n)$ to the upper limit. In any case, if $\bar{y} \leq 1/(2n)$, the lower limit is taken to be 0; and if $1 - \bar{y} \leq 1/(2n)$, the upper limit is taken to be N. But in these cases the nominal approximation is surely not appropriate (because either n is small, or p is close to 0 or 1).

$1/(2n)$ to $\bar{y}$ wherever $\bar{y}$ appears in the upper limit; by using this "continuity correction" we attempt to adjust for the fact that Φ_n is not a continuous function of t while Φ is.) For sampling without replacement, we replace $P(1 - P)$ by $\{(N - n)/(N - 1)\}P(1 - P)$. This replacement gives the same interval as given above, except that t is replaced by $t\{(N - n)/(N - 1)\}^{\frac{1}{2}}$. Analogous results are obtained for confidence bounds. Exact results are discussed under 15.4 below.

15.3. USE OF NORMAL TABLES IN THE GENERAL CASE

In the above case it was possible to obtain an interval computed from the statistic $\bar{y}$ alone; in general this is not possible. More often we therefore estimate $V(\bar{y})$ and replace $V(\bar{y})$ by its estimator $v(\bar{y})$.

When $\mathscr{Y}, \mathscr{P}$ and n are such that $\Phi_n(t)$ is close to $\Phi(t)$ in the relevant range of t-values, we get as a confidence interval for Y with approximate 0.95 confidence coefficient:

$$[N\bar{y} - Ntv(\bar{y})^{\frac{1}{2}}, N\bar{y} + Ntv(\bar{y})^{\frac{1}{2}}].$$

This method is often also a fair approximation for the case in which $v(\bar{y})$ is a function of $\bar{y}$ alone. In particular, if $Y = NP$ is a frequency in $\mathscr{P}$, $v(\bar{y})$ equals $\bar{y}(1 - \bar{y})/(n - 1)$ (or in sampling without replacement

$$\{(N - n)/N\}\bar{y}(1 - \bar{y})/(n - 1));$$

so that, if we substitute $v(\bar{y})$ for $V(\bar{y})$, we obtain that, in simple random sampling with replacement, $\bar{y} - P$ is with probability approximately 0.95 between $-t\{\bar{y}(1 - \bar{y})/(n - 1)\}^{\frac{1}{2}}$ and $t\{\bar{y}(1 - \bar{y})/(n - 1)\}^{\frac{1}{2}}$, or NP in

$$[N\bar{y} - tN\{\bar{y}(1 - \bar{y})/(n - 1)\}^{\frac{1}{2}}, N\bar{y} + tN\{\bar{y}(1 - \bar{y})/(n - 1)\}^{\frac{1}{2}}].$$

But the approximation methods of 15.2 are preferable in this case.

15.4. EXACT INTERVALS IN THE CASE OF FREQUENCIES OF PROPERTIES

With Y a frequency, we may obtain exact confidence intervals or limits as follows. Consider first sampling with replacement, so that

$$\mathrm{P}\{y = k\} = \binom{n}{k} P^k(1 - P)^{n-k}.$$

Then, given α_1 and α_2 (each less than $\frac{1}{2}$), we can find the largest integer $Nc_1(y)$ and the smallest integer $Nc_2(y)$ such that for all $P > c_2(y)$

$$\sum_{k \leq y} \binom{n}{k} P^k(1 - P)^{n-k} \leq \alpha_2$$

when $y < n$, and for all $P < c_1(y)$

$$\sum_{k \geq y} \binom{n}{k} P^k (1-P)^{n-k} \leq \alpha_1$$

when $y > 0$. The reason that such integers exist is that

$$\sum_{k \leq y} \binom{n}{k} P^k (1-P)^{n-k} = 1 - \frac{n!}{y!(n-y-1)!} \int_0^P t^y (1-t)^{n-y-1} \, dt$$

is decreasing in P (from 1 to 0) for $y < n$ and

$$\sum_{k \geq y} \binom{n}{k} P^k (1-P)^{n-k} = \frac{n!}{(y-1)!(n-y)!} \int_0^P t^{y-1} (1-t)^{n-y} \, dt$$

is increasing in P (from 0 to 1) for $y > 0$. We also define $c_1(0) = 0$, $c_2(n) = 1$.

The latter integral is called the incomplete beta function ratio $I_P(y, n-y+1)$ and has been tabulated (a convenient short table and graph appears in PEARSON and HARTLEY, Biometrika Tables for Statisticians, Vol. I (Cambridge University Press, 1956)). A solution, for continuous $c_i(y)$ can be obtained from these, or more conveniently from tables of the F-distribution, which are more widely available. Let $F = F_{\nu_1 \nu_2}(\alpha)$ be the upper α-point of the F-distribution with ν_1 and ν_2 degrees of freedom. To find $c_2(y)$, set $\nu_1 = 2(y+1)$, $\nu_2 = 2(n-y)$, and $\alpha = \alpha_2$; then $c_2(y) = \nu_1 F/(\nu_2 + \nu_1 F)$. To find $c_1(y)$, set $\nu_1 = 2(n-y+1)$, $\nu_2 = 2y$, and $\alpha = \alpha_1$; then $c_1(y) = \nu_2/(\nu_2 + \nu_1 F)$. If $\mathscr{Y}$ is a Poisson variable, the upper confidence bound for $\lambda = nP$ may be conveniently read from the widely available chi-square tables as half the upper α_2-point of chi-square with $\nu_1 = 2(y+1)$ degrees of freedom; the lower confidence bound for λ is half the lower α_1-point of chi-square with $\nu_2 = 2y$ degrees of freedom. Also Table 40 in PEARSON and HARTLEY, Biometrika Tables for Statisticians, Vol. I (Cambridge University Press, 1956) gives confidence intervals for the Poisson parameter.

A confidence interval of confidence coefficient (at least) $1 - \alpha_1 - \alpha_2$ is given by the closed interval from $Nc_1(y)$ to $Nc_2(y)$.

For let $t_1(P)$ be the largest integer t for which

$$\sum_{k \leq t} \binom{n}{k} P^k (1-P)^{n-k} \leq \alpha_2,$$

and $t_2(P)$ the smallest integer t for which

$$\sum_{k \geq t} \binom{n}{k} P^k (1-P)^{n-k} \leq \alpha_1.$$

Then

$$\mathrm{P}\{t_1(P) < y < t_2(P)\} \geq 1 - \alpha_1 - \alpha_2.$$

But

$$t_1(P) < y < t_2(P)$$

if and only if

$$\sum_{k \leq y} \binom{n}{k} P^k(1-P)^{n-k} > \alpha_2$$

and

$$\sum_{k \geq y} \binom{n}{k} P^k(1-P)^{n-k} > \alpha_1,$$

that is, if and only if

$$c_1(y) \leq P \leq c_2(y).$$

Tables are given by MAITLAND, HERRERA and SUTCLIFFE, Tables for Use with Binomial Samples (New York University College of Medicine, 1956).

In sampling without replacement

$$\binom{n}{k} P^k(1-P)^{n-k}$$

has to be replaced by

$$\binom{NP}{k}\binom{N-NP}{n-k}\Big/\binom{N}{n},$$

the hypergeometric frequency function.† Charts for sampling fractions of $\frac{1}{20}$ and more, and for N at least 500 are found in CHUNG and DE LURY, Confidence Limits for the Hypergeometric Distribution (University of Toronto Press, 1950). Consult also ARMSEN, Biometrika, **42** (1955) 494–511, LIEBERMAN and OWEN, Tables of the Hypergeometric Probability Distribution (Stanford University Press, 1961) and FINNEY, LATSCHA, BENNETT and HSU, Tables for Testing Significance in a 2 × 2 Contingency Table (Cambridge University Press, 1963).

REMARK. If we are interested in an upper or a lower confidence bound of confidence coefficient $1-\alpha_2$ or $1-\alpha_1$, respectively, we put α_1 or α_2, respectively, equal to zero. Otherwise, we have various ways of choosing α_1 and α_2 such that $1-\alpha_1-\alpha_2$ is at least equal to the desired confidence coefficient β. One choice often adopted‡ is to take both α_1 and $\alpha_2 \leq \frac{1}{2}(1-\beta)$, but as close as possible to it. In discrete distributions this may give rather wide

† Given a binomial confidence bound, the corresponding hypergeometric one is well approximated by shrinking the distance of that bound from $N\bar{y}$ by $[(N-n)/(N-1)]^{\frac{1}{2}}$.

‡ The tables of Maitland *et al.* follow this approach.

intervals.† There is perhaps more sense in choosing α_1 and α_2 such that, for such P, the terms $P(y = k)$ with $k \leq t_1(P)$ and $k \geq t_2(P)$ are the smallest (least likely to occur for that value of P). Systems of confidence intervals for the binomial distribution with this property have been computed among others by CROW, Biometrika, **43** (1956) 423–435 [New Statistical Tables XXIII] for $n \leq 30$;‡ and for the Poisson distribution with $y \leq 300$ by CROW and GARDNER, Biometrika, **46** (1959) 441–453 [New Statistical Tables XXVIII].

15.5. EXAMPLE

A sample of 20 is taken by simple random sampling with replacement from a population $\mathscr{P} = \{u_1, \ldots, u_{100}\}$. We consider five examples of the values of the characteristic $\mathscr{Y}$:

(i) $y_1 = \ldots = y_4 = 1;\ y_5 = \ldots = y_{100} = 0$; that is, $P = 0.04$.
(ii) $y_1 = \ldots = y_{10} = 1;\ y_{11} = \ldots = y_{100} = 0$; that is, $P = 0.10$.
(iii) $y_1 = \ldots = y_{30} = 1;\ y_{31} = \ldots = y_{100} = 0$; that is, $P = 0.30$.
(iv) $y_1 = \ldots = y_{40} = 1;\ y_{41} = \ldots = y_{100} = 0$; that is, $P = 0.40$.
(v) $y_1 = \ldots = y_{50} = 1;\ y_{51} = \ldots = y_{100} = 0$; that is, $P = 0.50$.

The sample space consists of 2^{20} points. For each possible sample we can compute a confidence interval for P with confidence coefficient 0.95. For example, for each point of the sample space for which $y = 5$, the interval in the table of Crow is [0.104, 0.467]; the tables of Maitland *et al.* give [0.086 8, 0.491 3]; the approximation formula of section 15.2 yields [0.112, 0.468] (with continuity correction§ [0.096, 0.494]). As $N = 100$, all these need to be rounded, to [0.10, 0.47], etc. We observe that, for instance, in case (iii) the value of P falls in this interval. Indeed, 0.30 falls in $[c_1(y), c_2(y)]$ given by

† For given α_1, the probability that $c_1(y)$ exceeds P is *less* than α_1 when P is not one of the numbers $c_1(1), \ldots, c_1(n)$, and, for small n, often is much less if α_1 is small; and, for given α_2, the probability that $c_2(y)$ is less than P is *less* than α_2 when P is not one of the numbers $c_2(0), \ldots, c_2(n-1)$, and, for small n, often is much less if α_2 is small. As these c's are generally all different, it follows that, for $c_1(y)$ and $c_2(y)$ corresponding to the indicated choice of α_1 and, α_2, the probability that $c_1(y) \leq P \leq c_2(y)$ will generally be *less* than $1 - \alpha_1 - \alpha_2$ for *all* values of P, and for small n, often considerably less for all values of P.

‡ In a private communication G. S. Walton recommends the following method of extrapolating this and similar tables up to values for which the normal or Poisson approximation is close enough. In the part of the table that relates to the maximum tabulated sample size, n_0, find the confidence interval for P with $y = n_0\bar{y}$ (this generally involves interpolation) and shrink this interval by multiplying the differences from $\bar{y}$ by $(n_0/n)^{\frac{1}{2}}$. Thus, for $n = 50$, $y = 39$ and $\beta = 0.99$, $30\bar{y} = 23.4$, giving the bounds 0.546 6 and 0.925 0, which differ from 0.78 by $-0.233\,4$ and 0.145 0, respectively. Multiplying by $(30/50)^{\frac{1}{2}}$ and adding back to 0.78 gives 0.60 and 0.89, which coincides with the result yielded by the formula of section 15.2.

§ If we follow the method of the footnote to section 15.2, we obtain [0.087, 0.493].

Crow (unrounded) when $y = 3, 4, \ldots, $ or 11, and not otherwise. But the probability that y equals $3, 4, \ldots,$ or 11, when $P = 0.30$, is 0.959. The intervals given in Maitland *et al.* cover 0.30 if and only if $y = 2, 3, \ldots,$ or 10, and the probability that y equals any of these values when $P = 0.30$ is as much as 0.975.

Consider now the same example in the case of sampling without replacement, for the case in which y takes on the value 5.

Let us first find the confidence interval for which α_1 and α_2 do not exceed 0.025, using Lieberman and Owen's tables. Denote by $a(P, k)$ the hypergeometric frequency function (with $N = 100$ and $n = 20$). Note that $a(P, k)$ for sample size n equals $a(n/N, k)$ for sample size NP. $A_1(P, l)$ denotes the sum of $a(P, k)$ for $k \leq l$, and $A_2(P, l)$ the sum for $k \geq l$. $A_1(P, 5) = 0.023\,9$ for $P = 0.47$, but more than 0.025 for $P = 0.46$; $A_2(P, 5) = 0.014\,7$ for $P = 0.09$, and $0.025\,46$ for $P = 0.10$. This leads to the confidence interval [0.10, 0.47].

Let us now find the Crow-type interval. We first try the upper bound 0.46. $A_1(0.46, 5) = 0.037\,492$ with $a(0.46, 5) = 0.022\,133$; values for $a(0.46, k)$ smaller than the latter are for $k < 5$ and for $k \geq 14$, and $A_2(0.46, 14) = 0.015\,252$. We now try the next smaller value, 0.45, for the upper confidence bound. $A_1(0.45, 5) = 0.037\,492$ with $a(0.45, 5) = 0.027\,125$; values for $a(0.45, k)$ smaller than the latter are for $k < 5$ and for $k \geq 14$, and $A_2(0.45, 14) = 0.011\,735$, which added to $A_1(0.45, 5)$ gives $0.049\,227$. So the upper bound is 0.45.

We try the lower bound 0.11. $A_2(0.11, 5) = 0.040\,417$ with $a(0.11, 5) = 0.032\,895$; values of $a(0.11, k)$ smaller than the latter are only for $k > 5$. Trying the next larger value, 0.12, for the lower confidence bound, we get $A_2(0.12, 5) = 0.059\,954$ which exceeds 0.05. So the lower bound is 0.11, and the interval is [0.11, 0.45].

Let us see what we would have obtained by the adjustment mentioned in the first footnote of the preceding section: $[(N - n)/(N - 1)]^{\frac{1}{2}} = 0.899$. The upper bound given by Maitland *et al.* differs from $\frac{5}{20}$ by 0.241 3. Adding 0.899 times this to $\frac{5}{20}$ gives 0.466 9, which we round to 0.47. Similarly we get the lower bound by rounding 0.103 3 to 0.10.

The upper bound given by Crow differs from $\frac{5}{20}$ by 0.217. Adding 0.899 times this to $\frac{5}{20}$ gives 0.445, which we round to 0.45. Similarly we get the lower bound by rounding 0.119 to 0.11 or 0.12. (Note that we should round without making the coverage probability fall below 0.95; this will usually imply rounding the upper bound up and the lower one down, unless the bound is very close to a multiple of N^{-1}.)

15.6. OTHER CASES IN WHICH Y IS NOT A FREQUENCY

If Y is not a frequency, and if F_N is not such that, for the size of the sample obtained, Φ_n can be well approximated by the normal or other known approximating function, then simple random sampling will not permit us obtaining an approximate confidence interval. Thus for the example in 15.1 of the stores we shall see in Chapter III that stratified sampling may enable us to obtain a more usable confidence interval.

In some cases our information on F_N is sufficiently detailed to make the method of transformation mentioned in section 15.1 dependable. Other ways of utilizing information on F_N are briefly discussed in section 10 of Chapter I and in Chapter VIII.

15.7. INTERVALS NOT BASED ON SAMPLE TOTALS

We can also apply the normal approximation to statistics other than $N\bar{y}$ and $\bar{y}$: in particular† to $\hat{Y}_R$ (see Chapter IIA, section 4). But, for given t and F_N, the number of observations needed for a certain degree of approximation may be larger than that required for y or $\bar{y}$.

15.8. EFFECT OF BIAS

We now examine the effect of bias on the usual confidence interval of confidence coefficient 0.95 in the case of a normal random variable with known variance σ_n^2. Suppose that x has mean $\mu + B$, where B is its *bias* for estimating μ. Then the probability of $[x - 1.96\sigma_n, x + 1.96\sigma_n]$ covering μ is still equal to the probability that $(x - \mu)/\sigma_n$ belongs to the interval from -1.96 to 1.96. However, now $(x - \mu)/\sigma_n - B/\sigma_n$ is standard normal; and the probability that $(x - \mu)/\sigma_n$ belongs to $[-1.96, 1.96]$ equals the probability that the standard normal variable $(x - \mu)/\sigma_n - B/\sigma_n$ belongs to $[-1.96 - B/\sigma_n, 1.96 - B/\sigma_n]$. For $B/\sigma_n = \pm 0.10$ this is 0.948 9 instead of 0.950 0, hardly any difference; for $B/\sigma_n = \pm 0.20$ it is 0.945 4; for $B/\sigma_n = \pm 0.30$, it is 0.939 6.

In the case of an upper or lower confidence bound, the effect of a negative (or, respectively, a positive) bias is much more pronounced. Thus the probability that the upper confidence bound $x + 1.645\sigma_n$ of confidence coefficient 0.95 exceeds μ is now equal the probability that the standard normal variable $(x - \mu)/\sigma_n - B/\sigma_n$ exceeds $-1.645 - B/\sigma_n$; already for $B/\sigma_n = -0.10$ this equals as little as 0.938 8 (for $B/\sigma_n = 0.10$: 0.959 5).

The effect may be larger if the sampling distribution of x is only approximately normal and if an estimated variance has to be used.

In many cases we know that the bias cannot exceed some small fraction of the standard error of the estimator (for examples see Chapter IIA, especially sections 3 and 7). However, in the absence of such a bound on the bias, the chance of not covering μ is quite uncontrolled.

† Using a method given in section 28.4 of CRAMÉR, Mathematical Methods of Statistics (Princeton University Press, 1946).

15.9. EFFECT OF CLUSTERS

The effect of clusters on the approximation to normality is not known. It may be substantial – see the example in section 3.2 of Chapter VII. The method of section 14.2 of Chapter III (see also section 11 of Chapter IX) will generally accelerate the approach to normality.

15.10. SAMPLE SIZE AND CONFIDENCE INTERVALS

In section 14 we discussed the determination of n so as to keep the variance of the estimator of a parameter such as $\bar{Y}$ below a given amount. In terms of confidence intervals $[L(\mathscr{s}), U(\mathscr{s})]$, it is often asked to determine n such that the mean length of the interval be smaller than a given amount. However, the latter is not necessarily a reasonable objective, since a short interval is not of much use and may even be misleading in cases in which it fails to cover the actual value of $\bar{Y}$.† Moreover, this objective does not extend to the case of upper (or lower) confidence limits.

Instead, one may consider some positive Δ and determine n such that

$$\mathrm{P}\{U(\mathscr{s}) - \bar{Y} > \Delta\} \quad \text{and} \quad \mathrm{P}\{\bar{Y} - L(\mathscr{s}) > \Delta\}$$

will not exceed some small number γ. In case we may use $\bar{y}$ and the normal approximation, these two are equal and are somewhat more‡ than

$$\mathrm{P}\{n^{\frac{1}{2}}(\bar{y} - \bar{Y})/\sigma > n^{\frac{1}{2}}\Delta/\sigma - t\mathscr{E}x\},$$

† Actually for normal y's the conditional expected length of the usual interval, given the interval covers $\bar{Y}$, will, for the usual confidence coefficients β, not be much different for the unconditional mean length when n is not very small; we find for it $2t\sigma n^{-\frac{1}{2}}\theta$, where

$$\beta\theta > -\left(\frac{2}{\pi}\right)^{\frac{1}{2}}\left(\frac{n-1}{t^2+n-1}\right)^{\frac{n-1}{2}} t^{-1} + \left(\frac{2}{n-1}\right)^{\frac{1}{2}} \Gamma\left(\frac{n}{2}\right)\Big/\Gamma\left(\frac{n-1}{2}\right)$$

$$= -\left(\frac{2}{\pi}\right)^{\frac{1}{2}} e^{-\frac{1}{2}t^2}\, t^{-1} + 1 + \mathrm{O}(n^{-1})$$

and

$$\beta\theta < -\left(\frac{2}{\pi}\right)^{\frac{1}{2}}\left(\frac{n-1}{t^2+n-1}\right)^{\frac{n-1}{2}}\left\{t^{-1} - \frac{t^2+n-1}{n-3}\,t^{-3}\right\} + \left(\frac{2}{n-1}\right)^{\frac{1}{2}} \Gamma\left(\frac{n}{2}\right)\Big/\Gamma\left(\frac{n-1}{2}\right)$$

$$= -\left(\frac{2}{\pi}\right)^{\frac{1}{2}} e^{-\frac{1}{2}t^2}(t^{-1} - t^{-3}) + 1 + \mathrm{O}(n^{-1}).$$

For $\beta = 0.95$ this gives θ between $0.99 - 0.5n^{-1}$ and $1.01 - 0.3n^{-1}$. When the normal approximation for the confidence interval may be used, but the y's are very different from normal, there may be a somewhat larger discrepancy between the conditional and unconditional mean lengths.

‡ If we compute this probability as the mean of the conditional probability given $v(\bar{y})$, and take into account the convexity of the function $1 - \Phi$ in the range here considered (where its value has mean γ), we see that it will exceed the amount given. For most n and F_N for which the above confidence intervals may be used, the excess will be small.

where $x^2 = nv(\bar{y})/\sigma^2$, and this is approximately† $1 - \Phi(n^{\frac{1}{2}}\Delta/\sigma - t)$. If the confidence coefficient is $\beta = 1 - \alpha$,

$$\Phi(t) = 1 - \frac{\alpha}{2},$$

which, if Ψ denotes the function inverse to Φ, may be written

$$t = \Psi\left(1 - \frac{\alpha}{2}\right).$$

We therefore have approximately

$$1 - \gamma = \Phi\left\{n^{\frac{1}{2}}\Delta/\sigma - \Psi\left(1 - \frac{\alpha}{2}\right)\right\}$$

or

$$\Psi(1 - \gamma) = n^{\frac{1}{2}}\Delta/\sigma - \Psi\left(1 - \frac{\alpha}{2}\right),$$

so that $n^{\frac{1}{2}}$ has to be about

$$\left\{\Psi\left(1 - \frac{\alpha}{2}\right) + \Psi(1 - \gamma)\right\}\sigma/\Delta,$$

and usually somewhat more. In the case of an upper confidence limit with confidence coefficient $1 - \alpha$, we have instead

$$\{\Psi(1 - \alpha) + \Psi(1 - \gamma)\}\sigma/\Delta.$$

15.11. EFFECT OF BIAS ON PROPERTIES OF CONFIDENCE INTERVALS OTHER THAN THEIR CONFIDENCE COEFFICIENT

The effect of bias on the confidence coefficient was discussed in 15.8. We now study the effect on certain other properties.

Consider an estimator of $\bar{Y}$ (alternative to the one in 15.10) for which again approximate normality holds, but which has bias B and standard error $\sigma'/n^{\frac{1}{2}}$. Calling the corresponding confidence interval $[L'(\mathscr{s}), U'(\mathscr{s})]$,

$$\mathrm{P}\{U'(\mathscr{s}) - \bar{Y} > \Delta\} \quad \text{and} \quad \mathrm{P}\{\bar{Y} - L'(\mathscr{s}) > \Delta\} \qquad (*)$$

† Due to concavity of the square-root function and the fact that $\mathscr{E}x^2 = (N - n)/(N - 1)$, $\mathscr{E}x$ is slightly less than 1; thus, for normal y's the mean of a sample standard deviation is

$$\sigma\left(\frac{2}{n - 1}\right)\Gamma\left(\frac{n}{2}\right)\Big/\Gamma\left(\frac{n - 1}{2}\right) \simeq \sigma\{1 - \tfrac{1}{4}(n - 1)^{-1}\},$$

and for general y's the coefficient of $(4n)^{-1}$ is half of the difference between their fourth standardized moment and 1.

will be about

$$1 - \Phi\{n^{\frac{1}{2}}(\Delta - B)/\sigma' - t\} \quad \text{and} \quad 1 - \Phi\{n^{\frac{1}{2}}(\Delta + B)/\sigma' - t\},$$

respectively. As an example, let $\Delta/\sigma' = 0.25$. In 15.10 we saw that for $\alpha = 0.05 = \gamma$, n needs to be

$$\{(1.96 + 1.645)/0.25\}^2 = 208.$$

If B/σ' is as small as 0.10, we already have for this n

$$1 - \Phi\{208^{\frac{1}{2}} \times 0.15 - 1.96\} = 1 - \Phi(0.20) = 0.42$$

instead of $\gamma = 0.05$.

If we want (*) to be at most γ, we have to take for n about the square of

$$\left\{\Psi\left(1 - \frac{\alpha}{2}\right) + \Psi(1 - \gamma)\right\}\sigma'/(\Delta - |B|).$$

(We assume $\Delta > |B|$; otherwise there is no n for which both probabilities (*) $\leq \gamma$.) In our example this is

$$\{(1.96 + 1.645)/0.15\}^2 = 577$$

instead of 208.

If $B/\sigma' = 0.014$, so that $B/\sigma'_n = 0.20$ (compare section 15.8) we have

$$1 - \Phi\{208^{\frac{1}{2}} \times 0.236 - 1.960\} = 1 - \Phi(1.44) = 0.075$$

instead of $\gamma = 0.050$.

16. Confidence intervals and bounds for fractiles, tolerance intervals and limits†

(a) Consider first the case in which the values $y(u_i)$ are all different. For any number P which is a multiple of $1/N$ and satisfies‡ $0 < P \leq 1$, we say that W_P is the *P-ile* (*P-fractile*, or 100*P percentile*) of $\mathscr{Y}$ in $\mathscr{P}$, if W_P is the smallest§ v

† The author briefly discussed the state of this subject to date, and the place the new results of this section occupies amidst the other literature, in a paper presented to the European meeting of the Institute of Mathematical Statistics, at Amsterdam in September, 1968; see abstract in Ann. Math. Statist., **39** (1968) 2171–2172.

‡ W_P is usually defined for all real numbers P between 0 and 1, viz., as that v for which $F_N(v) \geq P$ but for which, for all $v' < v$, $F_N(v') < P$ (and similarly w_p). It is possible, but somewhat cumbersome, to generalize the presentation of the present section to such W_P (and w_p with a similarly extended definition).

§ Under alternative definitions often encountered, W_P is the average of the smallest v for which $F_N(v) = P$ and the smallest v' for which $F_N(v') > P$; or is the set of all numbers between these extremes.

for which $F_N(v) = P$. We define $W_0 = -\infty$. Thus, if $P = \frac{1}{2}$, W_P is the *median* of $\mathscr{Y}$ in $\mathscr{P}$, i.e., the smallest value of $\mathscr{Y}$ such that, as closely as possible, an equal number of elements have $\mathscr{Y}$-values above and below it.† The lower *quartile* has three times as many elements with $\mathscr{Y}$-values above it as below, and the upper quartile three times as many below as above; the median is the second quartile.

In the sample we define, for each p which is a multiple of $1/n$ and satisfies $0 < p \leq 1$, w_p as the smallest v for which $F_n(v) = p$; and define $w_0 = -\infty$, $w_{(n+1)/n} = \infty$. The entire set $\{w_p : p = 1/n, 2/n, \ldots, 1\}$ is called the set of *order statistics* of the sample; $w_{1/n}$ is the smallest observed value of $\mathscr{Y}$ (also: the smallest order statistic), and $w_{n/n}$ the largest. If n is not very small, one may expect w_p to furnish a good estimator of W_P when p is as close as possible to P. It is not usually an unbiased estimator.

An upper *P-tolerance limit* v_n with confidence (or tolerance) coefficient β is defined as a statistic such that the (random variable) $F_N(v_n)$, the fraction of units in $\mathscr{P}$ for which the values of $\mathscr{Y}$ are *less than or equal* to v_n, is at least P with probability (at least) β.‡ We similarly define lower tolerance limits and tolerance intervals. A practical example is given under (ii) below.

(i) Consider first samples of n taken without replacement and let k be fixed between 1 and n. We shall show that the probability that $w_{k/n}$ equals $W_{t/N}$ for any t between k and $N - n + k$, equals

$$\binom{t-1}{k-1}\binom{N-t}{n-k}\Big/\binom{N}{n}.$$

This follows from the fact that $w_{k/n}$ equals $W_{t/N}$ if and only if the sample contains $k - 1$ of the $t - 1$ elements in the population with $\mathscr{Y}$-values smaller than $W_{t/N}$, and $n - k$ of the $N - t$ elements with values exceeding $W_{t/N}$.

† If N is even, the first definition given in the previous footnote makes the median the average of those two successive values of $\mathscr{Y}$ such that, as closely as possible, an equal number of elements have $\mathscr{Y}$-values above and below them.

‡ So, for (at least) a fraction β of the points $\mathfrak{s}$ of the sample space, $v_n(\mathfrak{s})$ will not be exceeded by the $\mathscr{Y}$-value of at least PN of the units of $\mathscr{P}$. Sometimes one considers a statistic v_n' such that $m(\mathfrak{s})$, the number of unsampled units of $\mathscr{P}$ whose $\mathscr{Y}$-values do not exceed $v_n'(\mathfrak{s})$, has *mean NP*, although $m(\mathfrak{s})$ may be smaller than NP for a substantial part of the $\mathfrak{s}$ in $\mathscr{S}$. Either statistic may be interpreted as a prediction about a future random drawing from $\mathscr{P}$, but only the latter has been given the name *prediction limit*, as it is, in fact, a confidence limit for the $\mathscr{Y}$-value, x, of a further unit drawn at random from $\mathscr{P}$. The following are the $n + 1$ different possible arrangements in sampling without replacement: $w_{1/n}, \ldots, w_{n/n}, x$; $w_{1/n}, \ldots, w_{(n-1)/n}, x, w_{n/n}$; $\ldots$; $x, w_{1/n}, \ldots, w_{n/n}$. By the symmetry of the method of sampling they are equally likely, and for k_2 of them x lies between $w_{k_1/n}$ and $w_{(k_1+k_2)/n}$. Therefore the probabilitity that x lies between $w_{k_1/n}$ and $w_{(k_1+k_2)/n}$ is $k_2/(n + 1)$, where $k_1 = 0, \ldots, n + 1 - k_2$.

Therefore the probability that, for fixed k, $w_{k/n} \geq W_P$ is, for any NP between k and $N - n + k$,

$$\sum_{t=NP}^{N-n+k} \binom{t-1}{k-1} \binom{N-t}{n-k} \Big/ \binom{N}{n}.$$

To get an *upper* confidence bound for W_P, find the smallest value of k for which this probability equals at least the desired confidence coefficient β. (Such k may not exist when N and n are small and at the same time β is large.) Then $w_{k/n}$ is an upper confidence bound for W_P with confidence coefficient β.

Now $w_{k/n} \geq W_P$ means that at least NP of the $\mathscr{Y}$-values in $\mathscr{P}$ are less than or equal to $w_{k/n}$; therefore $w_{k/n}$ is *also* an upper P-tolerance limit for F_N with tolerance coefficient β:

$$\mathrm{P}\{F_N(w_{k/n}) \geq P\} \geq \beta.$$

Similarly, we find a *lower* confidence bound $w_{k/n}$ using the largest value of k such that

$$\mathrm{P}\{w_{k/n} \leq W_P\} = \sum_{t=k}^{NP} \binom{t-1}{k-1} \binom{N-t}{n-k} \Big/ \binom{N}{n}$$

$$= \sum_{i=N-NP+1}^{N-k+1} \binom{N-i}{k-1} \binom{i-1}{n-k} \Big/ \binom{N}{n}$$

equals at least the desired confidence coefficient β.

Since $w_{k/n} \leq W_P$ means that at least $N(1 - P) + 1$ of the values of $\mathscr{Y}$ in $\mathscr{P}$ exceed or equal $w_{k/n}$, or that at least $N(1 - P)$ of them exceed $w_{k/n}$, we have†

$$\mathrm{P}\left\{1 - F_N(w_{k/n} - 0) \geq 1 - P + \frac{1}{N}\right\} = \mathrm{P}\{1 - F_N(w_{k/n}) \geq 1 - P\} \geq \beta.$$

So $w_{k/n}$ is also a lower $(1 - P + (1/N))$ - tolerance limit for F_N with tolerance coefficient β. If we wish to obtain a lower $(1 - P)$ - tolerance limit for F_N, we must find the largest k for which

$$\sum_{t=k}^{NP+1} \binom{t-1}{k-1} \binom{N-t}{n-k} \Big/ \binom{N}{n} \geq \beta.$$

To obtain confidence *intervals* for W_P, we first find, in a way similar to that followed above, that

$$\mathrm{P}\{w_{k_1/n} = W_{t_1/N},\ w_{(k_1+k_2)/n} = W_{(t_1+t_2)/N}\}$$

$$= \binom{t_1-1}{k_1-1} \binom{t_2-1}{k_2-1} \binom{N-t_1-t_2}{n-k_1-k_2} \Big/ \binom{N}{n}.$$

† If $\varepsilon_1, \varepsilon_2, \ldots$ is a sequence of positive numbers converging to 0, and if there exist a function g and a number t such that the sequence of numbers $g(t - \varepsilon_1), g(t - \varepsilon_2), \ldots$ converges, we denote the limit by $g(t - 0)$. It is easy to see that $NF_N(t - 0)$ is the number of units whose $\mathscr{Y}$-value is less than t.

Therefore,

$$\mathrm{P}\{w_{k_1/n} \leq W_P \leq w_{(k_1+k_2)/n}\}$$
$$= \sum_{t_1=k_1}^{NP} \sum_{t_2=NP-t_1}^{N-t_1-n+k_1+k_2} \binom{t_1-1}{k_1-1}\binom{t_2-1}{k_2-1}\binom{N-t_1-t_2}{n-k_1-k_2}\Big/\binom{N}{n}.$$

For all sets of values of k_1 and k_2 for which this is at least equal to β, the interval from $w_{k_1/n}$ to $w_{(k_1+k_2)/n}$ forms a confidence interval for W_P with confidence coefficient β.

In order to obtain a P-tolerance interval with tolerance coefficient β, we first derive the frequency function of the number (r) of elements in $\mathscr{P}$ for which $y(u_i)$ exceeds $w_{k_1/n}$ but is less than $w_{(k_1+k_2)/n}$, by summing

$$\binom{t_1-1}{k_1-1}\binom{t_2-1}{k_2-1}\binom{N-t_1-t_2}{n-k_1-k_2}\Big/\binom{N}{n}$$

over all possible values of t_1, when $t_2 - 1$ is fixed at r. The sum of

$$\binom{t_1-1}{k_1-1}\binom{N-t_1-1-r}{n-k_1-k_2}$$

over these values of t_1 equals

$$\binom{N-r-1}{n-k_2},$$

since the ratio is the probability that $w_{k_1/(n-k_2)}$ equals $W_{t_1/(N-1-r)}$ and so sums to 1. Therefore, the required frequency function is

$$\binom{r}{k_2-1}\binom{N-r-1}{n-k_2}\Big/\binom{N}{n};$$

i.e., it is the same expression as we obtained for the probability that $w_{k_2/n}$ equals $W_{(r+1)/N}$. The result is valid for $1 \leq k_1 \leq n - k_2$ (e.g., for $k_1 = 0$, we would talk about the probability that the number of elements in $\mathscr{P}$ with $\mathscr{Y}$-values not exceeding $w_{(k_1+k_2)/n}$ and for which "$y(u_i) \geq -\infty$" instead of "for which $y(u_i) > -\infty$", since $w_0 = -\infty$).

From the frequency function we now obtain at once

$$\mathrm{P}\left\{F_N(w_{(k_1+k_2)/n}) - F_N(w_{k_1/n} - 0) \geq \frac{t+2}{N}\right\}$$
$$= \sum_{r=t}^{N-n+k_2-1} \binom{r}{k_2-1}\binom{N-r-1}{n-k_2}\Big/\binom{N}{n} \qquad (1 \leq k_1 \leq n-k_2),$$

so that the closed interval $[w_{k_1/n}, w_{(k_1+k_2)/n}]$ defines a $(t+2)/N$ - tolerance interval of confidence coefficient equal to the right-hand side (β). We can use the above expression also for $k_1 = 0$ and for $k_1 = n - k_2 + 1$ by considering half-open intervals instead of closed intervals, while changing $t + 2$ to $t + 1$. Then, for $k_1 = 0$, the left-hand side becomes instead

$$\mathrm{P}\left\{F_N(w_{k_2/n}) - F_N(w_{0/n}) \geq \frac{t+1}{N}\right\} = \mathrm{P}\left\{F_N(w_{k_2/n}) \geq \frac{t+1}{N}\right\},$$

so that $w_{k_2/n}$ is an upper $(t + 1)/N$ - tolerance limit of confidence coefficient β; and for $k_1 = n - k_2 + 1$

$$\mathrm{P}\left\{1 - F_N(w_{k_1/n} - 0) \geq \frac{t+1}{N}\right\},$$

so that $w_{(n-k_2+1)/n}$ is a lower $(t + 1)/N$ - tolerance limit of that coefficient. These results check with those given earlier.

EXAMPLE. $\mathscr{P}$ consists of 7 elements with different $\mathscr{Y}$-values. Let us find an interval which with approximate probability 0.90 will contain at least 5 of these values when the sample size n is 4. The answer is unique in this case, viz., the sample *range*† (defined as $w_{1/n}$ to $w_{n/n}$):

$$\mathrm{P}\{F_N(w_{4/n}) - F_N(w_{1,n} - 0) \geq \tfrac{5}{7}\} = \sum_{r=3}^{5} \binom{r}{2}\binom{6-r}{1}\Big/\binom{7}{4} = \tfrac{31}{35} = 0.89.$$

To clarify the meaning of the tolerance coefficient, let us enumerate the 4 points in the sample space (which has a total of 35 points) for which the sample range embraces *fewer* than 5 of the 7 elements of $\mathscr{P}$, denoting the $\mathscr{Y}$-values of the elements of $\mathscr{P}$ by $y^{(1)}, y^{(2)}, \ldots, y^{(7)}$ (with $y^{(1)} < y^{(2)} \ldots < y^{(7)}$); these points are

$$\{y^{(1)}, y^{(2)}, y^{(3)}, y^{(4)}\},\quad \{y^{(2)}, y^{(3)}, y^{(4)}, y^{(5)}\},\quad \{y^{(3)}, y^{(4)}, y^{(5)}, y^{(6)}\},\quad \{y^{(4)}, y^{(5)}, y^{(6)}, y^{(7)}\}.$$

For $k_1 = 0$ we obtain the upper $\frac{4}{7}$ - tolerance bound $w_{3/n}$ with the same tolerance coefficient. The 4 points in the sample space for which *fewer* than 4 $\mathscr{Y}$-values in $\mathscr{P}$ are less than or equal to $w_{3/n}$ (the third smallest observation) are

$$\{y^{(1)}, y^{(2)}, y^{(3)}, y^{(4)}\},\quad \{y^{(1)}, y^{(2)}, y^{(3)}, y^{(5)}\},\quad \{y^{(1)}, y^{(2)}, y^{(3)}, y^{(6)}\},\quad \{y^{(1)}, y^{(2)}, y^{(3)}, y^{(7)}\}.$$

The student should also find the 4 points in the sample space for which *fewer* than 4 $\mathscr{Y}$-values in $\mathscr{P}$ are larger than or equal to $w_{2/n}$.

(ii)‡ For sampling with replacement consider the probability, $Q_P\{l_1, l_2\}$, of the event that at least l_1 of the observations on $\mathscr{Y}$ are less than or equal to W_P and at least $n - l_1 - l_2 + 1$ exceed or equal W_P. The complementary event is that at most $l_1 - 1$ are less than or equal to W_P *or* at most $n - l_1 - l_2$ exceed or equal W_P (i.e., *or* at least $l_1 + l_2$ are less than W_P). Since the prob-

† The word "range" has two meanings: the interval between the minimum and the maximum, and the difference between the maximum and the minimum. Here we refer to the first meaning.

‡ For what follows, keep in mind that we interpret $\sum_{l=a}^{a-1} f(l) = 0$ for any function f.

ability that a random observation is less than or equal to W_P is by definition P, and that a random observation is less than W_P is $P - (1/N)$, we have

$$Q_P\{l_1, l_2\} = 1 - \sum_{l=0}^{l_1-1} \binom{n}{l} P^l(1-P)^{n-l} - \sum_{l=l_1+l_2}^{n} \binom{n}{l} \left(P - \frac{1}{N}\right)^l \left(1 - P + \frac{1}{N}\right)^{n-l}.$$

By choosing l_1 positive and l_2 nonnegative such that $l_1 + l_2 \leq n$, and such that the formula gives an amount at least equal to the desired confidence coefficient β, we obtain a confidence *interval* for W_P with at least this confidence coefficient as follows:

In a particular sample let k_1' be the smallest integer k for which $w_{k/n} = w_{k_1/n}$ and let $(k_1 + k_2)''$ be the largest integer k for which $w_{k/n} = w_{(k_1+k_2)/n}$. Then

$$\begin{aligned} \text{P}\{w_{k_1/n} \leq W_p \leq w_{(k_1+k_2)/n}\} &= \text{P}\{w_{k\ \ /n} \leq W_P \leq w_{(k_1+k_2)''/n}\} \\ &= Q_P\{k_1', (k_1 + k_2)'' - k_1'\}. \end{aligned}$$

The reason for this is that, if k_1' or more of the observations are less than or equal to W_P, $w_{k_1/n} = w_{k_1'/n}$ is less than or equal to W_P, while, if $w_{k_1/n} = w_{k_1'/n}$ is less than or equal to W_P, k_1' or more of the observations are less than or equal to W_P; and that similarly $W_P \leq w_{(k_1+k_2)/n}$ if and only if at least $n - (k_1 + k_2)'' + 1$ of the observations exceed or equal W_P. Therefore, if k_1 and k_2 are such that $Q_P\{k_1, k_2\} \geq \beta$, then

$$\text{P}\{w_{k_1/n} \leq W_P \leq w_{(k_1+k_2)/n}\} = Q_P\{k_1', (k_1 + k_2)'' - k_1'\} \geq Q_P\{k_1, k_2\} \geq \beta.$$

By the same reasoning we obtain that the probability that at least l_1 of the observations are less than or equal to W_P and at least $n - l_1 - l_2 + 1$ *exceed* W_P is

$$1 - \sum_{l=0}^{l_1-1} \binom{n}{l} P^l(1-P)^{n-l} - \sum_{l=l_1+l_2}^{n} \binom{n}{l} P^l(1-P)^{n-l} = \sum_{l=l_1}^{l_1+l_2-1} \binom{n}{l} P^l(1-P)^{n-l}.$$

By choosing l_1 positive and l_2 nonnegative such that $l_1 + l_2 \leq n + 1$, we can, by the above method, derive from this $\text{P}\{w_{k_1/n} \leq W_P < w_{(k_1+k_2)/n}\}$, which in particular, for $k_2 = n + 1 - k_1$, yields *lower* confidence and tolerance bounds. Thus for $0 \leq k \leq n$

$$\text{P}\{w_{k/n} \leq W_P\} = \sum_{l=k'}^{n} \binom{n}{l} P^l(1-P)^{n-l} \geq \sum_{l=k}^{n} \binom{n}{l} P^l(1-P)^{n-l}.$$

so that $w_{k/n}$ is a lower confidence bound for W_P and a lower $(1 - P + (1/N))$-tolerance limit for F_N with confidence coefficient at least equal to the right-hand side.

The probability that at least l_1 of the observations are *less* than W_P and at least $n - l_1 - l_2 + 1$ exceed or equal W_P is

$$1 - \sum_{l=0}^{l_1-1} \binom{n}{l} \left(P - \frac{1}{N}\right)^l \left(1 - P + \frac{1}{N}\right)^{n-l} - \sum_{l=l_1+l_2}^{n} \binom{n}{l} \left(P - \frac{1}{N}\right)^l \left(1 - P + \frac{1}{N}\right)^{n-l}$$
$$= \sum_{l=l_1}^{l_1+l_2-1} \binom{n}{l} \left(P - \frac{1}{N}\right)^l \left(1 - P + \frac{1}{N}\right)^{n-l}.$$

By choosing l_1 and l_2 nonnegative and such that $l_1 + l_2 \leq n$, we can, by the above method, derive from this $\mathrm{P}\{w_{k_1/n} < W_P \leq w_{(k_1+k_2)/n}\}$, which in particular, for $k_1 = 0$, yields *upper* confidence and tolerance bounds. Thus for $1 \leq k \leq n$

$$\mathrm{P}\{W_P \leq w_{k/n}\} = \sum_{l=0}^{k''-1} \binom{n}{l} \left(P - \frac{1}{N}\right)^l \left(1 - P + \frac{1}{N}\right)^{n-l}$$
$$\geq \sum_{l=0}^{k-1} \binom{n}{l} \left(P - \frac{1}{N}\right)^l \left(1 - P + \frac{1}{N}\right)^{n-l},$$

so that $w_{k/n}$ is an upper confidence bound for W_P and an upper P-tolerance limit for F_N with confidence coefficient at least equal to the right-hand side. For the P-tolerance interval we shall only state a result which holds approximately for large N (proof below):

$$\mathrm{P}\left\{F_N(w_{(k_1+k_2)/n}) - F_N(w_{k_1/n} - 0) \geq \frac{t+1}{N}\right\} \approx \frac{1}{N^n} \sum_{l=0}^{k_2-1} \binom{n}{l} t^l (N-t)^{n-l}.$$

REMARK. One may check that for $k_1 = 0$ and $(t + 1)/N = P$ this is consistent with the upper P-tolerance limit given above, and that for

$$k_2 = n - k_1 + 1 \quad \text{and} \quad (t + 1)/N = 1 - P + (1/N)$$

it is consistent with the lower $(1 - P + (1/N))$-tolerance limit given above. A table for finding n such that $[w_{1/n}, w_{n/n}]$ constitutes a P-tolerance interval with given P and given confidence coefficient (when N is large) is given by HARMAN, Sankhya, **A29** (1967) 215–218. SOMERVILLE, Ann. Math. Statist., **29** (1958) 599–601 gives a table for finding $m = r + s$ such that $[w_{r/n}, w_{(n-s+1)/n}]$ constitutes a P-tolerance interval with given P and given confidence coefficient (when N is large). MURPHY, Ann. Math. Statist., **19** (1948) 581–589 gives graphs (for N large).

EXAMPLE. On the basis of a simple random sample of 100, the Army's office in charge of procurement of soldiers' clothing wishes to estimate, with confidence coefficient 0.95, an interval such that (at least) 90% of the soldiers have a girth between the limits of this interval. (For other soldiers the clothing will have to be made to measure.) From a binomial table we find

$$\sum_{j=0}^{95} \binom{100}{j} 0.9^j 0.1^{n-j} = 0.976\,29, \qquad \sum_{j=0}^{94} \binom{100}{j} 0.9^j 0.1^{n-j} = 0.942\,42,$$

so $k_2 = 96$. Any of the intervals from $w_{k_1/n}$ to $w_{(k_1+96)/n}$ with $k_1 = 1, 2, 3, 4$, will serve as a 0.90 - tolerance interval with confidence coefficient 0.95. For $k_1 = 0$ or 5 we get an upper and lower tolerance limit, respectively. Note that, if alteration of clothing that is too wide is less costly than making to measure, the office may decide on *two numbers*, Q_1 and Q_2, with $Q_1 > Q_2$, and ask for an interval such that with probability 0.95, not more than NQ_1 of the soldiers have a girth less than its lower limit and not more than NQ_2 a girth exceeding its upper limit.

REMARK. If F_N is known to be approximately normal, a better solution is the following: Compute $\bar{y}$ and $v(\bar{y})$ from the sample, and from this obtain the interval $\bar{y} - b\{nv(\bar{y})\}^{\frac{1}{2}}$ to $\bar{y} + b\{nv(\bar{y})\}^{\frac{1}{2}}$, where $b = b_{P,\beta,n}$ is read from special tables for tolerance intervals for normal distributions, e.g., in EISENHART *et al.* (eds.), Selected Techniques of Statistical Analysis (McGraw Hill, 1947). Thus $b_{0\cdot 90, 0\cdot 95, 100} = 1.874$.

Proof of the above result for tolerance intervals when N is large: First note that for sampling without replacement $\mathrm{P}\{F_N(w_{(k_1+k)/n}) - F_N(w_{k_1/n} - 0) \geq t/N\}$ equals the sum over l from $t-1$ to $N-n+k$ of

$$\binom{l-1}{k-1}\binom{N-l}{n-k}\Big/\binom{N}{n} = \frac{n!}{(k-1)!(n-k)!}\,\frac{l-1}{N}\cdots\frac{l-k+1}{N}\,\frac{N-l}{N}$$

$$\cdots\frac{N-l-n+k+1}{N}\left(\frac{N}{N}\cdots\frac{N-n+1}{N}\right)^{-1}\frac{1}{N}$$

which converges to

$$\frac{n!}{(k-1)!(n-k)!}\int_P^1 x^{k-1}(1-x)^{n-k}\,\mathrm{d}x = \sum_{j=0}^{k-1}\binom{n}{j}P^j(1-P)^{n-j}$$

as $N \to \infty$ in such a way that $(t-1)/N$ goes to P. But, as N increases, results for sampling without replacement approach those for sampling with replacement, and so the limiting result given applies also to sampling with replacement.

(b) Consider now the case in which not all $y(u_i)$ are different. Then there are values of P, multiples of $1/N$, such that there is no v for which $F_N(v) = P$. A suitable generalization of our previous definition of W_P is† the smallest v for which $F_N(v) \geq P$ and $1 - F_N(v-0) \geq 1 - P$.

† If we extend the definition to all real numbers P in the interval between 0 and 1, we obtain the definition of the first footnote of part (a) of the present section. The definition of sample fractiles may be similarly generalized.

Let NP' be the smallest integer such that $W_{P'} = W_P$ and NP'' the largest integer such that $W_{P''} = W_P$. Then for fixed k, in sampling without replacement,

$$\mathrm{P}\{w_{k/n} = W_P\} = \sum_{t=NP'}^{NP''} \binom{t-1}{k-1}\binom{N-t}{n-k}\Big/\binom{N}{n};$$

and so we have, for example,

$$\mathrm{P}\{w_{k/n} \geq W_P\} = \sum_{t=NP'}^{N-n+k} \binom{t-1}{k-1}\binom{N-t}{n-k}\Big/\binom{N}{n}.$$

The result is that, if we ignore the possibility of $y(u_i)$ being equal for several population units, the probability that the upper confidence bound so determined covers W_P, or the probability that at least NP of the units in $\mathscr{P}$ are smaller than the upper P-tolerance bound so determined, is possibly larger than required. If the location in $\mathscr{P}$ of multiple values of $y(u_i)$ is known, there are occasions in which it is possible to lower the k for which $w_{k/n}$ is an upper confidence or tolerance bound with the required confidence coefficient. In most cases, however, we do not know the location of multiple values, and it follows that then our bounds are conservative (i.e., on the safe side). The same is seen to hold for lower bounds and intervals.

These conclusions also hold in sampling with replacement. Thus, the probability that at least l_1 observations are less than or equal to W_P and at least $n - l_1 - l_2 + 1$ larger than or equal to W_P is, clearly,

$$1 - \sum_{l=0}^{l_1-1} \binom{n}{l} (P'')^l (1 - P'')^{n-l} - \sum_{l=l_1+l_2}^{n} \binom{n}{l} \left(P' - \frac{1}{N}\right)^l \left(1 - P' + \frac{1}{N}\right)^{n-l}.$$

This is at least as large as the expression without primes, as was shown above by expressing binomial sums in terms of incomplete beta-functions.

17. Appendix: Confidence bounds for $\bar{Y}$ based on order statistics

We saw in section 15.1 that, if F_N is very skew, confidence intervals computed by the usual approximation with $t = 2$ often have a confidence coefficient only slightly smaller than the normal one, but that this does not hold for confidence bounds. We here propose a different method of constructing confidence intervals which also appears to work well for confidence bounds. The point of departure is the discussion of section 15 of Chapter III for the case of a single stratum. It is shown there how to construct a confidence interval $[w_{p_1}, w_{p_2}]$ for the P-ile W_P, where $p_1 = P - tv^{\frac{1}{2}}$, $p_2 = P + tv^{\frac{1}{2}}$, and v is an estimator of

$$V\{F_n(W_P)\} = \left(\frac{1}{n} - \frac{1}{N}\right) P(1-P) \frac{N}{N-1}.$$

If we would know the value of P for which $\bar{Y}$ equals W_P (or equals it approximately), we could use this interval. We generally do not, but we may use the fact that $\{P(1-P)\}^{\frac{1}{2}}$ changes very little for P in the interval 0.3 to 0.7, and that for most skew distributions of interest P would fall in that interval. Thus, even for the very skew lognormal distribution discussed in section 15.1, $P = 0.69$. We propose therefore to replace $P \pm tv^{\frac{1}{2}}$ by $F_N(\bar{y}) \pm tv^{\frac{1}{2}}$ and to replace v here by the maximum of $V\{F_N(W_P)\}$ over P, thus building in a slight safety factor to adjust for the error† in $F_n(\bar{y})$ in estimating $F_N(\bar{Y}) = P$. This amounts to using as upper and lower confidence bounds the observations about

$$\tfrac{1}{2}t\left(\frac{1}{n} - \frac{1}{N}\right)^{\frac{1}{2}} n$$

to the left and right of $\bar{y}$.

The method has been applied to the lognormal case discussed in section 15.1. The percentages of cases in which $\bar{Y}$ fell to the left or right of the confidence interval with $t = 2$ were‡, respectively

for $n =$	320	2.2 and 3.0	(0.42)
for $n =$	640	2.3 and 2.7	(0.30)
for $n =$	960	2.0 and 2.2	(0.24)
for $n =$	1 280	1.7 and 2.3	(0.21).

In brackets is given for each case the average of the half-widths of the intervals; they appear to be only slightly higher than those of section 15.1.

It may be expected that the error due to replacing $F_N(W_P)$ by $F_n(\bar{y})$ is generally smaller, the larger the amount of symmetry. It will not always be possible to adequately correct for this error: If $\bar{Y}$ is near a region of slow increase of F_N, a large fraction of the

$$\tfrac{1}{2}t\left(\frac{1}{n} - \frac{1}{N}\right)^{\frac{1}{2}} n$$

† Even in this very skew case $F_n(\bar{y}) - P$ is only about 0.002 for $n = 320$.

‡ Note that $\frac{1}{2}tn^{-\frac{1}{2}}$ is not an integer; in particular for $n = 640$ it was rounded down, and for $n = 1\,280$ it was rounded up significantly. This, and the overconservatism of the method for large n, are reflected in the figures. It appears that even for confidence *intervals* the present method requires fewer observations to attain the theoretical confidence coefficient than the standard method. The data in the table are based on 11 000 trials.

observations one side (or even on both sides) of $\bar{y}$ may lie very far apart from each other, leading to excessively conservative confidence bounds. (However, such a case may be recognized from the data.)

This was, in fact, the case with an attempt to apply the proposed method, suitably modified,† to *stratified* sampling from the lognormal population mentioned above. In that case, taking 90 observations from the part of the population with $\mathcal{Y}$-values less than 5.32, and 60 from the other part, the large amount of skewness together with the smallness of the sample (relative to this skewness) caused the relevant observations preceding $\bar{y}_{st}$ and especially the ones following $\bar{y}_{st}$ to be very far apart in many cases. The method would therefore appear to require for this population an excessive number of observations as compared with the ordinary method for stratified sampling.

† Denoting N_h/N by Q_h and n_h/n by q_h, neglecting the factor $N_h/(N_h - 1)$, and assuming the stratification is according to the $\mathcal{Y}$-values, v is replaced by

$$\frac{1}{n}\left(\frac{1}{q_h} - \frac{1}{Q_h}\right)\left(\frac{1}{2}Q_h\right)^2.$$

Here h denotes that stratum which contains W_P. So, when using this method, we get the smallest half-widths if the stratification and allocation are such that for any of the strata, h, which *may* contain W_P, $Q_h^2(q_h^{-1} - Q_h^{-1})$ is small and the observations in any such strata fall relatively close together.

CHAPTER IIA

RATIO AND RELATED ESTIMATORS IN SIMPLE RANDOM SAMPLING (FURTHER RESULTS)

Ratio estimators have already been discussed in sections 8–10 of Chapter II and applied in other parts of that chapter. Here we give additional results and also discuss difference estimators, and, very briefly (in sections 6 and 13), regression estimators.

1. Notation

For any characteristic $\mathscr{Y}$ we already defined Y, $\bar{Y}$, $\sigma^2(\mathscr{Y})$, $S^2(\mathscr{Y})$. If $\bar{Y} \neq 0$, we also define

the *relative variance* of $\mathscr{Y}$:

$$\sigma^2(\mathscr{Y})/\bar{Y}^2 \quad (\text{or } \sigma(\mathscr{Y}\mathscr{Y})/\bar{Y}^2);$$

the *adjusted relative variance* of $\mathscr{Y}$:

$$S^2(\mathscr{Y})/\bar{Y}^2 \quad (\text{or } S(\mathscr{Y}\mathscr{Y})/\bar{Y}^2);$$

the *coefficient of variation* of $\mathscr{Y}$:

$$\sigma(\mathscr{Y})/|\bar{Y}|;$$

the *adjusted coefficient of variation* of $\mathscr{Y}$:

$$S(\mathscr{Y})/|\bar{Y}|.$$

For any two characteristics $\mathscr{X}$, $\mathscr{Y}$, let us define:

the *covariance* of $\mathscr{Y}$ and $\mathscr{X}$:

$$\sigma(\mathscr{Y}\mathscr{X}) = \frac{1}{N} \sum_{i=1}^{N} (y_i - \bar{Y})(x_i - \bar{X});$$

the *adjusted covariance* of $\mathscr{Y}$ and $\mathscr{X}$:

$$S(\mathscr{Y}\mathscr{X}) = \frac{1}{N-1} \sum_{i=1}^{N} (y_i - \bar{Y})(x_i - \bar{X});$$

and, if $\bar{X}$ and $\bar{Y}$ are different from 0,

the *relative covariance* of $\mathscr{Y}$ and $\mathscr{X}$:

$$\sigma(\mathscr{Y}\mathscr{X})/\bar{Y}\bar{X};$$

the *adjusted relative covariance* of $\mathscr{Y}$ and $\mathscr{X}$:

$$C(\mathscr{Y}\mathscr{X}) = S(\mathscr{Y}\mathscr{X})/\bar{Y}\bar{X}$$

[accordingly $C(\mathscr{Y}\mathscr{Y})$ and $C(\mathscr{X}\mathscr{X})$ will designate the squares of the adjusted coefficients of variation of $\mathscr{Y}$ and $\mathscr{X}$];

the *correlation coefficient* of $\mathscr{Y}$ and $\mathscr{X}$:

$$\rho(\mathscr{Y}\mathscr{X}) = \sigma(\mathscr{Y}\mathscr{X})/\sigma(\mathscr{Y})\,\sigma(\mathscr{X}) = S(\mathscr{Y}\mathscr{X})/S(\mathscr{Y})\,S(\mathscr{X}).$$

If

$$d_i = y_i - Rx_i,$$

since $\bar{D} = 0$,

$$d_i = y_i - \bar{Y} - R(x_i - \bar{X}),$$

and so

$$S^2(\mathscr{D}) = S(\mathscr{Y}\mathscr{Y}) - 2R\,S(\mathscr{X}\mathscr{Y}) + R^2 S(\mathscr{X}\mathscr{X}),$$

which may be estimated by

$$s^2(\hat{\mathscr{D}}) = s^2(\mathscr{Y}) - 2\hat{R}\,s(\mathscr{Y}\mathscr{X}) + \hat{R}^2\,s^2(\mathscr{X}),$$

or, simpler, since $\hat{\bar{d}} = 0$, by

$$s^2(\hat{\mathscr{D}}) = \frac{1}{n-1}\{\sum_{i=1}^{N} a_i y_i^2 - 2\hat{R}\sum_{i=1}^{N} a_i x_i y_i + \hat{R}^2 \sum_{i=1}^{N} a_i x_i^2\}.$$

We can use the above notations in order to express $V(\hat{R})$ in another form:

$$V(\hat{R}) \approx \frac{N-n}{N}\,\frac{1}{n}\,\frac{1}{\bar{X}^2}\,S^2(\mathscr{D})$$

$$= \frac{N-n}{N}\,\frac{1}{n}\,\frac{1}{\bar{X}^2}(S(\mathscr{Y}\mathscr{Y}) - 2R\,S(\mathscr{Y}\mathscr{X}) + R^2 S(\mathscr{X}\mathscr{X}))$$

$$= \frac{N-n}{N}\,\frac{1}{n}\,R^2(C(\mathscr{Y}\mathscr{Y}) - 2C(\mathscr{Y}\mathscr{X}) + C(\mathscr{X}\mathscr{X})).$$

EXAMPLE. Let us compute once more the quantities of section 10 according to the last formulas, where we already obtained

$$\hat{R} = \frac{5\,611}{10\,360} = 0.541\,602\,31,$$

$$\sum_{i=1}^{N} a_i y_i^2 = 1\,992\,151,$$

$$\sum_{i=1}^{N} a_i x_i y_i = 3\,663\,300,$$

$$\sum_{i=1}^{N} a_i x_i^2 = 6\,749\,000,$$

$$s^2(\hat{\mathscr{D}}) = \frac{1}{15}\{1\,992\,151 - 2 \times 0.541\,602\,31 \times 3\,663\,300$$

$$+ 0.541\,602\,31^2 \times 6\,749\,000\} = \frac{3\,752.357}{15} = 250.157,$$

$$v(\hat{Y}_R) = \frac{80 - 16}{80}\,\frac{250.157}{16} = 12.507\,9,$$

$$v(\hat{Y}_R)^{\frac{1}{2}} = 3.537, \quad v(\hat{R})^{\frac{1}{2}} = \frac{3.537}{650} = 0.005\,45.$$

NOTE. If $\bar{X}$ is not known, we shall estimate $V(\hat{R})^{\frac{1}{2}}$ by

$$3.536/647.50 = 0.005\,47;$$

as mentioned before, this estimator may be preferable even if $\bar{X}$ is known.

The quantities which we obtained here are not exactly equal to the quantities in section 10. Actually here they are more precise, because we rounded numbers at fewer stages than in section 10.

2. Asymptotic formula for the bias of $\hat{R}$

If $\bar{X} \neq 0$ (so that $R = \bar{Y}/\bar{X}$ is well defined) and if† $|(\bar{x} - \bar{X})/\bar{X}| < 1$, then from

$$\hat{R} - R = \frac{\bar{y} - R\bar{x}}{\bar{x}} = \frac{\bar{y} - R\bar{x}}{\bar{X}(1 + (\bar{x} - \bar{X})/\bar{X})}$$

$$= \frac{\bar{y} - R\bar{x}}{\bar{X}}\left[1 - \frac{\bar{x} - \bar{X}}{\bar{X}} + \left(\frac{\bar{x} - \bar{X}}{\bar{X}}\right)^2 - \dots\right]$$

we get

$$\mathscr{E}\hat{R} - R = -\frac{\mathscr{E}(\bar{y} - R\bar{x})(\bar{x} - \bar{X})}{\bar{X}^2} + \frac{\mathscr{E}(\bar{y} - R\bar{x})(\bar{x} - \bar{X})^2}{\bar{X}^3}$$
$$- \frac{\mathscr{E}(\bar{y} - R\bar{x})(\bar{x} - \bar{X})^3}{\bar{X}^4} + \dots$$

We shall see that, if we evaluate the first 2 terms‡ on the right-hand side of the expansion of $\mathscr{E}\hat{R} - R$, we arrive at

† Often the condition is not fulfilled for some values of $\bar{x}$ in the sample space; one would have to show that these contribute negligibly to the expected value of $\hat{R} - R$. Fortunately it is possible to prove the result we obtain without this condition, provided $|(X - x)/X| < 1$. See Koop, Bull. Intern. Statist. Inst. **33**, Pt. II (Intern. Statist. Conf., India, 1951) 141–146. However, his proof tacitly *assumes* that the series containing moments converges absolutely (for n not too small this is likely to hold for many distributions arising in practice, namely those whose moments do not grow rapidly with order). The latter condition is easily seen to be equivalent with

$$-1 < \frac{\bar{x} - \bar{X}}{\bar{X}} < \frac{2N - n}{n} = 1 + 2\frac{N - n}{n} \qquad \text{rather than} \qquad -1 < \frac{\bar{x} - \bar{X}}{\bar{X}} < 1.$$

In particular, if all x_i are nonnegative (or nonpositive), $x/X > 0$, so $(X - x)/X < 1$ and we have

$$(X - x)/X = \Sigma_{i \notin \sigma}\, x_i/X > 0 \; (> -1).$$

In that case $(\bar{x} - \bar{X})/\bar{X}$ is not necessarily < 1 (unless $n \geq N/2$). Even if only a few x_i in $\mathscr{P}$ differ in sign from the others, but are not much larger in absolute value, samples for which $x/X < 0$ or

$$\Sigma_{i \notin \sigma}\, x_i/X < 0$$

will occur only with very small probability when n and $N - n$ are fairly large.

‡ It is at least conceivable that further terms in the expansion contribute additional terms of order $1/n$.

$$\mathscr{E}\hat{R} - R = -\frac{N-n}{N}\frac{1}{n}\frac{1}{\bar{X}^2}S(\mathscr{X}\mathscr{D}) + O(1/n^2)$$

$$= \frac{N-n}{N}\frac{1}{n}\frac{1}{\bar{X}^2}(RS^2(\mathscr{X}) - S(\mathscr{X}\mathscr{Y})) + O(1/n^2)$$

$$= R\frac{N-n}{N}\frac{1}{n}\{C(\mathscr{X},\mathscr{X}) - C(\mathscr{X},\mathscr{Y})\} + O(1/n^2),$$

where $O(1/n^2)$ designates a function of n, whose ratio to n^2 is bounded as $n \to \infty$. It has to be carefully noted that the term $O(1/n^2)$ need not be small compared to the previous terms for small n. The meaning of "n small" cannot be stated without further investigation.

To show the above formula, we compute

$$\mathscr{E}\bar{y}(\bar{x} - \bar{X}) = \mathscr{E}(\bar{y} - \bar{Y})(\bar{x} - \bar{X}) = \frac{N-n}{N}\frac{1}{n}S(\mathscr{Y}\mathscr{X})$$

exactly the way we found $\mathscr{E}(\bar{x} - \bar{X})^2$ in Chapter II. We also need for this purpose

$$\mathscr{E}(\bar{y} - \bar{Y})(\bar{x} - \bar{X})^2 = \frac{N-n}{N}\frac{1}{n^2}\frac{N-2n}{N-2}\frac{\sum_{i=1}^{N}(y_i - \bar{Y})(x_i - \bar{X})^2}{N-1},$$

$$\mathscr{E}(\bar{x} - \bar{X})^3 = \frac{N-n}{N}\frac{1}{n^2}\frac{N-2n}{N-2}\frac{\sum_{i=1}^{N}(x_i - \bar{X})^3}{N-1},$$

which may be obtained by an approach similar to that followed in Remark (vi) of section 5 of Chapter II.†

EXAMPLE. In the above example we may estimate the bias of $\hat{R}$ by

$$\frac{N-n}{N}\frac{1}{n}\frac{1}{\bar{x}^2}(\hat{R}s^2(\mathscr{X}) - s(\mathscr{Y}\mathscr{X}))$$

$$= \frac{N-n}{N}\frac{1}{n}\frac{1}{\bar{x}^2}\frac{1}{n-1}\{\hat{R}\sum_{i=1}^{N}a_i x_i^2 - \sum_{i=1}^{N}a_i x_i y_i\}$$

$$= \frac{80-16}{80}\frac{1}{16}\frac{1}{647.50^2}\frac{1}{15}\{0.541\,602\,31 \times 6\,749\,000 - 3\,663\,300\}$$

$$= \frac{64}{80}\frac{1}{16}\frac{1}{647.50^2}\frac{1}{15}(-8\,026.08) = -0.000\,063\,8,$$

† SUKHATME, Sampling Theory of Surveys with Applications (Indian Society of Agricultural Statistics and Iowa State College, 1954) gives another derivation on p. 189. He also obtains a further approximation to $\mathscr{E}\hat{R} - R$ by including 3 terms in the expansion of $\mathscr{E}\hat{R} - R$ instead of 2.

and we estimate the bias of $\hat{Y}_R$ by 650 times this amount, that is by $-0.041\,5$. (The calculation of the estimates of bias and of bias adjusted estimates of $\hat{R}$ etc. in later sections have been carried to more places than is justified by the amount of sampling variation present; in order to illustrate more clearly the different methods.)

It is clear that estimators of the bias which we obtain in this way may, for small n, have large standard errors.

For the relative bias we obtain

$$\frac{\mathscr{E}\hat{R} - R}{R} \approx \frac{N-n}{N}\frac{1}{n}\{C(\mathscr{X}\mathscr{X}) - C(\mathscr{Y}\mathscr{X})\}$$

$$= \frac{N-n}{N}\frac{1}{n}\{C(\mathscr{X}\mathscr{X}) - s\rho(\mathscr{Y}\mathscr{X})(C(\mathscr{X}\mathscr{X})\,C(\mathscr{Y}\mathscr{Y}))^{\frac{1}{2}}\},$$

where

$$s = \begin{cases} 1 & \text{if } R > 0, \\ -1 & \text{if } R < 0. \end{cases}$$

If $\mathscr{X}$ is the characteristic $\mathscr{Y}$ as a previous point of time, then it may be that $C(\mathscr{X}\mathscr{X})$ is about equal to $C(\mathscr{Y}\mathscr{Y})$, and then

$$\frac{\mathscr{E}\hat{R} - R}{R} \approx \frac{N-n}{N}\frac{1}{n}C(\mathscr{Y}\mathscr{Y})\,(1 - \rho(\mathscr{Y}\mathscr{X})).$$

3. Bound for the absolute value of the bias of $\hat{Y}_R$

Since

$$-N\,\mathrm{Cov}\,(\hat{R}, \bar{x}) = -N\left(\mathscr{E}\frac{\bar{y}}{\bar{x}}\bar{x} - \mathscr{E}\hat{R}\,\mathscr{E}\bar{x}\right) = -N(\bar{Y} - \bar{X}\mathscr{E}\hat{R})$$

$$= X\mathscr{E}\hat{R} - Y = \mathscr{E}\,\hat{Y}_R - Y,$$

we have

$$\frac{\mathscr{E}\,\hat{Y}_R - Y}{V(\hat{Y}_R)^{\frac{1}{2}}} = \frac{-N\,\mathrm{Cov}\,(\hat{R}, \bar{x})}{V(\hat{Y}_R)^{\frac{1}{2}}} = -\frac{1}{|\bar{X}|}V(\bar{x})^{\frac{1}{2}}\rho' = -\rho'\left(\frac{N-n}{N}\frac{1}{n}C(\mathscr{X}\mathscr{X})\right)^{\frac{1}{2}},$$

where ρ' is the correlation coefficient between $\hat{R}$ and $\bar{x}$ (or between $\hat{Y}_R$ and $\bar{x}$). This result does not depend on approximations!

Since $|\rho'| \leq 1$,

$$\left|\frac{\mathscr{E}\hat{Y}_R - Y}{V(\hat{Y}_R)^{\frac{1}{2}}}\right| \leq \left(\frac{N-n}{N}\frac{1}{n}C(\mathscr{X}\mathscr{X})\right)^{\frac{1}{2}} = \mathrm{CV}(\bar{x});$$

that is the bias of $\hat{Y}_R$ relative to its standard deviation cannot exceed the coefficient of variation of $\bar{x}$ in absolute value.

EXAMPLE. In our example the bias of $\hat{Y}_R$ divided by its standard error is in absolute value sure to be less than $((0.8/16)C(\mathscr{X}\mathscr{X}))^{\frac{1}{2}}$ which, since

$$s^2(\mathscr{X}) = \frac{6\,749\,000 - 10\,360^2/16}{15} = 2\,726.67 = 54.098\,7^2,$$

may be estimated by

$$16^{-\frac{1}{2}} \times 650^{-1} \times 0.8^{\frac{1}{2}} \times 54.098\,7 = 0.018\,610.$$

This indicates that the biases of the estimators $\hat{Y}_R$ and $\hat{R}$ are not important compared with their variances.

In practice we often have some *advance* knowledge of $C(\mathscr{X}\mathscr{X})$ and thus may be able to give an upper bound for the relative bias.

4. Confidence intervals and bounds using ratio estimators

4.1. GENERAL RESULTS

Confidence limits for ratio estimators may be computed like for other estimators.† Sometimes Fieller's method is used which is based on the following argument. If n is sufficiently large for $\bar{y}$ and $\bar{x}$ to have approximately a joint normal distribution,‡

$$f(R) = (\bar{y} - R\bar{x})\Big/\left\{\frac{N-n}{N}\frac{1}{n}(s^2(\mathscr{Y}) + R^2s^2(\mathscr{X}) - 2Rs(\mathscr{Y}\mathscr{X}))\right\}^{\frac{1}{2}}$$

will be approximately standard normal. Therefore, e.g.,

$$\mathrm{P}\{|f(R)| \leq 1.96\} \approx 0.95,$$

from which quadratic inequality we may try to solve for R in order to find confidence limits on R. However, the solution is not always real, and the units do not always span an interval – and if they do, the interval is not

† COCHRAN, Sampling Techniques (Wiley, 2nd ed., 1963), states that the distribution of $\hat{R}$ tends to have positive skewness in the kinds of populations for which it is most often used. He recommends to use the usual normal approximation when $n \geq 30$ and the coefficients of variation of $\bar{x}$ and of $\bar{y}$ are less than 0.1.

‡ Often such n is much smaller than that required for approximate normality of $\hat{R}$.

always finite. If the (sample) coefficient of variation of $\bar{x}$ is less than 1 over the t used in the confidence interval, these difficulties do not arise, but any confidence interval so obtained is not an improvement, since, as shown by Hájek,† the interval

(i) is never shorter than the one obtained by the usual method;‡ and

(ii) in the important case discussed in Chaper VIII, section 8, the approximate probability of the interval covering R is nonetheless never larger than that of the usual confidence interval.

Moreover, Fieller's method can only be used for the simplest methods of sampling.

4.2. SOME EMPIRICAL RESULTS

Consider the ordinary confidence interval for R with $t = 2$, based on the assumption that the estimator is approximately normally distributed. Denote by l the percentage of 10 000 repetitions in which R falls to the left of the lower confidence bound, by r the percentage in which it falls to the right of the upper confidence bound, and by $\bar{w}$ the average of the half-widths of the intervals.

In section 7.3.1 we consider 3 *bivariate normal* populations with correlation coefficients $\rho = 0.4$, 0.6 and 0.8 respectively. Already for samples of size 50 the percentage of cases in which R fell outside the confidence interval based on $\hat{R}$ was close to the theoretical 4.5, even though the coefficients of variation of $\bar{y}$ and $\bar{x}$ are 0.2:

	$\rho = 0.4$			$\rho = 0.6$			$\rho = 0.8$		
	l	r	$\bar{w}$	l	r	$\bar{w}$	l	r	$\bar{w}$
$n = 50$	0.3	4.1	1.43	0.7	3.6	1.16	1.0	2.3	0.82
$n = 200$	1.2	3.3	0.67	1.6	3.0	0.55	1.9	2.4	0.39
$n = 1\,000$	1.7	2.7	0.30	2.1	2.6	0.24	1.9	2.9	0.17

† HÁJEK, Apl. Matem., 3 (1958) 384–398.

‡ Cochran notes that its length and center come close to those of the usual interval if the sample coefficients of variation of $\bar{x}$ and $\bar{y}$ are both much less than $1/t$. Inspection of the formulas shows, however, that the center may be displaced by a relative amount which is not necessarily negligible. Thus, if the sample coefficients of variation of $\bar{x}$ and $\bar{y}$ are 0.10 and 0.03 respectively, with $\bar{x}\bar{y} > 0$ and $t = 2$, and if the sample correlation coefficient between the x's and y's is 0.9, the usual confidence interval for R is $\hat{R}$ (1 ± 0.148 3), and Fieller's is $\hat{R}$(1.030 4 ± 0.154 4). With correlation 0.6, the usual interval is $\hat{R}$(1 ± 0.170 9) and Fieller's $\hat{R}$(1.034 2 ± 0.177 7). Indeed, without some displacement of the center, statement (ii) would appear to be contradictory (except for allowing for cases in which Fieller's confidence set is empty or the complement of an interval).

In section 7.3.2 we consider 3 corresponding *very skew lognormal* populations. The results for confidence intervals based on $\hat{R}$ are as follows (for some results using stratification, see section 14.1 of Chapter III):†

	$\rho = 0.4$			$\rho = 0.6$			$\rho = 0.8$		
	l	r	$\bar{w}$	l	r	$\bar{w}$	l	r	$\bar{w}$
$n = 50$	1.4	7.2	1.16	1.6	6.9	0.92	2.0	7.3	0.63
$n = 200$	1.3	4.9	0.62	1.2	4.8	0.50	1.3	5.0	0.35
$n = 1\,000$	1.4	3.5	0.29	1.4	3.9	0.23	1.2	4.6	0.17

It appears that, for obtaining the sum $l + r$ close to the theoretical value 4.5, a sample of size 1 000 was still insufficient,‡ especially for high ρ; even though, for high ρ, a much smaller sample sufficed to obtain an average half-width of less than 10% of R.

In addition it was found that for $n = 50$ the average of the sample estimates of the standard error of $\hat{R}$ fell some 12% below the true standard error, for $n = 200$ some 5%, and for $n = 1\,000$ some $2\frac{1}{2}$%.

For $\hat{R}_{(1)}$ and $\hat{R}_Q$ (defined in section 7.1), using the same estimator of variance as for $\hat{R}$ (and so obtaining the same $\bar{w}$), r is generally somewhat bigger than for $\hat{R}$:

	$\rho = 0.4$		$\rho = 0.6$		$\rho = 0.8$	
	l	r	l	r	l	r
$n = 50$						
$\hat{R}_{(1)}$	1.3	8.3	1.6	7.9	2.2	7.9
$\hat{R}_Q$	1.4	8.8	1.7	8.4	2.4	8.3
$n = 200$						
$\hat{R}_{(1)}$	1.3	5.4	1.2	5.2	1.3	5.3
$\hat{R}_Q$	1.3	5.4	1.2	5.1	1.3	5.3
$n = 1\,000$						
$\hat{R}_{(1)}$	1.4	3.7	1.4	4.1	1.1	4.7
$\hat{R}_Q$	1.4	3.8	1.3	4.1	1.2	4.8

† KOKAN, Bull. Calcutta Statist. Ass., **12** (1963) 149–158 computes the coefficient of variation of $s(\hat{R}\bar{X})$ for samples from lognormal populations. He finds that, for $C(\mathscr{Y})/C(\mathscr{X})$ equal to some $\frac{3}{4}$ or less and large ρ, it is much larger than that of $\bar{y}$ when $C(\mathscr{X}) \geq 0.7$ (say). For $C(\mathscr{Y})/C(\mathscr{X})$ equal to 1 as well as 10 they are not very far apart (unless $C(\mathscr{Y})$ is very large), but for some intermediate values they are more separated. In our case they would differ little, being equal to about $7n^{-\frac{1}{2}}$. In the normal case they are $(\frac{1}{2} + C(\mathscr{X},\mathscr{X}))^{\frac{1}{2}}n^{-\frac{1}{2}}$ and $(2n)^{-\frac{1}{2}}$, respectively (one fourth and one tenth of the lognormal case).

‡ The slowness of its decrease as the sample size increases beyond 1 000 led us not to pursue much further our search for "an adequate minimum sample size".

Computation of $\hat{R}_Q$ with $g > 2$ for the case $n = 50$ and $\rho = 0.6$ did not yield results much different from the case $g = 2$ presented in the table.

Finally, we consider a *finite population of real data*. The population consists of all 845 U.S. cities which, according to the 1960 Census, had over 25 000 inhabitants. However, we exclude the 54 cities which in 1965 had over 250 000 inhabitants (since in practice all of these would be included in a sample). The data for each city are for the fiscal year that ended somewhere between July 1, 1966 and June 30, 1967.

$\mathcal{Y}$ is total long term debt in thousands of dollars and $\mathcal{X}$ is total general revenue in thousands of dollars. The data have been published in Table IV of The Municipal Yearbook 1969, published by International City Management Association (Washington, 1969). The following are summary data

$$\bar{Y} = 11\,670.705 \qquad \bar{X} = 7\,498.334 \qquad R = 1.556\,439\,846$$

$$\sigma(\mathcal{Y}) = 16\,437.626 \qquad \sigma(\mathcal{X}) = 8\,340.197 \qquad \rho = 0.548$$

$$\gamma_1(\mathcal{Y}) = 4.78 \qquad \gamma_1(\mathcal{X}) = 3.16 \qquad \mathcal{E}\hat{R} - R \approx \left(\frac{1}{n} - \frac{1}{N}\right) 0.588\,51$$

$$\gamma_2(\mathcal{Y}) = 37.64 \qquad \gamma_2(\mathcal{X}) = 13.76 \qquad V(\hat{R}) \approx \left(\frac{1}{n} - \frac{1}{N}\right) 3.640\,55.$$

For any reasonable sample size the bias is negligible.

The following results were obtained for this population (based, this time, on 20 000 replications):

	l	r	$\bar{w}$
Samples of 300	1.1	5.1	0.172
Samples of 400	1.1	4.5	0.134
Samples of 500	1.2	4.3	0.103
Samples of 600	1.3	4.2	0.077
Samples of 700	1.2	4.0	0.049.

One may compare the above data with those obtained for $\bar{y}/\bar{X}$, also based on 20 000 replications:

	l	r	$\bar{w}$
Samples of 300	0.8	5.1	0.200
Samples of 400	0.9	4.9	0.154
Samples of 500	0.9	4.6	0.119
Samples of 600	0.9	4.4	0.088
Samples of 700	0.8	4.5	0.056.

Note that the values of $l + r$ are about equal in both cases, but that the skewness is less pronounced for the ratio estimator.

The average of the sample estimates of the standard error of $\hat{R}$ is extremely close to the true standard error of $\hat{R}$.

Stratified sampling from this population is discussed in section 14.1 of Chapter III, cluster sampling in section 3.2 of Chapter VII.

4.3. THE CASE OF RELATIVE OR TOTAL FREQUENCIES OF PROPERTIES IN A DOMAIN

We saw in section 11.5 (ii) of Chapter II that the relative frequency, $\bar{Y}_1$, of a characteristic in a domain may be estimated by the ratio estimator $\bar{y}_1$, whose variance is approximately

$$\frac{1}{n}\frac{N}{N_1}\bar{Y}_1(1-\bar{Y}_1)$$

in sampling with replacement, and $(N-n)/(N-1)$ times this in sampling without replacement. In section 15.2 of Chapter II, on approximating a binomial by a normal distribution, we found for P the confidence interval $[a-tw, a+tw]$, where

$$a = (\bar{y}n + \tfrac{1}{2}t^2)/(n+t^2), \quad w = \{n\bar{y}(1-\bar{y}) + \tfrac{1}{4}t^2\}^{\frac{1}{2}}/(n+t^2)$$

when sampling with replacement, by solving $(\bar{y}-P)^2 = t^2P(1-P)/n$ for P.

It follows that, if the normal approximation continues to be applicable in our case of domain estimation, a confidence interval is obtained from the above formulas by replacing in them $\bar{y}$ by $\bar{y}_1$ and t^2 by t^2N/N_1. In case N_1 is unknown (and perhaps even if it is known) we may replace N_1 by its estimator $(n_1/n)N$; the result is then $[a-tw, a+tw]$ with

$$a = (\bar{y}_1n_1 + \tfrac{1}{2}t^2)/(n+t^2), \quad w = \{n_1\bar{y}_1(1-\bar{y}_1) + \tfrac{1}{4}t^2\}^{\frac{1}{2}}/(n+t^2).$$

The corresponding confidence interval for Y' when N_1 is known is found by multiplying by N_1. If N_1 is not known, but Y is, we use the estimator $(y'/y)Y$ for Y'. An estimator for its variance is derived in section 11.3 of Chapter II; entirely analogously we find for its variance in sampling with replacement the approximate formula

$$n^{-1}NY'\{1-(Y'/Y)\}$$

and for $V(y'/y)$

$$n^{-1}\bar{Y}^{-1}(Y'/Y)\{1-(Y'/Y)\}.$$

A confidence interval for Y'/Y is therefore obtained from that for P by replacing $\bar{y}$ by y'/y and t^2 by $t^2/\bar{Y}$, giving $[a-tw, a+tw]$ with

$$a = \{(y'/y)n + \tfrac{1}{2}(t^2/\bar{Y})\}/\{n+(t^2/\bar{Y})\},$$
$$w = \{n\bar{Y}^{-1}(y'/y)(1-(y'/y)) + \tfrac{1}{4}(t^2/\bar{Y}^2)\}^{\frac{1}{2}}/\{n+t^2/\bar{Y}\}$$

and so for Y' we get $[a - tw, a + tw]$ with

$a = \{(y'/y)\,Yn + \frac{1}{2}t^2N\}/\{n + (t^2/\bar{Y})\}$,

$$w = N^{\frac{1}{2}}\{nY(y'/y)(1 - (y'/y)) + \tfrac{1}{4}t^2N\}^{\frac{1}{2}}/\{n + (t^2/\bar{Y})\}.$$

5. More precise formulas for $V(\hat{Y}_R)$

By the method of section 2, if n is not too small, it is possible to obtain a closer approximation for $V(\hat{Y}_R)$ than the one obtained before, which was of order $1/n$. In this approximation the term of order $1/n^2$ appears often to be positive, indicating a tendency of the previously given approximation to be too low. (Further detail is contained in the books of Cochran and Sukhatme. According to Cochran, Sampling Techniques (Wiley, 2nd ed., 1963) pp. 163, 164, the bias in the estimator† $v(\hat{Y}_R)$ is of order $1/n$ and will often be negative.)

6. Comparison between the variance of $\hat{Y}_R$ and that of $N\bar{y}$

Under the condition under which we computed the approximation for $V(\hat{Y}_R)$, we obtained

$$V(\hat{Y}_R) \approx \frac{N-n}{N}\frac{1}{n} N^2(S^2(\mathscr{Y}) + R^2S^2(\mathscr{X}) - 2R\rho(\mathscr{Y}\mathscr{X})\, S(\mathscr{Y})\, S(\mathscr{X})).$$

On the other hand,

$$V(N\bar{y}) = N^2\frac{N-n}{N}\frac{1}{n}S^2(\mathscr{Y}),$$

which is smaller than the expression for $V(\hat{Y}_R)$ if and only if (for $R > 0$)

$$\rho(\mathscr{Y}\mathscr{X}) < \tfrac{1}{2}RS(\mathscr{X})/S(\mathscr{Y}),$$

that is, if and only if (for $R > 0$)

$$\rho(\mathscr{Y}\mathscr{X}) < \tfrac{1}{2}(C(\mathscr{X}\mathscr{X})/C(\mathscr{Y}\mathscr{Y}))^{\frac{1}{2}};$$

† It may be remarked in that connection that there is no clear justification for the divisor $n - 1$ (rather than n), in this estimator. In the ordinary variance estimator this divisor is caused essentially by the fact that then $\bar{Y}$ is not known, so that the variance is estimated from sample deviations from the estimator $\bar{y}$ of $\bar{Y}$; but, in $s^2(\mathscr{D})$, $\bar{D}$ is known (it is zero). In using the divisor $n - 1$ we follow convention and, incidentally, stay on the "safe" side.

and, therefore, in particular if $C(\mathscr{X}\mathscr{X}) > 4C(\mathscr{Y}\mathscr{Y})$ (since a correlation coefficient is necessarily ≤ 1) or if $\rho(\mathscr{X}\mathscr{Y}) \leq 0$.† (To use these results, keep in mind what is said in the latter part of section 5.)

We know that in many populations the standard deviation of the total expenditures is not very much different from that of expenditures for food. In our example, the families receive above-average incomes, and we know that in such populations $100R$ must be between 35 and 60 and that $\rho(\mathscr{Y}\mathscr{X})$ is at least 0.40. Therefore we may expect in advance that $\hat{Y}_R$ is more efficient than $\bar{y}$ as an estimator of $\bar{Y}$.

REMARKS. The variance of the estimator is important for deciding what the sample size will be; therefore, if there is a large difference between $V(\hat{Y}_R)$ and $V(N\bar{y})$, it is important to know in advance which of these variances is the smaller. Beside variance and bias one should also take into account that usually, for any given expenditure of time and effort, $\hat{Y}_R$ has to be based on *fewer observations* than $N\bar{y}$ due to the cost or effort to obtain $\mathscr{X}$-observations (if not already obtained), of obtaining X, and of computing $s(\mathscr{Y}\mathscr{X})$ and $s(\mathscr{X}\mathscr{X})$.

If $\mathscr{X}$ is the characteristic $\mathscr{Y}$ at a previous date, then possibly

$$C(\mathscr{X}\mathscr{X}) \approx C(\mathscr{Y}\mathscr{Y});$$

and then, if there is a strong correlation between them, the variance of $\hat{Y}_R$ will be smaller than that of $N\bar{y}$. More generally, as already remarked at the end of section 9 in Chapter II, if y_i is approximately a multiple of x_i,‡ the variance of $\hat{Y}_R$ will be smaller than that of $N\bar{y}$, since then this multiple must be close to R and so Σd_i^2 relatively small. If y_i is approximately a constant plus a (constant) multiple of x_i for all i, it is more precise to estimate Y by fitting a linear regression on $\mathscr{X}$. Such an estimator is called a *regression estimator*. We shall not discuss that kind of situation here, but will mention only that for n large enough, its variance is about $(1 - \rho^2(\mathscr{X}\mathscr{Y}))$ times that of $N\bar{y}$ (see more in section 13). If there is between $\mathscr{Y}$ and $\mathscr{X}$ an approximate relation of a different kind we may be able to make efficient use of $\mathscr{X}$ as a stratifying variable (see the chapter on stratified sampling).

There are cases in which a sample may be meant to be a simple random

† When $\mathscr{X}$ and $\mathscr{Y}$ are negatively correlated, one may consider using the *product estimator* $N\bar{y}\bar{x}/\bar{X}$, which is readily found to have relative bias $\{(N-n)/Nn\}\ C(\mathscr{X}\mathscr{Y})$ and variance approximately equal to $\{(N-n)/Nn\}N^2(S^2(\mathscr{Y}) + R^2S^2(\mathscr{X}) + 2R\rho(\mathscr{X}\mathscr{Y})S(\mathscr{Y})S(\mathscr{X}))$. Further on this in MURTHY, Sankhya **26A** (1964) 69–74.

‡ The student should note that in the example of Chapter II, section 10, this is indeed the case for the given *sample*. Sometimes we are able to define $\mathscr{Y}$ and $\mathscr{X}$ or to determine (in advance of sampling) constants to be subtracted from the $\mathscr{Y}$- and/or $\mathscr{X}$-values, such that the situation mentioned may be expected to hold in the population. The definition of $\mathscr{X}$ in the example referred to (see the footnote there) was a step in this direction. – It may be shown that another factor favoring a ratio estimator is a tendency for the absolute value of the deviation of y_i from a multiple of x_i to increase with the value of x_i, especially if this increase is less than proportional.

sample from a population $\mathscr{P}$, but is not, for unavoidable or unforeseen reasons; see section 1 of Chapter X. This will often show up in that, for some subpopulation $\mathscr{P}_1$ and some characteristic $\mathscr{X}$, a significant discrepancy appears between the average value for $\mathscr{X}$ in the sample and the (exactly or approximately) known average value for *all* members of $\mathscr{P}_1$. There may, however, be evidence that y'/x' is very much closer to Y'/X' than is $N_1\bar{y}_1$ to Y'. In that case $(y/x)X$ may have a much smaller bias than $N\bar{y}$. (If X' and $X'' = X - X'$ are known, the estimator $(y'/x')X' + (y''/x'')X''$ may be preferable.)

For example, the sample may be such that poor families ($\mathscr{P}_1$) are seriously underrepresented, but available data may indicate that there is no appreciable difference between the poor and others in the ratio of the time they watch sports telecasts to their total television viewing time, whereas $\mathscr{Y}$, the amount of time a family devotes to looking at sports telecasts during the period under investigation, may generally be much larger for poor families. We may then use $(y/x)X$, or, if the number of poor families in the sample is very small, $(y''/x'')X$, for estimating Y, assuming total viewing time (X) is known.

A similar situation, involving regression estimators, is discussed in section 7 of Chapter I.

7. Approximate corrections for bias of a ratio estimator, and unbiased ratio estimators

7.1. BIAS CORRECTION

We shall discuss three bias correction methods. Each of these eliminates the term of $O(1/n)$ in the above expression for the bias. As remarked above, there is no certainty that for small n the actual bias is diminished, but, on the basis of knowledge about the higher order terms, we have reason to expect that the bias will be diminished when n is not extremely small and when the joint frequencies of the $\mathscr{Y}$ and $\mathscr{X}$ values in the population are not extremely irregular.

(i) The most obvious sort of correction is to subtract from $\hat{R}$ an estimator of the first term in the bias expansion given in section 2; that gives

$$\hat{R}_{(i)} = \hat{R}\left\{1 + \frac{N-n}{N}\,\frac{1}{n}\left(\frac{s(\mathscr{Y}\mathscr{X})}{\bar{x}\bar{y}} - \frac{s^2(\mathscr{X})}{\bar{x}^2}\right)\right\}.$$

(ii) Beale has suggested the very similar expression

$$\hat{R}_{(ii)} = \hat{R}\left\{1 + \frac{N-n}{N}\,\frac{1}{n}\,\frac{s(\mathscr{Y}\mathscr{X})}{\bar{x}\bar{y}}\right\}\left\{1 + \frac{N-n}{N}\,\frac{1}{n}\,\frac{s^2(\mathscr{X})}{\bar{x}^2}\right\}^{-1}.$$

(iii) Quenouille's estimator $\hat{R}_Q$, which we shall discuss below.

(iv) Some other estimators which we shall not explicitly mention here.

TIN, J. Amer. Statist. Ass., **60** (1965) 294–307 gives expansions for the bias and variance of these estimators. Subject to the previously mentioned qualifications as to the unreliability of such expansions, his results lead to a preference for (i) and (ii) over (iii) (with $g = 2$ for $\hat{R}_Q$), the reason being that the variances are in that order. As to the comparison of biases, little can be said in general.†

Up to order $1/n$, the variances of the first three estimators are the same as that of $\hat{R}$. The variances of $\hat{R}_{(i)}$ and $\hat{R}_{(ii)}$ are the same up to order $1/n^2$, but (for large N) the term of order $1/n^2$ is twice as large for $\hat{R}_Q$ as for $\hat{R}_{(i)}$. If the (y_i, x_i) are approximately joint normal, the variance of $\hat{R}$ is to that (and the next higher) order larger than that of $\hat{R}_Q$, and to order $1/n^3$ the variance of $\hat{R}_{(ii)}$ exceeds that of $\hat{R}_{(i)}$. In that case also the nonnormality measures $|\gamma_1|$ and γ_2 are up to order $1/n$ equal for $\hat{R}_{(i)}$ and $\hat{R}_{(ii)}$, larger for $\hat{R}_Q$, and still larger for $\hat{R}$.‡

We devote some space to $\hat{R}_Q$ because this method is of wide applicability in other contexts, even though here it may (at least for $g = 2$) give somewhat worse results:

Supposing that we are estimating some parameter T by some $\hat{T}$ which satisfies

$$\mathscr{E}\hat{T} - T = \sum_{k=1}^{\infty} b_k n^{-k}.$$

Let us divide the sample at random into g parts, each of size $m = n/g$ (assuming m is an integer); let $\hat{T}_{)j(}$ be the statistic $\hat{T}$ computed for the entire sample excluding the jth part. Then

$$\mathscr{E}\hat{T}_{)j(} - T = \sum_{k=1}^{\infty} b_k\{(g-1)m\}^{-k}$$

† Tin gives some conclusions about bias (and also about variance), which do not appear to be deducible from his formulae. One may state, however, that up to order $1/n$, $\hat{R}_{(i)}$ and $\hat{R}_{(ii)}$ are free from bias, while that of $\hat{R}_Q/R$ is $-\{C^2(\mathscr{X}) - C(\mathscr{X},\mathscr{Y})\}/N$. (It is clear from comparisons that the sign of the last factor in the expectation of $\hat{R}_Q$ is misprinted in (5) of Tin.) In all references given to Tin's work, $\hat{R}_Q$ relates to the case $g = 2$.

‡ Tin presents some results of sampling from certain bivariate normal populations. In section 7.3 new results are presented to replace Tin's, since, as explained there, the latter are not reliable.

and so

$$\mathscr{E}[g\hat{T} - (g-1)\hat{T}_{)j(}] - T = \sum_{k=1}^{\infty} b_k\{gn^{-k} - (g-1)^{-k+1}m^{-k}\}$$

$$= \sum_{k=1}^{\infty} b_k\{g^{-k+1} - (g-1)^{-k+1}\}m^{-k}$$

$$= -\frac{1}{g(g-1)}\frac{b_2}{m^2} - \frac{2g-1}{g^2(g-1)^2}\frac{b_3}{m^3} - \ldots$$

$$= -\frac{g}{g-1}\frac{b_2}{n^2} - \frac{g(2g-1)}{(g-1)^2}\frac{b_3}{n^3} - \ldots .$$

What is important here is that the term of order $1/n$ has fallen out. We now define the so-called pseudo-values

$$\hat{T}(j) = g\hat{T} - (g-1)\hat{T}_{)j(},$$

and define $\hat{T}_Q$ as their average:

$$\hat{T}_Q = g\hat{T} - (g-1)g^{-1}\sum_{j=1}^{g}\hat{T}_{)j(}.$$

No investigation has been published as to the best choice of g in *general*; in some cases it appears to be $g = n$. When $g = n$, the statistic also has the advantage of being defined without reference to a set of random numbers, so that all statisticians would come out with the same estimate for R if they use the same formula (estimator) on the same data.

REMARK. There has been a suggestion that the pseudo-values are "nearly" independent and that therefore

$$\frac{1}{g}\sum_{j=1}^{g}\{\hat{T}(j) - \hat{T}_Q\}^2/(g-1)$$

should be a good estimator of $V(\hat{T}_Q)$ (in the case of sampling with replacement,† presumably).

† For the case without replacement, JONES (Chapter 9 in Proceedings of the IBM Scientific Computing Symposium on Statistics, October 21–23, 1963, IBM 1963) has suggested (for the case $m = 1$) to define the pseudovalues by using ω instead of g, where

$$\omega = \frac{n}{N-n}\bigg/\left(\frac{n}{N-n} - \frac{n-m}{N-n+m}\right) = g\frac{N-n+m}{N}.$$

He gets as an estimator of the variance of the average of these new pseudovalues $\hat{T}'(j)$

$$\frac{N-g}{N}\frac{1}{g}\frac{g-1}{\omega-1}\Sigma\left\{\hat{T}'(j) - \frac{1}{g}\Sigma\hat{T}'(j)\right\}^2\bigg/(\omega-1) = \frac{N-g}{N}\frac{1}{g}\frac{1}{g-1}\Sigma\left\{\hat{T}(j) - \frac{1}{g}\Sigma\hat{T}(j)\right\}^2,$$

as found below, and the average of the $\hat{T}'(j)$ is $\hat{T}'_Q = a\hat{T}_Q + (1-a)\hat{T}$ with $a = (\omega-1)/(g-1)$.

It is not clear what "nearly" independent means in numerical terms or how it would affect the validity of this estimator of $V(\hat{T}_Q)$. In our case, we can make a comparison between the above expression and $v(\hat{T})$ when $g = n$.† If we denote the sample by $u_{i_1}, \ldots, u_{i_n}$,

$$\hat{R}(j) = g\frac{y}{x} - (g-1)\frac{(1-(y_{i_j}/y))y}{(1-(x_{i_j}/x))x}$$

$$= \frac{y}{x}\left\{g - (g-1)\left(1 - \frac{y_{i_j}}{y} + \frac{x_{i_j}}{x}\right)\right\} - \frac{y}{x}(g-1)\left(\frac{x_{i_j}}{x} - \frac{y_{i_j}}{y}\right)\sum_{k=1}^{\infty}\left(\frac{x_{i_j}}{x}\right)^k$$

$$= \frac{y}{x}\left\{1 + \frac{g-1}{y}\hat{d}_{i_j}\sum_{k=0}^{\infty}\left(\frac{x_{i_j}}{x}\right)^k\right\},$$

so, by neglecting terms with $k > 0$,

$$\hat{R}(j) - \frac{1}{g}\Sigma\hat{R}(j) \approx \frac{g-1}{g}\frac{1}{\bar{\bar{x}}}\hat{d}_i,$$

and

$$\frac{N-g}{N}\frac{1}{g}\Sigma\{\hat{R}(j) - \hat{R}_Q\}^2/(g-1) \approx \left(\frac{g-1}{g}\right)^2 v(\hat{R}).$$

The approximation is good if none of the x_{i_j} is close to the sum x; otherwise the expression is likely to be larger. For cross-products occurring in the sums of squares are of the sort: x^k times

$$\Sigma\hat{d}_{i_j}\left\{\hat{d}_{i_j}(x_{i_j})^k - \frac{1}{g}\sum_j\hat{d}_{i_j}(x_{i_j})^k\right\} = \Sigma\hat{d}_{i_j}^2(x_{i_j})^k \quad (k > 0),$$

which are >0, and of the sort: x^k times

$$\Sigma\left\{\hat{d}_{i_j}(x_{i_j})^k - \frac{1}{g}\sum_j\hat{d}_{i_j}(x_{i_j})^k\right\}\left\{\hat{d}_{i_j}(x_{i_j})^l - \frac{1}{g}\sum_j d_{i_j}(x_{i_j})^l\right\} \quad (0 < k < l),$$

most of which are likely to be positive unless very small (we assume $x > 0$).

Finally note that

$$\frac{1}{g}\Sigma\{\hat{R}(j) - \hat{R}_Q\}^2/(g-1) = \frac{g-1}{g}\Sigma\left\{\hat{R}_{(j)} - \frac{1}{g}\Sigma\hat{R}_{(j)}\right\}^2.$$

† One would not expect the above estimator to be good if g is small. Computations for $n = 50$ for the populations discussed in section 7.3.1 and in section 7.3.2 ($\rho = 0.6$) with $g = 2, 5, 10, 25$, and 50 show, indeed, that reasonable results are obtained for $g = 25$ and $g = 50$ only.

The same expansion can be used to compare $\hat{R}_Q$ with (i), by neglecting terms with $k > 1$:

$$\hat{R}_Q \approx \frac{y}{x}\left\{1 + \frac{g-1}{yx}\sum_j d_{i_j}x_{i_j}\right\} = \hat{R}\left\{1 + \frac{1}{n}\left(\frac{n-1}{n}\right)^2\left(\frac{s(\mathcal{YX})}{\bar{y}\bar{x}} - \frac{s^2(\mathcal{X})}{\bar{x}^2}\right)\right\}.$$

EXAMPLE. In section 2 we already found $-0.000\,063\,8$ for the estimator of the bias of $\hat{R}$ in our example, giving

$$\hat{R}_{(\mathrm{i})} = 0.541\,602\,3 - (-0.000\,063\,8) = 0.541\,666\,1.$$

$$\hat{R}_{(\mathrm{ii})} = 0.541\,602\,3\left\{1 + \frac{80-16}{80}\,\frac{1}{16}\,\frac{1}{15}\,\frac{3\,663\,000 - (10\,360 \times 5\,611/16)}{10\,360 \times 5\,611/16^2}\right.$$

$$\times\left\{1 + \frac{80-16}{80}\,\frac{1}{16}\,\frac{1}{15}\,\frac{6\,749\,000 - (10\,360^2/16)}{10\,360^2/16^2}\right\}^{-1}$$

$$= 0.541\,602\,3\left[1 + \frac{80-16}{80}\,\frac{1}{15}\left\{\frac{3\,663\,300 \times 16}{10\,360 \times 5\,611} - 1\right\}\right]$$

$$\times\left[1 + \frac{80-16}{80}\,\frac{1}{15}\left\{\frac{6\,749\,000 \times 16}{10\,360^2} - 1\right\}\right]^{-1}$$

$$= 0.541\,606\,66.$$

To compute Quenouille's estimates, we first arrange the observations at random by finding a random arrangement of the numbers 1 to 16:

10, 9, 1, 5, 4, 2, 16, 8, 3, 15, 6, 13, 11, 12, 7, 14.

$g = 2$. If the first 8 in the random arrangement constitute the first group and the last 8 the second,

	$\mathcal{Y}$	$\mathcal{X}$	Ratio
Sum of the last 8 values	2 818	5 200	$\hat{R}_{)1(} = 0.541\,923\,0$
Sum of the first 8 values	2 793	5 160	$\hat{R}_{)2(} = 0.541\,279\,0$
	5 611	10 360	$\hat{R} = 0.541\,602\,3$

$$\hat{R}_Q = 2\hat{R} - \tfrac{1}{2}\{\hat{R}_{)1(} + \hat{R}_{)2(}\} = 0.541\,603\,6.$$

$g = 4$. If the first 4 in the random arrangement constitute the first group, etc.,

	$\mathscr{Y}$	$\mathscr{X}$	Remainder $\mathscr{Y}$	Remainder $\mathscr{X}$	Ratio
Sum of the first 4 values	1 356	2 550	4 255	7 810	$\hat{R}_{)1(} = 0.544\ 814\ 3$
Sum of the next 4 values	1 437	2 610	4 174	7 750	$\hat{R}_{)2(} = 0.538\ 580\ 6$
Sum of the next 4 values	1 405	2 620	4 206	7 740	$\hat{R}_{)3(} = 0.543\ 410\ 8$
Sum of the next 4 values	1 413	2 580	4 198	7 780	$\hat{R}_{)4(} = 0.539\ 588\ 6$
	5 611	10 360			

$$\hat{R}_Q = 4\hat{R} - \tfrac{3}{4}\{\hat{R}_{)1(} + \ldots + \hat{R}_{)4(}\} = 0.541\ 613\ 5.$$

$g = 8$. If the first 2 in the random arrangement constitute the first group, etc.,

	$\mathscr{Y}$	$\mathscr{X}$	Remainder $\mathscr{Y}$	Remainder $\mathscr{X}$	Ratio
Sum of the first 2 values	700	1 330	4 911	9 030	$\hat{R}_{)1(} = 0.543\ 853\ 8$
Sum of the next 2 values	656	1 220	4 955	9 140	$\hat{R}_{)2(} = 0.542\ 122\ 5$
Sum of the next 2 values	634	1 210	4 977	9 150	$\hat{R}_{)3(} = 0.543\ 934\ 4$
Sum of the next 2 values	803	1 400	4 808	8 960	$\hat{R}_{)4(} = 0.536\ 667\ 1$
Sum of the next 2 values	718	1 340	4 893	9 020	$\hat{R}_{)5(} = 0.542\ 461\ 1$
Sum of the next 2 values	687	1 280	4 924	9 080	$\hat{R}_{)6(} = 0.542\ 290\ 7$
Sum of the next 2 values	711	1 280	4 900	9 080	$\hat{R}_{)7(} = 0.539\ 647\ 5$
Sum of the next 2 values	702	1 300	4 909	9 060	$\hat{R}_{)8(} = 0.541\ 832\ 2$
	5 611	10 360			

$$\hat{R}_Q = 8\hat{R} - \tfrac{7}{8}\{\hat{R}_{)1(} + \ldots + \hat{R}_{)8(}\} = 0.541\ 662\ 8.$$

$g = 16$. Here there is no random arrangement. We find

$$\Sigma\hat{R}_{)j(} = 8.665\ 552, \quad \hat{R}_Q = 16\hat{R} - \frac{15}{16}\Sigma\hat{R}_{)j(} = 0.541\ 682\ 0,$$

$$\Sigma\hat{R}^2_{)j(} = 4.693\ 267\ 200\ 64,$$

so

$$\frac{N-g}{N}\,\frac{g-1}{g}\{\Sigma\hat{R}^2_{)j(} - (\Sigma\hat{R}_{)j(})^2/g\} = 0.000\ 022\ 675\ 568$$

with square root 0.004 762 (compare with the previous estimator 0.005 45).

7.2. UNBIASED ESTIMATORS

Besides estimators partially corrected for bias, the literature contains also unbiased ratio estimators. If the probability for any x_i to vanish is zero, such an estimator may be constructed as a function of the sample average of the ratios $r_i = y_i/x_i$; see, e.g., GOODMAN and HARTLEY, J. Amer. Statist. Ass., **53** (1958) 491–508. Since the average of the r_i is usually subject to much larger sampling fluctuations than the ratio $\bar{y}/\bar{x}$ of the average $\mathscr{Y}$ and $\mathscr{X}$ values, one should be reluctant to use such an estimator without clear evidence of its preferability.

This estimator may also require a modification in the method of sampling to ensure that, for no u_i in the sample, x_i vanishes. A less objectionable procedure has been proposed in a paper by MICKEY, J. Amer. Statist. Ass., **54** (1959) 594–612, namely†

$$\frac{1}{g}\Sigma\hat{R}_{)j(} + \omega\frac{\bar{x}}{\bar{X}}\left(\hat{R} - \frac{1}{g}\Sigma\hat{R}_{)j(}\right) = \omega\frac{\bar{x}}{\bar{X}}\hat{R}_{Q} + \left(1 - \omega\frac{\bar{x}}{\bar{X}}\right)\frac{1}{g}\Sigma\hat{R}_{)j(}$$

(or rather the average over all permutations of the sample), where

$$\omega = g(1 - (n - m)/N);$$

and a generalization yielding unbiased regression estimators. For $g = n$, Mickey's estimate is, in our example, 0.541 689 7.

7.3. SOME EMPIRICAL RESULTS ON THE DISTRIBUTION OF VARIOUS RATIO ESTIMATORS

7.3.1. *Bivariate normal populations.* In the paper of Tin referred to above, the results of sampling from three bivariate normal populations are reported. In all these populations $\bar{Y} = 15$, $\bar{X} = 5$, $S^2(\mathscr{Y}) = 500$, $S^2(\mathscr{X}) = 45$; the correlations $\rho = \rho(\mathscr{Y}, \mathscr{X})$ are 0.4, 0.6 and 0.8, so that the values of $S(\mathscr{Y}, \mathscr{X})$ are 60, 90 and 120, respectively. Thus the coefficients of variation of $\bar{y}$ and $\bar{x}$ are 0.21 and 0.19 for $n = 50$, 0.10 and 0.09 for $n = 200$, and 0.05 and 0.04 for $n = 1\,000$.

For various reasons Tin's results are not reliable: the second factor in the simulation of $s(\mathscr{Y}, \mathscr{X})/n$ has to be (in his notation) $G_3(\beta^2 + \frac{1}{2}\delta/S_x^2)^{\frac{1}{2}}(1 + \zeta^2)^{-\frac{1}{2}}$; the normal approximations he uses for $s^2(\mathscr{X})$ give negative results in many instances, and these cannot, of course, be simply discarded; the number of repetitions he uses, 1 000, is far too small to give reliable results; his results for γ_1 (and, of course, those for γ_2) are often exceedingly different from the values yielded by his asymptotic expansions, especially for $\rho = 0.8$.

New calculations have therefore been performed, based on 10 sets of 1 000 repetitions. As is seen below, even this number was often insufficient to yield

† That is, we take Mickey's α to be $n(1 - (1/g))$. Some more of Mickey's estimators may be considered in case it is desired to have $g < n$ but n is not divisible by the desired group size. WILLIAMS, J. Amer. Statist. Ass., **57** (1962) 184–186, by using interpenetrating samples (see, e.g., Chapter III, section 14.3) includes, apart from the case $g = 2$, the less desirable of Mickey's estimators, namely the ones for which α is less than $\frac{1}{2}n$.

values of reasonable precision for several of the characteristics; such values are given below in parentheses. Asterisks indicate that the values were exceedingly unstable; those include, not surprisingly, all γ_2 values. The reasonably precise values may be seen to be quite close to the values attained by the asymptotic approximations in Tin.

Approximation to the characteristics $\mathscr{E}$, V and γ_1 of the sampling distribution of three estimators for R from samples of sizes $n = 50$, 200 and 1 000 drawn from 3 bivariate normal populations

	$\rho = 0.4$			$\rho = 0.6$			$\rho = 0.8$		
	$\mathscr{E}$	V	γ_1	$\mathscr{E}$	V	γ_1	$\mathscr{E}$	V	γ_1
$n = 50$									
$\hat{R}$	3.06	0.53	(0.90)	3.04	0.36	(0.59)	3.02	0.17	(0.33)
$\hat{R}_{(1)}$	2.98	0.44	(0.58)	2.99	0.31	(0.42)	3.00	0.15	(0.18)
$\hat{R}_Q$	2.94	*	*	2.99	*	*	3.00	*	*
$n = 200$									
$\hat{R}$	3.02	0.11	0.29	3.01	0.078	0.20	3.00	0.038	(0.10)
$\hat{R}_{(1)}$	3.00	0.11	0.28	3.00	0.076	0.19	3.00	0.037	(0.10)
$\hat{R}_Q$	3.00	0.11	0.28	3.00	0.077	0.19	3.00	0.038	(0.10)
$n = 1\,000$									
$\hat{R}$	3.00	0.022	(0.10)	3.00	0.015	(0.09)	3.00	0.008	(0.02)
$\hat{R}_{(1)}$	3.00	0.022	(0.10)	3.00	0.015	(0.09)	3.00	0.008	(0.02)
$\hat{R}_Q$	3.00	0.022	(0.10)	3.00	0.015	(0.09)	3.00	0.008	(0.02)

The values of the bias (B), and of γ_1 and γ_2 attained by the asymptotic approximations given by Tin are:

	$\rho = 0.4$			$\rho = 0.6$			$\rho = 0.8$		
	B	γ_1	γ_2	B	γ_1	γ_2	B	γ_1	γ_2
$n = 50$									
$\hat{R}$	0.076	0.69	1.52	0.046	0.51	1.22	0.015	0.24	0.94
$\hat{R}_{(1)}$	−0.010	0.63	1.10	−0.006	0.46	0.86	−0.002	0.21	0.64
$\hat{R}_Q$	−0.020	0.66	1.28	−0.012	0.49	1.02	−0.004	0.23	0.79
$n = 200$									
$\hat{R}$	0.015 4	0.29	0.38	0.009 3	0.21	0.31	0.003 1	0.10	0.23
$\hat{R}_{(1)}$	−0.000 5	0.28	0.28	−0.000 3	0.21	0.22	−0.000 1	0.10	0.16
$\hat{R}_Q$	−0.000 9	0.28	0.32	−0.000 6	0.21	0.26	−0.000 2	0.10	0.20
$n = 1\,000$									
$\hat{R}$	0.003 0	0.12	0.08	0.001 8	0.09	0.06	0.000 6	0.04	0.05
$\hat{R}_{(1)}$	0.000 0	0.12	0.06	0.000 0	0.09	0.04	0.000 0	0.04	0.03
$\hat{R}_Q$	0.000 0	0.12	0.06	0.000 0	0.09	0.05	0.000 0	0.04	0.04

For $n = 50$ the asymptotic approximations for the variance of $\hat{R}_Q$ are 0.48, 0.32 and 0.16 for $\rho = 0.4$, 0.6 and 0.8, respectively. For the other n and for the other estimators the asymptotic approximations for the variance are very close to the values found by sampling.

Because of the closeness of $\hat{R}_{(ii)}$ to $\hat{R}_{(i)}$, no calculations have been made for the former.

7.3.2. *Bivariate lognormal populations.* Calculations have also been performed on bivariate populations with the same mean and variance and the same 3 correlation coefficients as the above bivariate normal populations. Thus $(\log_e \mathscr{Y}, \log_e \mathscr{X})$ are joint normal with means 2.123 0 and 1.094 6, standard deviations 1.078 5 and 1.014 7; and with correlations 0.537 108, 0.720 477 and 0.873 128, respectively.

For $\mathscr{Y}$ the skewness γ_1 is 7.8, and γ_2 is 199.9; for $\mathscr{X}$ the skewness γ_1 is 6.4, and γ_2 is 122.9. The values of the variance of $\hat{R}$ are very close to those implied by Tin's formulas up to order n^{-2}, except that the bias is about 0.01 to 0.02 higher for $n = 50$.

Approximation to the characteristics $\mathscr{E}$, V and γ_1 of the sampling distributions of three estimators for R from samples of sizes $n = 50$, 200 and 1 000 drawn from 3 bivariate lognormal populations

	$\rho = 0.4$			$\rho = 0.6$			$\rho = 0.8$		
	$\mathscr{E}$	V	γ_1	$\mathscr{E}$	V	γ_1	$\mathscr{E}$	V	γ_1
$n = 50$									
$\hat{R}$	3.05	0.43	0.88	3.02	0.27	0.76	3.00	0.13	(0.65)
$\hat{R}_{(1)}$	3.00	0.44	0.87	2.99	0.28	0.77	2.99	0.15	(0.75)
$\hat{R}_Q$	2.99	0.45	0.87	2.99	0.29	0.72	2.99	0.15	(0.72)
$n = 200$									
$\hat{R}$	3.01	0.108	(0.47)	3.00	0.069	0.35	2.99	0.035	(0.43)
$\hat{R}_{(1)}$	3.00	0.110	(0.46)	2.99	0.071	0.35	2.99	0.037	(0.47)
$\hat{R}_Q$	3.00	0.111	(0.44)	2.99	0.071	0.34	2.99	0.037	(0.47)
$n = 1\,000$									
$\hat{R}$	2.99	0.022	(0.30)	2.99	0.014	(0.18)	2.99	0.007 2	(0.17)
$\hat{R}_{(1)}$	2.99	0.022	(0.30)	2.99	0.015	(0.19)	2.99	0.007 2	(0.17)
$\hat{R}_Q$	2.99	0.022	(0.29)	2.99	0.015	(0.18)	2.99	0.007 2	(0.19)

8. Weighted averages when the weights are obtained from the same sample

If $x_i = w_i$ and $y_i = z_i w_i$,

$$R = \sum_{i=1}^{N} w_i z_i \Big/ \sum_{i=1}^{N} w$$

is a weighted mean.

In this case we obtain

$$V(\hat{R}) \approx \frac{N-n}{N}\frac{1}{n}\frac{1}{\bar{W}^2}\frac{\sum_{i=1}^{N} w_i^2(z_i - R)^2}{N-1},$$

since

$$d_i = w_i z_i - R w_i = w_i(z_i - R).$$

Similarly

$$\mathscr{E}(\hat{R}) - R \approx -\frac{N-n}{N}\frac{1}{n}\frac{1}{\bar{W}^2}\frac{\sum_{i=1}^{N} w_i^2(z_i - R)^2}{N-1}.$$

9. Double ratios

Sometimes we wish to estimate ratios of ratios. For X_0 and X_1 positive, let $R_0 = Y_0/X_0$, $R_1 = Y_1/X_1$; we may be interested in R_1/R_0 if also Y_0 is different from 0. The subscripts may refer, e.g., to observations at different times; in the following it is assumed that they are all obtained from the same sample (as otherwise no special formulas are needed).

Assuming $|(\bar{x}_1 - \bar{X}_1)/\bar{X}_1|$ and $|(\bar{y}_0 - \bar{Y}_0)/\bar{Y}_0|$ to be less than 1, Rao and Pereira, Sankhya, **A30** (1968) 83–90 obtained by the method of section 2, if $r_0 = y_0/x_0$ and $r_1 = y_1/x_1$,

$$\frac{\mathscr{E}r_1/r_0 - R_1/R_0}{R_1/R_0}$$

$$\approx \frac{N-n}{N}\frac{1}{n}\{C(\mathscr{Y}_0\mathscr{Y}_0) + C(\mathscr{X}_1\mathscr{X}_1) + C(\mathscr{Y}_0\mathscr{X}_1) + C(\mathscr{Y}_1\mathscr{X}_0)$$
$$- C(\mathscr{Y}_0\mathscr{Y}_1) - C(\mathscr{X}_0\mathscr{X}_1) - C(\mathscr{Y}_0\mathscr{X}_0) - C(\mathscr{Y}_1\mathscr{X}_1)\},$$

$$V(r_1/r_0)/(R_1/R_0)^2$$

$$\approx \frac{N-n}{N}\frac{1}{n}\{(C(\mathscr{Y}_1\mathscr{Y}_1) - 2C(\mathscr{Y}_1\mathscr{X}_1) + C(\mathscr{X}_1\mathscr{X}_1))$$
$$- 2(C(\mathscr{Y}_0\mathscr{Y}_1) - C(\mathscr{Y}_0\mathscr{X}_1) - C(\mathscr{Y}_1\mathscr{X}_0) + C(\mathscr{X}_0\mathscr{X}_1))$$
$$+ (C(\mathscr{Y}_0\mathscr{Y}_0) - 2C(\mathscr{Y}_0\mathscr{X}_0) + C(\mathscr{X}_0\mathscr{X}_0))\}.$$

Note that the latter equals simply

$$V\left(\frac{r_1}{R_1} - \frac{r_0}{R_0}\right).$$

For estimation of the variance we may replace the C's by c's and the R's by r's, and obtain

$$v(r_1/r_0) = \frac{N-n}{N}\frac{1}{n}\frac{1}{r_0^2}\sum_{i=1}^{N} a_i\left\{\frac{y_{1i}-r_1x_{1i}}{\bar{x}_1} - \frac{r_1}{r_0}\frac{y_{0i}-r_0x_{0i}}{\bar{x}_0}\right\}^2\Big/(n-1).$$

One is often interested in expressions of the form $\Sigma w_k R_{1k}/R_{0k}$, where the w_k are known numbers. It follows from the above that we may estimate the variance of $\Sigma w_k r_{1k}/r_{0k}$ by

$$\frac{N-n}{N}\frac{1}{n}\Sigma a_i(\Sigma w_k f_{ik}/r_{0k})^2/(n-1),$$

when f_i is the expression in the curly brackets above.

Occasionally one may wish to estimate $\bar{Y}_1$ by $(r_1/r_0)R_0\bar{X}_1$ or R_1 by $(r_1/r_0)R_0$, especially if $y_{0i} = 1$ for all i or $x_{0i} = 1$ for all i.

10. Specialization of ratio estimators to frequencies†

If $x_i = 1$ when u_i belongs to a subpopulation $\mathscr{C}$ (containing $N(\mathscr{C})$ elements) and 0 otherwise, and if $y_i = 1$ if u_i belongs to a subpopulation $\mathscr{D}$ (containing $N(\mathscr{D})$ elements) and 0 otherwise, we have $R = N(\mathscr{D})/N(\mathscr{C})$. Then, by substitution in our previously obtained formulae,

$$\frac{\mathscr{E}\hat{R}-R}{R} \approx \frac{N-n}{N}\frac{1}{n}\frac{N}{N-1}\left(\frac{N(\mathscr{C})(N-N(\mathscr{C}))}{N^2(\mathscr{C})} - \frac{NN(\mathscr{C}\mathscr{D})-N(\mathscr{C})N(\mathscr{D})}{N(\mathscr{C})N(\mathscr{D})}\right)$$

$$= \frac{N-n}{N-1}\frac{1}{n}\frac{N}{N(\mathscr{C})}\left(1-\frac{N(\mathscr{C}\mathscr{D})}{N(\mathscr{D})}\right),$$

$$V(\hat{R}/R) = V(\hat{R})/R^2$$

$$\approx \frac{N-n}{N}\frac{1}{n}\frac{N}{N-1}\left(\frac{N(\mathscr{D})(N-N(\mathscr{D}))}{N^2(\mathscr{D})} - 2\frac{NN(\mathscr{C}\mathscr{D})-N(\mathscr{C})N(\mathscr{D})}{N(\mathscr{C})N(\mathscr{D})} + \frac{N(\mathscr{C})(N-N(\mathscr{C}))}{N^2(\mathscr{C})}\right)$$

$$= \frac{N-n}{N-1}\frac{1}{n}\frac{N^2}{N(\mathscr{C})N(\mathscr{D})}\left(\frac{N(\mathscr{D})}{N} + \frac{N(\mathscr{C})}{N} - 2\frac{N(\mathscr{C}\mathscr{D})}{N}\right),$$

† If, in $\hat{R} = y/x$, x is a frequency, we cannot obtain unbiased estimators if $\mathrm{P}(x = 0) > 0$. Assume that the probability of x being zero is very small. Even so one has to modify the sampling plan so as to exclude the possibility that $x = 0$. (For example, we may require continuation of sampling beyond n observations until the sum of the x_i in the sample will be positive.) The formulas that follow (except those in the Note) neglect the small changes entailed by such a modification in the sampling plan.

where $N(\mathscr{C}\mathscr{D})$ is the number of elements belonging to both $\mathscr{C}$ and $\mathscr{D}$. The student should consider different cases ($\mathscr{C}$ and $\mathscr{D}$ overlapping or not, $\mathscr{C}$ and $\mathscr{D}$ exhausting the population or not, $\mathscr{D}$ contained in $\mathscr{C}$) and their application. Why does the formula given here for the approximate relative bias not contradict the result mentioned at the beginning of Chapter II, section 11.2? Note that for small $N(\mathscr{C})$ the size of n required to make the relative bias negligible may be considerable.

The student should also consider the mixed case in which $\mathscr{Y}$ is a quantitative characteristic and $\mathscr{X}$ is as above.

The special case $\mathscr{D} = \mathscr{P}$ is also important: then $R = 1/\bar{X}$ and

$$\mathscr{E}\,\frac{1}{x} \approx \frac{1}{n\bar{X}}\left\{1 + \frac{N-n}{N-1}\,\frac{1}{n\bar{X}}(1-\bar{X})\right\},$$

$$\mathscr{E}\,\frac{1}{x} - \frac{1}{\bar{X}} \approx \frac{N-n}{N-1}\,\frac{1}{n\bar{X}}\left\{1 + \frac{1}{n\bar{X}}(1-\bar{X})\right\},$$

$$V(1/x) \approx \frac{N-n}{N-1}(1-\bar{X})\,(n\bar{X})^{-3}.$$

NOTE. For n much smaller than N and $x > 0$, GRAB and SAVAGE, J. Amer. Statist. Ass., **49** (1954) 169–177, give for $\mathscr{E}(1/x)$ the approximation

$$\{n\bar{X} - (1-\bar{X})\}^{-1}.$$

MENDENHALL and LEHMAN, Technometrics, **2** (1960) 227–242 give the approximation

$$\frac{n-2}{n}\{(n-1)\,\bar{X} - 1\}^{-1},$$

which is good for $n\bar{X} > 3\frac{1}{2}$ and $\bar{X} \geq 0.10$ (and better than the Grab-Savage approximation). The latter also give approximations for $\mathscr{E}(1/x^k)$ for natural numbers k, and tables; they obtain, for example, for the variance of $1/x$ the approximation

$$\frac{n-2}{n}\,\frac{n-a-1}{n}\{(a-1)^2(a-2)\}^{-1}$$

$$= \frac{1}{n^2(n-1)}\,\frac{n-2}{n-1}\,\frac{1-\bar{X}}{(\bar{X}-(1/(n-1))^2(\bar{X}-(2/(n-1))},$$

where $a = (n-1)\bar{X}$.

Tables for $\mathscr{E}(1/x)$ in sampling without replacement are given by GOVINDARAJULU, Sankhya, **B26** (1964) 217–236.

Another special case of ratio estimation is one in which we estimate the *total number N in the population* by $\hat{N}_R = (n/x)X$, where $\mathscr{X}$ is any characteristic for which we know the population *total*. In this case $\mathscr{Y}$ is identically equal to 1, and so we get at once, since $R = 1/\bar{X}$ and $S(\mathscr{X}\mathscr{Y}) = 0$, that

$$(\mathscr{E}\hat{N}_R - N)/N \approx \left(1 - \frac{n}{N}\right)\frac{1}{n}\frac{1}{\bar{X}}RS^2(\mathscr{X}) = \left(1 - \frac{n}{N}\right)\frac{1}{n}C(\mathscr{X}\mathscr{X}),$$

$$V(\hat{N}_R)/N^2 \approx \left(1 - \frac{n}{N}\right)\frac{1}{n}\Sigma\left(1 - \frac{1}{\bar{X}}x_i\right)^2\Big/(N-1) = \left(1 - \frac{n}{N}\right)\frac{1}{n}C(\mathscr{X}\mathscr{X}).$$

From this it follows that we should try to find a characteristic with small coefficient of variation; in case X is a frequency, this means that $\bar{X}$ should be as close to $\bar{Y} = 1$ as possible.

The variance of $\hat{N}_R = \hat{R}X$ (with $\hat{R} = 1/\bar{x}$) may be estimated by

$$\hat{N}_R(\hat{N}_R - n)\frac{1}{n}c(\mathscr{X}\mathscr{X}),$$

and a simple correction for the bias consists in subtracting from $\hat{N}_R$ the amount $(\hat{N}_R - n)(1/n)c(\mathscr{X}\mathscr{X})$.

Since, in order to draw a simple random sample from a population by means of random numbers, the units need to be numbered, one may well wonder how the above problem could arise.

Sometimes a random sample is already at hand, but N is unknown to the analyst. In other cases, the numbering of the units in the population may have taken place on an earlier occasion, but since then an unknown number of units may have dropped out of the population and we have a list of additions. Then we know an upper bound for N, and evidently still can use random numbers to obtain a simple random sample from the present population. The above estimation method is not necessarily the best available in this case, but may be so if it is expensive or time consuming to draw specified units from the population and if we also require a random sample to estimate some other characteristic of the population.†

On the other hand, as remarked in the footnote to section 6 of Chapter I, random sampling is not always carried out by the use of random numbers. Suppose we wish to ascertain the number N of fish of a certain kind in a lake. Catch a large number X of them, tag them, and release them right away. We now set up traps at randomly placed locations and eventually catch n fish of

† In cases in which $(y/z)Z$ would efficiently estimate $\bar{Y}$, but N and Z are unknown, whereas some X is known, it may be that $\hat{Y} = (y/z)Z\hat{N}_R$ is an efficient estimator of Y. Being ZX times a double ratio, with $R_1 = \bar{Y}/\bar{Z}$, $R_0 = \bar{X}/1$, we find from the previous section

$$(\mathscr{E}\hat{Y} - Y)/Y \approx (\mathscr{E}r_1 - R_1)/R_1 + (\mathscr{E}\hat{N}_R - N)/N + C(\mathscr{X}(\mathscr{Z} - \mathscr{Y})),$$
$$V(\hat{Y}/Y) \approx V(r_1/R_1) + V(\hat{N}_R/N) + 2C(\mathscr{X}(\mathscr{Z} - \mathscr{Y})).$$

the given kind of which x turn out to be tagged, and use the above estimator for N.

11. Multivariate ratio estimators

Sometimes there are several auxiliary variables, for example, $\mathscr{X}_1$ and $\mathscr{X}_2$. If $\hat{Y}_{1R} = (y/x_1)X_1$ and $\hat{Y}_{2R} = (y/x_2)X_2$, we may consider using a linear combination of the two:

$$\hat{Y}_{MR} = W\hat{Y}_{1R} + (1 - W)\hat{Y}_{2R},$$

with that W minimizing $V(\hat{Y}_{MR})$. A simple calculation gives as minimizing W

$$\frac{V(\hat{Y}_{2R}) - \text{Cov}(\hat{Y}_{1R}, \hat{Y}_{2R})}{V(\hat{Y}_{1R}) - 2\,\text{Cov}(\hat{Y}_{1R}, \hat{Y}_{2R}) + V(\hat{Y}_{2R})},$$

and as minimized variance

$$\frac{V(\hat{Y}_{1R})V(\hat{Y}_{2R}) - \{\text{Cov}(\hat{Y}_{1R}, \hat{Y}_{2R})\}^2}{V(\hat{Y}_{1R}) - 2\,\text{Cov}(\hat{Y}_{1R}, \hat{Y}_{2R}) + V(\hat{Y}_{2R})}.$$

We find an approximately unbiased estimator of the variance by replacing the variances by the estimators we already found for them, and the covariance by

$$\text{cov}(\hat{Y}_{1R}, \hat{Y}_{2R}) = N^2\frac{N-n}{N}\frac{1}{n}(s^2(\mathscr{Y}) + \hat{R}_1\hat{R}_2 s(\mathscr{X}_1\mathscr{X}_2) - \hat{R}_1 s(\mathscr{Y}\mathscr{X}_1) - \hat{R}_2 s(\mathscr{Y}\mathscr{X}_2))$$

with, for example,

$$s(\mathscr{Y}\mathscr{X}_1) = \frac{1}{n-1}\sum_{i=1}^{N} a_i(y_i - \bar{y})(x_{1i} - x_1).$$

EXAMPLE. (OLKIN, Biometrika, **45** (1958) 154–165.) Suppose we wish to estimate the average number of inhabitants in 1950 of the 200 largest cities in the U.S.A. The population of these cities is subdivided into 2 strata: the 5 largest, and the others. We draw a simple random sample of 50 cities from the second stratum and obtain complete data for the first stratum. We shall here examine only the second stratum. Let $\mathscr{Y}$ be the number of inhabitants according to the census of 1950, $\mathscr{X}_1$ the number in 1940, and $\mathscr{X}_2$ the number in 1930. The following is a summary for *all* the cities for the second stratum

$\bar{Y} = 169\,900$	$\bar{X}_1 = 148\,200$	$\bar{X}_2 = 142\,000$
$\sigma(\mathscr{Y}) = 174\,000$	$\sigma(\mathscr{X}_1) = 155\,400$	$\sigma(\mathscr{X}_2) = 150\,900$
$C(\mathscr{Y}\mathscr{Y}) = 1.049$	$C(\mathscr{X}_1\mathscr{X}_1) = 1.098$	$C(\mathscr{X}_2\mathscr{X}_2) = 1.131$
$C(\mathscr{Y}\mathscr{X}_1) = 1.059$	$C(\mathscr{Y}\mathscr{X}_2) = 1.056$	$C(\mathscr{X}_1\mathscr{X}_2) = 1.108, \quad W = 2$

and for the cities in the sample from this stratum

$$\begin{array}{lll} \bar{y} = 189\,600 & \bar{x}_1 = 169\,300 & \bar{x}_2 = 164\,300 \\ s(\mathscr{Y}) = 208\,800 & s(\mathscr{X}_1) = 193\,200 & s(\mathscr{X}_2) = 193\,100 \\ c(\mathscr{Y}\mathscr{Y}) = 1.213 & c(\mathscr{X}_1\mathscr{X}_1) = 1.302 & c(\mathscr{X}_2\mathscr{X}_2) = 1.381 \\ c(\mathscr{Y}\mathscr{X}_1) = 1.241 & c(\mathscr{Y}\mathscr{X}_2) = 1.256 & c(\mathscr{X}_1\mathscr{X}_2) = 1.335, \quad w = 2.38 \end{array}$$

(w is the estimated value of W, etc.).

We compare:

(a) The estimator $\bar{y}$ (=189 600) with its estimated standard error

$$((195 - 50)/(195 \times 50))^{\frac{1}{2}} \times 208\,800 = 25\,470.$$

(b) $(\bar{y}/\bar{x}_1)\bar{X}_1$ (=166 000) with estimated standard error

$$((195 - 50)/(195 \times 50))^{\frac{1}{2}} \times 30\,200 = 3\,680.$$

(c) $w(\bar{y}/\bar{x}_1)\bar{X}_1 + (1 - w)(\bar{y}/\bar{x}_2)\bar{X}_2$ (=168 900) with estimated standard error

$$((195 - 50)/(195 \times 50))^{\frac{1}{2}} \times 15\,200 = 1\,850.$$

12. Double sampling to obtain an estimate of $\bar{X}$ (for use in ratio estimators)

If $\bar{X}$ is not known in advance, and if it is inexpensive to obtain sample information on it, we may consider a *double* (or *two-phase*) sampling procedure. Consider drawing a large preliminary random sample of n' in order to obtain an estimate $\bar{x}'$ for $\bar{X}$, and then estimate $\bar{Y}$ by $(y/x)\bar{x}'$ on the basis of a sample of n. We consider two cases:

(i) the second sample is a subsample from the first;
(ii) the two samples are drawn independently from the same population.

12.1. BIAS

We write for the bias:

$$\frac{y}{x}\bar{x}' - \bar{Y} = \left(\frac{y}{x}\bar{X} - \bar{Y}\right) + \frac{y}{x}(\bar{x}' - \bar{X}).$$

Thus the bias of our estimator equals that of $(y/x)\bar{X}$ plus the mean of $\hat{R}(\bar{x}' - \bar{X})$, which is 0 under (ii), and which under (i) is

$$\frac{1}{\bar{X}}\mathscr{E}\bar{y}(\bar{x}' - \bar{X}) - \frac{1}{\bar{X}^2}\mathscr{E}\bar{y}(\bar{x} - \bar{X})(\bar{x}' - \bar{X}) + \ldots$$

by the method of section 2. We now proceed to evaluate the terms of the latter expansion.

Now

$$\mathscr{E}\{(\bar{x}' - \bar{X})\bar{y}|u_{i_1}, \ldots, u_{i_{n'}}\} = (\bar{x}' - \bar{X})\mathscr{E}\{\bar{y}|u_{i_1} \ldots, u_{i_{n'}}\}$$
$$= (\bar{x}' - \bar{X})\bar{y}',$$

so

$$\frac{1}{\bar{X}}\mathscr{E}\bar{y}(\bar{x}' - \bar{X}) = \frac{1}{\bar{X}}\mathscr{E}\bar{y}'(\bar{x}' - \bar{X}) = \frac{N - n'}{N}\frac{1}{n'}\frac{1}{\bar{X}}S(\mathscr{X}\mathscr{Y}).$$

Also

$$\mathscr{E}\{(\bar{x}' - \bar{X})\bar{y}(\bar{x} - \bar{X})|u_{i_1}, \ldots, u_{i_{n'}}\}$$
$$= (\bar{x}' - \bar{X})\mathscr{E}\{(\bar{y} - \bar{Y})(\bar{x} - \bar{X})|u_{i_1}, \ldots, u_{i_{n'}}\}$$
$$+ (\bar{x}' - \bar{X})\,\bar{Y}\mathscr{E}\{(\bar{x} - \bar{X})|u_{i_1}, \ldots, u_{i_{n'}}\}.$$

But

$$\mathscr{E}\{(\bar{x} - \bar{X})|u_{i_1}, \ldots, u_{i_{n'}}\} = \bar{x}' - \bar{X}$$

and

$$\mathscr{E}\{(\bar{y} - \bar{Y})(\bar{x} - \bar{X})|u_{i_1}, \ldots, u_{i_{n'}}\}$$
$$= \mathscr{E}\{(\bar{y} - \bar{y}')(\bar{x} - \bar{x}')|u_{i_1}, \ldots, u_{i_{n'}}\}$$
$$+ (\bar{y}' - \bar{Y})\,\mathscr{E}\{(\bar{x} - \bar{x}')|u_{i_1}, \ldots, u_{i_{n'}}\}$$
$$+ (\bar{x}' - \bar{X})\,\mathscr{E}\{(\bar{y} - \bar{y}')|u_{i_1}, \ldots, u_{i_{n'}}\} + (\bar{y}' - \bar{Y})(\bar{x}' - \bar{X})$$
$$= \frac{n' - n}{n'}\frac{1}{n}\frac{1}{n' - 1}\sum_{j=1}^{n'}(y_{i_j} - \bar{y}')(x_i - \bar{x}') + (\bar{y}' - \bar{Y})(\bar{x}' - \bar{X})$$
$$= \frac{n' - n}{n' - 1}\frac{1}{n}\frac{1}{n'}\left\{\sum_{j=1}^{n'}(y_{i_j} - \bar{Y})(x_i - \bar{X}) - n'(\bar{y}' - \bar{Y})(\bar{x}' - \bar{X})\right\}$$
$$+ (\bar{y}' - \bar{Y})(\bar{x}' - \bar{X})$$
$$= \frac{n' - n}{n' - 1}\frac{1}{n}\frac{1}{n'}\sum_{j=1}^{n'}(y_{i_j} - \bar{Y})(x_{i_j} - \bar{X}) + \frac{n'(n - 1)}{n(n' - 1)}(\bar{x}' - \bar{X})(\bar{y}' - \bar{Y}).$$

If

$$z_i = (x_i - \bar{X})(y_i - \bar{Y})$$

and

$$S(\mathscr{X}\mathscr{X}\mathscr{Y}) = \sum_{i=1}^{N}(x_i - \bar{X})^2(y_i - \bar{Y})/(N - 1),$$

then

$$\mathscr{E}(\bar{x}' - \bar{X})\sum_{j=1}^{n'}(x_i - \bar{X})(y_i - \bar{Y})/n' = \mathscr{E}(\bar{x}' - \bar{X})\bar{z}' = \frac{N - n'}{N}\frac{1}{n'}S(\mathscr{X}\mathscr{X}\mathscr{Y}),$$

and, from section 2,

$$\mathscr{E}(\bar{x}' - \bar{X})^2(\bar{y}' - \bar{Y}) = \frac{N - n'}{N} \frac{1}{(n')^2} \frac{N - 2n'}{N - 2} S(\mathscr{X}\mathscr{X}\mathscr{Y}).$$

Therefore,

$$\frac{1}{\bar{X}^2} \mathscr{E}(\bar{x}' - \bar{X})\bar{y}(\bar{x} - \bar{X})$$

$$= \frac{1}{\bar{X}^2} \frac{n' - n}{n' - 1} \frac{1}{n} \mathscr{E}(\bar{x}' - \bar{X}) \sum_{j=1}^{n'} (x_{i_j} - \bar{X})(y_{i_j} - \bar{Y})/n'$$

$$+ \frac{1}{\bar{X}^2} \frac{n'(n - 1)}{n(n' - 1)} \mathscr{E}(\bar{x}' - \bar{X})^2(\bar{y}' - \bar{Y}) + \frac{\bar{Y}}{\bar{X}^2} \mathscr{E}(\bar{x}' - \bar{X})^2$$

$$= \frac{1}{\bar{X}^2} \left(\frac{n' - n}{n' - 1} \frac{1}{n} \frac{N - n'}{N} \frac{1}{n'} + \frac{n'(n - 1)}{n(n' - 1)} \frac{N - n'}{N} \frac{1}{(n')^2} \frac{N - 2n'}{N - 2} \right) S(\mathscr{X}\mathscr{X}\mathscr{Y})$$

$$+ \frac{R}{\bar{X}} \frac{N - n'}{N} \frac{1}{n'} S^2(\mathscr{X}) \approx \frac{1}{\bar{X}} \frac{N - n'}{N} \frac{1}{n'} (RS^2(\mathscr{X}) + \frac{1}{n\bar{X}} S(\mathscr{X}\mathscr{X}\mathscr{Y})).$$

If the second expression in the bracket is negligible compared with the first, the above expansion for the expected value of $\hat{R}(\bar{x}' - \bar{X})$ in case (i) yields approximately

$$\frac{N - n'}{N} \frac{1}{n'} \frac{1}{\bar{X}} \{S(\mathscr{Y}\mathscr{X}) - RS^2(\mathscr{X})\}.$$

Consequently, under (ii) the first order approximation to the bias of our estimator is

$$\frac{N - n}{N} \frac{1}{n} \frac{1}{\bar{X}} (RS^2(\mathscr{X}) - S(\mathscr{Y}\mathscr{X})),$$

and under (i)

$$\frac{n' - n}{n'} \frac{1}{n} \frac{1}{\bar{X}} (RS^2(\mathscr{X}) - S(\mathscr{Y}\mathscr{X})).$$

As a check, note that, if in the latter expression $n' = N$, we get the former expression; if $n' = n$, we get 0 (which is correct since in this case the estimator becomes $\bar{y}$). Like in section 7, replacement in these expressions of parameters by estimators leads to an estimator of $\bar{Y}$ of reduced bias.

12.2. VARIANCE

To approximate the variance, we write the deviation from $\bar{Y}$ of our estimator:

$$\frac{\bar{X}}{\bar{x}}(\bar{y} - R\bar{x}) + \frac{y}{x}(\bar{x}' - \bar{X})$$

as

$$(\bar{y} - R\bar{x}) + R(\bar{x}' - \bar{X})$$

plus smaller terms (which, on squaring and taking expected value, will lead to terms of order $1/n^2$). Squaring and taking the expected value, this gives under (ii),

$$\frac{N-n}{N}\frac{1}{n}(S^2(\mathscr{Y}) - 2RS(\mathscr{X}\mathscr{Y}) + R^2S^2(\mathscr{X})) + \frac{N-n'}{N}\frac{1}{n'}R^2S^2(\mathscr{X}).$$

By using the expression for $\mathscr{E}\bar{y}(\bar{x}' - \bar{X})$ above (also for $\bar{y} = \bar{x}$), we obtain under (i), if $n^{-1}S(\mathscr{X}\mathscr{X}\mathscr{Y})/(\bar{Y}S^2(\mathscr{X}))$ is negligible, the approximation

$$\frac{N-n}{N}\frac{1}{n}(S^2(\mathscr{Y}) - 2RS(\mathscr{Y}\mathscr{X}) + R^2S^2(\mathscr{X}))$$

$$- \frac{N-n'}{N}\frac{1}{n'}(-2RS(\mathscr{Y}\mathscr{X}) + R^2S^2(\mathscr{X}))$$

$$= \frac{N-n'}{N}\frac{1}{n'}S^2(\mathscr{Y}) + \frac{n'-n}{n'}\frac{1}{n}(S^2(\mathscr{Y}) - 2RS(\mathscr{X}\mathscr{Y}) + R^2S^2(\mathscr{X})).$$

(If $n' = n$, $(\bar{y}/\bar{x})\bar{x}' = \bar{y}$, and so we should get $\{(N-n)/Nn\}S^2(\mathscr{Y})$, which we do. If $n' = N$, our estimator is the ordinary ratio estimator, and our formula reduces to the approximate formula for its variance.) Note that (if $R > 0$) the approximate variance of $(y/x)\bar{x}'$ (like that of $\hat{Y}_R$) is smaller than $V(\bar{y})$ if and only if

$$\rho(\mathscr{X}\mathscr{Y}) > \tfrac{1}{2}(C(\mathscr{X}\mathscr{X})/C(\mathscr{Y}\mathscr{Y}))^{\frac{1}{2}}$$

under (i), and if

$$\rho(\mathscr{X}\mathscr{Y}) > \tfrac{1}{2}(C(\mathscr{X}\mathscr{X})/C(\mathscr{Y}\mathscr{Y}))^{-\frac{1}{2}}(n' + n)/n'$$

under (ii).

13. Difference estimators and their generalizations

13.1. DIFFERENCE ESTIMATOR

Let $\mathscr{Y}$ and $\mathscr{X}$ be correlated characteristics and suppose we wish to estimate Y. If from a simple random sample we obtain the unbiased estimators $\hat{Y} = N\bar{y}$ and $\hat{X} = N\bar{x}$ of Y and X, respectively, then

$$\hat{Y}_D = \hat{Y} + (X - \hat{X})$$

is also an unbiased estimator of Y, and its variance is

$$V(\hat{Y}_D) = V(\hat{Y}) - 2\,\mathrm{Cov}\,(\hat{Y}, \hat{X}) + V(\hat{X}) = V(\hat{Y})\{1 - 2\rho\theta + \theta^2\},$$

where $\rho = \rho(\mathscr{Y}\mathscr{X})$ and $\theta = S(\mathscr{X})/S(\mathscr{Y})$.

From this we see that $\hat{Y}_{\mathrm{D}}$ has a smaller variance than $\hat{Y}$ if ρ exceeds $\frac{1}{2}\theta$. The approximate variance of the ratio estimator $\hat{Y}_R$ is

$$V(\hat{Y})\{1 - 2\rho s\theta_0 + \theta_0^2\},$$

where $\theta_0^2 = C(\mathscr{X}\mathscr{X})/C(\mathscr{Y}\mathscr{Y})$, and $s = 1$ if $R > 0$, and -1 if $R < 0$. So, within this approximation, $\hat{Y}_{\mathrm{D}}$ has a smaller variance than $\hat{Y}_R$ if ρ exceeds $\frac{1}{2}\theta(1 + R)$ (for $R < 1$) or if ρ is less than this (for $R > 1$); in the special case $R > 0$, this means: ρ exceeds $\frac{1}{2}(\theta + \theta_0)$ (for $R < 1$) or ρ less than this (for $R > 1$).

The approximate variance of the *regression estimator* mentioned in section 6 is, for sufficiently large n, $V(\hat{Y})\{1 - \rho^2\}$, so that, within this approximation, $\hat{Y}_{\mathrm{D}}$ has a larger variance than this estimator unless $\rho = \theta$.

EXAMPLE 1. $\mathscr{X}$ is the property $\mathscr{Y}$ but at an earlier time, and X is already known. If not much time lapsed, we often have that either θ or θ_0 is close to 1, and R only fractionally different from 1. (We often have approximately $S(\mathscr{Y}) = |R|^\alpha S(\mathscr{X})$, with $0 \leq \alpha \leq 1$.) Therefore, in these cases:

(i) $\hat{Y}_{\mathrm{D}}$ has a smaller variance than $\hat{Y}$ if ρ well exceeds $\frac{1}{2}$.

(ii) For sufficiently large n, the variance of the regression estimator would be smaller than that of $\hat{Y}_{\mathrm{D}}$ (unless $\rho = \theta$).

REMARK. In practice there are usually changes in the composition of the population from period to period (in addition to changes in $\mathscr{Y}$-values) which have to be taken into account.

EXAMPLE 2. $\mathscr{X}$ is a crude version of $\mathscr{Y}$, i.e., is obtained by a quicker and/or cheaper method of measurement. Then usually θ and θ_0 exceed 1 and R is close to 1. So:

(i) For $\hat{Y}_{\mathrm{D}}$ to be have a variance smaller than that of $\hat{Y}$, ρ has to exceed a number well above $\frac{1}{2}$.

(ii) $\hat{Y}_{\mathrm{D}}$ will usually have a larger variance than $\hat{Y}_R$ when $R < 1$, and a smaller variance when $R > 1$, as long as our formula for approximating the variance of $\hat{Y}_R$ is applicable.

(iii) The variance of the regression estimator would, for sufficiently large n, be smaller than that of $\hat{Y}_{\mathrm{D}}$.

REMARK. An important advantage of $\hat{Y}_{\mathrm{D}}$ is that it is particularly simple, and always unbiased.

13.2. DOUBLE SAMPLING

If in the formula for the difference estimator we use instead of X an estimator from a large sample, of size n', we obtain as variance

$$N^2 \left\{ \frac{N-n'}{Nn'} S^2(\mathscr{Y}) + \frac{n'-n}{n'n} (S^2(\mathscr{Y}) - 2S(\mathscr{X}\mathscr{Y}) + S^2(\mathscr{X})) \right\}$$
$$= N^2 \frac{1}{n} \left[\frac{N-n}{N} S^2(\mathscr{Y}) - \frac{n'-n}{n'} \{2S(\mathscr{X}\mathscr{Y}) - S^2(\mathscr{X})\} \right]$$

if the sample of n is a subsample of the sample of n'; and

$$\frac{N^2}{n} \left\{ \frac{N-n}{N} (S^2(\mathscr{Y}) - 2S(\mathscr{Y}\mathscr{X}) + S^2(\mathscr{X})) + \frac{N-n'}{N} \frac{n}{n'} S^2(\mathscr{X}) \right\}$$
$$= N(N-n) \frac{1}{n} \{S^2(\mathscr{Y}) - 2S^2(\mathscr{Y}\mathscr{X}) + S^2(\mathscr{X})\} + N(N-n') \frac{n}{n'} S^2(\mathscr{X})$$

if the sample of n is drawn independently of the sample of n'.

13.3. GENERALIZATIONS

Note that also $\hat{Y} + k(X - \hat{X})$ is unbiased if k is a constant or is uncorrelated with $\hat{X}$. Its variance† is

$$V(\hat{Y}) - 2k \operatorname{Cov}(\hat{Y}, \hat{X}) + k^2 V(\hat{X}).$$

The optimum value of k is evidently $S(\mathscr{X}\mathscr{Y})/S^2(\mathscr{X})$.

The *regression estimator* uses for k the analogous function of the sample, namely

$$\frac{\sum_{i=1}^{N} a_i(y_i - \bar{y})(x_i - \bar{x})}{\sum_{i=1}^{N} a_i(x_i - \bar{x})^2};$$

it is not in general unbiased.‡

† In all the results above $S(\mathscr{X}\mathscr{Y})$ has then to be multiplied by k and $S^2(\mathscr{X})$ by k^2.

‡ If, in analogy with the d_i in the case of ratio estimation, we define e_i as the residuals from the (population) regression of $\mathscr{Y}$ on $\mathscr{X}$, we can show, by an elaboration of the argument given in section 7.4 of COCHRAN, Sampling Techniques (Wiley, 2nd ed., 1963), that the bias of the regression estimator is

$$-\frac{N-n}{N} \frac{1}{n} S(\mathscr{E}\mathscr{X}\mathscr{X})/S^2(\mathscr{X}) + O(n^{-3/2}),$$

and that its variance is

$$\frac{N-n}{N} \frac{1}{n} S^2(\mathscr{E}) + O(n^{-3/2}).$$

(*continued on opposite page*)

If we have a good independent estimator of this k, we can by its use improve the difference estimator.

EXAMPLE. Consider the Example 2 above and let θ be known to be between 2 and 4. Since $\frac{1}{2}\theta$ is at least 1, the ordinary difference estimator and the ratio estimator are not recommended. Let $S(\mathscr{Y}\mathscr{X})$ be between 0.7 and 0.9. Since

$$S(\mathscr{Y}\mathscr{X})/S(\mathscr{X}\mathscr{X}) = \rho/\theta,$$

we would take for k some number between 0.18 and 0.45, say 0.3. Then

$$V\{\hat{Y} + k(X - \hat{X})\} = \{N(N - n)/n\}S^2(\mathscr{Y})\{1 - 2k\rho\theta + k^2\theta^2\}$$

would be between 0.28 and 0.76 times $\{N(N - n)/n\}S^2(\mathscr{Y})$.

On the other hand if θ is known to be between 1.2 and 1.6 and $\rho(\mathscr{Y}\mathscr{X})$ between 0.85 and 0.95, there is sense in comparing the case $k = 1$ and $k = 0.65$ (say). For the former the variance of the estimator will be a multiple of $\{N(N - n)/n\}S^2(\mathscr{Y})$ lying between 0.16 and 0.84, for the latter between 0.13 and 0.37.

Note that, even if there is a mistake in these ranges for $\rho(\mathscr{Y}\mathscr{X})$ or θ, the estimators are unbiased.

If we use $\hat{R}$ for k, the estimator becomes $\hat{Y}_R$, the ratio estimator of Y. Note that if we substitute $k = R$ in $V\{\hat{Y} + k(X - \hat{X})\}$ we obtain $V(\hat{Y}_R)$, and if we substitute $k = S(\mathscr{Y}\mathscr{X})/S^2(\mathscr{X})$ we obtain the usual approximation for the variance of the regression estimator. It is also readily seen that in general in

---

If ρ'' denotes the correlation coefficient between e_i and $(x_i - \bar{X})^2$, and γ_2 the kurtosis of $\mathscr{X}$ (3 less than the fourth moment divided by the fourth power of the standard deviation),

$$S(\mathscr{E}\mathscr{X}\mathscr{X})/S^2(\mathscr{X}) = \rho''S(\mathscr{E})(\gamma_2 + 2)^{\frac{1}{2}},$$

so that we get for the bias of the regression estimator relative to its standard deviation

$$-\left(\frac{N - n}{N}\frac{1}{n}\right)^{\frac{1}{2}} \rho''(\gamma_2 + 2)^{\frac{1}{2}} + O(n^{-\frac{3}{4}}).$$

To obtain in advance of sampling a notion about the magnitude of this expression, it may help to write it

$$-\left(\frac{N - n}{N}\frac{1}{n}\right)^{\frac{1}{2}} \{\rho'''(\gamma_2 + 2)^{\frac{1}{2}} - \rho\gamma_1\}(1 - \rho^2)^{-\frac{1}{2}} + O(n^{-\frac{3}{4}}),$$

where ρ'' is the correlation coefficient between y_i and $(x_i - \bar{X})^2$, and γ_1 is the skewness of $\mathscr{X}$ (the third moment divided by the third power of the standard deviation, which is 0 in the normal case but often is much larger). Since we apply the regression estimator if ρ^2 is large, the bias can easily be sizeable in absolute value, especially when the distribution of $\mathscr{X}$ is very skew, unless n is very large. Moreover, the contribution of the terms of order $n^{-\frac{3}{4}}$ may not be negligible.

Example 2 it is best to take for k a value less than 1 ($S(\mathscr{Y}\mathscr{X})/S^2(\mathscr{X}) = \rho/\theta$ is less than 1 for $\theta > 1$); in that case the variance of $\hat{Y}_D$ can become less than that of $\hat{Y}_R$ and close to that of the regression estimator.

As in section 11 one can also consider using *more than one auxiliary variable* with weights, e.g., two auxiliary characteristics, $\mathscr{X}_1$ and $\mathscr{X}_2$; the value of W minimizing the variance of

$$\hat{Y} + Wk_1(X_1 - \hat{X}_1) + (1 - W)k_2(X_2 - \hat{X}_2)$$

is exactly that for the bivariate ratio estimator with Cov ($\hat{Y}_{iR}$, $\hat{Y}_{jR}$) replaced by Cov ($\hat{Y}_{iD}$, $\hat{Y}_{jD}$) (writing $\hat{Y}_{iD} = \hat{Y} + k_i(X_i - \hat{X}_i)$).

Another generalization: $\hat{Y}$ or $\hat{X}$ (or both) in the formula are estimators other than $N\bar{y}$ or $N\bar{x}$, respectively.

13.4. EXAMPLE

Take the example used to illustrate the bivariate ratio, $\hat{Y}_{MR}$, estimator of $\bar{Y}$, and put $k_1 = k_2 = 1$. W becomes

$$\frac{S^2(\mathscr{X}_2) - S(\mathscr{X}_1\mathscr{X}_2) + S(\mathscr{Y}\mathscr{X}_1) - S(\mathscr{Y}\mathscr{X}_2)}{S^2(\mathscr{X}_1) + S^2(\mathscr{X}_2) - 2S(\mathscr{X}_1\mathscr{X}_2)},$$

which is 2.36, and which we estimate by substituting in the formula $s^2(\mathscr{X}_1)$ for $S^2(\mathscr{X}_1)$ etc.; this gives $w = 2.59$. Then the bivariate difference estimator takes on the value

$$189\,600 + 2.59 \times (148\,200 - 169\,300) + (-1.59)(142\,000 - 164\,300) = 170\,400$$

and has estimated variance

$$\frac{N-n}{N}\frac{1}{n}\{s^2(\mathscr{Y}) + w^2s^2(\mathscr{X}_1) + (1-w)^2s^2(\mathscr{X}_2) - 2ws(\mathscr{Y}\mathscr{X}_1) - 2(1-w)s(\mathscr{Y}\mathscr{X}_2) + 2w(1-w)s(\mathscr{X}_1\mathscr{X}_2)\}$$

equal to

$$\{(195 - 50)/(195 \times 50)\} \times 416\,217\,000 = 6\,189\,900,$$

so the estimated standard error is 2 490. Compare this with the results on ratio estimators given previously and with the difference estimator using $\mathscr{X}_1$ only (with value 189 600 + (148 200 − 169 300) = 16 500), which has an estimated variance of

$$\frac{N-n}{N}\frac{1}{n}\{s^2(\mathscr{Y}) - 2s(\mathscr{Y}\mathscr{X}_1) + s^2(\mathscr{X}_1)\}$$

equal to

$$\{(195 - 50)/(195 \times 50)\}(1\,253\,267\,000) = 18\,638\,300,$$

so an estimated standard error of 4 320. (Unfortunately, there appear to be printing errors in the data given in the source mentioned in section 11; as a result the values of the sample variances and covariances had to be reconstructed from the given summary statistics $c(\mathscr{Y}\mathscr{Y})$ etc. and $\bar{y}$ etc., which affects the outcome of the above calculations, especially the value of w.)

There appears to be a considerable reduction in the standard error due to the use of two auxiliary variables, both for the ratio and for the difference estimator.

The *regression* estimator using $\mathscr{X}_1$, has the value

$$189\,600 + 1.067 \times (148\,200 - 169\,300) = 167\,100.$$

Its standard error is estimated at $25\,470 \times 0.093\,8 = 2\,390$ (using an estimated value of the correlation between $\mathscr{Y}$ and $\mathscr{X}_1$ of 0.990 5).

13.5. MULTIPLE REGRESSION ESTIMATOR

It is also possible to consider a multiple regression estimator (here: using $\mathscr{X}_1$ and $\mathscr{X}_2$). The formula for its standard error differs from that of the one for the standard error of the regression estimator with a single auxiliary variable in that the squared correlation coefficient is replaced by the squared multiple correlation coefficient.

In our example the standard error of the multiple regression estimator comes out as $25\,470 \times 0.005\,2 = 130$. (The use of reconstructed statistics may lead to larger discrepancies in this estimator than in some previously given results.) The multiple correlation coefficient comes out as 0.999 94, a figure which is exceptionally high for surveys.

The multiple regression estimator should, moreover, be used with the greatest care when the regressors (here: $\mathscr{X}_1$ and $\mathscr{X}_2$) are highly correlated, as is the case in our example. For one can readily show that the term of order $1/n$ in its *bias* is

$$\frac{N-n}{N}\,\frac{1}{n}\,\frac{1}{1-\rho^2(\mathscr{X}_1\mathscr{X}_2)}\,[\{2S_{e12}/(S(\mathscr{X}_1)(S\mathscr{X}_2)\rho^{-1}(\mathscr{X}_1\mathscr{X}_2))\}$$

$$-\,(S_{e11}/S^2(\mathscr{X}_1)) - (S_{e22}/S^2(\mathscr{X}_2))],$$

where, e.g.,

$$S_{e12} = \frac{1}{N-1}\,\Sigma e_i(x_{1i} - \bar{X}_1)(x_{2i} - \bar{X}_2)$$

and (with B_1 and B_2 standing for the population regression coefficients)

$$e_i = y_i - \bar{Y} - B_1(x_{1i} - \bar{X}_1) - B_2(x_{2i} - \bar{X}_2).$$

[This is a generalization of the result shown in Cochran's book for the bias of the simple regression estimators:

$$-\frac{1}{n}S_e(\mathscr{X}\mathscr{X})/S^2(\mathscr{X})$$

with

$$e_i = y_i - \bar{Y} - B(x_i - \bar{X}).]$$

Indeed, in our example, the multiple regression estimator has the value

$$189\,600 + 0.011\,6 \times (148\,200 - 169\,300) + (-0.006\,9)(142\,000 - 164\,300) = 189\,500,$$

which differs from $\bar{Y} = 169\,900$ by a very large multiple of the estimated standard error (130)!

14. Difference and ratio of ratio estimators for two domains in simple random sampling

We consider two *nonoverlapping* domains and use the notation and assumptions introduced in the sections on domains; for example $\mathscr{Y}''$ will denote the characteristic $\mathscr{Y}$ for units in domain 2, and will be 0 for other units.† The units are, for convenience of notation, supposed to be arranged in the population according to domain.

We shall first find an approximate formula for the covariance of ratios. This can most easily be obtained by replacing, as in section 9 of Chapter II, $(\bar{y}/\bar{x}) - (\bar{Y}/\bar{X})$ by $(\bar{y} - \bar{X}^{-1}\bar{Y}\bar{x})/\bar{X}$. In particular, if

$$c = (N - n)/(N - 1)$$

for sampling without replacement and 1 for sampling with replacement,

$$\text{Cov}\left(\frac{y'}{x'}, \frac{y''}{x''}\right) \approx \frac{c}{n}\frac{1}{\bar{X}'\bar{X}''}\Sigma\, d_i'\, d_i''/N = 0,$$

$$\text{Cov}\left(\frac{y'}{x}, \frac{y''}{x}\right) \approx \frac{c}{n}\frac{1}{\bar{X}^2}\left\{\sigma(\mathscr{Y}'\mathscr{Y}'') - \frac{\bar{Y}'}{\bar{X}}\sigma(\mathscr{X}\mathscr{Y}'') - \frac{\bar{Y}''}{\bar{X}}\sigma(\mathscr{X}\mathscr{Y}') + \frac{\bar{Y}'\bar{Y}''}{\bar{X}^2}\sigma^2(\mathscr{X})\right\}$$

$$= \frac{c}{n}\frac{1}{N\bar{X}^3}\{-\bar{Y}'\Sigma x_i y_i'' - \bar{Y}''\Sigma x_i y_i' + \bar{Y}'\bar{Y}''\bar{X}^{-1}\Sigma x_i^2\}.$$

† In the following we thus use, e.g., $\mathscr{D}'$ and $\mathscr{D}''$, where $d_i' = y_i' - R'x_i'$, etc. (for the case of different denominators in the two domains). If we denote $\mathscr{D}' - \mathscr{D}''$ by $\mathscr{D}^*$, we can immediately get expressions for the variance of the difference of ratios or of ratio estimators in these terms, and for estimators of these variances, though not always in the most convenient form. The dispersion of the d_i' in the first domain (i.e., for $i = 1, \ldots, N_1$) will be denoted by $\sigma^2(\mathscr{D}, 1)$, whereas $\sigma^2(\mathscr{D}')$ represents the dispersion of d_i' in the entire population (i.e., for $i = 1, \ldots, N$).

When $x_i = 0$ or 1, being 1 for all units in the first as well as the second domain, the latter becomes

$$-\frac{c}{n}\frac{1}{\bar{X}^3}\bar{Y}'\bar{Y}''.$$

Now

$$V\left(\frac{y'}{x'}\right) \approx \frac{c}{n}\frac{1}{\bar{X}'^2}\sigma^2(\mathscr{D}') = \frac{c}{n}\frac{1}{\bar{X}'^2}\frac{N_1}{N}\sigma^2(\mathscr{D}, 1) = \frac{c}{n}\frac{1}{\bar{X}_1^2}\frac{N}{N_1}\sigma^2(\mathscr{D}, 1),$$

which, in case $x_i' = 1$ for all units in the first domain, becomes

$$V(\bar{y}_1) \approx \frac{N}{n}c\frac{1}{N_1}\sigma_1^2.$$

Therefore we get in general

$$V\left(\frac{y'}{x'} - \frac{y''}{x''}\right) \approx \frac{c}{n}\left\{\frac{1}{\bar{X}_1^2}\frac{N}{N_1}\sigma^2(\mathscr{D}, 1) + \frac{1}{\bar{X}_2^2}\frac{N}{N_2}\sigma^2(\mathscr{D}, 2)\right\},$$

$$V\left(\frac{\bar{y}_1}{\bar{x}_1}\bar{X}_1 - \frac{\bar{y}_2}{\bar{x}_2}\bar{X}_2\right) \approx \frac{c}{n}\left\{\frac{N}{N_1}\sigma^2(\mathscr{D}, 1) + \frac{N}{N_2}\sigma^2(\mathscr{D}, 2)\right\},$$

and in particular

$$V(\bar{y}_1 - \bar{y}_2) \approx \frac{N}{n}c\left\{\frac{1}{N_1}\sigma_1^2 + \frac{1}{N_2}\sigma_2^2\right\}.$$

Similarly we obtain

$$v\left(\frac{y'}{x'} - \frac{y''}{x''}\right) = c'\frac{n}{n-1}\left(\frac{1}{\bar{x}_1^2}\frac{1}{n_1}s^2(\hat{\mathscr{D}}, 1)\right)\frac{n_1 - 1}{n_1} + \frac{1}{\bar{x}_2^2}\frac{1}{n_2}s^2(\hat{\mathscr{D}}, 2)\frac{n_2 - 1}{n_2}\Bigg),$$

$$v\left(\frac{\bar{y}_1}{\bar{x}_1}\bar{X}_1 - \frac{\bar{y}_2}{\bar{x}_2}\bar{X}_2\right) = c'\frac{n}{n-1}\left(\frac{1}{n_1}s^2(\hat{\mathscr{D}}, 1)\frac{n_1 - 1}{n_1} + \frac{1}{n_2}s^2(\hat{\mathscr{D}}, 2)\frac{n_2 - 1}{n_2}\right),$$

where $c' = (N - n)/N$ for sampling without replacement and 1 for sampling with replacement.

On the other hand, we find for differences of ratios with the same denominator:

$$V\left(\frac{y'}{x} - \frac{y''}{x}\right) \approx \frac{c}{n}\frac{1}{\bar{X}^2}\frac{1}{N}\left\{\sum_{i=1}^{N_1+N_2} y_i^2 - 2\frac{Y' - Y''}{X}\sum_{i=1}^{N_1+N_2} x_i(y_i' - y_i'') + \left(\frac{Y' - Y''}{X}\right)^2 \sum_{i=1}^{N} x_i^2\right\}$$

$$= \frac{c}{n}\frac{1}{\bar{X}^2}\frac{1}{N}\sum_{i=1}^{N}\left\{y_i' - y_i'' - \frac{Y' - Y''}{X}x_i\right\}^2,$$

where $(N-1)^{-1}$ times the last sum may be estimated by

$$\sum_{i=1}^{N} a_i \left\{ y_i' - y_i'' - \frac{y'-y''}{x} x_i \right)^2 \Big/ (n-1).$$

If x_i is 0 or 1, being 1 for all units in the first or second domain, we obtain

$$V\left(\frac{y'}{x} - \frac{y''}{x}\right) \approx \frac{c}{n} \frac{1}{\bar{X}^2} \frac{1}{N} \left\{ \sum_{i=1}^{N_1+N_2} y_i^2 - \frac{(Y'-Y'')^2}{X} \right\};$$

on the other hand, if $\mathscr{X} = \mathscr{Y}$, we obtain

$$V\left(\frac{y'}{y} - \frac{y''}{y}\right) \approx \frac{c}{n} \frac{1}{\bar{Y}^2} \frac{1}{N} \sum_{i=1}^{N} \left\{ y_i' - y_i'' - \frac{Y'-Y''}{Y} y_i \right\}^2.$$

If $\mathscr{Y}$ has value 1 or 0 only, and $x_i = 0$ outside the 2 domains but 1 inside, the former formula becomes

$$V\left(\frac{n_1p_1 - n_2p_2}{n_1+n_2}\right) \approx \frac{c}{n} \frac{N}{(N_1+N_2)} \left\{ \frac{N_1P_1}{N_1+N_2}\left(1 - \frac{N_1P_1}{N_1+N_2}\right) + \frac{N_2P_2}{N_1+N_2}\left(1 - \frac{N_2P_2}{N_1+N_2}\right) + 2\frac{N_1N_2P_1P_2}{(N_1+N_2)^2} \right\}.$$

The formula for the estimator of this variance has n, p_1, p_2 and

$$c' \frac{N-1}{N} \frac{n}{n-1}$$

in place of N, P_1, P_2 and c.

The above results immediately give us approximations for the variance of ratio estimators of differences between totals: $X((y'/x) - (y''/x))$, $Y(y'-y'')/y$ and $N_1\bar{y}_1 - N_2\bar{y}_2$. For the exactly unbiased estimator $N(\bar{y}' - \bar{y}'')$, we have

$$V(N\bar{y}' - N\bar{y}'') = \frac{c}{n} N^2\{\sigma^2(\mathscr{Y}') + \sigma^2(\mathscr{Y}'') - 2\sigma(\mathscr{Y}'\mathscr{Y}'')\},$$

with

$$\sigma(\mathscr{Y}'\mathscr{Y}'') = N^{-1}\Sigma(y_i' - \bar{Y}')(y_i'' - \bar{Y}'') = -\bar{Y}'\bar{Y}'' = -\frac{N_1}{N}\frac{N_2}{N}\bar{Y}_1\bar{Y}_2,$$

which is small if N_1/N and N_2/N are small; and we have

$$v(N\bar{y}' - N\bar{y}'') = \frac{c'}{n} N^2\{s^2(\mathscr{Y}') + s^2(\mathscr{Y}'') - 2s(\mathscr{Y}'\mathscr{Y}'')\},$$

with

$$s(\mathscr{Y}'\mathscr{Y}'') = (n-1)^{-1}\Sigma a_i(y_i' - \bar{y}')(y_i'' - \bar{y}'') = -\frac{n}{n-1}\bar{y}'\bar{y}''.$$

Using (i) of section 11.5, Chapter II, we may also write

$$V(N\bar{y}' - N\bar{y}'') = \frac{N^2}{n} c \left\{ \frac{N_1}{N} \sigma_1^2 + \frac{N_2}{N} \sigma_2^2 + \frac{N_1}{N} \bar{Y}_1^2 + \frac{N_2}{N} \bar{Y}_2^2 - \left(\frac{Y' - \bar{Y}''}{N} \right)^2 \right\}$$

$$= \frac{N^2}{n} c \left\{ \frac{1}{N} \sum_{i=1}^{N_1+N_2} y_i^2 - (\bar{Y}' - \bar{Y}'')^2 \right\}.$$

Often the contribution of $(\bar{Y}' - \bar{Y}'')^2$ is relatively negligible.

In case $\mathscr{Y}$ has values 1 and 0, the formula becomes

$$V\left\{ \frac{N}{n} (n_1 p_1 - n_2 p_2) \right\} = N^2 \frac{c}{n} \left\{ \sum \frac{N_i P_i}{N} \left(1 - \frac{N_i P_i}{N} \right) + 2 \frac{N_1 N_2 P_1 P_2}{N^2} \right\}.$$

Compare the formula for

$$V\{(n_1 p_1 - n_2 p_2)/n\}$$

with that given for

$$V\{(n_1 p_1 - n_2 p_2)/(n_1 + n_2)\};$$

one obtains the estimator for the former by the same replacement as we applied to the latter.

REMARK. Cov $(y'/x', y''/x'')$ actually vanishes. This follows from the facts that the conditional covariance of y'/x' and y''/x'', given the samples from the first and second domains, is zero (because of conditional independence), and that the conditional expectations of y'/x' and y''/x'' are the constants (Y'/X') and (Y''/X''), as it is assumed that x' and x'' are assured to be different from zero. For those facts imply that the unconditional covariance of y'/x' and y''/x'' is also 0. Also y' and y'' are conditionally independent, but the conditional expectations of y' and y'' are $n_1 \bar{Y}_1$ and $n_2 \bar{Y}_2$, which are not constants, so that the unconditional covariance of y' and y'' is not zero.

We now readily obtain that both for

$$(y'/x') - (y''/x'')$$

and for

$$(y'/x')\bar{X}_1 - (y''/x'')\bar{X}_2$$

the bias (if any) divided by the standard deviation is bounded by

$$k = ((1/n) - (1/N))^{\frac{1}{2}} \max (C^{\frac{1}{2}}(\mathscr{X}'\mathscr{X}'), C^{\frac{1}{2}}(\mathscr{X}''\mathscr{X}'')),$$

provided the correlation between y'/x' and x' has the same sign as the correlation between y''/x'' and x'',† and by $2^{\frac{1}{2}}k$ in any case;‡ and that for $(y' - y'')/x$ the bias divided by the standard deviation is bounded by

$$((1/n) - (1/N))^{\frac{1}{2}} C^{\frac{1}{2}}(\mathscr{X}\mathscr{X}).$$

The latter formula may lead to a very wide bound for the bias of $(y' - y'')/x$ in cases in which the covariance between y'/x and y''/x is negative and not small in absolute value compared with their variances.

From section 9 we may also obtain estimators of the bias and variance of *ratios* of two domain estimators. If the subscripts in section 9 refer to two non-overlapping domains, the relative covariances relating to two different domains are all -1. Thus we find

$$\frac{\mathscr{E}\bar{y}_1/\bar{y}_2 - \bar{Y}_1/\bar{Y}_2}{\bar{Y}_1/\bar{Y}_2} \approx \frac{N}{n} c\, N_2^{-1}\sigma_2^2/\bar{Y}_2^2,$$

$$V(\bar{y}_1/\bar{y}_2)/(\bar{Y}_1/\bar{Y}_2)^2 \approx \frac{N}{n} c\{N_1^{-1}\sigma_1^2/\bar{Y}_1^2 + N_2^{-1}\sigma_2^2/\bar{Y}_2^2\}.$$

† Denote these correlation coefficients by ρ_1 and ρ_2, and the coefficient of variation of $\bar{x}'$ by k_1, of $\bar{x}''$ by k_2. The bound continues to hold when $\rho_1\rho_2 < 0$, provided $-\rho_1\rho_2$ is sufficiently small. This is seen as follows. The bias of $(y'/x') - (y''/x'')$ equals $-\rho_1 k_1 d_1 + \rho_2 k_2 d_2$, where $d_1^2 = V(y'/x')$, $d_2^2 = V(y''/x'')$. Write $r_1 = |\rho_1|$, $r_2 = |\rho_2|$, and suppose without loss of generality that $r_1 \geq r_2$. Then $(-\rho_1 k_1 d_1 + \rho_2 k_2 d_2)^2 \leq k^2\{r_1^2(d_1^2 + d_2^2) + 2r_1 r_2 d_1 d_2\} \leq k^2(d_1^2 + d_2^2)(r_1^2 + r_1 r_2) = k^2 V\{(y'/x') - (y''/x'')\}(r_1^2 + r_1 r_2)$. The last factor ≤ 1 for any $r_2 \leq 1$ if $r_1 \leq \frac{1}{2}(5^{\frac{1}{2}} - 1) = 0.618$, and in general if $-\rho_1\rho_2 \leq 1 - \max(\rho_1^2, \rho_2^2)$. (Thus, for $\max(|\rho_1|, |\rho_2|)$ as large as 0.90, the inequality still holds if $-0.90 \leq -\rho_1\rho_2 \leq 0.19$.) It seems most unlikely that in any case of practical interest this condition is not fulfilled.

‡ Referring to the previous footnote, $r_1^2 + r_1 r_2 \leq 2r_1^2 \leq 2$.

CHAPTER III

STRATIFIED SAMPLING

1. Introduction

Often we divide the population into several disjoint parts (named strata) and draw, from *each* stratum *separately*, a sample – independently from the sampling in the other strata. In some of the strata the units of sampling may themselves be clusters.

The reasons for stratified sampling may be one or more of the following:

(a) The organization conducting the survey has several branches, each of which is in full administrative control of recruiting, instruction and supervision of interviewers in its region.

(b) The characteristics subject to the research vary in different parts of the population (the population is very "heterogeneous"), but vary much less within these parts (the strata are more "homogeneous"). For instance, consumption habits of men serving in the army, people living in hotels, homes for the aged, etc. are quite different from those of people living with their families. Therefore it is desirable to draw separate samples for soldiers, hotel guests, etc. (Here and in many other instances, the reason for stratified sampling is not only (b) but also (d).)

(c) Not only the characteristics under study, but also related characteristics of the units that can be utilized for efficient design or estimation, are different in different parts of the population.†

† Also ratio and related estimators as well as "pps" and "ppes" sampling defined in section 3 of Chapter VII, utilize such related characteristics. Moreover these methods are not mutually exclusive (see, e.g., section 11 of Chapter VII). Note, however, that, when several characteristics are of interest, stratification and allocation of the number of units to be sampled to the different strata are matters of compromise (see the third footnote of section 8); whereas we can choose different related characteristics for estimating different characteristics of interest by ratio estimators (or use the same related characteristics with different k for difference or regression estimators). The efficiency of estimating different characteristics of interest may also be differently affected by simple *vs* "pps" or "ppes" sampling (and by different choices of measures of size), but usually much less so than by the choice of stratification cum allocation.

Because of the reasons stated under (b) and (c), the variances of estimators in stratified samples are usually less than those in unstratified samples of the same size (see section 8). Also stratification usually enables us to obtain better estimators of the variances of our estimators.

(d) Survey costs and practical problems connected with sampling vary in different parts of the population.

(e) Partial lists of the population are available; it may then be best to draw a part of the sample directly from lists, and the rest in the field. More generally, prior information available is not the same in different parts of the population.

(f) Effective stratification offers some protection against obtaining by chance a sample which gives very poor estimates – see section 12.1.

(g) Often it is required to obtain from a survey not only estimates of certain overall parameters, such as Y, but also estimates for a set A of subdivisions of the population which are to have (approximately) specified standard errors. Suppose these requirements imply that the sample should contain at least a fraction f_h from subpopulation h (of size N_h, with $\Sigma N_h = N$); then a simple random sample would have to be of size at least $N \max \{f_h(h \in A)\}$, which may much exceed the size needed for sufficiently precise estimation of the overall parameters.

2. Estimation of $\bar{Y}$ by $\bar{y}_{\text{st}}$

2.1. THE ESTIMATOR AND ITS VARIANCE

Suppose we divide the population into L strata, where the respective sizes are N_h ($h = 1, . . ., L$). Since the sampling in the strata is done separately and independently in each of the strata, we may use the formulae previously obtained to obtain estimators of the components of the mean and variance from the different strata.† Hence, if $\bar{Y}_h$ is the mean of the characteristic $\mathscr{Y}$ in the hth stratum, we may estimate it by $\bar{y}_h$ – the mean of $\mathscr{Y}$ for the sample of size $n_h > 0$ drawn in the h stratum.

The general mean of $\mathscr{Y}$ in the population is

$$\bar{Y} = \sum_{h=1}^{L} N_h \bar{Y}_h / N,$$

where

$$N = \sum_{h=1}^{L} N_h.$$

We shall estimate $\bar{Y}$ by

$$\bar{y}_{\text{st}} = \Sigma N_h \bar{y}_h / N$$

(st – short for "stratified").

We have also

$$V(\bar{y}_{\text{st}}) = \sum_{h=1}^{L} N_h^2 V(\bar{y}_h)/N^2 \quad \left(V(\bar{y}_h) = \frac{S_h^2}{n_h} \frac{N_h - n_h}{N_h} \right),$$

† In this chapter we assume simple random sampling without replacement in each stratum.

and

$$v(\bar{y}_{st}) = \sum_{h=1}^{L} N_h^2 v(\bar{y}_h)/N^2 \qquad \left(v(\bar{y}_h) = \frac{s_h^2}{n_h}\frac{N_h - n_h}{N_h}\right).$$

Here

$$S_h^2 = \sum_{i=1}^{N_h} (y_{hi} - \bar{Y}_h)^2/(N_h - 1)$$

if $N_h > 1$, and 0 otherwise; and s_h^2 is the corresponding sample expression when $n_h > 1$, 0 when $n_h = N_h = 1$, and undefined if $n_h = 1$ and $N_h > 1$.

With allocation of the sample to the strata proportional to the N_h, these formulas reduce to

$$\frac{N-n}{N}\frac{1}{n}\sum_{h=1}^{L}\frac{N_h}{N}S_h^2, \qquad \frac{N-n}{N}\frac{1}{n}\sum_{h=1}^{L}\frac{N_h}{N}s_h^2.$$

2.2. EXAMPLE

Consider the example of section 9 of Chapter II with stratification according to total expenditures.

$\mathscr{S}_1$: total expenditures not exceeding 600 pounds,
$\mathscr{S}_2$: total expenditures exceeding 600 pounds but not exceeding 650 pounds,
$\mathscr{S}_3$: total expenditures exceeding 650 pounds but not exceeding 700 pounds,
$\mathscr{S}_4$: total expenditures exceeding 700 pounds.

It is given that $N_1 = 20$, $N_2 = 20$, $N_3 = 25$, $N_4 = 15$. Desired are an estimate of $\bar{Y}$ and of its variance, and of $\bar{X}$ and of its variance.

x_{1i}	y_{1i}	x_{2i}	y_{2i}	x_{3i}	y_{3i}	x_{4i}	y_{4i}
580	300	620	332	690	383	720	412
590	302	640	356	660	340	740	400
600	318	610	330	670	360	710	420
600	325			670	381		
580	290			680	362		
2 950	1 535	1 870	1 018	3 370	1 826	2 170	1 232

$\bar{x}_1 = 590$ $\bar{y}_1 = 307$ $\bar{x}_2 = 623\frac{1}{3}$ $\bar{y}_2 = 339\frac{1}{3}$ $\bar{x}_3 = 674$ $\bar{y}_3 = 365\frac{1}{5}$ $\bar{x}_4 = 723\frac{1}{3}$ $\bar{y} = 410\frac{2}{3}$

$s_1^2(\mathscr{X}) = \frac{1}{4}[(-10)^2 + 0^2 + 10^2 + 10^2 + (-10)^2] = 100,$
$s_2^2(\mathscr{X}) = \frac{1}{2}[0^2 + 20^2 + (-10)^2 - \frac{10^2}{3}] = 233\frac{1}{3}$ (using arbitrary origin 620),

$$s_3^2(\mathscr{X}) = \tfrac{1}{4}[16^2 + (-14)^2 + (-4)^2 - (-4)^2 + 6^2] = 130,$$
$$s_4^2(\mathscr{X}) = \tfrac{1}{2}[0^2 + 20^2 + (-10)^2 - \tfrac{10^2}{3}] = 233\tfrac{1}{3},$$
$$s_1^2(\mathscr{Y}) = \tfrac{1}{4}[0^2 + 2^2 + 18^2 + 25^2 + (-10)^2 - \tfrac{35^2}{5}] = 202,$$
$$s_2^2(\mathscr{Y}) = \tfrac{1}{2}[0^2 + 24^2 + (-2)^2 - \tfrac{22^2}{3}] = 209\tfrac{1}{3},$$
$$s_3^2(\mathscr{Y}) = \tfrac{1}{4}[23^2 + (-20)^2 + 0^2 + (21)^2 + 2^2 - \tfrac{26^2}{5}] = 309\tfrac{7}{10},$$
$$s_4^2(\mathscr{Y}) = \tfrac{1}{2}[12^2 + 0^2 + 20^2 - \tfrac{32^2}{3}] = 101\tfrac{1}{3}.$$

For future reference we also compute $s_h(\mathscr{X}\mathscr{Y})$

$$s_1(\mathscr{X}\mathscr{Y}) = \tfrac{1}{4}\Big[(-10)(0) + (0)(2) + (10)(18) + (10)(25) + (-10)(-10) - \frac{(0)(35)}{5}\Big] = 132\tfrac{1}{2},$$

$$s_2(\mathscr{X}\mathscr{Y}) = \tfrac{1}{2}\left[(0)(0) + (20)(24) + (-10)(-2) - \frac{(10)(22)}{3}\right] = 213\tfrac{1}{3},$$

$$s_3(\mathscr{X}\mathscr{Y}) = \tfrac{1}{4}\Big[(16)(23) + (-14)(-20) + (-4)(0) + (-4)(21) + (6)(2) - \frac{(0)(26)}{5}\Big] = 144,$$

$$s_4(\mathscr{X}\mathscr{Y}) = \tfrac{1}{2}\left[(0)(12) + (20)(0) + (-10)(20) - \frac{(10)(32)}{3}\right] = -153\tfrac{1}{3},$$

$$\bar{x}_{st} = (1/80)[(20)(590) + (20)(623\tfrac{1}{3}) + (25)(674) + (15)(723\tfrac{1}{3})] = 649.58,$$
$$\bar{y}_{st} = (1/80)[(20)(307) + (20)(339\tfrac{1}{3}) + (25)(365\tfrac{1}{5}) + (15)(723\tfrac{1}{3})] = 352.71,$$
$$v(\bar{x}_{st}) = (1/80)^2[(60)(100) + (113\tfrac{1}{3})(233\tfrac{1}{3}) + (100)(130) + (60)(233\tfrac{1}{3})] = 9.29,$$
$$v(\bar{y}_{st}) = (1/80)^2[(60)(202) + (113\tfrac{1}{3})(209\tfrac{1}{3}) + (100)(309\tfrac{7}{10}) + (60)(101\tfrac{1}{3})] = 11.31.$$

Compare these results with

$$\bar{x} = 647.50, \quad \bar{y} = 350.69,$$
$$v(\bar{x}) = 170.42, \quad v(\bar{y}) = 81.48.$$

Also compare these with $\hat{Y}_R = 352.04$ (using $\bar{X} = 650$) and $v(\hat{Y}_R) = 12.51$. This stratification by $\mathscr{X}$ is therefore estimated to be somewhat more effective than ratio estimation. One can achieve still greater precision by combining stratification and ratio estimation; see section 6.

3. Illustration of the sampling distribution of $\bar{y}_{st}$ and $v(\bar{y}_{st})$

We take again the population

$$\mathscr{P} = \{u_1, u_2, u_3, u_4\},$$

and the property $\mathscr{Y}$ with values $y_1 = 3$, $y_2 = 4$, $y_3 = 3$, $y_4 = 5$, as in the example of Chapter II, section 6. Recall that there we found that $\bar{Y} = 3\frac{3}{4}$ and $S^2 = \frac{11}{12}$, and that for simple random samples of two taken without replacement from the population $\mathscr{P}$, $V(\bar{y}) = \frac{11}{48}$. We also gave an unbiased estimator $v(\bar{y})$ of $V(\bar{y})$ based on the sample. Now let us divide $\mathscr{P}$ into two strata, $\mathscr{S}_1$ and $\mathscr{S}_2$, of sizes N_1 and N_2 respectively, and let us obtain the samples of size two from $\mathscr{P}$ by drawing simple random samples of size one from each stratum separately. The average $\bar{y}_{st}$ is the weighted average of the two observations, using as weights N_1/N and N_2/N, the relative frequencies in the strata. Let us compute $V(\bar{y}_{st})$ directly, and let us compare the results with the formula

$$V(\bar{y}_{st}) = \left(\frac{N_1}{N}\right)^2 \frac{S_1^2}{n_1} \frac{N_1 - n_1}{N_1} + \left(\frac{N_2}{N}\right)^2 \frac{S_2^2}{n_2} \frac{N_2 - n_2}{N_2}.$$

Since here is only one observation in each stratum, it is not possible to compute s_1^2 and s_2^2; indeed, there does not exist a reasonable estimator of $V(\bar{y}_{st})$ which is based on the sample. Nonetheless, as we shall see in the next section, it is possible to give a reasonable estimator if some approximate evidence permits making suitable assumptions.

(1) Let these two strata be

$$\mathscr{S}_1 = \{u_1, u_2, u_3\}, \quad \mathscr{S}_2 = \{u_4\};$$

then

$$\bar{Y}_1 = 3\tfrac{1}{3}, \quad \bar{Y}_2 = 5,$$

$$S_1^2 = \tfrac{1}{2}\{(\tfrac{1}{3})^2 + (\tfrac{2}{3})^2 + (\tfrac{1}{3})^2\} = \tfrac{1}{3}, \quad S_2^2 = 0$$

(subpopulation of 1 element!),

$$\bar{y}_{st} = \tfrac{3}{4}\bar{y}_1 + \tfrac{1}{4}\bar{y}_2.$$

The formula gives

$$V(\bar{y}_{st}) = (\tfrac{3}{4})^2 \times \tfrac{1}{3} \times \tfrac{2}{3} + (\tfrac{1}{4})^2 \times 0 \times \tfrac{0}{1} = \tfrac{1}{8}.$$

Possible samples	Values of $\mathscr{Y}$	$\bar{y}_{st}$	
$\{u_1, u_4\}$	$\{3, 5\}$	$3\frac{1}{2}$	
$\{u_2, u_4\}$	$\{4, 5\}$	$4\frac{1}{4}$	
$\{u_3, u_4\}$	$\{3, 5\}$	$3\frac{1}{2}$	
		$3\frac{3}{4}$	Average

Direct calculation

Distribution of $\bar{y}_{st}$			$(\bar{y}_{st} - 3\frac{3}{4})^2$	f	$f(\bar{y}_{st} - 3\frac{3}{4})^2$	
$\bar{y}_{st}$	f	$f\bar{y}_{st}$				
$3\frac{1}{2}$	$\frac{2}{3}$	$2\frac{1}{3}$	$\frac{1}{16}$	$\frac{2}{3}$	$\frac{1}{24}$	
$4\frac{1}{4}$	$\frac{1}{3}$	$1\frac{5}{12}$	$\frac{1}{4}$	$\frac{1}{3}$	$\frac{1}{12}$	
	1	$\mathscr{E}\bar{y}_{st} = 3\frac{3}{4}$		1	$V(\bar{y}_{st}) = \frac{1}{8}$	Total

(2) Let the two strata be

$$\mathscr{S}_1 = \{u_1, u_3\}, \quad \mathscr{S}_2 = \{u_2, u_4\};$$
$$\bar{Y}_1 = 3, \quad \bar{Y}_2 = 4\tfrac{1}{2},$$
$$S_1^2 = 0, \quad S_2^2 = \tfrac{1}{2},$$
$$\bar{y}_{st} = \tfrac{1}{2}\bar{y}_1 + \tfrac{1}{2}\bar{y}_2$$

$$V(\bar{y}_{st}) = (\tfrac{1}{2})^2 \times 0 \times \frac{2-1}{2} + (\tfrac{1}{2})^2 \times \tfrac{1}{2} \times \frac{2-1}{2} = \tfrac{1}{16}.$$

Possible samples	Values of $\mathscr{Y}$	$\bar{y}_{st}$	
$\{u_1, u_2\}$	$\{3, 4\}$	$3\frac{1}{2}$	
$\{u_1, u_4\}$	$\{3, 5\}$	4	
$\{u_3, u_2\}$	$\{3, 4\}$	$3\frac{1}{2}$	
$\{u_3, u_4\}$	$\{3, 5\}$	4	
		$3\frac{3}{4}$	Average

Direct calculation

Distribution of $\bar{y}_{st}$			$(\bar{y}_{st} - 3\frac{3}{4})^2$	f	$f(\bar{y}_{st} - 3\frac{3}{4})^2$	
$\bar{y}_{st}$	f	$f\bar{y}_{st}$				
$3\frac{1}{2}$	$\frac{1}{2}$	$1\frac{3}{4}$	$\frac{1}{16}$	$\frac{1}{2}$	$\frac{1}{32}$	
4	$\frac{1}{2}$	2	$\frac{1}{16}$	$\frac{1}{2}$	$\frac{1}{32}$	
	1	$\mathscr{E}\bar{y}_{st} = 3\frac{3}{4}$		1	$V(\bar{y}_{st}) = \frac{1}{16}$	Total

(3) Let the two strata be

$$\mathscr{S}_1 = \{u_1, u_2\}, \quad \mathscr{S}_2 = \{u_3, u_4\};$$
$$\bar{Y}_1 = 3\tfrac{1}{2}, \quad \bar{Y}_2 = 4,$$
$$S_1^2 = \tfrac{1}{2}, \quad S_2^2 = 2,$$
$$\bar{y}_{st} = \tfrac{1}{2}\bar{y}_1 + \tfrac{1}{2}\bar{y}_2,$$

$$V(\bar{y}_{st}) = (\tfrac{1}{2})^2 \times \tfrac{1}{2} \times \frac{2-1}{2} + (\tfrac{1}{2})^2 \times 2 \times \frac{2-1}{2} = \tfrac{5}{16}.$$

Possible samples	Values of $\mathscr{Y}$	$\bar{y}_{st}$	
$\{u_1, u_3\}$	$\{3, 3\}$	3	
$\{u_1, u_4\}$	$\{3, 5\}$	4	
$\{u_2, u_3\}$	$\{4, 3\}$	$3\frac{1}{2}$	
$\{u_2, u_4\}$	$\{4, 5\}$	$4\frac{1}{2}$	
		$3\frac{3}{4}$	Average

Direct calculation

Distribution of $\bar{y}_{st}$					
$\bar{y}_{st}$	f	$f\bar{y}_{st}$	$(\bar{y}_{st} - 3\frac{3}{4})^2$	$f(\bar{y}_{st} - 3\frac{3}{4})^2$	
3	$\frac{1}{4}$	$\frac{3}{4}$	$\frac{9}{16}$	$\frac{9}{64}$	
4	$\frac{1}{4}$	1	$\frac{1}{16}$	$\frac{1}{64}$	
$3\frac{1}{2}$	$\frac{1}{4}$	$\frac{7}{8}$	$\frac{1}{16}$	$\frac{1}{64}$	
$4\frac{1}{2}$	$\frac{1}{4}$	$\frac{9}{8}$	$\frac{9}{16}$	$\frac{9}{64}$	
	1	$\mathscr{E}\bar{y}_{st} = 3\frac{3}{4}$		$V(\bar{y}_{st}) = \frac{5}{16}$	Total

REMARK. In (1) and in (2) the strata are more homogeneous and therefore $V(\bar{y}_{st}) < V(\bar{y}_{srs})$; in (3), for a bad choice of the strata, $V(\bar{y}_{st}) > V(\bar{y}_{srs})$.

4. Collapsed strata

If the number of strata is large, but also if the units sampled within the strata are large clusters which are themselves subsampled, we may have $n_h = 1$ for some of the h's. For such h, $v(\bar{y}_h)$ is not defined. A way of estimating $V(\bar{y}_{st})$ in such a case, and more in general when some of the n_h are very small, is the method of collapsed strata.

Let us first consider the case in which we wish to combine an even number of strata. Assume that these are the first in line ($h = 1, 2, \ldots, 2k$). Let us group the $2k$ strata in pairs, such that in each pair (h, h') we could reasonably suppose that the $N^{-2}(Y_h - Y_{h'})^2$ are small relative to the contributions of strata h and h' to $V(\bar{y}_{st})$. Since

$$V(\bar{y}_h) = \frac{N_h - n_h}{N_h}\frac{1}{n_h}S_h^2, \quad V(\bar{y}_{h'}) = \frac{N_{h'} - n_{h'}}{N_{h'}}\frac{1}{n_{h'}}S_{h'}^2,$$

we obtain

$$V(N_h\bar{y}_h - N_{h'}\bar{y}_{h'}) = N_h^2\left(1 - \frac{n_h}{N_h}\right)\frac{1}{n_h}S_h^2 + N_{h'}^2\left(1 - \frac{n_{h'}}{N_{h'}}\right)\frac{1}{n_{h'}}S_{h'}^2,$$

and therefore

$$\mathscr{E}(N_h\bar{y}_h - N_{h'}\bar{y}_{h'})^2$$
$$= N_h^2\left(1 - \frac{n_h}{N_h}\right)\frac{1}{n_h}S_n^2 + N_{h'}^2\left(1 - \frac{n_{h'}}{N_{h'}}\right)\frac{1}{n_{h'}}S_{h'}^2 + (Y_h - Y_{h'})^2.$$

Hence, roughly, we may estimate

$$N_h^2\left(1 - \frac{n_h}{N_h}\right)\frac{1}{n_h}S_h^2 + N_{h'}^2\left(1 - \frac{n_{h'}}{N_{h'}}\right)\frac{1}{n_{h'}}S_{h'}^2$$

by $(N_h\bar{y}_h - N_{h'}\bar{y}_{h'})^2$, and

$$V(\bar{y}_{st}) = \sum_{(h,h')\leq 2k}\left(N_h^2\frac{N_h - n_h}{N_h}\frac{1}{n_h}S_h^2 + N_{h'}^2\frac{N_{h'} - n_{h'}}{N_{h'}}\frac{1}{n_{h'}}S_{h'}^2\right)\Big/N^2$$
$$+ \sum_{h>2k} N_h^2V(\bar{y}_h)/N^2$$

by

$$\hat{V}_c(\bar{y}_{st}) = \sum_{(h,h')\leq 2k}(N_h\bar{y}_h - N_{h'}\bar{y}_{h'})^2/N^2 + \sum_{h>2k}N_h^2v(\bar{y}_h)/N^2,$$

where the first sum contains k *terms*, corresponding to the k pairs (h, h'). When $N_h = N_{h'}$,

$$(N_h\bar{y}_h - N_{h'}\bar{y}_{h'})^2/N^2 = (N_h/N)^2(\bar{y}_h - \bar{y}_{h'})^2.$$

WARNING. The grouping has to be done without knowledge of the sample $\bar{y}_h$'s. To see this, suppose we had grouped the strata with 1 observation such that not the $(N_h\bar{Y}_h - N_{h'}\bar{Y}_{h'})^2/N^2$, but the $(N_h\bar{y}_h - N_{h'}\bar{y}_{h'})^2/N^2$ are relatively small. The statistic computed with this grouping from the same formula as that for $\hat{V}_c(\bar{y}_{st})$ generally will take a smaller value than $\hat{V}_c(\bar{y}_{st})$ computed with a grouping decided on prior to sampling. Therefore there is a danger of it underestimating $V(\bar{y}_{st})$.

Thus consider the following *example*, in which there are 4 strata, each of size 3, which contribute only one observation each to the sample:

$$\begin{array}{llllll} y_{11} = 11, & y_{12} = 12, & y_{13} = 13, & \text{so} & Y_1 = 36, & S_1^2 = 1; \\ y_{21} = 8, & y_{22} = 13, & y_{23} = 15, & \text{so} & Y_2 = 36, & S_2^2 = 13; \\ y_{31} = 11, & y_{32} = 12, & y_{33} = 17, & \text{so} & Y_3 = 40, & S_3^2 = 10\tfrac{1}{3}; \\ y_{41} = 9, & y_{42} = 13, & y_{43} = 18, & \text{so} & Y_4 = 40, & S_4^2 = 20\tfrac{1}{3}. \end{array}$$

The contribution of these strata to $V(N\bar{y}_{st})$ is

$$9(1 - \tfrac{1}{3})(44\tfrac{2}{3}) = 268.$$

For each point in the sample space we may compute the minimum value of

$$t = 9\{(y_h - y_{h'})^2 + (y_{h''} - y_{h'''})^2\}.$$

E.g., if the sample from the 4 strata consists of u_{11}, u_{21}, u_{31} and u_{41}, the minimum value of t is 9, attained by combining $\mathscr{S}_1$ with $\mathscr{S}_3$ and $\mathscr{S}_2$ with $\mathscr{S}_4$. The average of the minimum values of t over the sample space is found to be $9(1\,222)/3^4 = 135.8$, an underestimate by $268 - 135.8 = 132.2$. On the other hand we obtain the following mean values of t for the 3 possible pairings:

$$\begin{array}{lll} \{\mathscr{S}_1, \mathscr{S}_2\} & \text{and} & \{\mathscr{S}_3, \mathscr{S}_4\}: 268 + (Y_1 - Y_2)^2 + (Y_3 - Y_4)^2 = 268, \\ \{\mathscr{S}_1, \mathscr{S}_3\} & \text{and} & \{\mathscr{S}_2, \mathscr{S}_4\}: 268 + (Y_1 - Y_3)^2 + (Y_2 - Y_4)^2 = 300, \\ \{\mathscr{S}_1, \mathscr{S}_4\} & \text{and} & \{\mathscr{S}_2, \mathscr{S}_3\}: 268 + (Y_1 - Y_4)^2 + (Y_2 - Y_3)^2 = 300. \end{array}$$

Occasionally we need to combine three strata. Let strata h, h', h'' be not much different in their total $\mathscr{Y}$-values, then we may use as contribution to the estimated variance from these strata

$$\{(N_h\bar{y}_h - N_{h'}\bar{y}_{h'})^2 + (N_h\bar{y}_h - N_{h''}\bar{y}_{h''})^2 + (N_{h'}\bar{y}_{h'} - N_{h''}\bar{y}_{h''})^2\}/(2N^2),$$

since its expected value is

$$\sum_{j=h,h',h''} N_j^2\left(1 - \frac{n_j}{N_j}\right)\frac{1}{n_j} S_j^2/N^2$$
$$+ \{(Y_h - Y_{h'})^2 + (Y_h - Y_{h''})^2 + (Y_{h'} - Y_{h''})^2\}/(2N^2).$$

REMARK. HANSEN, HURWITZ and MADOW, Sample Survey Methods and Theory (Wiley, 1953) Vol. II, Chapter 9, section 5, give an estimator of $V(\bar{y}_{st})$ for collapsed strata, which in the case $N_h \neq N_{h'}$ is different from the one given above: the term for a pair of combined strata in their estimator is

$$4\{(1 - W_h)N_h\bar{y}_h - W_hN_{h'}\bar{y}_{h'}\}^2/N^2,$$

with $0 < W_h < 1$, where W_h would usually be chosen to be $N_h/(N_h + N_{h'})$, but may be chosen differently. The authors assume N_h and $N_{h'}$ are not very small and recommend their estimator for the case in which W_h is not very different from $\frac{1}{2}$ (i.e., in which N_h and $N_{h'}$ do not differ much, if W_h is proportional to N_h), since then the bias contributed by the pair (h, h') is close to

$$4N^{-2}\{(1 - W_h)Y_h - W_h Y_{h'}\}^2.$$

But then the bias of the estimator we gave above is also close to this amount when W_h is not far from $\frac{1}{2}$.

More generally, if for any given positive quantities A_h, $A_{h'}$

$$W_h = A_h/(A_h + A_{h'}), \quad W_{h'} = 1 - W_h, \quad \tilde{Y}_h = Y_h/A_h, \quad \tilde{Y}_{h'} = Y_{h'}/A_{h'},$$

we have that $4N^{-2}\{W_{h'}Y_h - W_h Y_{h'}\}^2$ equals

$$N^{-2}A^2(h, h')(\tilde{Y}_h - \tilde{Y}_{h'})^2,$$

where $A(h, h')$ is the harmonic mean of A_h and $A_{h'}$. For arbitrary A_h (and negligible N_h^{-1}, $N_{h'}^{-1}$) it is readily shown that the contribution to the bias neglected above (for the case W_h close to $\frac{1}{2}$) is

$$C_{hh'} = N^{-2}\left[\left\{1 - \frac{A(h, h')}{\frac{1}{2}(A_h + A_{h'})}\right\}(N_h^2 S_h^2 + N_{h'}^2 S_{h'}^2) - \frac{A_h - A_{h'}}{\frac{1}{2}(A_h + A_{h'})}(N_h^2 S_h^2 - N_{h'}^2 S_{h'}^2)\right],$$

where the term in curly brackets is non-negative (since the harmonic mean of positive quantities cannot exceed their arithmetic mean).

There may be occasions when we cannot pair strata with a single observation such that the sum of the $N^{-2}(Y_h - Y_{h'})^2$ is relatively small, but can pair them such that the sum of the $N^{-2}A^2(h, h')(\tilde{Y}_h - \tilde{Y}_{h'})^2 + C_{hh'}$ is relatively small. Then we may want to use the estimator of Hansen *et al.* (who give the more general formula from which we may obtain the contribution from a pooled triple as well).

5. Illustration of the sampling distribution of our estimator of variance for collapsed strata

In the previous examples in (2) and (3) there are only 2 strata. Out of each we have one observation in the sample, which we pair. Since $N_1 = N_2 = 2$, our estimator reduces to

$$\hat{V}_c(\bar{y}_{st}) = \frac{1}{4}(y^{(1)} - y^{(2)})^2,$$

where $y^{(1)}$ is the observation obtained from the first element of the pair, and $y^{(2)}$ from the second.

In the partition (2) we obtain

Possible samples	$\frac{1}{4}(y^{(1)} - y^{(2)})^2$	f	Product
$\{u_1, u_2\}$	$\frac{1}{4}$	$\frac{1}{4}$	$\frac{1}{16}$
$\{u_1, u_4\}$	1	$\frac{1}{4}$	$\frac{1}{4}$
$\{u_3, u_2\}$	$\frac{1}{4}$	$\frac{1}{4}$	$\frac{1}{16}$
$\{u_3, u_4\}$	1	$\frac{1}{4}$	$\frac{1}{4}$
		1	$\mathscr{E}(\hat{V}_c(\bar{y}_{st})) = \frac{5}{8}$ Total (instead of $\frac{1}{16}$)

In partition (2)

$$Y_1 = 3 + 3 = 6, \quad Y_2 = 4 + 5 = 9, \quad (Y_1 - Y_2)^2/N^2 = \tfrac{9}{16}.$$

In partition (3) we obtain

Possible samples	$\frac{1}{4}(y^{(1)} - y^{(2)})^2$	f	Product
$\{u_1, u_3\}$	0	$\frac{1}{4}$	0
$\{u_1, u_4\}$	1	$\frac{1}{4}$	$\frac{1}{4}$
$\{u_2, u_3\}$	$\frac{1}{4}$	$\frac{1}{4}$	$\frac{1}{16}$
$\{u_2, u_4\}$	$\frac{1}{4}$	$\frac{1}{4}$	$\frac{1}{16}$
		1	$\mathscr{E}(\hat{V}_c(\bar{y}_{st})) = \frac{6}{16}$ Total (instead of $\frac{5}{16}$)

In partition (3)

$$Y_1 = 3 + 4 = 7, \quad Y_2 = 3 + 5 = 8, \quad (Y_1 - Y_2)^2/N^2 = \tfrac{1}{16}.$$

The estimator with the method of collapsed sampling was less precise in (2) than in (3), because in the former the difference between the totals for the two collapsed strata was larger in absolute value than in the latter.

6. Ratio and difference estimators

6.1. SEPARATE AND COMBINED RATIO ESTIMATORS

In stratified sampling, there are different ways of estimating $R = Y/X$ and for obtaining a ratio estimator for Y. One way is the method of *separate estimation* (we shall use the subscript s in that case), and another – the *combined estimation* (we shall use the subscript c in that case).

In *separate estimation* $\hat{R}_h$ is computed separately for each stratum. Hence the estimator for Y is

$$\hat{Y}_{R_s} = \Sigma \hat{R}_h X_h = \Sigma \hat{Y}_{R_h}.$$

Since each $\hat{R}_h$ is generally a biased estimator for R_h, and the bias of $\hat{Y}_{R_s}$ is the sum of the biases of the $\hat{Y}_{R_h}$, it is not desirable to use $\hat{Y}_{R_s}$ (or the corresponding estimator for R) unless we have priori information which justifies it – for instance:

(a) the sum of the upper bounds of absolute values of the individual biases is relatively small (this may be the case, among others, if each stratum contains many observations, or if an approximate bias correction has been applied);
(b) we know with certainty that the individual biases cancel each other.

The *combined estimator* of R is $\hat{R}_c = \bar{y}_{st}/\bar{x}_{st}$ and of Y is $\hat{Y}_{R_c} = \hat{R}_c X$, so that for computing this estimator of Y only knowledge of X is required (the computation of $\hat{Y}_{R_s}$ requires knowledge of all the X_h's!).† In the section on ratio estimators (simple random sampling), we wrote

$$\hat{Y}_R - Y = N \frac{\bar{X}}{\bar{x}} (\bar{y} - R\bar{x}) \approx N(\bar{y} - R\bar{x})$$

and obtained correspondingly the approximate expression for the variance (using $\mathscr{E}\,(\bar{y} - R\bar{x}) = 0$):

$$\begin{aligned} V(\hat{Y}_R) &\approx N^2 \mathscr{E}(\bar{y} - R\bar{x})^2 \\ &= N^2 V\{\bar{y} - R\bar{x}\} \\ &= N^2 \frac{N-n}{N} \frac{1}{n} \sum_{i=1}^{N} (y_i - Rx_i - \bar{Y} + R\bar{X})^2/(N-1); \end{aligned}$$

now, instead of the approximation $\hat{Y}_R - Y \approx N(\bar{y} - R\bar{x})$, we use the approximation

$$\begin{aligned} \hat{Y}_{R_c} - Y &= N \frac{\bar{X}}{\bar{x}_{st}} (\bar{y}_{st} - R\bar{x}_{st}) \\ &\approx N(\bar{y}_{st} - R\bar{x}_{st}) \\ &= \Sigma N_h \Sigma a_{hi} (y_{hi} - Rx_{hi})/n_h \end{aligned}$$

† Another advantage of $\hat{R}_c$ is that it is not required that each $\bar{x}_h$ will be guaranteed to turn out different from zero, but only that $\bar{x}_{st}$ will have this property.

and, using $\mathscr{E}(\bar{y}_{\text{st}} - R\bar{x}_{\text{st}}) = 0$, obtain

$$V(\hat{Y}_{R_c}) \approx N^2 \sum_{h=1}^{L} \frac{N_h^2}{N^2} \frac{N_h - n_h}{N_h} \frac{1}{n_h} \frac{\sum_{i=1}^{N_h} (y_{hi} - Rx_{hi} - \bar{Y}_h + R\bar{X}_h)^2}{N_h - 1}$$

$$= N^2 \sum_{h=1}^{L} \frac{N_h^2}{N^2} \frac{N_h - n_h}{N_h} \frac{1}{n_h} (S_h^2(\mathscr{Y}) - 2RS_h(\mathscr{X}\mathscr{Y}) + R^2 S_h^2(\mathscr{X}))$$

where N_h^2/N^2 appears because of the weights N_h/N in the formulae for $\bar{y}_{\text{st}}$ and $\bar{x}_{\text{st}}$. As

$$\hat{R}_c - R = (\bar{X}/\bar{x}_{\text{st}})\bar{X}^{-1}(\bar{y}_{\text{st}} - R\bar{x}_{\text{st}}) \approx \bar{X}^{-1}(\bar{y}_{\text{st}} - \bar{x}_{\text{st}}),$$

the approximation for $V(\hat{R}_c)$ corresponding to that for $V(\hat{Y}_{R_c})$ may be obtained from the latter by multiplication by X^{-2}.

For the estimation of $V(\hat{Y}_{R_c})$ we put, instead of $S_h^2(\mathscr{Y})$, $S_h^2(\mathscr{X}\mathscr{Y})$ and $s_h^2(\mathscr{X})$, their estimators; and, instead of R, $\hat{R}_c$. For estimating $V(\hat{R}_c)$, divide the result by $(N\bar{x}_{\text{st}})^2$.

The formula for the approximate variance of $\hat{Y}_{R_s}$ is the same as that given for $V(\hat{Y}_{R_c})$, except that R_h comes in the place of R, since $V(\hat{Y}_{R_s}) = \Sigma V(\hat{Y}_{R_h})$. Therefore, the difference between the approximate variance of $\hat{Y}_{R_c}$ and of $\hat{Y}_R$ is

$$\Sigma N_h(N_h - n_h)n_h^{-1}\{(R^2 - R_h^2)S_h^2(\mathscr{X}) + 2(R_h - R)S_h(\mathscr{X}\mathscr{Y})\}$$
$$= \Sigma N_h(N_h - n_h)n_h^{-1}\{(R - R_h)^2 + 2(R_h - R)(\beta_h - R_h)\}S_h^2(\mathscr{X}),$$

where $\beta_h = S_h(\mathscr{X}\mathscr{Y})/S_h^2(\mathscr{X})$. Under circumstances under which $\hat{Y}_{R_h}$ is a very good estimator of Y_h, β_h is generally close to R_h.† Consequently, under such circumstances, and if, moreover, the R_h vary much from stratum to stratum, $\hat{Y}_{RS}$ may have substantially smaller variance than $\hat{Y}_{R_c}$.‡

The strata in which we sample 2 units (say u_i and u_j) make a contribution to the estimator of $V(\hat{Y}_{R_c})$ which simplifies to

$$\hat{Y}^2 N_h(N_h - 2)\left\{\frac{1}{2\hat{Y}}(y_{hi} - y_{hj}) - \frac{1}{2\hat{X}}(x_{hi} - x_{hj})\right\}^2.$$

When strata h and h' have been collapsed, their contribution to the estimator of variance of $\hat{Y}_{R_c}$ may, if $(Y_h - RX_h - Y_{h'} + RX_{h'})^2$ is small relative to this variance, be taken to be

$$\{N_h(\bar{y}_h - R\bar{x}_h) - N_{h'}(\bar{y}_{h'} - R\bar{x}_{h'})\}^2.$$

† Cf. the end of section 9 of Chapter II or the end of section 6 of Chapter IIA.

‡ Note also that from the approximation methods we used it is clear that, if some n_h are small, the approximate formula for $V(\hat{Y}_{R_s})$ may not be as reliable as that for $V(\hat{Y}_{R_c})$.

Exactly like† in section 3 of Chapter IIA, $|\hat{Y}_{R_c}/V(\hat{Y}_{R_c})^{\frac{1}{2}}|$ does not exceed the coefficient of variation of $\bar{x}_{st}$. To obtain an approximate expression for the bias of $\hat{Y}_{R_c}$, write the expected value of

$$\bar{X}^{-1}(\hat{Y}_{R_c} - \bar{Y}) = \bar{x}_{st}^{-1}(\bar{y}_{st} - R\bar{x}_{st}),$$

with

$$\bar{x}_{st} = \bar{X}\left(1 + \frac{\bar{x}_{st} - \bar{X}}{\bar{X}}\right),$$

as

$$\mathscr{E}\frac{\bar{y}_{st} - R\bar{x}_{st}}{\bar{X}}\left\{1 - \frac{\bar{x}_{st} - \bar{X}}{\bar{X}} + \left(\frac{\bar{x}_{st} - \bar{X}}{\bar{X}}\right)^2 - \dots\right\}$$

$$= -\frac{1}{\bar{X}^2}\mathscr{E}\bar{y}_{st}(\bar{x}_{st} - \bar{X}) + \frac{R}{\bar{X}^2}\mathscr{E}(\bar{x}_{st} - \bar{X})^2 - \dots .$$

The second term on the right is

$$R\Sigma\left(\frac{N_h}{N}\right)^2 \frac{N_h - n_h}{N_h}\frac{1}{n_h} S_h^2(\mathscr{X})/\bar{X}^2,$$

and similarly the first is

$$-\Sigma\left(\frac{N_h}{N}\right)^2 \frac{N_h - n_h}{N_h}\frac{1}{n_h} S_h(\mathscr{Y}\mathscr{X})/\bar{X}^2.$$

Hence we may give as a first approximation to the bias of $\hat{Y}_{R_c}$:

$$X^{-1}\Sigma N_h(N_h - n_h)\frac{1}{n_h}(-S_h(\mathscr{Y}\mathscr{X}) + RS_h^2(\mathscr{X})).$$

Under proportional allocation this reduces to

$$X^{-1}N(N - n)\frac{1}{n}\Sigma\frac{N_h}{N}\{-S_h^2(\mathscr{Y}\mathscr{X}) + RS_h^2(\mathscr{X})\};$$

and the next terms will be approximately of order n^{-2}. In other circumstances the bias may not be of order n^{-1}.

In general bias may be diminished or avoided‡ by using the methods of section 7 of Chapter IIA; in dividing the sample into groups of equal size, the allocation over the strata should be preserved in the subsamples as much as possible (cf. section 14.2 of this chapter).

† Replace $\hat{R}$ by $\hat{R}_c$ and $\bar{x}$ by $\bar{x}_{st}$.

‡ See also in this connection NIETO DE PASCUAL, J. Amer. Statist. Ass., **56** (1961) 70–87, **59** (1964) 1298–1299, especially for the case of a large number of strata.

NOTE. Sometimes there may be occasion to incorporate additional information into a ratio estimator. For example, if X is known and if $\mathscr{Z}$ is a further characteristic with known stratum totals Z_h, and if the biases of

$$\hat{Y}_R = \Sigma(y_h/z_h)Z_h \quad \text{and} \quad \hat{X}_R = \Sigma(x_h/z_h)Z_h$$

are small, we may use $(\hat{Y}_R/\hat{X}_R)X$ as an estimator of Y.

Now in

$$X(\hat{Y}_R/\hat{X}_R) - Y = \frac{X}{\hat{X}_R}\left(\hat{Y}_R - \frac{Y}{X}\hat{X}_R\right) \approx \hat{Y}_R - \frac{Y}{X}\hat{X}_R$$

substitute

$$\hat{Y}_R = Y + \Sigma\frac{Z_h}{z_h}\left(y_h - \frac{Y_h}{Z_h}z_h\right) \approx Y + \Sigma\frac{N_h}{n_h}\left(y_h - \frac{Y_h}{Z_h}z_h\right),$$

$$\hat{X}_R = X + \Sigma\frac{Z_h}{z_h}\left(x_h - \frac{X_h}{Z_h}z_h\right) \approx X + \Sigma\frac{N_h}{n_h}\left(x_h - \frac{X_h}{Z_h}z_h\right),$$

giving

$$X(\hat{Y}_R/\hat{X}_R) - Y \approx \Sigma N_h\left\{y_h - \frac{Y_h}{Z_h}z_h - \frac{Y}{X}\left(x_h - \frac{X_h}{Z_h}z_h\right)\right\}\Big/ n_h.$$

The expression on the right-hand side has variance

$$\sum_{h=1}^{L} N_h^2\frac{N_h - n_h}{N_h}\frac{1}{n_h}\sum_{i=1}^{N_h}\left\{y_{hi} - \frac{Y_h}{Z_h}z_{hi} - \frac{Y}{X}\left(x_{hi} - \frac{X_h}{Z_h}z_{hi}\right)\right\}^2\Big/(N_h - 1),$$

which under the given conditions may be used as an approximation to the variance of $(\hat{Y}_R/\hat{X}_R)X$. In estimating this variance, the ratios Y_h/Z_h and X_h/Z_h are, of course, replaced by y_h/z_h and x_h/z_h, respectively.

6.2. EXAMPLE

Consider the example of section 2. If it is given that $\bar{X}_1 = 565$, $\bar{X}_2 = 640$, $\bar{X}_3 = 690$, $\bar{X}_4 = 710$, we can compute the separate ratio estimator.

$$\hat{\bar{Y}}_{R_s} = N^{-1}\Sigma N_h\hat{\bar{Y}}_{Rh} = N^{-1}\Sigma N_h\hat{R}_h\bar{X}_h = N^{-1}\Sigma N_h\frac{y_h}{x_h}\bar{X}_h.$$

This gives:

$$(1/80)\left[20\frac{1\,535}{2\,950}565 + 20\frac{1\,018}{1\,870}640 + 25\frac{1\,826}{3\,370}690 + 15\frac{1\,232}{2\,170}710\right] = 353.01.$$

$$\begin{aligned} v(\hat{\bar{Y}}_{R_s}) &= N^{-2}\Sigma N_h^2 v(\hat{\bar{Y}}_{Rh}) \\ &= N^{-2}\Sigma N_h(N_h - n_h)n_h^{-1}\{s_h^2(\mathscr{Y}) - 2\hat{R}_h s_h(\mathscr{X}\mathscr{Y}) + \hat{R}_h^2 s_h^2(\mathscr{X})\}.\end{aligned}$$

This gives:

$$(1/80)^2[(60)(91.184\,5) + (113\tfrac{1}{3})(46.211\,9) + (100)(191.811\,8) + (60)(193.954\,6)] = 6.49.$$

Compare this with

$$v(\hat{Y}_R) = 12.51.$$

An estimated upper bound on the bias $\hat{Y}_{Rh}$ is

$$\{v(\hat{Y}_{Rh})v(\bar{x}_h)\}^{\frac{1}{2}}/|\bar{X}_h| = \frac{N_h - n_h}{N_h}\frac{1}{n_h} s_h(\mathscr{X})\{(s_h^2(\mathscr{Y}) - 2\hat{R}_h s_h(\mathscr{X}\mathscr{Y}) + \hat{R}_h^2 s_h^2(\mathscr{X})\}^{\frac{1}{2}}/|\bar{X}_h|$$

and so an estimated upper bound for the bias of $\hat{Y}_{R_s}$ is the weighted sum of these expressions (weighted by N_h/N):

$$(1/80)[(3)\{(100)(91.184\,5)\}^{\frac{1}{2}}/(565) + (5\tfrac{2}{3})\{(233\tfrac{1}{3})(46.211\,9)\}^{\frac{1}{2}}/(640) + (4)\{(130)(191.811\,8)\}^{\frac{1}{2}}/(690) + (4)\{(233\tfrac{1}{3})(193.954\,6)\}^{\frac{1}{2}}/(710)] = 0.044,$$

which is relatively negligible.

If the $\bar{X}_h$ are unknown, we may compute the combined ratio estimator

$$\hat{Y}_{R_c} = \frac{\bar{y}_{st}}{\bar{x}_{st}}\bar{X} = \frac{352.71}{649.58}(650) = 352.94,$$

$$v(\hat{Y}_{R_c}) = N^{-2}\Sigma N_h(N_h - n_h)n^{-1}\{s_h^2(\mathscr{Y}) - 2\hat{R}_c s_h(\mathscr{X}\mathscr{Y}) + \hat{R}_c^2 s_h^2(\mathscr{X})\}.$$

This gives

$$(1/80)^2[(60)(87.593) + (113\tfrac{1}{3})(46.455) + (100)(191.680) + (60)(336.641)] = 7.79,$$

somewhere higher than $v(\hat{Y}_{R_s})$.

A bound for the bias is estimated from

$$\bar{X}^{-1}\{v(\bar{y}_{st})v(\bar{x}_{st})\}^{\frac{1}{2}} = (1/650)\{(11.31)(9.29)\}^{\frac{1}{2}} = 0.016.$$

If desired, the bias of $\hat{Y}_{R_c}$ may be estimated from

$$\bar{X}^{-1}N^{-2}\Sigma N_h(N_h - n_h)n_h^{-1}(-s_h(\mathscr{X}\mathscr{Y}) + \hat{R}_c s_h^2(\mathscr{X})).$$

This gives

$$(1/650)(1/80)^2[(60)(-78.202) + (113\tfrac{1}{3})(-86.638) + (100)(-73.413) + (60)(280.029)] = -0.001.$$

6.3. DIFFERENCE ESTIMATORS; DOUBLE SAMPLING

Difference estimators in stratified sampling are weighted combinations of difference estimators within the strata. In generalized difference estimators one may use a different k for each stratum, or use one k for all of them; in the latter case the use of a k based on the sample (such as an estimator of Y/X) introduces little bias.

In stratified sampling, in order to use ratio, regression, and difference estimators, we may estimate the values of the required X_h from large preliminary samples (*double sampling*).

6.4. APPENDIX: A FORMULA LIKE THE SEPARATE RATIO ESTIMATOR IN NONSTRATIFIED SAMPLING

The formula given for $\hat{Y}_{R_s}$ also occurs in a different situation: a nonstratified random sample of n units is obtained from a population, each of whose units is subdivided into L subunits by some criterion. Then

$$y_i = \sum_h y_{hi}$$

for all i, and the average of the y_{hi} is

$$\bar{Y}(h) = Y(h)/N = \sum_{i=1}^{N} y_{hi}/N,$$

with a corresponding sample average

$$\bar{y}(h) = y(h)/n = \sum_{i=1}^{N} a_i y_{hi}/n.$$

The estimator in question is $\Sigma X(h)\hat{R}(h)$.

Now

$$\Sigma X(h)\hat{R}(h) - Y = \Sigma X(h)\frac{y(h)}{x(h)} - Y = \sum \frac{X(h)}{x(h)}\{y(h) - R(h)\,x(h)\}$$

is the sum of

$$\sum_h \frac{X(h)}{x(h)}\{y_{hi} - R(h)x_{hi}\}$$

for the units n_i which appear in the sample. For large n, we may expect $\bar{x}(h)$ to be close to $\bar{X}(h)$ for each h, so that the above terms are approximately equal to

$$\frac{N}{n}\sum_h \{y_{hi} - R(h)x_{hi}\}.$$

Therefore, the mean square error of $\Sigma\hat{R}(h)X(h)$ is approximately

$$\frac{N-n}{N}\,n\left(\frac{N}{n}\right)^2 \sum_i [\sum_h \{y_{hi} - R(h)x_{hi}\}]^2/(N-1),$$

which may be estimated by

$$N(N-n)\,\frac{1}{n}\sum_i a_i[\sum_h\{y_{hi} - \hat{R}(h)x_{hi}\}]^2/(n-1),$$

or, perhaps better (cf. end of section 8, Chapter VIII) by

$$N(N-n)\,\frac{1}{n}\,\Sigma a_i\left[\sum_h \frac{\bar{X}(h)}{\bar{x}(h)}\{y_{hi} - \hat{R}(h)x_{hi}\}\right]^2\Big/(n-1).$$

Using the first term in the power series

$$\frac{\bar{X}(h)}{\bar{x}(h)} = 1 - \frac{\bar{x}(h) - \bar{X}(h)}{\bar{x}(h)} + \ldots$$

we get

$$\mathscr{E}\,\frac{N}{n}\sum \frac{\bar{X}(h)}{\bar{x}(h)}\{y(h) - R(h)\,x(h)\}$$

$$\approx -\,N\Sigma\bar{X}(h)^{-1}\mathscr{E}[\{\bar{x}(h) - \bar{X}(h)\}\bar{y}(h) - R(h)\{\bar{x}(h) - \bar{X}(h)\}\bar{x}(h)]$$

$$= -(N-n)\,\frac{1}{n}\,\Sigma\bar{X}(h)^{-1}\{S(\mathscr{X}(h)\mathscr{Y}(h)) - R(h)S^2(\mathscr{X}(h))\},$$

so that the relative bias of $\Sigma X(h)\hat{R}(h)$ is approximately

$$\frac{N-n}{N}\,\frac{1}{n}\,\frac{\Sigma\,\bar{Y}(h)\{-C(\mathscr{X}(h)\mathscr{Y}(h)) + C^2(\mathscr{X}(h))\}}{\Sigma\,\bar{Y}(h)},$$

which may sometimes be roughly estimated from the data.

One may also give an analogue to the estimator for the case discussed in the Note above, namely

$$\Sigma X(g)\,\frac{\hat{Y}(g)}{\hat{X}(g)},$$

where g refers to one of the subunits into which the units are subdivided in addition to the division into L parts, and

$$\hat{Y}(g) = \sum \frac{Z(h)}{z(h)}\,y(g,h), \quad \hat{X}(g) = \sum \frac{Z(h)}{z(h)}\,x(g,h).$$

Thus

$$\sum X(g)\frac{\hat{Y}(g)}{\hat{X}(g)} - Y = \sum \frac{X(g)}{\hat{X}(g)}\left[\hat{Y}(g) - Y(g) - \frac{Y(g)}{X(g)}\{\hat{X}(g) - X(g)\}\right]$$

$$= \sum \frac{X(g)}{\hat{X}(g)} \sum_h \frac{Z(h)}{z(h)}\left[y(g, h) - \frac{Y(g, h)}{Z(h)} z(h) - \frac{Y(g)}{X(g)}\left\{x(g, h) - \frac{X(g, h)}{Z(h)} z(h)\right\}\right].$$

This equals $(N/n)\Sigma a_i d_i$ with

$$d_i = \sum \frac{X(g)}{\hat{X}(g)} \sum_h \frac{\bar{Z}(h)}{\bar{z}(h)}\left[y_{ghi} - \frac{Y(g, h)}{Z(h)} z_{hi} - \frac{Y(g)}{X(g)}\left\{x_{ghi} - \frac{X(g, h)}{Z(h)} z_{hi}\right\}\right],$$

satisfying $\Sigma d_i = 0$. Therefore, if we write d_i' for d_i in which $\bar{Z}(h)/\bar{z}(h)$ and $X(g)/\hat{X}(g)$ are replaced by 1, the variance of our estimator will be approximately

$$\{N(N - n)/n\}\Sigma(d_i')^2/(N - 1)$$

under circumstances in which this replacement has small effect. As estimator of this variance we may use

$$\{N(N - n)/n\}\Sigma a_i \hat{d}_i^2/(n - 1),$$

with

$$\hat{d}_i = \sum \frac{X(g)}{\hat{X}(g)} \sum \frac{\bar{Z}(h)}{\bar{z}(h)}\left[y_{ghi} - \frac{y(g, h)}{z(h)} z_{hi} - \frac{y(g)}{x(g)}\left\{x_{ghi} - \frac{x(g, h)}{z(h)} z_{hi}\right\}\right].$$

7. Domains of study

Let the numbering of units in stratum h be such that the first N_{h1} among them fall in domain $\mathscr{D}_1$. In this section, if $N_{h1} > 0$, the number n_{h1} of observations in stratum h falling in domain $\mathscr{D}_1$ is assumed positive (and more than 1, if we estimate variances), and the probability for n_{h1} to be zero is assumed to be negligible. However, the estimators sub (b) and (c) below have the advantage that they require this property only for the sum over h, that is, for the total number n' of observations falling in $\mathscr{D}_1$.

We remember that for simple random sampling the sum Y' of $\mathscr{Y}$ for domain $\mathscr{D}_1$ was estimated

by $N'\bar{y}_1$ if N' is known and n' is sure to be not zero, by $(y'/y)\,Y$ if Y is known and N' is not,

by $N\bar{y}'$ (if N is known) when neither N' nor Y are known.†

The corresponding formulae for estimating Y' in stratified sampling in the case of domains of study are the sums over the strata of the above expressions:

(a)‡

$$\sum_{h=1}^{L} N_{h1}\bar{y}_{h1}$$

if the N_{h1} are known.

(b)§

$$\sum_{h=1}^{L} (y_h'/y_h)\,Y_h$$

if the Y_h are known, but not the N_{h1}.

(c)

$$\sum_{h=1}^{L} N_h\bar{y}_h'$$

if none of the above are known.

We now obtain the variance of these estimators:

(a) Since $\bar{y}_{h1}$ is a ratio y_h'/x_h' of random quantities (where $x_{hi}' = 1$ if u_{hi} is in $\mathscr{D}_1$ and zero otherwise),

$$V(\bar{y}_{h1}) \approx \frac{N_h - n_h}{N_h}\,\frac{1}{n_h}\,\frac{1}{(\bar{X}_h')^2}\sum_{i=1}^{N_h}\left(y_{hi}' - \frac{Y_h'}{X_h'}\,x_{hi}'\right)^2\Big/(N_h - 1),$$

where $\bar{X}_h' = X_h'/N_h$ equals N_{h1}/N_h. Therefore

$$V(\Sigma N_{h1}\bar{y}_{h1}) \approx \Sigma N_{h1}{}^2\,\frac{N_h - n_h}{N_h}\,\frac{1}{n_h}\left(\frac{N_h}{N_{h1}}\right)^2\sum_{i=1}^{N_h}\left(y_{hi}' - \frac{Y_h'}{X_h'}\,x_{hi}'\right)^2\Big/(N_h - 1)$$

$$= \Sigma N_h(N_h - n_h)\,\frac{1}{n_h}\sum_{i=1}^{N_{h1}}(y_{hi} - \bar{Y}_{h1})^2(N_h - 1),$$

which may be estimated by

$$\Sigma N_h(N_h - n_h)\,\frac{1}{n_h}\sum_{i=1}^{N_{h1}} a_{hi}(y_{hi} - \bar{y}_{h1})^2/(n_h - 1).$$

† In this section, in order to avoid confusion between the number in domain 1 and the number in stratum 1, we write the former as $N'(n')$. If there is a danger of confusing the average for $\mathscr{Y}$ in domain 1 with its average for stratum 1, we write the former as Y'/N'.

‡ If some of $\mathscr{E}n_{h1}$ are small, the bias of this estimator may be unacceptably large.

§ For reasons explained in the previous section, the bias of this separate ratio estimator may be large. An alternative estimator is $(\Sigma N_h\bar{y}_h'/\Sigma N_h\bar{y}_h)\,Y$ (which, moreover, requires only knowledge of Y, not of the Y_h). For an approximation to its bias see the result on the bias of a combined ratio estimator in the previous section.

Note that if the n_{h1} were not random (so that the intersections of the original L strata and $\mathscr{D}_1$ form a new set of strata), $V(\Sigma N_{h1}\bar{y}_{h1})$ and its estimator are as above if

$$n_{h1} = (n_h/N_h)N_{h1}$$

and the

$$\{N_h/(N_h - 1)\}\{(N_{h1} - 1)/N_{h1}\} \quad \text{or} \quad \{n_h/(n_h - 1)\}\{(n_{h1} - 1)/n_{h1}\},$$

respectively, are negligible (compare end of section 11.5, Chapter II; with stratification the n_{h1} may become so small that these approximations are too rough).

(b) We obtain like above

$$V\left(\sum Y_h \frac{y'_h}{y_h}\right) \approx \sum N_h(N_h - n_h)\frac{1}{n_h}\sum_{i=1}^{N_h}\left(y'_{hi} - \frac{Y'_h}{Y_h}y_{hi}\right)^2\bigg/(N_h - 1),$$

which may be estimated by

$$\sum N_h(N_h - n_h)\frac{1}{n_h}\sum_{i=1}^{N_h} a_{hi}\left(y'_{hi} - \frac{y'_h}{y_h}y_{hi}\right)^2\bigg/(n_h - 1).$$

For the combined ratio estimator given in the footnote we obtain at once from the previous section the following approximate variance formula:

$$\sum N_h(N_h - n_h)\frac{1}{n_h}\sum_{i=1}^{N_h}\{y'_{hi} - \bar{Y}'_h - (Y'/Y)(y_{hi} - \bar{Y}_h)\}^2/(N_h - 1).$$

(c) As $\bar{y}'_h$ is not a ratio of random quantities, we get at once

$$V(\Sigma N_h\bar{y}'_h)$$
$$= \sum N_h(N_h - n_h)\frac{1}{n_h}\sum_{i=1}^{N_h}(y'_{hi} - \bar{Y}'_h)^2/(N_h - 1)$$
$$= \sum N_h(N_h - n_h)\frac{1}{n_h}\frac{1}{N_h - 1}\left\{(N_{h1} - 1)S^2_{h1} + N_{h1}\left(1 - \frac{N_{h1}}{N_h}\right)\bar{Y}^2_{h1}\right\}$$

(compare section 11.5 (i), Chapter II) with estimator

$$\sum N_h(N_h - n_h)\frac{1}{n_h}\sum_{i=1}^{N_h} a_{hi}(y'_{hi} - \bar{y}'_h)^2/(n_h - 1)$$
$$= \sum N_h(N_h - n_h)\frac{1}{n_h}\frac{1}{n_h - 1}\left\{(n_{h1} - 1)s^2_{h1} + n_{h1}\left(1 - \frac{n_{h1}}{n_h}\right)\bar{y}^2_{h1}\right\}.$$

The corresponding estimators of $\bar{Y}_1$ are:

(a)† $\Sigma N_{h1}\bar{y}_{h1}/N'$ if the N_{h1} are known; otherwise

(b)‡ $\Sigma(y_h'/y_h)Y_h/N'$ if the Y_h are known – which is modified to

$$\Sigma(y_h'/y_h)Y_h/\hat{N}' \quad (\text{with } \hat{N}' = \Sigma N_h n_{h1}/n_h)$$

if N' is unknown;

(c) $\Sigma N_h\bar{y}_h'/N'$ or

$$\Sigma N_h\bar{y}_h'/\hat{N}' = \Sigma\hat{N}_{h1}\bar{y}_h/\Sigma\hat{N}_{h1}$$

if neither the N_{h1} nor the Y_h are known; here

$$\hat{N}' = \Sigma\hat{N}_{h1} = \Sigma N_h n_{h1}/n_h = \Sigma N_h\bar{x}_h',$$

where x_{hi}' is 1 if u_{hi} belongs to the hth stratum and also to $\mathscr{D}_1$; and zero otherwise. If the n_h are proportional to the N_h, i.e., if $n_h/N_h = f$, then $\hat{N}'$ equals

$$n'/f \quad (\text{with } n' = \sum_h n_{h1});$$

and $\Sigma N_h\bar{y}_h'/\hat{N}'$ becomes $y'/n' = \bar{y}_1$, the mean of $\mathscr{Y}$ for the u's that are in the sample and in $\mathscr{D}_1$. If the n_h are not proportional to N_h, the latter is of the form $\Sigma N_h\bar{y}_h'/\Sigma N_h\bar{x}_h'$, which is a combined ratio estimator $\hat{R}_c$ as defined in the previous section. Thus, from the results given there with $X' = N'$, we obtain, if we call the estimator $\hat{Y}_{R1}$:

$$\begin{aligned}
v(\hat{Y}_{R1}) &= \frac{1}{\hat{N}'^2}\sum_{h=1}^{L} N_h^2\frac{N_h - n_h}{N_h}\frac{1}{n_h}\frac{1}{n_h - 1} \\
&\quad\times \sum_{i=1}^{N_h} a_{hi}[y_{hi}' - \hat{Y}_{R1}x_{hi}' - (\bar{y}_h' - \hat{Y}_{R1}\bar{x}_h')]^2 \\
&= \frac{1}{\hat{N}'^2}\sum_{h=1}^{L} N_h^2\frac{N_h - n_h}{N_h}\frac{1}{n_h}\frac{1}{n_h - 1} \\
&\quad\times \left[\sum_{i=1}^{N_{h1}} a_{hi}(y_{hi} - \hat{Y}_{R1})^2 - \frac{n_{h1}^2}{n_h}(\bar{y}_{h1} - \hat{Y}_{R1})^2\right] \\
&= \frac{1}{\hat{N}'^2}\sum_{h=1}^{L} N_h^2\frac{N_h - n_h}{N_h}\frac{1}{n_h}\frac{1}{n_h - 1} \\
&\quad\times \left[\sum_{i=1}^{N_{h1}} a_{hi}(y_{hi} - \bar{y}_{h1})^2 + n_{h1}\left(1 - \frac{n_{h1}}{n_h}\right)(\bar{y}_{h1} - \hat{Y}_{R1})^2\right];
\end{aligned}$$

† If some of $\mathscr{E}n_{h1}$ are small, the bias of this estimator may be unacceptably large.

‡ For reasons explained in the previous section, the estimators under (b) may have large biases. Alternative estimators are $(N^1)^{-1}$ or $(\hat{N}^1)^{-1}$ times $(\Sigma N_h\bar{y}_h'/\Sigma N_h\bar{y}_h)Y$ (these estimators, moreover, require only knowledge of Y, not of the Y_h).

and similarly

$$V(\hat{\bar{Y}}_{R1}) \approx \frac{1}{N'^2} \sum_{h=1}^{L} N_h^2 \frac{N_h - n_h}{N_h} \frac{1}{n_h} \frac{1}{N_h - 1}$$

$$\times \left[\sum_{i=1}^{N_{h1}} (y_{hi} - \bar{Y}_{h1})^2 + N_{h1}\left(1 - \frac{N_{h1}}{N_h}\right)(\bar{Y}_{h1} - \bar{Y}_1)^2 \right].$$

The estimator of $\bar{Y}_1$ under (b) above with N' estimated may be written in the form

$$\Sigma(y'_h/y_h)Y_h/\Sigma(x'_h/x_h)X_h,$$

which is like the expression discussed in the Note of section 6. Therefore its variance is approximately†

$$\frac{1}{(N')^2} \Sigma N_h(N_h - n_h) \frac{1}{n_h} \sum_{i=1}^{N_h} \left[y'_{hi} - \frac{Y'_h}{Y_h} y_{hi} - \frac{Y'}{X'}\left(x'_{hi} - \frac{X'_h}{X_h} x_{hi} \right)^2 \right] \Big/ (N_h - 1)$$

$$= \frac{1}{(N')^2} \Sigma N_h(N_h - n_h) \frac{1}{n_h} \left[\sum_{i=1}^{N_h} \left(y'_{hi} - \frac{Y'_h}{Y_h} y_{hi} \right)^2 \Big/ (N_h - 1) \right.$$

$$\left. + \frac{1}{N_h - 1} \bar{Y}_1 \left\{ \frac{N}{N'}\left(Y' \frac{N_h}{N} \right)\left(1 - \frac{N_{h1}}{N_h} \right) - 2Y'_h\left(1 - \frac{Y'_h}{Y_h} \right) \right\} \right]$$

estimated by

$$\frac{1}{(\hat{N}')^2} \Sigma N_h(N_h - n_h) \frac{1}{n_h} \left[\sum_{i=1}^{N_h} a_{hi} \left(y'_{hi} - \frac{y'_h}{y_h} y_{hi} \right)^2 \Big/ (n_h - 1) \right.$$

$$\left. + \frac{1}{n_h - 1} \hat{\bar{Y}}_1 \left\{ \hat{\bar{Y}}_1(n_h - n_{h1}) - 2y'_h\left(1 - \frac{y'_h}{y_h} \right) \right\} \right].$$

The last of the estimators in the footnote to (b) (with N' estimated) is of a more complicated character.

It is also possible to use an auxiliary characteristic $\mathscr{X}$ (more general than the one used above, which can only be 0 or 1) for ratio estimation of Y' or $\bar{Y}_1$, as well as for estimation of Y'/X'. For example, Y'/X' may be estimated by

$$\hat{R}' = \Sigma N_h \bar{y}'_h / \Sigma N_h \bar{x}'_h$$

† In many cases $Y'N_h/N$ will, for most strata h that contribute significantly to the variance, differ relatively little from Y'_h, and N_{h1}/N_h relatively little from Y'_h/Y_h; then the expression in curly brackets will be about

$$\left(\frac{N}{N'} - 2 \right) Y'_h \left(1 - \frac{Y'_h}{Y_h} \right).$$

with estimated variance

$$v(\hat{R}') = \frac{1}{\hat{X}'^2} \sum_{h=1}^{L} N_h^2 \frac{N_h - n_h}{N_h} \frac{1}{n_h} \frac{1}{n_h - 1}$$

$$\times \sum_{i=1}^{N_h} a_{hi} \left[y'_{hi} - \bar{y}'_h - \frac{y'_{\text{st}}}{x'_{\text{st}}} (x'_{hi} - \bar{x}'_h) \right]^2$$

$$= \frac{1}{\hat{X}'^2} \sum_{h=1}^{L} N_h^2 \frac{N_h - n_h}{N_h} \frac{1}{n_h} \frac{1}{n_h - 1}$$

$$\times \left[\sum_{i=1}^{N_{h1}} a_{hi} \left(y_{hi} - \frac{y'_{\text{st}}}{x'_{\text{st}}} x_{hi} \right)^2 - \frac{n_{h1}^2}{n_h} \left(\bar{y}_{h1} - \frac{y'_{\text{st}}}{x'_{\text{st}}} \bar{x}_{h1} \right)^2 \right];$$

and, for known X' (or $\bar{X}_1$), a ratio estimator of Y' (or $\bar{Y}_1$) is $\hat{R}'X'$ (or $\hat{R}'\bar{X}_1$), which has an estimated variance equal to the same expression as above without the divisor $(\hat{X}')^2$ (and, for $\hat{R}'\bar{X}_1$, divided by $(N')^2$ or $(\hat{N}')^2$ instead of $(\hat{X}')^2$).

If in the formula for estimating Y', given in the footnote to (b), the factor Y is left out, we obtain an estimator of Y'/Y.

REMARK. Like in section 11.6 of Chapter II, one sometimes wishes to estimate a total $Y'(z)$ for a subdomain $\mathscr{D}_{1z}$ (see the notation there). In the estimator

$$\Sigma N_h y'_h(z)/n_h$$

the denominators are necessarily positive. On the other hand, in many cases the probability for some denominator in

$$\Sigma N'_h(z) y'_h(z)/n'_h(z)$$

to be zero is nonnegligible, without this being so for the estimators

$$\sum_h N_h(z) y'_h(z)/n_h(z), \quad \sum_h N'_h y'_h(z)/n'_h.$$

The variances of these estimators are derived at once from the expressions given in the section referred to.

For estimating $Y'(z)/X'(z)$ [such as: the average for $\mathscr{Y}$ over $\mathscr{D}_{1z}$ if $X'(z) = N'(z)$], we may use the combined ratio estimator

$$\frac{\Sigma N_h y'_h(z)/n_h}{\Sigma N_h x'_h(z)/n_h},$$

or one of the more complicated estimators

$$\frac{\Sigma N_h(z) y'_h(z)/n_h(z)}{\Sigma N_h(z) x'_h(z)/n_h(z)}, \quad \frac{\Sigma N'_h y'_h(z)/n'_h}{\Sigma N'_h x'_h(z)/n'_h},$$

with the choice depending on the ease with which information is obtainable on the N_h, $N_h(z)$ and N'_h, and on evaluation of the biases and variances.

8. Size of sample and its allocation among the strata; comparison with simple random sampling

A simple way of allocating the number n of observations among the strata is *proportional* allocation. Under it the *sampling fractions* $f_h = n_h/N_h$ are equal for all strata, and so are all equal to $f = n/N$. It follows that then $\bar{y}_{st}$ is equal to the unweighted sample mean over the whole sample,

$$\bar{y} = \sum_{h=1}^{L} n_h \bar{y}_h/n = \Sigma\Sigma y_{hi}/n$$

(we say the sample is *self-weighting*); this greatly simplifies calculations. The variance of $\bar{y}_{st}$ becomes simply

$$\frac{1-f}{n} \sum_{h=1}^{L} \frac{N_h}{N} S_h^2,$$

which differs from the variance of $\bar{y}$ under simple random sampling of n elements from the population by

$$\frac{1-f}{n} \frac{L-1}{N-1} \left[N \frac{\sum_{h=1}^{L} (N_h/N)(\bar{Y}_h - \bar{Y})^2}{L-1} - \frac{\sum_{h=1}^{L} \{(N-N_h)/N\} S_h^2}{L-1} \right]$$

(proof in Remark (i) at the end of this section), which may be negative only if N times the variation between the stratum means (as measured by the first term in brackets) is less than the average† of the adjusted variances within strata (as measured by the second term in brackets). This result is not unexpected; in fact, one of the main aims of stratification is to diminish variability of the estimator by selecting strata in such a way‡ as to increase homogeneity within strata, and this generally leads to large variation between the stratum means. If we succeed well in this aim, we may therefore obtain a substantially reduced variance by stratification. Nonetheless, proportional allocation may be improved upon if the S_h^2 are known (at least approximately) or if the cost of sampling vary among the strata (apart from a fixed cost of c_0 for the entire survey).

† If all N_h are equal, it is an unweighted average; otherwise it is a weighted average with weights $(N - N_h)/N$.

‡ Sometimes we are in a position to formulate rules for choosing stratum boundaries that make the variance of $\bar{y}_{st}$ as small as possible. The first requirement for this is that there must be good prior information on F_N for $\mathscr{Y}$ (or for some closely related characteristic $\mathscr{X}$). On the basis of this we then compute stratum boundaries for $\mathscr{Y}$ (or $\mathscr{X}$), see e.g. COCHRAN, Bull. Intern. Statist. Inst., **38**, Pt. II (1961) 345–358. The second requirement is that it be in fact possible to use, exactly or approximately, these stratum boundaries. – Cochran's paper also has some remarks on the choice of L.

Suppose† that the S_h^2 are known. Let us find optimum numbers $n_1, \ldots, n_L$ in two cases:

(a) The unit cost c_h of sampling of stratum h is the same for each stratum: $c_1 = \ldots = c_L = c$, so that the cost function is $c_0 + c\Sigma n_h$.

(b) The cost function is $c_0 + \Sigma c_h n_h$, and not all c_h are equal.

Case (a): Let us attempt to choose $n_1, \ldots, n_L$ such as to minimize $V(\bar{y}_{\text{st}})$ subject to a given upper bound C_0 on $c_0 + c\Sigma n_h$ (or, equivalently, an upper bound on Σn_h equal to $(C_0 - c_0)/c = n$).

An approximate solution is obtained by first treating the arguments $n_1, \ldots, n_L$ of $V(\bar{y}_{\text{st}})$ and $n_1 + \ldots + n_L \leq n$ as magnitudes which can be varied continuously over the real line (in which case it is evident that the optimum is obtained for $n_1 + \ldots + n_L$ *equal* to n), and then searching for an integer solution with $0 < n_h \leq N_h$ in the neighborhood of the obtained solution.

If in our tentative solution some n_h are bigger than the corresponding N_h, we take them to equal these N_h, and we shall allocate the remaining observations to the remaining strata by the method outlined above.

Since

$$V(\bar{y}_{\text{st}}) = \frac{1}{N^2} \sum_{h=1}^{N} \frac{N_h^2 S_h^2}{n_h} - \frac{1}{N} \sum_{h=1}^{L} \frac{N_h}{N} S_h^2,$$

it suffices to minimize

$$U = U(n_1, \ldots, n_L) = \Sigma((N_h^2 S_h^2)/n_h)$$

under the condition $n_1 + \ldots + n_L = n$.

For this purpose we shall find all the L-tuples $(n_1, \ldots, n_L)$ which minimize

$$W = (n_1 + \ldots + n_L) U(n_1, \ldots, n_L),$$

and among them look for $(n_1, \ldots, n_L)$ such that also $n_1 + \ldots + n_L = n$.

† If we measure several characteristics, we take into account the S_h^2 for the characteristic of greatest interest. Often the relative S_h^2 values for the most important characteristics do not differ greatly; if they do we may use some sort of best compromise allocation. For further details on this topic see YATES, Sampling Methods for Censuses and Surveys (Griffin, 3rd ed., 1960). Methods similar to those discussed there are applicable if there is more than one cost function, e.g., one in terms of money, one in terms of certain types of personnel, and one in terms of time. A closely related question is the following: Extra funds are made available for the purpose of including in a survey observations on a further characteristic; how should these extra funds be allocated? Rather unexpectedly, the answer is provided by a technique designed for a Bayesian approach to allocation; see p. 767 of ERICSON, J. Amer. Statist. Ass., **60** (1965) 750–771. The cost or effort involved in measuring different characteristics may also vary.

Since $n > 0$, for such $(n_1, \ldots, n_L)$, and only for such, U attains its smallest value over the set of all L-tuples whose sum is n.†

Let us use‡ the Schwarz-Cauchy or Bunyakovski inequality for the real numbers $a_1, \ldots, a_L$ and $b_1, \ldots, b_L$:

$$(\Sigma a_h^2)(\Sigma b_h^2) \geq (\Sigma a_h b_h)^2$$

where the equality holds if and only if $a_1 : b_1 = \ldots = a_L : b_L$.§

In our case, putting $a_h = n_h^{\frac{1}{2}}$, $b_h = (N_h S_h / n_h^{\frac{1}{2}})$ with $W = \Sigma a_h^2 \Sigma b_h^2$, we obtain the minimum if and only if

$$n_1 : N_1 S_1 = \ldots = n_L : N_L S_L$$

or

$$n_h : \Sigma n_h = N_h S_h : \Sigma N_h S_h ;$$

and the minimum value of U for n_h satisfying $\Sigma n_h = n$ is

$$\frac{1}{\Sigma a_h^2} (\Sigma a_h b_h)^2 = \frac{1}{n} (\Sigma N_h S_h)^2. \qquad (*)$$

† Clearly the method involved is a general one for the problem of minimizing a function U for the values of the argument for which a function g takes on a given positive value. In general, however, it may be that the problem has a solution, but that none of the minimizers of the product, W, of g and U give to g the specified value, so that the method does not work. E.g., $f(x_1, x_2) = 4x_1 + 4x_2$, $g(x) = x_1 + x_2$. If S_c denotes the set of (x_1, x_2) for which $g(x_1, x_2) = c$, f is constant on S_5 and so its minimum over S_5 is attained at *all* points of S_5, whereas W attains its minimum on S_0 and so at *no* point of S_5.

‡ Since the n_h appear in the numerator of the first term of W and in the denominator in the second term, we can also proceed directly as follows

$$\begin{aligned} W &= \sum_{h}\sum_{h'} n_h N_{h'}^2 S_{h'}^2 / n_h \\ &= \sum\sum_{h>h'} \{(n_h N_{h'}^2 S_{h'}^2 / n_{h'}) + (n_{h'} N_h^2 S_h^2 / n_h)\} + \Sigma N_h^2 S_h^2 \\ &= \sum\sum_{h>h'} \{(n_h/n_{h'})^{\frac{1}{2}} N_{h'} S_{h'} - (n_{h'}/n_h)^{\frac{1}{2}} N_h S_h\}^2 + 2\sum\sum_{h>h'} N_{h'} N_h S_{h'} S_h + \Sigma N_h^2 S_h^2. \end{aligned}$$

So $W = W_{\min}$ if and only if the expression in brackets is equal to zero, i.e.,

$$n_h : n_{h'} = N_h S_h : N_{h'} S_{h'}.$$

Then

$$W_{\min} = 2\sum\sum_{h>h'} N_{h'} N_h S_{h'} S_h + \Sigma N_h^2 S_h^2 = (\Sigma N_h S_h)^2.$$

§ Proof:

$$\Sigma a_h^2 \Sigma b_h^2 - (\Sigma a_h b_h)^2 = \tfrac{1}{2} \Sigma\Sigma \{a_h b_{h'} - b_h a_{h'}\}^2$$

is positive unless all brackets on the right-hand side vanish.

Since we also require $\Sigma n_h = n$, we obtain

$$n_h = n \frac{N_h S_h}{\sum_h N_h S_h}.$$

This allocation is called the *Neyman-Tshuprov allocation.*

If we substitute these n_h in the formula for the variance, or from (*), we obtain for $V(\bar{y}_{st})$ its "minimal" value†

$$V' = \frac{1}{n}\left(\sum \frac{N_h}{N} S_h\right)^2 - \frac{1}{N}\sum \frac{N_h}{N} S_h^2.$$

REMARK. We can also use the method of Lagrange multipliers in order to find the minimum of U under the restraint $\Sigma n_h = n$. We take the derivative of the auxiliary function

$$\phi = \Sigma N_h^2 S_h^2 / n_h + \lambda(\Sigma n_h - n)$$

with respect to n_h ($h = 1, 2, \ldots, L$) and with respect to λ, and by setting the derivatives equal to 0 we obtain the equations

$$\frac{\partial \phi}{\partial n_h} = -N_h^2 S_h^2 / n_h^2 + \lambda = 0 \quad (h = 1, 2, \ldots, L)$$

$$\frac{\partial \phi}{\partial \lambda} = \Sigma n_h - n = 0.$$

The first equation gives us the optimal proportion between the n_h (with proportionality factor $\lambda^{-\frac{1}{2}}$). The disadvantage of this method is that we still have to prove that the point $(n_1, \ldots, n_L)$ thus obtained is indeed a minimizing point, and not a maximizing point or a saddle point.

Let us find now the n_h which minimize the cost $c_0 + c\Sigma n_h$ (or Σn_h) if we are given in advance a bound‡ V_0 for the variance $V(\bar{y}_{st})$. For this purpose we may use the set of L-tuples for minimizing W that we already found,§ and among them look for $(n_1, \ldots, n_L)$ satisfying

$$\Sigma N_h^2 S_h^2 / n_h = \sum_{h=1}^{L} N_h S_h^2 + N^2 V_0.$$

† Usually the actual solution (in integers satisfying $0 < n_h \leq N_h$), as well as any reasonable approximation to it arrived at by making small adjustments in the solution for the continuous problem, give a value of $V(\bar{y}_{st})$ close to V', especially if n is not very small. In such a case extensive search for the actual solution is not worthwhile. Moreover, it may be helpful to point out that one should not round down the smallest allocations unless the corresponding $N_h S_h$ are relatively large.

‡ In practice we also have a bound C_0 on the budget; if calculation shows that, for the minimum cost C', $C' > C_0$, we do not sample.

§ This can be seen by interchanging the expression defining g and U, respectively; their product, W, is left unchanged.

REMARK. If we use Lagrange multipliers, we obtain the first L equations (with proportionality factor $\mu^{\frac{1}{2}}$), because now we take derivatives of the auxiliary function

$$\psi = \Sigma n_h + \mu(\Sigma N_h^2 S_h^2/n_h - \Sigma N_h S_h^2 - N^2 V_0).$$

The "minimal" value for Σn_h is

$$n' = \frac{(\Sigma(N_h/N)S_h)^2}{V_0 + (1/N)\,\Sigma(N_h/N)S_h^2};$$

and therefore the "minimal" expenditure is $c' = c_0 + cn'$.

We can easily obtain an expression for the reduction in the variance obtained by using Neyman-Tshuprov allocation as compared with proportional allocation:

$$V_{\text{prop}} - V'$$
$$= \left\{\frac{1}{n}\sum\frac{N_h}{N}S_h^2 - \frac{1}{N}\sum\frac{N_h}{N}S_h^2\right\} - \left\{\frac{1}{n}\left(\sum\frac{N_h}{N}S_h\right)^2 - \frac{1}{N}\sum\frac{N_h}{N}S_h^2\right\}$$
$$= \frac{1}{nN}\Sigma N_h(S_h - \bar{S})^2 \qquad \left(\bar{S} = \sum\frac{N_h}{N}S_h\right).$$

From this we see that the difference is caused by the variation between the S_h. When the characteristic of interest is a proportion and is not too close to 0 or 1, calculations show that V_{prop} is close to V'. But in other cases, this shows that, even if the variation among the strata means is small, V_{st} may be much smaller than V_{srs} if there are marked differences among the S_h and an allocation close to the Neyman-Tshuprov allocation may be achieved.

Usually the minimal variance is somewhat larger than V', since in the first place the values of n_h which are obtained in the approximate solution are not whole numbers, and are not always between 1 and N_h; and since in the second place the adjusted variances S_h^2 which we used are in practice only estimates.† This may influence V' very much only if for certain h's we obtain n_h very much larger than the corresponding N_h.

Case (b): The cost function is $c_0 + \Sigma c_h n_h$; then we obtain by any of the previous methods‡ that the n_h have to satisfy

$$\frac{n_h}{n} = \frac{N_h S_h / c_h^{\frac{1}{2}}}{\Sigma N_h S_h / c_h^{\frac{1}{2}}}.$$

† Note the fact that, even if the estimated values of the adjusted variances S_h^2 are very far from their actual values, $\bar{y}_{\text{st}}$ is still an unbiased estimator of $\bar{Y}$.

‡ Then take $a_h = (c_h n_h^{\frac{1}{2}})$ and b_h as before, as we minimize $\Sigma c_h n_h U(n_1, \ldots, n_L)$; it is evident that c_o has no effect on the allocation of n among the strata.

If we substitute these n_h/n in $V(\bar{y}_{st})$, we obtain its "minimal" value

$$V''(n) = \frac{1}{n}\left(\sum \frac{N_h}{N} S_h c_h^{\frac{1}{2}}\right)\left(\sum \frac{N_h}{N} S_h/c_h^{\frac{1}{2}}\right) - \frac{1}{N}\sum \frac{N_h}{N} S_h^2.$$

In order to find the "best" n for given C_0, it suffices to substitute in the equation $C_0 = c_0 + \Sigma c_h n_h$ the values which we obtained for n_h/n:

$$n'' = \frac{(C_0 - c_0)\Sigma N_h S_h/c_h^{\frac{1}{2}}}{\Sigma N_h S_h c_h^{\frac{1}{2}}}.$$

(This formula of course reduces to $n'' = (C_0 - c_0)/c$ if $c_1 = \ldots = c_L = c$.) Therefore the "minimal" variance, with a budget of C_0, is

$$V'' = \frac{1}{C_0 - c_0}\left(\sum \frac{N_h}{N} S_h c_h^{\frac{1}{2}}\right)^2 - \frac{1}{N}\sum \frac{N_h}{N} S_h^2.$$

If V_0 is fixed in advance, again the n_h are proportional to the N_h, S_h and $c_h^{-\frac{1}{2}}$; if we now equate $V''(n)$ to V_0 and solve for n, we obtain

$$n'' = \frac{(\Sigma(N_h/N)S_h c_h^{\frac{1}{2}})(\Sigma(N_h/N)S_h/c_h^{\frac{1}{2}})}{V_0 + N^{-1}\Sigma(N_h/N)S_h^2}.$$

Let us now substitute the "optimal" n_h/n and this n'' in the cost function; we then obtain the "minimal" cost if the variance is fixed in advance to be V_0:

$$C'' = c_0 + \frac{n''\Sigma(N_h/N)S_h c_h^{\frac{1}{2}}}{\Sigma(N_h/N)S_h/c_h^{\frac{1}{2}}} = c_0 + \frac{(\Sigma(N_h/N)S_h c_h^{\frac{1}{2}})^2}{V_0 + N^{-1}\Sigma(N_h/N)S_h^2}.$$

EXAMPLE. In sampling firms to obtain an average net selling price of a certain product, the population was divided into three strata as follows:

	N_h	c_h	S_h
firms of large size	10	1	50
firms of intermediate size	60	5	20
firms of small size	100	10	10

For each firm we observe the appropriate total proceeds from sales of the product divided by the average number of units sold per firm in the population. The given S_h and c_h are rough estimates, the c_h are given in units of 10 hours of a professional's time. (The high cost for small firms is due to their not keeping books, so that the information has to be compiled from documents on individual transactions. Most firms of intermediate size keep books, but only for the large firms can the net selling price be readily obtained from the accounts.) Find:

(i) the sample sizes required to assure that $V(\bar{y}_{st})$ not exceed 4,

(ii) the sample sizes required to keep direct time cost below 3 500 hours; and in both cases assess the total direct time involved and $V(\bar{y}_{st})$.

Solution: As a first approximation we have

$$n_1:n_2:n_3 = N_1S_1c_1^{-\frac{1}{2}}:N_2S_2c_2^{-\frac{1}{2}}:N_3S_3c_3^{-\frac{1}{2}} = 500:537:316,$$

giving

$$n_1/n = \frac{500}{1\,353}, \quad n_2/n = \frac{537}{1\,353}, \quad n_3/n = \frac{316}{1\,353};$$

$$c_1n_1/\Sigma c_hn_h = \frac{500}{6\,345}, \quad c_2n_2/\Sigma c_hn_h = \frac{2\,685}{6\,345}, \quad c_3n_3/\Sigma c_hn_h = \frac{3\,160}{6\,345}.$$

(i) As $\Sigma N_hS_h^2 = 59\,000$ and $N = 170$, we have

$$n'' = \frac{6\,345 \times 1\,353}{4 \times 170^2 + 59\,000} = 49.2,$$

and so $(500/1\,353)n'' > 10 = N_1$, whereas N_2 and N_3 exceed n''. So we take $n_1 = 10$ and start the calculations afresh without stratum 1:

$$n_2/(n_2+n_3) = \frac{537}{853}, \quad c_2n_2/(c_2n_2+c_3n_3) = \frac{2\,685}{5\,845}, \quad N_2S_2^2 + N_3S_3^2 = 34\,000;$$

$$n'' = \frac{5\,845 \times 853}{4 \times 170^2 + 34\,000} = 33.3,$$

so take $n_2 + n_3 = 34$. Now $\frac{537}{853}\,34 = 21.4$. If $n_2 = 21$ and $n_3 = 13$. we assess $V(\bar{y}_{st})$ at $(N_2^2S_2^2n_2^{-1} + N_3^2S_3^2n_3^{-1} - N_2S_2^2 - N_3S_3^2)/N^2 = 3.9$. So take $n_1 = 10$, $n_2 = 21$, $n_3 = 13$. The total direct time involved is assessed at 2 450 hours.

(ii) Since here the direct time allowed is larger, we shall surely include all the units of the first stratum (with $c_1n_1 = 10$);

$$n'' = \frac{340 \times 853}{5\,845} = 49.6$$

(as the time allowed for strata 2 and 3 is $350 - 10 = 340$), so take $n_2 + n_3 = 50$. Now $\frac{537}{853}\,50 = 31.48$. If $n_2 = 31$ and $n_3 = 19$, $c_2n_2 + c_3n_3 = 345 > 340$; but if $n_2 = 32$ and $n_3 = 18$, $c_2n_2 + c_3n_3 = 340$. So take $n_1 = 10$, $n_2 = 32$, $n_3 = 18$. Then $V(\bar{y}_{st})$ is assessed at $(N_2^2S_2^2n_2^{-1} + N_3^2S_3^2n_3^{-1} - N_2S_2^2 - N_3S_3^2)/N^2 = 3.5$, and the total direct time at 3 500 hours.

If the cost of travel between strata is high, a better approximation to the cost function is $c_0 + \Sigma c_hn_h^{\frac{1}{2}}$; see HANSEN, HURWITZ and MADOW, Sample Survey Methods and Theory (Wiley, 1953) Vol. I, Chapter VI, section 12.

For arbitrary allocation ratios $w_h = n_h/n$, if $V(\bar{y}_{st})$ is to be smaller than V_0, n must exceed

$$\frac{\Sigma(N_h^2/N^2)S_h^2/w_h}{V_0 + N^{-1}\Sigma(N_h/N)S_h^2}.$$

This gives as a first approximation for n

$$n_0 = \frac{1}{V_0} \sum \frac{N_h^2}{N^2} S_h^2/w_h.$$

If the w_h are not very different from N_h/N, n_0 is approximately equal to $V_0^{-1}\Sigma(N_h/N)S_h^2$ and n approximately equal to $n_0/(1 + (n_0/N))$. Therefore in this case n_0 is close to n if n_0/N is small. In other cases it is worthwhile to improve n_0 with the help of the formula

$$n = \frac{n_0}{1 + (V_0 N)^{-1}\Sigma(N_h/N)S_h^2}$$

or (with proportional allocation) with the help of the formula

$$n = n_0/(1 + (n_0/N)).$$

REMARKS. (i) In the beginning of the section we gave a formula for $V(\bar{y}_{\text{srs}}) - V(\bar{y}_{\text{st}})$ for the case of stratified sampling with proportional allocation. This is obtained by noting that because of

$$N\sigma^2(\mathscr{Y}) = \Sigma(N_h - 1)S_h^2 + \Sigma N_h(\bar{Y}_h - \bar{Y})^2,$$

and

$$\frac{N}{N-1} = 1 + \frac{1}{N-1},$$

we have

$$\begin{aligned} V(\bar{y}_{\text{srs}}) &= \frac{N-n}{N-1}\frac{1}{n}\sigma^2 \\ &= \frac{N-n}{N}\frac{1}{n}\left(1 + \frac{1}{N-1}\right)\sum\frac{N_h}{N}S_h^2 - \frac{N-n}{N}\frac{1}{n}\frac{1}{N-1}\sum S_h^2 \\ &\quad + \frac{N-n}{N}\frac{1}{n}\frac{1}{N-1}\Sigma N_h(\bar{Y}_h - \bar{Y})^2 \\ &= \frac{N-n}{N}\frac{1}{n}\left\{\sum\frac{N_h}{N}S_h^2 - \frac{1}{N-1}\sum\frac{N-N_h}{N}S_h^2 \right. \\ &\quad \left. + \frac{1}{N-1}\Sigma N_h(\bar{Y}_h - \bar{Y})^2\right\}. \end{aligned}$$

(ii) If N/L is large, the middle term in the brackets is small in absolute value compared with the first,† so that $V_{\text{srs}} > V_{\text{prop}}$ and

$$V_{\text{srs}} - V_{\text{prop}} \approx \frac{N-n}{N}\frac{1}{n}\Sigma N_h(\bar{Y}_h - \bar{Y})^2/(N-1).$$

† If $N/N_h \approx L$, the middle term is about $-\dfrac{L-1}{N-1}V_{\text{prop}}$.

If N/L is not large, it may be that $V_{\text{srs}} < V_{\text{prop}}$. From the formula for $V_{\text{prop}} - V'$ it is clear that it may even occur that $V_{\text{srs}} < V'$. (For instance, if, for all h, $\bar{Y}_h \approx \bar{Y}$ and $S_h \approx \bar{S}$ with $N/N_h \approx L$, this will be so.) Moreover, note that $V_{\text{srs}} - V_{\text{prop}}$ and $V_{\text{prop}} - V'$ are (if positive) less for big n than for small n (the relative differences in variance according to the two modes of allocation

$$\frac{V_{\text{srs}} - V_{\text{prop}}}{V_{\text{prop}}} \quad \text{and} \quad \frac{V_{\text{prop}} - V'}{V'}$$

do not depend on n).

(iii) If the c_h vary from stratum to stratum, we may wish to compare the minimum variance of $\bar{y}_{\text{st}}$ under optimal allocation given the budget C_0:

$$\frac{1}{C_0 - c_0}\left(\Sigma \frac{N_h}{N} S_h c_h^{\frac{1}{2}}\right)^2 + \frac{1}{N} \Sigma \frac{N_h}{N} S_h^2$$

with the minimum variance of $\bar{y}_{\text{st}}$ under Neyman-Tshuprov allocation given the budget C_0:

$$\frac{1}{C_0 - c_0}\left(\Sigma \frac{N_h}{N} S_h c_h\right)\left(\Sigma \frac{N_h}{N} S_h\right) - \frac{1}{N} \Sigma \frac{N_h}{N} S_h^2.$$

The latter formula is obtained if we recall that under Neyman-Tshuprov allocation

$$n_h = nN_hS_h/\Sigma N_hS_h,$$

so that

$$C_0 - c_0 = \Sigma c_h n_h = n\Sigma c_h N_h S_h/\Sigma N_h S_h,$$

which gives the value of n to be substituted in the formula for V'.

(iv) In the case of the domain estimator $\hat{\bar{Y}}_{R1}$ of section 7, we find at once that the "optimal" ratios n_h/N_h are proportional to the square root of

$$\frac{N_{h1} - 1}{N_h - 1}(S_{h1}^2 + T_{h1}^2)/c_h,$$

where

$$T_{h1}^2 = (1 - N_{h1}/N_h)(\bar{Y}_{h1} - \bar{Y}_1)^2/(N_{h1}/(N_{h1} - 1)).$$

In many cases T_{h1}^2 is a fairly small or not very far from constant proportion of S_{h1}^2; the "optimal" n_h are then proportional to $N_h^{\frac{1}{2}}$, $(N_{h1} - 1)^{\frac{1}{2}}$, S_{h1} and $c_h^{-\frac{1}{2}}$.

9. Double sampling for stratification

Often we wish to stratify but find it impracticable to prepare lists of all the elements classified by stratum. In this case we may take a simple random sample of size n' from $\mathscr{P}$, classify the units in this sample according to the strata to which they belong, and use this classified listing to draw a stratified subsample. (As an alternative, to be discussed in section 11, we may, instead of further subsampling, weight the simple random sample, according to the strata weights N_h/N, provided we know these weights; in the present method, the N_h/N need not be known.) An extension to triple sampling (for double stratification) to estimate a relative frequency, with application to estimation of magazine readership, is given by ROBSON and KING, J. Amer. Statist. Ass., **47** (1952) 203–215, **48** (1953) 911.

9.1. NOTATION; ESTIMATOR FOR $\bar{Y}$

Let us denote the sample of size n' by $\mathscr{s}'$, and denote those of its members that belong to stratum h by $\mathscr{s}'_h$. Let n'_h denote the number of elements in $\mathscr{s}'_h$. We assume that n' is so large that the probability that any n'_h takes on the value 1 or 0 is negligible.† (If under these circumstances some n'_h is 1 or 0, we may increase n' until this is no more the case, or we may merge strata such that all n'_h exceed 1.)

In subsampling we take, separately and independently for each h, a sample $\mathscr{s}_h$ of size n_h from $\mathscr{s}'_h$. Let $\bar{y}_h$ and s^2_h be the sample mean of the $\mathscr{Y}$-values and the sample estimator of S^2_h based on $\mathscr{s}_h$.

Our estimator for $\bar{Y}$ will be

$$\bar{y}_{\mathrm{st,d}} = \sum_{h=1}^{L} \frac{n_h}{n'} \bar{y}_h.$$

This estimator is unbiased, since its conditional mean, given $n'_1, \ldots, n'_L$ (all positive), is

$$\sum_{h=1}^{L} (n'_h/n') \bar{Y}_h,$$

which has expected value

$$\sum_{h=1}^{L} (N_h/N) \bar{Y}_h = \bar{Y}.$$

Occasionally the N_h/N are known and our estimator will then be

$$\sum_{h=1}^{L} (N_h/N) \bar{y}_h.$$

The various formulas given below will then simplify in an obvious way.

† These conditions are sufficient for the theory given below to be applicable. However, it is clear that the method could only be efficient if n' is large and most n'_h not small.

9.2. EXAMPLE

The population is

$$\mathscr{P} = \{u_1, u_2, u_3, u_4, u_5\},$$

the strata are

$$\mathscr{S}_1 = \{u_1, u_2\} \quad \text{and} \quad \mathscr{S}_2 = \{u_3, u_4, u_5\}.$$

$\mathscr{Y}$ is defined by $y(u_i) = i$, so $\bar{Y}_1 = 1\frac{1}{2}$, $\bar{Y}_2 = 4$, $S_1^2 = \frac{1}{2}$, $S_2^2 = 1$, $\bar{Y} = 3$, $S^2 = 2\frac{1}{2}$. Take a preliminary sample s' of size $n' = 4$; s_1' constitutes those of its elements belonging to $\mathscr{S}_1$, and s_2' those belonging to $\mathscr{S}_2$. (Note that in this case no modification of the method of drawing s' is needed to ensure that the n_h' are positive.)

Then take a sample of size $n_1 = 1$ from s_1' and $n_2 = 2$ from s_2'.

Possible different samples s' of size 4	n_1'	n_2'	Possible different subsamples	Probability	$(\bar{y}_1, \bar{y}_2)$	$\bar{y}_{st,d}$	$\mathscr{E}\bar{y}_{st,d}$ (conditional)
$\{u_1, u_2; u_3, u_4\}$	2	2	$\{u_1\}, \{u_3, u_4\}$	$\frac{1}{10}$	$(1, 3\frac{1}{2})$	$2\frac{1}{4}$	$2\frac{1}{2}$
			$\{u_2\}, \{u_3, u_4\}$	$\frac{1}{10}$	$(2, 3\frac{1}{2})$	$2\frac{3}{4}$	
$\{u_1, u_2; u_3, u_5\}$	2	2	$\{u_1\}, \{u_3, u_5\}$	$\frac{1}{10}$	$(1, 4)$	$2\frac{1}{2}$	$2\frac{3}{4}$
			$\{u_2\}, \{u_3, u_5\}$	$\frac{1}{10}$	$(2, 4)$	3	
$\{u_1, u_2; u_4, u_5\}$	2	2	$\{u_1\}, \{u_4, u_5\}$	$\frac{1}{10}$	$(1, 4\frac{1}{2})$	$2\frac{3}{4}$	3
			$\{u_2\}, \{u_4, u_5\}$	$\frac{1}{10}$	$(2, 4\frac{1}{2})$	$3\frac{1}{4}$	
$\{u_1; u_3, u_4, u_5\}$	1	3	$\{u_1\}, \{u_3, u_4\}$	$\frac{1}{15}$	$(1, 3\frac{1}{2})$	$2\frac{7}{8}$	
			$\{u_1\}, \{u_3, u_5\}$	$\frac{1}{15}$	$(1, 4)$	$3\frac{1}{4}$	$3\frac{1}{4}$
			$\{u_1\}, \{u_4, u_5\}$	$\frac{1}{15}$	$(1, 4\frac{1}{2})$	$3\frac{5}{8}$	
$\{u_2; u_3, u_4, u_5\}$	1	3	$\{u_2\}, \{u_3, u_4\}$	$\frac{1}{15}$	$(2, 3\frac{1}{2})$	$3\frac{1}{8}$	
			$\{u_2\}, \{u_3, u_5\}$	$\frac{1}{15}$	$(2, 4)$	$3\frac{1}{2}$	$3\frac{1}{2}$
			$\{u_2\}, \{u_4, u_5\}$	$\frac{1}{15}$	$(2, 4\frac{1}{2})$	$3\frac{7}{8}$	
						Mean	3

Variance

$$\begin{aligned}\{(2\tfrac{1}{4} - 3)^2 + (2\tfrac{3}{4} - 3)^2\}\tfrac{1}{10} &+ \{(2\tfrac{1}{2} - 3)^2 + (3 - 3)^2\}\tfrac{1}{10} \\ &+ \{(2\tfrac{3}{4} - 3)^2 + (3\tfrac{1}{4} - 3)^2\}\tfrac{1}{10} \\ &+ \{(2\tfrac{7}{8} - 3)^2 + (3\tfrac{1}{4} - 3)^2 + (3\tfrac{5}{8} - 3)^2\}\tfrac{1}{15} \\ &+ \{(3\tfrac{1}{8} - 3)^2 + (3\tfrac{1}{2} - 3)^2 + (3\tfrac{7}{8} - 3)^2\}\tfrac{1}{15} = \tfrac{1}{5}.\end{aligned}$$

On the other hand, the variance of $\bar{y}$ in a simple random sample of 3 is $\frac{1}{3}$, and the variance of $\bar{y}_{st}$ in a stratified (single phase) sample (with $n_1 = 1$ and $n_2 = 2$) is $\frac{1}{10}$. The method of section 11 gives 0.136.

9.3. THE VARIANCE OF THE ESTIMATOR

We now note that the conditional variance of $\bar{y}_{\text{st,d}}$, given the n'_h, is obtained from our formula for $V(\bar{y}_{\text{st}})$ by replacing the weights N_h/N by n'_h/n':

$$\sum \left(\frac{n'_h}{n'}\right)^2 \frac{N_h - n_h}{N_h} \frac{1}{n_h} S_h^2.$$

This would be obvious if the samples $\mathscr{s}_1, \ldots, \mathscr{s}_h$ would have been drawn from the strata independently of the sample $\mathscr{s}'$. (Actually, all results given below hold for this case, which is, however, of little practical interest.) For the case of subsampling, this statement may be proved as follows: let $(n'_h - 1)s_h'^2$ be the sum of squares of deviations of the $\mathscr{Y}$ values in $\mathscr{s}'_h$ from their average $\bar{y}'_h$. Then

$$\mathscr{E}\{\bar{y}_h|\mathscr{s}'_h\} = \bar{y}'_h \quad \text{and} \quad V\{\bar{y}_h|\mathscr{s}'_h\} = ((1/n_h) - (1/n'_h))s_h'^2.$$

Therefore

$$\begin{aligned} V\{\bar{y}_h|n'_h\} &= \mathscr{E}[V\{\bar{y}_h|\mathscr{s}'_h\}|n'_h] + V[\mathscr{E}\{\bar{y}_h|\mathscr{s}'_h\}|n'_h] \\ &= \left(\frac{1}{n_h} - \frac{1}{n'_h}\right) \mathscr{E}\{s_h'^2|n'_h\} + V\{\bar{y}'_h|n'_h\} \\ &= \left(\frac{1}{n_h} - \frac{1}{n'_h}\right) S_h^2 + \left(\frac{1}{n'_h} - \frac{1}{N_h}\right) S_h^2 \\ &= \left(\frac{1}{n_h} - \frac{1}{N_h}\right) S_h^2. \end{aligned}$$

Therefore the variance of $\Sigma(n'_h/n')\bar{y}_h$ given the n'_h is as stated above. Since, given the n'_h, the conditional bias of $\bar{y}_{\text{st,d}}$ equals $\Sigma((n'_h/n') - (N_h/N))\bar{Y}_h$, we get as conditional mean square error

$$\sum \left(\frac{n'_h}{n'}\right)^2 \frac{N_h - n_h}{N_h} \frac{1}{n_h} S_h^2 + \left\{\sum \left(\frac{n'_h}{n'} - \frac{N_h}{N}\right) \bar{Y}_h\right\}^2.$$

Case (i)†: the n_h/n do not depend on the n'_h/n'. Because

$$V(n'_h/n') = \frac{N - n'}{N - 1} \frac{1}{n'} \frac{N_h}{N}\left(1 - \frac{N_h}{N}\right),$$

and similarly

$$\text{Cov}\,(n'_h/n', n'_k/n') = -\frac{N - n'}{N - 1} \frac{1}{n'} \frac{N_h}{N} \frac{N_k}{N} \qquad (h \neq k),$$

† Strictly speaking it is not possible to choose $n_h > 1$ independently of n'_h when we subsample; but we can, by choosing n' sufficiently large, ensure that $n_h(>1)$ chosen in advance will not exceed n'_h with a probability very close to 1, provided we have some notion of the N_h.

we obtain

$$\mathscr{E}\left(\frac{n_h'}{n'}\right)^2 = \frac{N-n'}{N-1}\frac{1}{n'}\frac{N_h}{N}\left(1-\frac{N_h}{N}\right) + \left(\frac{N_h}{N}\right)^2$$

$$\mathscr{E}\left\{\sum\left(\frac{n_h'}{n'} - \frac{N_h}{N}\right)\bar{Y}_h\right\}^2$$

$$= \frac{N-n'}{N-1}\frac{1}{n'}\left\{\sum\frac{N_h}{N}\left(1-\frac{N_h}{N}\right)\bar{Y}_h^2 - \sum_{h\neq k}\sum\frac{N_h}{N}\frac{N_k}{N}\bar{Y}_h\bar{Y}_k\right\}$$

$$= \frac{N-n'}{N-1}\frac{1}{n'}\left(\sum\frac{N_h}{N}\bar{Y}_h^2 - \bar{Y}^2\right)$$

$$= \frac{N-n'}{N-1}\frac{1}{n'}\sum\frac{N_h}{N}(\bar{Y}_h - \bar{Y})^2$$

and

$$V(\bar{y}_{\mathrm{st,d}}) = V(\bar{y}_{\mathrm{st}}) + \frac{N-n'}{N-1}\frac{1}{n'}\sum\frac{N_h}{N}$$

$$\times\left\{\left(1-\frac{N_h}{N}\right)\frac{N_h-n_h}{N_h}\frac{S_h^2}{n_h} + (\bar{Y}_h - \bar{Y})^2\right\}.$$

Case (ii): the n_h/n depend on the n_h'/n'. This is the more usual case. Let us say that

$$n_h/n \approx \lambda_h n_h'/\Sigma\lambda_h n_h',$$

where the λ_h are known in advance (for instance, in proportional allocation $\lambda_h = 1$). Then the conditional MSE is

$$\frac{1}{n}\sum\frac{n_h'}{n'}S_h^2\frac{\Sigma\lambda_k n_k'}{\lambda_h n'} - \sum\left(\frac{n_h'}{n'}\right)^2 S_h^2\frac{1}{N_h} + \left\{\sum\left(\frac{n_h'}{n'} - \frac{N_h}{N}\right)\bar{Y}_h\right\}^2.$$

The expectation of

$$n^{-1}(n_h'/n')\sum_{k=1}^{L}\lambda_k n_k'/(n'\lambda_h)$$

is equal to

$$\frac{1}{n}\left[\frac{N-n'}{N-1}\frac{1}{n'}\frac{N_h}{N}\left(1-\frac{N_h}{N}\right) + \left(\frac{N_h}{N}\right)^2\right.$$

$$\left. + \left(1 - \frac{N-n'}{N-1}\frac{1}{n'}\right)\frac{N_h}{N}\frac{Q - \lambda_h N_h/N}{\lambda_h}\right]$$

$$= \frac{1}{n}\frac{N_h}{N}\left\{\frac{Q}{\lambda_h} + \frac{N-n'}{N-1}\frac{1}{n'}\left(1-\frac{Q}{\lambda_h}\right)\right\},$$

where

$$Q = \Sigma \frac{N_h}{N} \lambda_h.$$

Also

$$\mathscr{E}\Sigma \left(\frac{n_h'}{n'}\right)^2 S_h^2 \frac{1}{N_h} = \frac{1}{N}\left\{\frac{N-n'}{N-1}\frac{1}{n'}\Sigma\left(1-\frac{N_h}{N}\right)S_h^2 + \Sigma\frac{N_h}{N}S_h^2\right\}.$$

Therefore†

$$V(\bar{y}_{\text{st,d}}) = \Sigma \frac{N_h}{N}\left[\left(\frac{1}{n}\frac{Q}{\lambda_h} - \frac{1}{N}\right) S_h^2 \right.$$

$$\left. + \frac{N-n'}{N-1}\frac{1}{n'}\left\{\left(\frac{1}{n}\left(1-\frac{Q}{\lambda_h}\right) + \frac{1}{N}\right) S_h^2 + (\bar{Y}_h - \bar{Y})^2\right\}\right]$$

$$- \frac{N-n'}{N-1}\frac{1}{n'}\frac{1}{N}\Sigma S_h^2$$

$$= \Sigma\frac{N_h}{N}\left[\left\{\frac{1}{n}\frac{Q}{\lambda_h} + \frac{1}{n'n}\left(1-\frac{Q}{\lambda_h}\right)\right\} S_h^2 + \frac{1}{n'}(\bar{Y}_h - \bar{Y})^2\right]$$

$$- \frac{A_1}{N-1}\left(\frac{1}{n'} - \frac{1}{N}\right) + \frac{A_2(n', n)}{N-1} - \frac{A_3}{N},$$

where

$$A_1 = \Sigma\left(1 - \frac{N_h}{N}\right) S_h^2,$$

$$A_2(n', n) = \left(\frac{1}{n'} - 1\right)\Sigma\frac{N_h}{N}\left\{\frac{1}{n}\left(1 - \frac{Q}{\lambda_h}\right) S_h^2 + (\bar{Y}_h - \bar{Y})^2\right\},$$

$$A_3 = \Sigma\frac{N_h}{N} S_h^2.$$

† Note that this result, which is exact if we can choose n_h/n precisely as indicated above, does not coincide with the approximations given in DEMING, Sample Design in Business Research (Wiley, 1960), for his plans H and I ($\lambda_h = 1$ and $\lambda_h = S_h$, respectively). Also the optimum n comes out different, viz.

$$n = n'\Sigma N_h S_h \{N_h(\bar{Y}_h - \bar{Y})^2\}^{-\frac{1}{2}}(c'/c)^{\frac{1}{2}}.$$

In Deming's first example (in Chapter 15, D) both formulae give the same answer; his second allows only approximate computation according to our formula, and that gives nearly the same answer.

9.4. ALLOCATION

In order to obtain a minimum for this expression (neglecting the term in A_2), we have to choose the λ_k such that $\Sigma N_k\lambda_k \Sigma N_h S_h^2/\lambda_h$ is minimum. This is achieved (by putting

$$a_h = (N_h\lambda_h)^{\frac{1}{2}}, \quad b_h = S_h(N_h/\lambda_h)^{\frac{1}{2}}$$

in the Schwarz-Cauchy inequality) if

$$\lambda_h = S_h/\Sigma S_h.$$

Then

$$\Sigma \frac{N_h}{N} \frac{1}{n} \frac{Q}{\lambda_h} S_h^2$$

becomes

$$\frac{1}{n}\left(\Sigma \frac{N_h}{N} S_h\right)^2.$$

If the cost function is $c_0 + c'n' + cn$, we can obtain approximations to the optimal n and n' by neglecting

$$\{A_1/(N-1)\}\{(1/n') - (1/N)\} \quad \text{and} \quad A_2(n', n)/(N-1).$$

That is to say, we have to minimize

$$B_0 = (1/n)(1 - (1/n'))\, B_1 + (1/nn')B_2 + (1/n')B_3$$

under the condition that

$$c_0 + c'n' + cn \leq C_0,$$

where

$$B_1 = \left(\Sigma \frac{N_h}{N} S_h\right)^2, \quad B_2 = A_3, \quad B_3 = \Sigma \frac{N_h}{N} (\bar{Y}_h - \bar{Y})^2.$$

Note that $B_2 \geq B_1$. If we take the derivatives of $B_0(n, n') + \mu(c'n' + cn)$, we obtain

$$\frac{(n')^2c'}{n^2c} = \frac{n'}{n} \frac{(B_2 - B_1) + B_3 n}{B_1(n'-1) + B_2}$$

or

$$\frac{n'c'}{nc} = \frac{(B_2 - B_1) + B_3 n}{B_1(n'-1) + B_2};$$

therefore

$$B_1 c'(n')^2 + (B_2 - B_1)(c'n' - cn) - B_3 cn^2 = 0$$

or

$$0 = \left\{\frac{B_1}{c'}(c'n' + cn) + (B_2 - B_1)\right\}(c'n' - cn) + \left(\frac{B_1}{c'} - \frac{B_3}{c}\right)c^2n^2$$

$$= B_4(C_0 - c_0) - 2B_4cn + \left(\frac{B_1}{c'} - \frac{B_3}{c}\right)c^2n^2,$$

where

$$C_0 - c_0 = c'n' + cn, \quad B_4 = \frac{B_1}{c'}(C_0 - c_0) + (B_2 - B_1) > 0.$$

Therefore

$$cn\left(\frac{B_1}{c'} - \frac{B_3}{c}\right) = B_4 \pm \left\{B_4^2 - B_4(C_0 - c_0)\left(\frac{B_1}{c'} - \frac{B_3}{c}\right)\right\}^{\frac{1}{2}}$$

$$= B_4 \pm \left[B_4\left\{(C_0 - c_0)\frac{B_3}{c} + (B_2 - B_1)\right\}\right]^{\frac{1}{2}},$$

because

$$0 \leq nn'B_0(n, n')$$

$$= B_4 - cn\left(\frac{B_1}{c'} - \frac{B_3}{c}\right)$$

$$= B_4 - \left[B_4 \pm \left\{B_4^2 - B_4(C_0 - c_0)\left(\frac{B_1}{c'} - \frac{B_3}{c}\right)\right\}^{\frac{1}{2}}\right].$$

The minus sign is required and therefore

$$cn\left(\frac{B_1}{c'} - \frac{B_3}{c}\right) = B_4 - \left[B_4\left\{(C_0 - c_0)\frac{B_3}{c} + (B_2 - B_1)\right\}\right]^{\frac{1}{2}}$$

(note that if $(B_1/c') - (B_3/c) \gtrless 0$, the right-hand side $\gtrless 0$), and

$$n' = (C_0 - c_0 - cn)/c'.$$

REMARK. The point we found is not a maximum or a saddle point, since

$$\det\begin{pmatrix} \left(\dfrac{\partial^2 B_0(n, n')}{\partial^2(n,n')}\right) & \begin{matrix} c \\ c' \end{matrix} \\ \begin{matrix} c & c' \end{matrix} & 0 \end{pmatrix}$$

$$= \frac{-2}{(nn')^3}\left[\{(nc - n'c')^2 + ncn'c'\}(B_2 - B_1) + n^3c^2B_3 + n'^3c'^2B_1\right] < 0.$$

9.5. ESTIMATOR FOR $V(\bar{y}_{st,d})$ IN CASE (i)

The expected value of

$$\sum \frac{n_h'}{n'} (\bar{y}_h - \bar{y}_{st,d})^2 = \sum \frac{n_h}{n'} \bar{y}_h^2 - \bar{y}_{st,d}^2$$

equals

$$\sum \frac{N_h}{N} \left(\frac{1}{n_h} - \frac{1}{N_h}\right) S_h^2 + \sum \frac{N_h}{N} (\bar{Y}_h - \bar{Y})^2 - V(\bar{y}_{st,d}).$$

Let us denote the difference between $V(\bar{y}_{st,d})$ and

$$\frac{N-n'}{N-1} \frac{1}{n'} \sum \frac{N_h}{N} (\bar{Y}_h - \bar{Y})^2$$

by K; then

$$V(\bar{y}_{st,d}) = \frac{1}{n'-1} \left[\frac{N-n'}{N} \mathscr{E} \left\{\sum \frac{n_h'}{n'} (\bar{y}_h - \bar{y}_{st,d})^2 - \sum \frac{n_h'}{n'} \frac{s_h^2}{n_h} + \frac{1}{N} \sum s_h^2\right\} + \frac{N-1}{N} n'K\right].$$

In order to estimate

$$K = \left(1 - \frac{N-n'}{N-1} \frac{1}{n'}\right) \left\{\sum \frac{N_h^2}{N^2} \frac{S_h^2}{n_h} - \frac{1}{N} \sum \frac{N_h}{N} S_h^2\right\} + \frac{N-n'}{N-1} \frac{1}{n'} \left\{\sum \frac{N_h}{N} \frac{S_h^2}{n_h} - \frac{1}{N} \sum S_h^2\right\},$$

note that an unbiased estimator of $(N_h/N)^2$ is

$$\frac{(N-1)n'(n_h'^2/n'^2) - (N-n')(n_h'/n')}{N(n'-1)};$$

therefore an unbiased estimator of K is

$$\hat{K} = \sum \frac{n_h'^2}{n'^2} \frac{s_h^2}{n_h} - \frac{(n'-1)}{(N-1)n'} \sum \frac{n_h'}{n'} s_h^2 - \frac{N-n'}{N-1} \frac{1}{n'} \frac{1}{N} \sum s_h^2$$

and of $V(\bar{y}_{st,d})$ is

$$\frac{n'}{n'-1} \left[\frac{N-n'}{N} \frac{1}{n'} \sum \frac{n_h'}{n'} (\bar{y}_h - \bar{y}_{st,d})^2 + \frac{N-1}{N} \sum \frac{n_h'^2}{n'^2} \frac{s_h^2}{n_h} - \frac{N-n'}{N} \frac{1}{n'} \sum \frac{n_h'}{n'} \frac{s_h^2}{n_h} - \frac{n'-1}{n'} \frac{1}{N} \sum \frac{n_h'}{n'} s_h^2\right].$$

9.6. ESTIMATOR FOR $V(\bar{y}_{st,d})$ IN CASE (ii)

Denote by K the difference between $V(\bar{y}_{st,d})$ and

$$\{(N - n')/(N - 1)\}n'^{-1}\Sigma(N_h/N)(\bar{Y}_h - \bar{Y})^2.$$

First let us try to estimate $\Sigma(N_h/N)(\bar{Y}_h - \bar{Y})^2$. Since

$$\mathscr{E}\sum\frac{n'_h}{n'}(\bar{y}_h - \bar{y}_{st,d})^2$$

$$= \mathscr{E}\sum\frac{n'_h}{n'}\bar{y}_h^2 - \bar{Y}^2 - V(\bar{y}_{st,d})$$

$$= \mathscr{E}\sum\frac{n'_h}{n'}\frac{S_h^2}{n_h} - \mathscr{E}\sum\frac{n'_h}{n'}\frac{S_h^2}{N_h} + \mathscr{E}\sum\frac{n'_h}{n'}(\bar{Y}_h - \bar{Y})^2 - V(\bar{y}_{st,d})$$

$$= \mathscr{E}\,\frac{1}{n}\sum S_h^2\hat{Q}/\lambda_h - \frac{1}{N}\sum S_h^2 + \sum\frac{N_h}{N}(\bar{Y}_h - \bar{Y})^2 - V(\bar{y}_{st,d}),$$

where $\hat{Q} = \Sigma n'_h\lambda_h/n'$, the expected value of

$$T_1 = \sum\frac{n'_h}{n'}(\bar{y}_h - \bar{y}_{st,d})^2 - \frac{1}{n}\,\Sigma s_h^2\hat{Q}/\lambda_h + \frac{1}{N}\,\Sigma s_h^2$$

equals

$$\sum\frac{N_h}{N}(\bar{Y}_h - \bar{Y})^2 - V(\bar{y}_{st,d}).$$

Therefore

$$V(\bar{y}_{st,d}) = \frac{1}{n' - 1}\left\{\frac{N - n'}{N}\,\mathscr{E}T_1 + \frac{N - 1}{N}\,n'K\right\}.$$

Now let us try to estimate K. We find that the expected value of

$$\frac{n'_h}{n'}\,s_h^2\,\frac{\hat{Q}}{\lambda_h} = \frac{n'^2_h}{n'^2}\,s_h^2 + \frac{n'_h}{n'}\sum_{k\neq h}\frac{n'_k}{n'}\,\lambda_k\,\frac{s_h^2}{\lambda_h}$$

is equal to

$$\frac{N_h}{N}\,S_h^2\,\frac{Q}{\lambda_h} + \frac{N - n'}{N - 1}\,\frac{1}{n'}\,\frac{N_h}{N}\,S_h^2\left(1 - \frac{Q}{\lambda_h}\right);$$

therefore the expected value of

$$T_2 = \frac{1}{n}\sum\frac{n'_h}{n'}\,s_h^2\,\frac{\hat{Q}}{\lambda_h} - \frac{n' - 1}{n'}\,\frac{1}{N - 1}\sum\frac{n'_h}{n'}\,s_h^2 - \frac{N - n'}{N - 1}\,\frac{1}{n'}\,\frac{1}{N}\,\Sigma s_h^2$$

is equal to K; and the expected value of

$$\frac{1}{n'-1}\left\{\frac{N-n'}{N}T_1+\frac{N-1}{N}n'T_2\right\}$$

is equal to $V(\bar{y}_{\mathrm{st,d}})$.

9.7. APPLICATION OF THIS ANALYSIS TO DOMAIN ESTIMATION

As mentioned in the first footnote of section 11.1, Chapter II, it may be worthwhile to use double sampling instead of the method for domain estimation used there. Let us again define $\mathscr{Y}'$ as the characteristic $\mathscr{Y}$ for elements in the domain and as 0 for other elements, and let us identify the domain with the first stratum in the present setup, and the rest of the population with stratum $2(=L)$. Since we use a prime (′) to denote the domain, we shall denote the number of units constituting the preliminary sample by n^*, and the number of these that turn out to belong to the domain by n_1^*. We subsample n_1 of these and observe their $\mathscr{Y}$-values, denoting their average by $\bar{y}_1$. We shall assume that n_1 is fixed in advance, and that n^* has been taken large enough to make it very unlikely for n_1 to exceed n_1^*. (It may be advisable to make n_1 depend on what n_1^* turns out to be, especially if we have very little advance information on N_1/N. Other methods mentioned in section 14 of Chapter II may well be appropriate.)

To estimate Y' we may use

$$N(n_1^*/n^*)\bar{y}_1.$$

Its conditional mean square error given n_1^* is

$$N^2\left\{\left(\frac{n_1^*}{n^*}\right)^2\frac{N_1-n_1}{N_1}\frac{1}{n_1}S_1^2+\left(\frac{n_1^*}{n^*}-\frac{N_1}{N}\right)^2\bar{Y}_1^2\right\},$$

which, since n_1 is fixed in advance, has expected value

$$N^2\left[\left(\frac{1}{n_1}-\frac{1}{N_1}\right)N_1S_1^2\left\{N_1+\frac{N}{N-1}\left(\frac{1}{n^*}-\frac{1}{N}\right)(N-N_1)\right\}\right.$$
$$\left.+\frac{N}{N-1}\left(\frac{1}{n^*}-\frac{1}{N}\right)N_1(N-N_1)\bar{Y}_1^2\right].$$

To estimate $\bar{Y}_1$ we may use $\bar{y}_1$, which has variance $((1/n_1)-(1/N_1))S_1^2$. The cost function would usually be of the form

$$c_0+(c_1+c)n^*+c_2n_1,$$

where c_0 is fixed cost, c is cost of classifying a unit as to whether or not it belongs to the domain, c_1 is the cost of drawing a unit from the population (we neglect the cost of drawing n_1 units from the subsample), and c_2 the cost of measuring $\mathscr{Y}$ on any one unit in the domain. Note that n^* does not appear in $V(\bar{y}_1)$, but, since we have to make n^* sufficiently large to make $n_1^* \geq n_1$ with probability close to 1, n^* should somewhat exceed n_1N/N_1.

The cost function with single stage sampling may be taken to be

$$c_0 + (c_1 + c)n + c_2n_1,$$

which has expectation

$$c_0 + \left(c_1 + c + \frac{N_1}{N}\,c_2\right)n.$$

The comparison of the double sampling estimator of Y' with the single stage sampling one is not easily carried out in general terms; for small c double sampling often involves a large saving. On the other hand, it is easy to make this comparison for the estimation of $\bar{Y}_1$:

In single sampling of n, the approximate variance of $\bar{y}_1$ is found to be:

$$\left(\frac{1}{n} - \frac{1}{N}\right)\frac{N}{N_1}\,S_1^2\left\{1 + \frac{1}{n}\,\frac{N}{N_1}\left(1 - \frac{N_1}{N}\right)\right\}.$$

For given total cost C,

$$n = (C - c_0)/\{c_1 + c + (N_1/N)c_2\},$$

so that the variance in single sampling is about

$$\left[\left\{\frac{(c_1 + c)(N/N_1) + c_2}{C - c_0} - \frac{1}{N_1}\right\}S_1^2\right]\left\{1 + \frac{1}{n}\,\frac{N}{N_1}\left(1 - \frac{N_1}{N}\right)\right\}.$$

The term in square brackets is the variance in double sampling if

$$n^* = n_1(N/N_1).$$

Usually $n^{-1}(N/N_1)(1 - (N_1/N))$ will be rather small compared with 1; and it is the smaller, the smaller is $(c_1 + c)$ compared with c_2. Moreover, in most cases much of the contribution to the advantage of double sampling of the term under discussion would be expected to be balanced out by the fact that n^* has to be taken to exceed somewhat n_1N/N_1, especially if N/N_1 is not known.

The extension to ratio estimation, stratification, etc. is not difficult to carry out.

10. Estimation of what the variance of an estimator would have been if a different method of sampling had been employed

Consider the usual estimator of Y under simple random sampling ($N\bar{y}$) and under stratified sampling

$$\hat{Y}_{\text{st}} = \sum_{h=1}^{L} N_h\bar{y}_h.$$

We examine 2 problems:

(a) Given a stratified sample of size n, estimate what would have been the variance of $N\bar{y}$ in a simple random sample of the same size.

(b) Given a simple random sample of size n, estimate what would have been the variance of $\hat{Y}_{\text{st}}$ in a stratified sample of the same size.

These questions arise in assessing whether or not in future research it will be worthwhile to stratify.

Sub (a).† Since

$$\mathscr{E} N_h \sum_{i=1}^{N_h} a_{hi} y_{hi}^2 / n_h = \sum_{i=1}^{N_h} y_{hi}^2$$

(where $a_{hi} = 1$ if u_{hi} is in the sample and 0 otherwise), and since

$$Y^2 = \mathscr{E} \hat{Y}_{\text{st}}^2 - V(\hat{Y}_{\text{st}}),$$

we can estimate

$$V(N\bar{y}) = N^2 \frac{N-n}{N-1} \frac{1}{n} \{ (\sum_{h=1}^{L} \sum_{i=1}^{N_h} y_{hi}^2 / N) - (Y^2/N^2) \}$$

without bias by

$$\frac{N(N-1)}{n(N-1)} \left[\sum_{h=1}^{L} \frac{N_h}{n_h} \sum_{i=1}^{N_h} a_{hi} y_{hi}^2 - \frac{1}{N} \{ \hat{Y}_{\text{st}}^2 - v(\hat{Y}_{\text{st}}) \} \right]$$

$$= \frac{N(N-n)}{n(N-1)} \left[\sum_{h=1}^{L} N_h \left\{ s_h^2 \left(1 - \frac{1}{n_h} \right) + \bar{y}_h^2 \right\} - \frac{1}{N} \{ \hat{Y}_{\text{st}}^2 - v(Y_{\text{st}}) \} \right].$$

Under proportionate allocation this equals

$$\hat{V} = N^2 \frac{N-n}{N-1} \frac{1}{n^2} [\sum_{h=1}^{L} \{ (n_h - 1) s_h^2 + n_h (\bar{y}_h - \bar{y}_{\text{st}})^2 \} + n v(\bar{y}_{\text{st}})]$$

$$= N^2 \frac{N-n}{N-1} \frac{1}{n^2} \sum_{h=1}^{L} \left[\left\{ \left(1 + \frac{N-n}{Nn} \right) n_h - 1 \right\} s_h^2 + n_h (\bar{y}_h - \bar{y}_{\text{st}})^2 \right]$$

and $\bar{y}_{\text{st}} = \bar{y}$.

Under proportionate allocation, a simple random sample will distribute itself among the strata approximately proportionately if n is not small. So in that case one would expect that a good estimator of the variance under simple random sampling would be

$$v_0^2 = N^2 \frac{N-1}{N} \frac{1}{n} s_0^2,$$

† SUKHATME, Sampling Theory of Surveys and Applications (Indian Soc. of Agric. Statistics and Iowa State University Press, 1954) section 3b.5 discusses the case of unequal probability sampling within strata.

where

$$s_0^2 = \sum_{h=1}^{L} \sum_{i=1}^{N_h} a_{hi}(y_{hi} - \bar{y})^2 / (\sum_{h=1}^{L} n_h - 1)$$
$$= (n-1)^{-1}\{\Sigma(n_h - 1)s_h^2 + \Sigma n_h(\bar{y}_h - \bar{y})^2\}$$

and

$$\bar{y} = \sum_{h=1}^{L} \sum_{i=1}^{N_h} a_{hi} y_{hi} / \sum_{h=1}^{L} n_h.$$

If $n_h/n = N_h/N$, s_0^2 equals

$$\sum_{h=1}^{L} \{(n_h - 1)s_h^2 + n_h(\bar{y}_h - \bar{y}_{\text{st}})^2\}/(n-1).$$

From the above it follows that under proportionate allocation

$$\frac{N}{N-1} \frac{n-1}{n} \{v_0^2 - v(\hat{Y}_{\text{st}})_{\text{prop}}\}$$

estimates the excess of the variance under simple random sampling over that under stratified sampling with proportional allocation of the estimator of Y. For

$$\frac{N-1}{N} \hat{V} = N^2 \frac{N-n}{N} \frac{1}{n^2} \{(n-1)s_0^2 + nv(\bar{y}_{\text{st}})_{\text{prop}}\},$$

but

$$N^2 \frac{N-n}{N} \frac{n-1}{n^2} s_0^2 = \frac{n-1}{n} v_0^2,$$

so

$$\frac{N-1}{N} \hat{V} = \frac{n-1}{n} v_0^2 + N^2 \frac{N-n}{Nn} v(\bar{y}_{\text{st}})_{\text{prop}}$$

and

$$\frac{N-1}{N} \frac{n}{n-1} \{\hat{V} - v(\hat{Y}_{\text{st}})_{\text{prop}}\}$$
$$= v_0^2 + N^2 \left(\frac{N-n}{Nn} \frac{n}{n-1} - \frac{N-1}{N} \frac{n}{n-1}\right) v(\bar{y}_{\text{st}})_{\text{prop}}$$
$$= v_0^2 - N^2 v(\bar{y}_{\text{st}})_{\text{prop}}$$
$$= v_0^2 - v(\hat{Y}_{\text{st}})_{\text{prop}}.$$

Sub (b). Let the population be $\{u_1, \ldots, u_N\}$, of which the first N_1 belong to the first stratum, etc. We have to estimate

$$\sum_{h=1}^{L} N_h^2 \frac{N_h - n_h}{N_h} \frac{1}{n_h} S_h^2$$

with, for example,

$$S_1^2 = \{(\sum_{i=1}^{N_1} y_i^2 - N_1 \bar{Y}_1^2)\}/(N_1 - 1).$$

If

$$\sum_{i=1}^{N_1} a_i > 1,$$

we can estimate S_1^2 by the familiar expression

$$\frac{1}{\sum_{i=1}^{N_1} a_i - 1} \{\sum_{i=1}^{N_1} a_i y_i^2 - (\sum_{i=1}^{N_1} a_i y_i)^2 / \sum_{i=1}^{N_1} a_i\},$$

which will be approximately unbiased if

$$\mathrm{P}\{\sum_{i=1}^{N_1} a_i \leq 1\}$$

is negligible.

11. Post-stratification

11.1. THE USUAL METHOD

Often we wish to use strata but (as in section 9) do not have listings for all the data, although we know the number N_h in each stratum. If it is feasible to classify each member of a simple random sample according to the stratum to which it belongs, we may compute

$$\bar{y}_{\text{post}} = \sum_{h=1}^{L} \frac{N_h}{N} \bar{y}_h.$$

NOTE. Like in the previous section , all n_h must be positive, and we assume that the sample is sufficiently large or the stratification such that the probability for this not to occur in a simple random sample is very small. We then modify the sampling procedure to ensure positivity of all n_h. Thus $\bar{y}_{\text{post}}$ is unbiased and the approximation for its variance given below a good one. FULLER, J. Amer. Statist. Ass., **61** (1966) 1172–1183, discusses alternative procedures for the case in which some n_h are 0 or small.

EXAMPLE. Consider again the example of section 9.2.

Possible different samples of 3	$(\bar{y}_1, \bar{y}_2)$	$\bar{y}_{\text{post}}$
$\{u_1, u_{,2}\, u_3\}$	$(1\frac{1}{2}, 3)$	2.4
$\{u_1, u_2, u_4\}$	$(1\frac{1}{2}, 4)$	3.0
$\{u_1, u_2, u_5\}$	$(1\frac{1}{2}, 5)$	3.6
$\{u_1, u_3, u_4\}$	$(1, 3\frac{1}{2})$	2.5
$\{u_1, u_3, u_5\}$	$(1, 4)$	2.8
$\{u_1, u_4, u_5\}$	$(1, 4\frac{1}{2})$	3.1
$\{u_2, u_3, u_4\}$	$(2, 3\frac{1}{2})$	2.9
$\{u_2, u_3, u_5\}$	$(2, 4)$	3.2
$\{u_2, u_4, u_5\}$	$(2, 4\frac{1}{2})$	3.5
$\{u_3, u_4, u_5\}$	$\begin{cases}(1, 4)\\(2, 4)\end{cases}$	$\left.\begin{matrix}2.8\\3.2\end{matrix}\right\}$†
	Mean	3.0

† Since no units from the first stratum fall in this sample, we take one more observation, which will be u_1 or u_2 with equal probability.

In this case the variance is

$$(2.4 - 3.0)^2 \tfrac{1}{10} + (3.0 - 3.0)^2 \tfrac{1}{10} + \ldots + (2.8 - 3.0)^2 \tfrac{1}{20} + \\ + (3.2 - 3.0)^2 \tfrac{1}{20} = 0.136.$$

Let us investigate whether it is worthwhile to compute $\bar{y}_{\text{post}}$ in place of $\bar{y}$; so we need to know $V(\bar{y}_{\text{post}})$.

Because of the unbiasedness of $\bar{y}_{\text{post}}$, the variance of $\bar{y}_{\text{post}}$ equals

$$\sum \frac{N_h^2}{N^2}\left(\mathscr{E}\frac{1}{n_h} - \frac{1}{N_h}\right)S_h^2.$$

We saw in section 10 of Chapter IIA (with $X = N_h$, $x = n_h$) that

$$\mathscr{E}\frac{1}{n_h} - \frac{1}{N_h} \approx \frac{N-n}{N-1}\frac{N}{nN_h}\left\{1 + \frac{N}{nN_h} - \frac{1}{n}\right\},$$

so that

$$V(\bar{y}_{\text{post}}) \approx \frac{N-n}{N-1}\frac{1}{n^2}\left\{(n-1)\sum_{h=1}^{L}\frac{N_h}{N}S_h^2 + \sum_{h=1}^{L}S_h^2\right\}.$$

So the contribution to the variance that can be ascribed to the fact that the sample is not a stratified sample with proportional allocation is approximately

$$\frac{N-n}{N-1}\frac{1}{n^2}\sum_{h=1}^{L}(1 - W_h)\,S_h^2$$

where $W_h = N_h/N$. Note that, if the S_h are approximately equal, this expression is approximately

$$\{(L - 1)\, n^{-1}\}\; V(\bar{y}_{st})_{prop},$$

which is relatively small if n is large compared with the number of strata. This is as may be expected, since, if n is large, the allocation in a simple random sample is approximately proportional. Moreover, using Remark (i) in section 8,

$$V(\bar{y}_{srs}) - V(\bar{y}_{post})$$
$$\approx \frac{N-n}{N}\frac{1}{n}\left\{\sum \frac{N_h}{n_h}(\bar{Y}_h - \bar{Y})^2 - \frac{1}{n}\frac{N+n}{N}\sum\left(1 - \frac{N_h}{N}\right)S_h^2\right\}.$$

Post-stratification may also give some of the protection mentioned under (f) in section 1.

In order to estimate $V(\bar{y}_{post})$, we may in the first formula for $V(\bar{y}_{post})$ replace the S_h^2 by their estimators discussed in section 10(b) and $\mathscr{E}n_h^{-1}$ by n_h^{-1}. This is so, because its conditional expectation given $n_1, \ldots, n_L$ will then be

$$\sum \frac{N_h^2}{N^2}\left(\frac{1}{n_h} - \frac{1}{N_h}\right)S_h^2,$$

and so its expectation $V(\bar{y}_{post})$. It may seem strange that the same formula

$$\sum \frac{N_h^2}{N^2}\left(\frac{1}{n_h} - \frac{1}{N_h}\right)s_h^2$$

may serve as an unbiased estimator of both $V(\bar{y}_{post})$ and $V(\bar{y}_{st})$, despite the fact that $V(\bar{y}_{st}) > V(\bar{y}_{post})$; the explanation lies in the well-known fact that $\mathscr{E}n_h^{-1}$ always exceeds $(\mathscr{E}n_h)^{-1}$.

11.2. OTHER METHODS

Another method of saving on the expense of classifying units, which, however, leads to classification of *some* units which ultimately will not be in the sample, is to fix, in advance, for each h, the number n_h of units desired from stratum h. (It may not be possible to achieve these numbers in a single drawing.) The formula for the variance of an estimator for this plan is the same as that with ordinary stratified sampling with these numbers (n_h) of observation. It is often not very convenient to sample this way. If one wishes to have the n_h proportional to the N_h, there is, moreover, rarely occasion to use this plan (rather than the previous one) when the sample is not very small, since then simple random sampling will automatically lead to approximately proportional allocation. If, however, one desires an allocation different from a proportional one (say in proportions $k_1 : \ldots : k_L$), the following plan may be considered as an alternative:

Let h_0 be such that $k_{h_0} \geq k_h$ for $h = 1, \ldots, L$. We suppose that the numbers k_{h_0}/k_h are all approximately whole numbers and round them to the nearest whole numbers l_h. Then we take a simple random sample and for each $h(\neq h_0)$, we drop units at random from that part of the sample which fell in stratum h, retaining 1 out of l_h only.

Like in the method above, a certain number of units will be classified without ultimately being included in the sample; also the formula for the variance is affected by the randomness of the sample sizes, like in 11.1.

YOUNG, Biometrika, **48** (1961) 333–342 discusses a sequential method of obtaining specified numbers of units in the different strata.

12. Multiple classification and quota sampling

12.1. DEFINITIONS AND NOTATION

A population $\mathscr{P}$ may be classified (partitioned) by two criteria, $\mathscr{C}'$ and $\mathscr{C}''$; there are, say, L' classes according to $\mathscr{C}'$: $\mathscr{P}'_1, \ldots, \mathscr{P}'_{L'}$, and L'' classes according to $\mathscr{C}''$: $\mathscr{P}''_1, \ldots, \mathscr{P}''_{L''}$. If $\mathscr{P}_{h'h''}$ consists of the units common to $\mathscr{P}'_{h'}$ and $\mathscr{P}''_{h''}$, then we have what is called a *product partition* (*cross classification*) of $\mathscr{P}$ according to both criteria $\mathscr{C}'$ and $\mathscr{C}''$; we call $\mathscr{P}_{h'h''}$ a *cell* of that partition.

The cell frequencies are

$$\#\{u_i \in \mathscr{P}_{h'h''}\} = N_{h'h''} \quad (h' = 1, \ldots, L'; h'' = 1, \ldots, L'').$$

The marginal frequencies are

$$\#\{u_i \in \mathscr{P}'_{h'}\} = N_{h'.} = \sum_{h''=1}^{L''} N_{h'h''} \quad (h' = 1, \ldots, L');$$

and

$$\#\{u_i \in \mathscr{P}''_{h''}\} = N_{.h''} = \sum_{h'=1}^{L'} N_{h'h''} \quad (h'' = 1, \ldots, L'').$$

For a sample of n, the numbers

$$\begin{aligned} &n_{h'h''} \quad (h' = 1, \ldots, L'; h'' = 1, \ldots, L''), \\ &n_{h'.} \quad (h' = 1, \ldots, L'), \\ &n_{.h''} \quad (h'' = 1, \ldots, L''), \end{aligned}$$

are analogously defined. How to extend all this to more than two criteria is obvious.†

The sample is said to be *representative* with respect to some criterion, $\mathscr{C}'$, if its relative frequencies $n_{h'.}/n$ are close to $N_{h'.}/N$; and representative with respect

† It is often expedient to stratify a population according to several criteria, without the classification constituting a cross classification. Thus, for an adult population, stratification according to sex, occupation and size of residential community, the occupational classification may well differ among the cells formed by the other two criteria, and may well be omitted altogether for the purely farm areas.

to two criteria, $\mathscr{C}'$ and $\mathscr{C}''$, if its relative cell frequencies $n_{h'h''}/n$ are close to $N_{h'h''}/N$. (Sometimes one calls the sample representative if its marginal relative frequencies $n_{h'.}/n$ and $n_{.h''}/n$ are close to $N_{h'.}/N$ and $N_{.h''}/N$, though the relative cell frequencies $n_{h'h''}/n$ may differ much from $N_{h'h''}/N$!) A small simple random sample may easily turn out to be quite unrepresentative, but even a large one may occasionally be quite unrepresentative, and, if the criteria bear a strong relation to the $\mathscr{Y}$-values, can give a poor estimate of Y. On the other hand, at least with respect to the criteria used for stratification, a stratified sample is always representative.†

12.2. CONTROLLED SELECTION

With several desirable criteria of classification it may be that the number of units to be sampled (especially if the units are themselves clusters) is less than the number of cells. This problem is studied in BRYANT, HARTLEY and JESSEN, J. Amer. Statist. Ass., **55** (1960) 105–124 and earlier papers quoted there. The reader should also consult a series of related papers by AVADHANI and SUKHATME, the latest of which to hand is in Austral. J. Statist., **10** (1968) 1–7, which deal in general terms with the problem of controlled selection, in which the aim may be to achieve better balance with respect to a number of characteristics than can be achieved by ordinary stratification, as well as to keep down travel and other costs.

12.3. POST-STRATIFICATION WITH MULTIPLE CLASSIFICATION

Sometimes we stratify one way and use post-stratification for a second criterion of classification. Consider the estimator $\Sigma\Sigma N_{hh'}\bar{y}_{hh'}$; this evidently is the sum over h of L post-stratified estimators of $Y_{h.}$ and so has approximate variance

$$\sum_h N_{h.}^2 \frac{(N_{h.} - n_{h.})}{N_{h.} - 1} \frac{1}{n_{h.}} \left\{\left(1 - \frac{1}{n_{h.}}\right) \sum_{h'} \frac{N_{hh'}}{N_{h.}} S_{hh'}^2 + \frac{1}{n_h} \sum_{h'} S_{hh'}^2\right\}.$$

If we do not know the $N_{hh'}$ but only the marginal frequencies $N_{h.}$, we may use instead of $\Sigma\Sigma N_{hh'}\bar{y}_{hh'}$ the estimator $\Sigma \hat{Y}_{.h'}$, where $\hat{Y}_{.h'}$ is the sum over the strata of the domain estimators $N_{h.}y_{hh'}/n_{h.}$ discussed in section 7(c). Now

$$\begin{aligned} V(\hat{Y}_{.h'}) &= \sum_h N_{h.}(N_{h.} - n_{h.}) \frac{1}{n_{h.}} \frac{1}{N_{h.} - 1} \sum_{i=1}^{N_{h.}} (y'_{hh'i} - \bar{Y}'_{hh'})^2 \\ &= \sum_h N_{h.}(N_{h.} - n_{h.}) \frac{1}{n_{h.}} \frac{1}{N_{h.} - 1} \\ &\quad \times \left\{(N_{hh'} - 1)S_{hh'}^2 + N_{hh'}\left(1 - \frac{N_{hh'}}{N_{h.}}\right) \bar{Y}_{hh'}^2\right\} \end{aligned}$$

† If allocation is nonproportional, this is true after taking into account the weighting applied in estimation.

and for $k' \neq l'$

$$\begin{aligned}\operatorname{Cov}(\hat{Y}_{.k'}, \hat{Y}_{.l'}) &= \sum_h N_{h.}(N_{h.} - n_{h.}) \frac{1}{n_{h.}} \frac{1}{N_{h.} - 1} \sum_{i=1}^{N_{h.}} (y'_{hk'i} - \bar{Y}'_{hk'})(y'_{hl'i} - \bar{Y}'_{hl'}) \\ &= \sum_h N_{h.}(N_{h.} - n_{h.}) \frac{1}{n_{h.}} \frac{1}{N_{h.} - 1} Y_{hk'} Y_{hl'}/N_{h.}.\end{aligned}$$

So we get

$$\begin{aligned}V(\Sigma \hat{Y}_{.h'}) = \sum_h N_{h.}(N_{h.} - n_{h.}) \frac{1}{n_{h.}} \frac{1}{N_{h.} - 1} \\ \times \{\sum_{h'}(N_{hh'} - 1)S^2_{hh'} + \sum_{h'} N_{hh'}(\bar{Y}_{hh'} - \bar{Y}_{h.})^2\},\end{aligned}$$

with a corresponding expression for an estimator of this variance.

If we do not stratify at all, we may, if we know only marginal frequencies, use the post-stratified estimator†

$$\hat{Y} = \Sigma\{N_{.h'} \hat{Y}_{.h'}/N_{.h'}\}.$$

An approximation to its variance may be obtained from

$$\hat{Y} - Y = \sum_{h'} \frac{N_{.h'}}{\hat{N}_{.h'}} \left[\sum \frac{N_{h.}}{n_{h.}} \left\{y_{hh'} - \frac{Y_{hh'}}{N_{h.}} n_{h.} - \frac{Y_{.h'}}{N_{.h'}} \left(n_{hh'} - \frac{N_{hh'}}{N_{h.}} n_{h.}\right)\right\}\right],$$

which is suggested by the development of section 7. When n is sufficiently *large* for the bias to be negligible, the sampling variation in $N_{.h'}/\hat{N}_{.h'}$ and $N_{h.}/n_{h.}$ may be taken to be very small, so that

$$\begin{aligned}V(\hat{Y}) &\approx V\left[\frac{N}{n} \sum_{h'} \sum_{h} \left\{y_{hh'} - \frac{Y_{hh'}}{N_{h.}} n_{h.} - \bar{Y}_{.h'} \left(n_{hh'} - \frac{N_{hh'}}{N_{h.}} n_{h.}\right)\right\}\right] \\ &= V\left[\frac{N}{n} \sum_{h'} \left\{\sum_h n_{hh'}(\bar{y}_{hh'} - \bar{Y}_{.h'}) - N_{hh'} \frac{n_{h.}}{N_{h.}} (\bar{Y}_{hh'} - \bar{Y}_{.h'})\right\}\right] \\ &\approx V\left\{\frac{N}{n} (y - \Sigma n_{.h'} \bar{Y}_{.h'})\right\}.\end{aligned}$$

The conditional variance of y is

$$\sum\sum n^2_{hh'} \left(\frac{1}{n_{hh'}} - \frac{1}{N_{hh'}}\right) S^2_{hh'},$$

† The same estimator may be used in the case of stratification on the first criterion and post-stratification on the second criterion when only the $N_{.h'}$ are known; of course, its variance will not be the same.

which, since

$$\mathscr{E}n_{hh'}^2 = nN_{hh'}(N - N_{hh'}) \frac{N-n}{N-1} N^{-2} + n^2 N_{hh'}^2 N^{-2},$$

has mean

$$\frac{N-n}{N-1} \frac{n}{N} \Sigma\Sigma(N_{hh'} - 1)S_{hh'}^2.$$

The conditional expectation of $y - \Sigma n_{.h'} \bar{Y}_{.h'}$ is

$$\Sigma\Sigma n_{hh'}(\bar{Y}_{hh'} - \bar{Y}_{.h'}),$$

which has variance

$$\frac{N-n}{N-1} \frac{n}{N^2} \{\Sigma\Sigma(\bar{Y}_{hh'} - \bar{Y}_{.h'})^2 N_{hh'}(N - N_{hh'})$$

$$- \underset{(k,k') \neq (l,l')}{\sum\sum \sum\sum} (\bar{Y}_{kk'} - \bar{Y}_{.k'})(\bar{Y}_{ll'} - \bar{Y}_{.l'})N_{kk'}N_{ll'}\}.$$

The latter term in the bracket equals

$$-\{\Sigma\Sigma(\bar{Y}_{hh'} - \bar{Y}_{.h'})N_{hh'}\}^2 + \Sigma\Sigma(\bar{Y}_{hh'} - \bar{Y}_{.h'})^2 N_{hh'}^2,$$

where the first part is 0, so that we get

$$\frac{N-n}{N-1} \frac{n}{N} \Sigma\Sigma(\bar{Y}_{hh'} - \bar{Y}_{.h'})^2 N_{hh'}.$$

Therefore

$$V(\hat{Y}) \approx \frac{N}{n} \{\Sigma\Sigma(N_{hh'} - 1)S_{hh'}^2 + \sum\sum N_{hh'}(\bar{Y}_{hh'} - \bar{Y}_{.h'})^2\}.$$

Different expressions for the approximate variances of the above estimators are given without proof in W. H. Williams, J. Amer. Statist. Ass., **59** (1964) 1054–1062.

12.4. QUOTA SAMPLING

In quota sampling we determine in advance how many units to interview for each cell of the classification – the *quota* – without further prescribing the method of sampling. Often the quotes are fixed so as to make the sample representative with respect to those criteria thought to effect the measured values most. In stratified sampling – with the cells as strata – we sample the prescribed number of units from a list of the $N_{h'h''}$ members of cell $(h'h'')$, and thus all members of $\mathscr{P}$ have a definite probability of being chosen, a probability which we have fixed in advance. In quota sampling the process of selection is such that, even if there are probabilities of selection for each unit in the population, they are entirely unknown to us. We have no probability model for the process of selection (or at least not one in which all quantities are fully specified in advance), which means that the process cannot be reduced *ultimately* to some definite scheme of drawing a random sample by means of a list of objects and a table of random numbers.† It follows that we cannot use the results of our theory – and that in particular the practice of some people to indiscriminately use formulae for stratified sampling in the case of quota sampling is quite unacceptable.‡

13. Specialization to frequencies

This presents no difficulties in general. We merely make some remarks on the estimation of F_N. If for a unit u_i and for any given number v we define the property $\mathscr{Z}(v)$ as having the value 1 when $y(u_i) \leq v$ and 0 otherwise, $\bar{Z}(v) = F_N(v)$ is naturally estimated by

$$\bar{\ }_{\mathrm{st}}(v) = \sum \frac{N_h}{N} \bar{z}_h(v).$$

As v takes on all values between $-\infty$ and ∞, this equation defines a function which we shall denote by F_n, being a generalization of the previously defined empirical distribution function. However, other estimators may also be considered, provided the resulting function F_n is nondecreasing and has values in the interval from zero to one.

14. Confidence intervals and bounds for Y and R

14.1. STANDARD CASE AND SOME EMPIRICAL RESULTS

Suppose we estimate Y by some estimator $\hat{Y}_n$, and denote the distribution of $(\hat{Y}_n - Y)\{v(\hat{Y}_n)\}^{-\frac{1}{2}}$ by Φ_n. For confidence intervals we are interested in Φ_n being close to the standard normal distribution function Φ for certain values

† More on models in Chapter VIII.

‡ STEPHAN and MCCARTHY, Sampling Opinions (Wiley, 1958), as well as MOSER and STUART, J. Roy. Statist. Soc., **116** (1953) 349–405, have studied the results of repeated application of the same quota sampling procedure. As the former authors say, these give "absolutely no information concerning the systematic error or bias of the procedure and estimate" (p. 220), but they do give some, not too readily interpretable, information on their variability. With regard to estimators relating to 19 out of 22 factual questions posed to individuals in 3 large British cities, the latter authors evaluated the ratios of the standard error for quota sampling (without callbacks) to that for unrestricted random sampling in the 3 cities (with 2 call backs) to range between 1.0 and 1.9. For so-called probability sampling with quotas see section 2.5 of Chapter X.

of the argument *only*. If this condition is fulfilled, we may replace Φ_n by Φ in the usual manner.

In section 15.1 of Chapter II we gave some examples where with simple random sampling this condition is not satisfied for moderate n. Let us here discuss the lognormal example, assuming that it is possible to stratify the population into two strata, one for which the $\mathscr{Y}$-values are less than 5.32 and the other for which they are 5.32 or more. In practice it will not usually be possible to stratify on the $\mathscr{Y}$-values themselves,† but only on some related characteristic, and thus the $\mathscr{Y}$-values of the two strata will usually overlap.

Numerical integration gives $\bar{Y}_1 = 1.72$, $\bar{Y}_2 = 9.62$, $\sigma_1 = 1.24$ and $\sigma_2 = 6.02$; and W_1 (the fraction in $\mathscr{P}$ less than 5.32) is 0.887. Therefore Neyman-Tshuprov allocation implies $n_1/(n_1 + n_2) = 0.62$. It was found that, for $n_1 = 90$ and $n_2 = 60$, $\bar{Y}$ fell outside the confidence interval in 4.8% of the trials (of which 69% on the right-hand side); the half-widths of the confidence intervals averaged 0.29. In practice insufficient knowledge of f_N often makes it impossible, and differences in costs among the strata often make it undesirable, to come very close to Neyman-Tshuprov allocation. For an even allocation it was found that three to four times‡ the total number of observations are needed to yield about the same percentage of trials with $\bar{Y}$ falling outside the confidence interval.§ For a 5:1 allocation the number of observations had again to be about doubled! For the mixed distribution, $\bar{Y}_1 = 1.81$, $\bar{Y}_2 = 8.25$, $\sigma_1 = 1.33$, $\sigma_2 = 4.86$, $N_1/N = 0.823$, so that Neyman-Tshuprov allocation gives $n_1/(n_1 + n_2) = 0.56$. Fewer trials were conducted for this case; they point to very similar results as for the lognormal case. Results on confidence intervals for $\bar{Y}$ based on $\bar{y}_{st}$ for a finite population (of 791 U.S. cities) are given below. For the less skewed characteristic $\mathscr{X}$, the percentages of 20 000 cases in which the confidence interval based on $\bar{x}_{st}$ failed to cover $\bar{X}$ were 5.1, 4.8, 5.3, 5.0, and 4.9 for samples of size 300, 400, 500, 600, and 700, respectively, their average half-widths declining from $5\frac{1}{2}$% to just over 1% of $\bar{X}$.

We have also investigated the estimation of $R = Y/X$ for the bivariate lognormal example of section 4.2 of Chapter IIA, when the population is divided into two strata, with 45% in one stratum and 55% in the other, in four different ways:

(i) The first stratum contains the units for which the $\mathscr{X}$-value is less than 2.629 5, the other stratum contains all other units.

(ii) The first stratum contains the units for which the $\mathscr{Y}$-value is less than 7.296 7, the other the remaining units.

(iii) The first stratum contains the units for which the $\mathscr{X}$-value is less than

† It is, however, possible to poststratify; the results must by very similar to proportional allocation, i.e. in the ratio N_1 to N_2.

‡ The cause of the indefiniteness is that again the approach, with increasing sample size, to the nominal percentage is slow. This made it impracticable to investigate the question exhaustively.

§ Such a number of observations is well above that needed to obtain a confidence interval whose half-width is 10% of $\bar{Y}$; in fact, the estimated half-width is 6 to 7% of $\bar{Y}$.

2.988 1 and the $\mathscr{Y}$-value less than 18.285 2, 13.704 3 and 10.152 9, for ρ equal to 0.4, 0.6 and 0.8, respectively, the second stratum the other units.

(iv) The first stratum contains the units for which the $\mathscr{Y}$-value is less than 8.356 3 and the $\mathscr{X}$-value less than 6.243 0, 4.759 1 and 3.589 0, for ρ equal to 0.4, 0.6 and 0.8, respectively, the second stratum contains the remaining units.

In each case we allocated 125 to the first and 875 to the second stratum. With prior knowledge of the population, one could minimize the variance of the combined ratio estimator

$$\hat{R}_c = \frac{0.45\bar{y}_1 + 0.55\bar{y}_2}{0.45\bar{x}_1 + 0.55\bar{x}_2}$$

that was used by allocating about the following percentage of the observations to the first stratum:

	$\rho = 0.4$	$\rho = 0.6$	$\rho = 0.8$
(i) and (ii)	22	15	12
(iii) and (iv)	10	9	8

The results were as follows:

	$\rho = 0.4$			$\rho = 0.6$			$\rho = 0.8$		
	l	r	$\bar{w}$	l	r	$\bar{w}$	l	r	$\bar{w}$
(i)	1.0	4.3	0.27	1.2	4.0	0.20	1.2	4.1	0.14
(ii)	1.9	3.1	0.26	1.6	3.5	0.20	1.4	4.1	0.14
(iii)	1.4	3.9	0.24	1.4	4.5	0.19	1.4	4.7	0.14
(iv)	1.5	3.8	0.24	1.6	4.2	0.19	1.4	4.4	0.14

It is seen that the intervals are narrower than in the nonstratified case, but that the fraction of cases in which they fail to cover Y/X is, except for $\rho = 0.8$, at least as large as in the unstratified case. Again, except for (i), the excess of this fraction over the theoretical fraction is larger for the larger ρ.

For the finite population reported on in that section, we also examined confidence intervals for $\hat{R}_c$ based on stratified samples. We divided the cities into 3 strata according to their 1965 population:

94 between 100 000 and 250 000 inhabitants,
231 between 50 000 and 100 000 inhabitants,
466 between 25 000 and 50 000 inhabitants.

Since

$$X^{-1}N_h\{S_h^2(\mathscr{Y}) - 2RS_h(\mathscr{X}\mathscr{Y}) + R^2S_h^2(\mathscr{X})\}^{\frac{1}{2}}$$

equals 0.50, 0.54 and 0.59, respectively, allocation that minimizes the variance of $\hat{R}_c$ consists in including all the cities from the first two strata in the sample when $n = 500$, 600 or 700, and including 175, 275 and 375 respectively, from the last stratum. When $n = 300$ or 400, this variance is minimized by including all or nearly all cities from the first stratum, and allocating the remaining required number of observations nearly equally between the other two strata; we allocated an equal number to these. The results were as follows (based on 20 000 repetitions):

	l	r	$\bar{w}$
Samples of 300	1.2	4.2	0.129
Samples of 400	1.4	3.8	0.093
Samples of 500	1.0	4.5	0.070
Samples of 600	1.2	3.8	0.046
Samples of 700	1.4	3.8	0.027

Again the intervals are narrower than without stratification. However, the percentage of cases in which the confidence intervals fail to cover Y/X is, if anything, lower than in the unstratified case (especially for samples of 300), and closer to the percentage that would prevail if the distribution were normal. The average of the sample estimates of the standard error of $\hat{R}_c$ is extremely close to the true standard error of $\hat{R}_c$. Compare the above results with those obtained for $\bar{y}_{st}/\bar{X}$, also based on 20 000 replications:

	l	r	$\bar{w}$
Samples of 300	1.1	4.6	0.121
Samples of 400	1.0	4.4	0.088
Samples of 500	1.0	4.9	0.069
Samples of 600	1.2	4.3	0.045
Samples of 700	1.0	4.2	0.026

14.2. USE OF INTERPENETRATING SAMPLES

An entirely different approach is that of *interpenetrating* (*or replicated*) *samples*. Before discussing this, we recall from section 7 of Chapter II that, if $\bar{y}$ is normally distributed, then $(\bar{y} - \bar{Y})v(\bar{y})^{-\frac{1}{2}}$ has a Student or t distribution. This result, which can only have an exact meaning if sampling is with replacement, is based on a theorem which states that, if the distribution of $\bar{y}$ is *exactly* normal, so is the distribution of each of the individual observations (which are independently distributed). The result is also *approximately* correct if the *individual observations* are only approximately normal and even if they are drawn without replacement. However, we saw that $\bar{y}$ may be *approximately* normal even if the individual observations are far from normal; and if the latter is the case, the distribution of $(\bar{y} - \bar{Y})v(\bar{y})^{-\frac{1}{2}}$ may be very different from Student's.

One way to partially overcome this difficulty is to subdivide the sample into k random subsets of size n' each. The averages $\bar{y}(1), \ldots, \bar{y}(k)$, of these sub-sets are much closer to normal than the individual observations,† and so, since $\bar{y}$ is the average of $\bar{y}(1), \ldots, \bar{y}(k)$,

$$v'(\bar{y}) = \frac{1}{k}\sum_{i=1}^{k}\{\bar{y}(i) - \bar{y}\}^2/(k-1)$$

is an unbiased estimator of $V(\bar{y})$ and

$$(\bar{y} - \bar{Y})v'(\bar{y})^{-\frac{1}{2}}$$

may be much closer to having a t distribution than $(\bar{y} - \bar{Y})v(\bar{y})^{-\frac{1}{2}}$. However, the tabulated t-value which is relevant here is not the one corresponding to n observations (called t_{n-1} or "t with $n-1$ degrees of freedom", and found in most statistics, textbooks and collections of tables), but the one corresponding to k observations (t_{k-1}), and therefore may be much larger. Thus with $\beta = 0.95$, if $k = 6$, we already saw that $t_{k-1} = 2.57$. For $k = 3$, $t_{k-1} = 4.30$, and for $k = 2$, t_{k-1} is as large as 12.71.‡

By just assuming the $\bar{y}(i)$ to have *any* continuous distribution F with a median μ (so that $F(\mu) = \frac{1}{2}$), one readily shows that the probability for μ to lie between the smallest and the largest of the $\bar{y}(i)$ is $1 - 2(\frac{1}{2})^k$, so that these form a confidence interval for μ of confidence coefficient $1-2^{-k+1}$.

The method outlined above may occasionally be useful in simple random sampling, with ordinary and with ratio estimation; but we described it primarily as a background for the discussion of interpenetrating sampling in the stratified case. Again we shall confine our discussion to a simple estimator, namely $\hat{Y}_n = \Sigma N_h \bar{y}_{h\cdot}$. Let us suppose that $n_1, \ldots, n_L$ are such that there exist whole numbers $n'_1, \ldots, n'_L$ and k, such that $n_h = n'_h k$ for $h = 1, \ldots, L$. If n is reasonably large, the theory that follows will give good approximate results even if this assumption is violated (and the obvious alterations are made in the required computation).

Consider first drawing the stratified sample as follows: take a stratified sample of sizes $n'_1, \ldots, n'_L$ (with $n'_1 + \ldots + n'_L = n'$) in the usual way, record the L sample means $\bar{y}_1(1), \ldots, \bar{y}_L(1)$, and compute $\hat{Y}_{n'}(1) = \Sigma N_h \bar{y}_h(1)$. Replace the units drawn and take once more a stratified sample of sizes $n'_1, \ldots, n'_L$, recording the L sample means $\bar{y}_1(2), \ldots, \bar{y}_L(2)$, and computing

† This results from one of the class of theorems mentioned in section 15.1 of Chapter II. In what follows, we shall also use the fact that it continues to hold for the sort of linear combinations of observations we shall consider, and also for the usual kinds of sampling with unequal probabilities. However, the convergence would usually be slower in these cases.

‡ It may be shown that

$$[V\{v'(\bar{y})\} - V\{v(\bar{y})\}]/V\{v(\bar{y})\} = (n-k)/(k-1)$$

when sampling from a normal distribution (or in general when sampling from any distribution for which the standardized fourth central moment is that of the normal distribution).

$\hat{Y}_{n'}(2) = \Sigma N_h \bar{y}_h(2)$, and so forth. We thus obtain k independent unbiased estimators $\hat{Y}_{n'}(1), \ldots, \hat{Y}_{n'}(k)$ of Y, which are, for n' sufficiently large,† approximately normally distributed, and so $(\hat{Y}_n - Y)v'(\hat{Y}_n)^{-\frac{1}{2}}$ with

$$v'(\hat{Y}_n) = \frac{1}{k}\sum_{i=1}^{k}\{\hat{Y}_{n'}(i) - \hat{Y}_n\}^2/(k-1)$$

has an approximate t distribution with $k - 1$ degrees of freedom. If we do not replace the sample each time, since sampling is done independently in each stratum, $v'(\hat{Y}_n)$ will give only a slight overestimate of $V(\hat{Y}_n)$ when n' is sufficiently large, particularly if L is not very small. (Sometimes an ordinary sample has been obtained first and then k subsamples are obtained from the observations from each of the strata in order to compute $v'(\hat{Y}_n)$.)

The method may also be used if sampling within the strata is *not* simple random sampling, for example, if within stratum h we sample clusters (with equal or unequal probabilities – in the latter case it may be difficult to apply the idea without replacement). Furthermore, it can be used for cluster sampling without stratification (see, e.g., Chapter IX, "*l* random starts" and section 11). In some of these cases the usual variance estimators lead to complicated calculations, which have often been considered prohibitive. – For another aspect of the method see Chapter X, section 3.2.‡

A variant of the idea of replicated samples is that of *pseudoreplication.* It has been partially worked out, under the name of *half-sample replication*, for the case in which there are two observations per stratum. As pointed out in the section on collapsed strata,§ this case often arises when there are many strata, or when the units sampled within the strata are clusters that are to be subsampled. The latter is usually the case when half-sample replication is considered, a major reason for such replication being the above-mentioned difficulties of estimating variances by the usual methods.

Let us denote the two units observed in stratum h by $u_h(0)$ and $u_h(1)$, with $\mathscr{Y}$-values $y_h(0)$ and $y_h(1)$, and let $d_h = y_h(0) - y(1)$. If sampling is with replacement or all N_h are large, the usual estimator of the variance of the estimator of Y in stratified sampling is

$$\Sigma N_h^2 \tfrac{1}{2}[\{y_h(0) - \bar{y}_h\}^2 + \{y_h(1) - \bar{y}_h\}^2] = \tfrac{1}{4}\Sigma N_h^2 d_h^2;$$

if sampling is without replacement and the N_h are approximately equal, the expression has to be multiplied by a constant approximately equal to

† See the first footnote of this section. The meaning of "sufficiently large" now also depends on the differences between the strata: other things being equal, the larger the differences, the larger will n' have to be, as a rule.

‡ Interpenerating samples have also been useful in checking for gross errors.

§ In fact, in many applications of half-sample replications, only one observation is made in each stratum, but pairs of strata are collapsed.

$(N_h - 2)/N_h$. If none of these conditions hold, the formula is still applicable if N_h is replaced by $\{N_h(N_h - n_h)\}^{\frac{1}{2}}$. For convenience we assume in the following sampling with replacement. In case of subsampling and other more complex methods, $y_h(0)$ and $y_h(1)$ in the formula stand for the estimators of the $\mathscr{Y}$-value of $u_h(0)$ and of $u_h(1)$.

For each h independently we define a random number z_h chosen from among 0 and 1 with equal probability, and refer to the sequence $(z_1, \ldots, z_L)$ as z. Define

$$\hat{Y}_{\text{st},z} = \Sigma N_h y_h(z_h).$$

For example, if $L = 3$ and $z_1 = 0$, $z_2 = 0$, $z_3 = 1$, then $z = (0, 0, 1)$ and

$$\hat{Y}_{\text{st},z} = N_1 y_1(0) + N_2 y_2(0) + N_3 y_3(1).$$

It is clear that the conditional expected value of $\hat{Y}_{\text{st},z}$ given $y_1(0), y_1(1), \ldots, y_L(0), y_L(1)$ is

$$\hat{Y}_{\text{st}} = \tfrac{1}{2}\Sigma N_h\{y_h(0) + y_h(1)\} = \Sigma N_h \bar{y}_h.$$

Also

$$\hat{Y}_{\text{st},z} - \hat{Y}_{\text{st}} = \Sigma N_h\{y_h(z_h) - \tfrac{1}{2}y_h(0) - \tfrac{1}{2}y_h(1)\} = \tfrac{1}{2}\Sigma N_h(-1)^{z_h} d_h,$$

since the term in curly brackets equals

$$\tfrac{1}{2}y_h(0) - \tfrac{1}{2}y_h(1) = \tfrac{1}{2}d_h = \tfrac{1}{2}(-1)^0 d_h,$$

if $z_h = 0$, and

$$\tfrac{1}{2}y_h(1) - \tfrac{1}{2}y_h(0) = -\tfrac{1}{2}d_h = \tfrac{1}{2}(-1)^1 d_h$$

if $z_h = 1$.

Squaring gives

$$(\hat{Y}_{\text{st},z} - \hat{Y}_{\text{st}})^2 = \tfrac{1}{4}N_h^2 d_h^2 + \tfrac{1}{2}\sum_{h<h'}\sum(-1)^{z_h + z_{h'}} d_h d_{h'}.$$

In all possible half-samples (that is, for all possible choices of z), the cross-product term connected with a given pair (h, h') of strata appears as many times with $z_h + z_{h'}$ equal to 0 or 2 as with $z_h + z_{h'}$ equal to 1, and therefore the conditional expected value of the above expression, given $d_1, \ldots, d_L$, equals just $\frac{1}{4}\Sigma N_h^2 d_h^2$, the expression given above. Therefore $(\hat{Y}_{\text{st},z} - \hat{Y}_{\text{st}})^2$ is an unbiased estimator of $V(\hat{Y}_{\text{st}})$, and so is its average for k random choices of z.

The thing to notice is that the cancelling of cross-product terms over the set of all 2^L possible choices of z also holds for much smaller sets of choices of z. Any set of z with this property is called (cross-product) *balanced*, and so is the corresponding set of half-samples. As an example with $L = 6$, consider the following set of 8 choices:

$$\begin{aligned} z^{(1)} &= (0, 1, 1, 0, 1, 0), & z^{(5)} &= (0, 1, 0, 0, 0, 1), \\ z^{(2)} &= (0, 0, 1, 1, 0, 1), & z^{(6)} &= (1, 0, 1, 0, 0, 0), \\ z^{(3)} &= (0, 0, 0, 1, 1, 0), & z^{(7)} &= (1, 1, 0, 1, 0, 0), \\ z^{(4)} &= (1, 0, 0, 0, 1, 1), & z^{(8)} &= (1, 1, 1, 1, 1, 1). \end{aligned}$$

One immediately obtains another such set by replacing every 0 by a 1 and every 1 by a 0; we shall refer to these as $z_*^{(1)}, \ldots, z_*^{(8)}$. Therefore

$$\tfrac{1}{8}\sum_{t=1}^{8} (\hat{Y}_{\text{st},z}(t) - \hat{Y}_{\text{st}})^2, \quad \tfrac{1}{8}\sum_{t=1}^{8} (\hat{Y}_{\text{st},z_*}(t) - \hat{Y}_{\text{st}})^2$$

are unbiased estimators of the variance of $\hat{Y}_{\text{st}}$ (the two are actually identical because of the linearity of our estimator). It may be shown that the minimum number, k, of half-samples required to obtain cross-product balance is at most $L + 3$. For L very large, this may still be more than desired; in that case one may wish to obtain partially balanced sets in which not all cross-product terms cancel. These and other matters are discussed in McCarthy, Replication: An Approach to the Analysis of Data from Complex Surveys (National Center for Health Statistics, 1966) Series 2, No. 14.

One can also apply the above to other estimators. For example, by replacing $\hat{Y}_{\text{st},z}$ by

$$\hat{R}_{\text{st},z} = \hat{Y}_{\text{st},z}/\hat{X}_{\text{st},z},$$

the estimated variance of $\hat{R}_c$ becomes

$$\sum_{t=1}^{k} (\hat{R}_{\text{st},z}(t) - \hat{R}_c)^2/k.$$

This method has been used by the U.S. Census Bureau. In this case

$$\sum_{t=1}^{k} (\hat{R}_{\text{st},z_*}(t) - \hat{R}_c)^2/k$$

is not necessarily identical with the expression without $*$, and one may decide at random whether to use the $z^{(t)}$ or the $z_*^{(t)}$, or use the average of the two estimators of variance. Another procedure is to combine half-samples with Quenouille's method, described in section 7.1 of Chapter IIA, with g, the number of groups, equal to the number of observations ($2L$). Thus

$$\hat{R}_{\text{st.})1(} = \frac{N_1 y_1(1) + N_2\bar{y}_2 + \ldots + N_L\bar{y}_L}{N_1 x_1(1) + N_2\bar{x}_2 + \ldots + N_L\bar{x}_L} = \frac{\hat{Y}_{\text{st}} - \frac{1}{2}N_1 d_1(\mathscr{Y})}{\hat{X}_{\text{st}} - \frac{1}{2}N_1 d_1(\mathscr{X})},$$

$$\hat{R}_{\text{st.})2(} = \frac{N_1 y_1(0) + N_2\bar{y}_2 + \ldots + N_L\bar{y}_L}{N_1 x_1(0) + N_2\bar{x}_2 + \ldots + N_L\bar{x}_L} = \frac{\hat{Y}_{\text{st}} + \frac{1}{2}N_1 d_1(\mathscr{Y})}{\hat{X}_{\text{st}} + \frac{1}{2}N_1 d_1(\mathscr{X})},$$

$$\hat{R}_{\text{st.})3(} = \frac{N_1\bar{y}_1 + N_2 y_2(1) + \ldots + N_L\bar{y}_L}{N_1\bar{x}_1 + N_2 x_2(1) + \ldots + N_L\bar{x}_L} = \frac{\hat{Y}_{\text{st}} - \frac{1}{2}N_2 d_2(\mathscr{Y})}{\hat{X}_{\text{st}} - \frac{1}{2}N_2 d_2(\mathscr{X})},$$

etc.

From these we compute the $2L$ pseudovalues $\hat{R}_{\text{st}}(j)$ and their average $\hat{R}_{\text{Q}}$. For estimating variances one can consider a cross-product balanced subset, replacing $\hat{R}_{\text{st.}z}(t)$ in the formula above by the corresponding pseudovalue.

15. Confidence intervals and bounds for fractiles, and tolerance intervals and limits

We use the definitions introduced in section 16 of Chapter II. By the non-decreasing nature of the empirical distribution function F_n, we have for any $0 < p_1 \leq p_2 \leq 1$, that $w_{p_1} \leq W_P \leq w_{p_2}$ if and only if $p_1 \leq F_n(W_P) \leq p_2$ (or more precisely, $p_1' \leq F_n(W_P) \leq p_2''$, where p_1' is the smallest p for which $w_p = w_{p_1}$ and p_2'' is the largest p for which $w_p = w_{p_2}$). So if $[p_1, p_2]$ covers $F_n(W_P)$ with given probability, $[w_{p_1}, w_{p_2}]$ covers W_P with this probability. Let $z_{hj}(t) = 1$ if y_{hj} is less than or equal to t, and 0 otherwise; and consider a given P, so that W_P is a fixed, though unknown, number.

Now

$$F_n(W_P) = \Sigma(N_h/N)\bar{z}_h(W_P)$$

has expected value

$$F_N(W_P) = P$$

and variance

$$\sum \frac{N_h^2}{N^2}\frac{N_h - n_h}{N_h - 1}\frac{1}{n_h}\bar{Z}_h(W_P)\{1 - \bar{Z}_h(W_P)\}.$$

If $\bar{z}_h(W_P)$ were a statistic, we would estimate this variance by

$$\sum \frac{N_h^2}{N^2}\frac{N_h - n_h}{N_h}\frac{1}{n_h - 1}\bar{z}_h(W_P)\{1 - \bar{z}_h(W_P)\};$$

since $\bar{z}_h(W_P)$ is not a statistic, we replace $\bar{z}_h(W_P)$ in this formula by $\bar{z}_h(w_P)$:†

$$v = v\{F_n(W_P)\} = \sum \frac{N_h^2}{N^2}\frac{N_h - n_h}{N_h}\frac{1}{n_h - 1}\bar{z}_h(w_P)\{1 - \bar{z}_h(w_P)\}.$$

† Recall that w_P has been defined on the basis of the function F_n, which here is defined for all v by $F_n(v) = N^{-1}\Sigma N_h\bar{z}_h(v)$.

Then we assume that n is sufficiently large for that replacement to have only a small effect on this estimator and for

$$\frac{F_n(W_P) - P}{v^{\frac{1}{2}}}$$

to be approximately standard normal. So for any given P, and confidence coefficient β, this leads to a "confidence interval" $[p_1, p_2]$ for $F_n(W_P)$ of the form $[P - tv^{\frac{1}{2}}, P + tv^{\frac{1}{2}}]$, i.e.,

$$p_1 = P - tv^{\frac{1}{2}}, \quad p_2 = P + tv^{\frac{1}{2}};$$

and from this we get the confidence interval $[w_{p_1}, w_{p_2}]$ for W_P.

We similarly get upper confidence bounds and lower confidence bounds, and from these derive tolerance limits as explained in Chapter II, section 16, part (a). The relation proved at the end of that section leads us to expect that, if w_{p_2} is an upper P-tolerance bound of approximately the required confidence coefficient, then $[w_{p_2+p_0}, w_{p_0}]$ is a P-tolerance interval of approximately that confidence coefficient, for all positive p_0 not exceeding $1 - p_2$.

REMARK. The same idea (here modified somewhat from that described in WOODRUFF, J. Amer. Statist. Ass., **47** (1952) 635–646†) is seen to be applicable, with obvious adaptations, to other forms of nonsimple sampling.

16. Difference of ratio estimators for two domains in stratified sampling

Like in section 14 of Chapter IIA, we consider two nonoverlapping domains and use the notation and assumptions of that section. In the summation over h we include only the strata which contain elements belonging to the domains in question.

The variances for differences of ratios are obtained directly from previous results. We get special formulae for particular cases; we shall examine the special case $V(\hat{Y}_{R1} - \hat{Y}_{R2})$ with

$$\hat{Y}_{R1} = \Sigma N_h \bar{y}'_h / \Sigma N_h \bar{x}'_h,$$

$x'_{hi} = 1$ for u_{hi} in domain 1, 0 otherwise, and with a similar definition for $\hat{Y}_{R2}$; and just discuss the without replacement case.

† For stratified sampling with proportional allocation a more exact method is given by McCARTHY, J. Amer. Statist. Ass., **60** (1965) 712–718.

$$V(\hat{Y}_{R1}) \approx \sum N_h^2 \left(1 - \frac{n_h}{N_h}\right) \frac{1}{n_h} \frac{(N')^{-2}}{N_h - 1}$$

$$\times \left\{ \sum_{i=1}^{N_{h1}} (y_{hi} - \bar{Y}_1)^2 - \frac{N_{h1}^2}{N_h} (\bar{Y}_{h1} - \bar{Y}_1)^2 \right\}$$

$$= \sum N_h^2 \left(1 - \frac{n_h}{N_h}\right) \frac{1}{n_h} (N')^{-2}$$

$$\times \left\{ \frac{N_{h1} - 1}{N_h - 1} S_{h1}^2 + \frac{N_{h1}}{N_h} \left(1 - \frac{N_{h1}}{N_h}\right) \frac{N_h}{N_h - 1} (\bar{Y}_{h1} - \bar{Y}_1)^2 \right\},$$

which may be estimated by

$$v(\hat{Y}_{R1}) = \sum N_h^2 \left(1 - \frac{n_h}{N_h}\right) \frac{1}{n_h} (\hat{N}')^{-2}$$

$$\times \left\{ \frac{n_{h1} - 1}{n_h - 1} s_{h1}^2 + \frac{n_{h1}}{n_h} \left(1 - \frac{n_{h1}}{n_h}\right) \frac{n_h}{n_h - 1} (\bar{y}_{h1} - \hat{Y}_{R1})^2 \right\}.$$

Since $\bar{X}_h' = N_{h1}/N_h$, $\bar{X}_h'' = N_{h2}/N_h$,

$$\frac{1}{N_h - 1} \sum_{i=1}^{N_h} \{y_{hi}' - \bar{Y}_1 x_{hi}' - (\bar{Y}_h' - \bar{Y}_1 \bar{X}_h')\}\{y_{hi}'' - \bar{Y}_2 x_{hi}'' - (\bar{Y}_h'' - \bar{Y}_2 \bar{X}_h'')\}$$

$$= \frac{1}{N_h - 1} \{ \sum_{i=1}^{N_h} (y_{hi}' - \bar{Y}_1 x_{hi}')(y_{hi}'' - \bar{Y}_2 x_{hi}'')$$

$$- N_h^{-1} N_{h1} N_{h2} (\bar{Y}_{h1} - \bar{Y}_1)(\bar{Y}_{h2} - \bar{Y}_2)\},$$

where the first sum on the right-hand side vanishes; it may be estimated by

$$- \frac{1}{n_h - 1} \{n_h^{-1} n_{h1} n_{h2} (\bar{y}_{h1} - \hat{Y}_{R1})(\bar{y}_{h2} - \hat{Y}_{R2})\}.$$

Therefore

$$V(\hat{Y}_{R1} - \hat{Y}_{R2}) \approx A - B$$

with†

$$A = \sum\left(1 - \frac{n_h}{N_h}\right)\frac{1}{n_h} N_h^2 \left[\left\{\frac{N_{h1} - 1}{N_h - 1}\frac{S_{h1}^2}{(N')^2} + \frac{N_{h2} - 1}{N_h - 1}\frac{S_{h2}^2}{(N'')^2}\right\}\right.$$
$$\left. + \frac{N_h}{N_h - 1}\left\{\frac{N_{h1}}{N_h}(\bar{Y}_{h1} - \bar{Y}_1)^2(N')^{-2} + \frac{N_{h2}}{N_h}(\bar{Y}_{h2} - \bar{Y}_2)^2(N'')^{-2}\right\}\right],$$

$$B = \sum\left(1 - \frac{n_h}{N_h}\right)\frac{1}{n_h} N_h^2 \frac{N_h}{N_h - 1}$$
$$\times \left\{\frac{N_{h1}}{N_h}(\bar{Y}_{h1} - \bar{Y}_1)(N')^{-1} - \frac{N_{h2}}{N_h}(\bar{Y}_{h2} - \bar{Y}_2)(N'')^{-1}\right\}^2.$$

For domain comparisons of interest one may expect that B is usually very small compared with A. Therefore, a reasonable estimator would also be provided by

$$\sum\left(1 - \frac{n_h}{N_h}\right)\frac{1}{n_h} N_h^2 \left[\left\{\frac{n_{h1} - 1}{n_h - 1}\frac{s_{h1}^2}{(\hat{N}')^2} + \frac{n_{h2} - 1}{n_h - 1}\frac{s_{h2}^2}{(\hat{N}'')^2}\right\}\right.$$
$$\left. + \frac{n_h}{n_h - 1}\left\{\frac{n_{h1}}{n_h}(\bar{y}_{h1} - \hat{Y}_{R1})^2(\hat{N}')^{-2} + \frac{n_{h2}}{n_h}(\bar{y}_{h2} - \hat{Y}_{R2})^2(\hat{N}'')^{-2}\right\}\right]$$
$$= \sum\left(1 - \frac{n}{N_h}\right)\frac{n_h}{n_h - 1}\frac{N_h^2}{n_h}\left\{\frac{n_{h1}}{n_h} g_{h1}^2(\hat{N}')^{-2} + \frac{n_{h2}}{n_h} g_{h1}^2(\hat{N}'')^{-2}\right\},$$

where, for example,

$$g_{h1}^2 = \sum_{i=1}^{N_{h1}} a_{hi}(y_{h1i} - \hat{Y}_{R1})^2/n_{h1}.$$

If desired, the amount of overestimation may be estimated by

$$\sum\left(1 - \frac{n_h}{N_h}\right)\frac{n_h}{n_h - 1}\frac{N_h^2}{n_h}$$
$$\times \left\{\frac{n_{h1}}{n_h}(\bar{y}_{h1} - \hat{Y}_{R1})(\hat{N}')^{-1} - \frac{n_{h2}}{n_h}(\bar{y}_{h2} - \hat{Y}_{R2})(\hat{N}'')^{-1}\right\}^2.$$

The variance of the difference between $\Sigma N_{h1}\bar{y}_{h1}/N_1$ and $\Sigma N_{h2}\bar{y}_{h2}/N_2$ equals the sum of their variances.

† One may check that, in case the domains are entire strata or combinations of entire strata, the second line of A equals B; e.g., if $\mathscr{D}_1$ consists of the first 2 strata and $\mathscr{D}_2$ of the third stratum, we get for $A - B$

$$\left(1 - \frac{n_1}{N_1}\right)\frac{1}{n_1}\left(\frac{N_1}{N_1 + N_2}\right)^2 S_1^2 + \left(1 - \frac{n_2}{N_2}\right)\frac{1}{n_2}\left(\frac{N_2}{N_1 + N^2}\right)^2 S_2^2 + \left(1 - \frac{n_3}{N_3}\right)\frac{1}{n_3} S_3^2,$$

which evidently is the correct formula for $V((y_1 + y_2)/(n_1 + n_2) - y_3/n_3)$.

CHAPTER IV

SAMPLING WITH REPLACEMENT AND UTILIZATION OF DISTINCT UNITS ONLY†

1. Computation of some basic probabilities connected with the sampling method under discussion

Let $A_1, \ldots, A_m$ be events defined over a given probability space and let

$$A = A_1 \cup \ldots \cup A_m.$$

Then

$$\begin{aligned} \mathrm{P}\{A_1 \cap \ldots \cap A_m\} \\ &= \mathrm{P}\{A\} - \sum_{t=1}^{m} \mathrm{P}\{A - A_t\} + \sum_{t_1 < t_2}\sum \mathrm{P}\{A - A_{t_1} - A_{t_2}\} \\ &\quad - \ldots + (-1)^{m-1} \sum_{t_1 < \ldots < t_{m-1}} \ldots \sum \mathrm{P}\{A - A_{t_1} - \ldots - A_{t_{m-1}}\}. \end{aligned}$$

Proof. It is easy to see that

$$(A - A_{i_1}) \cap \ldots \cap (A - A_{i_k}) = A - A_{i_1} - \ldots - A_{i_k}, \tag{1}$$

since

$$A_{i_1} \subset A, \ldots, A_{i_k} \subset A.$$

† The method of this chapter, largely due to Basu and Pathak, are interesting but not of major importance. Omitted altogether are inverse methods, i.e., those in which sampling is continued until a preassigned number of distinct units has been obtained (see, e.g., RAJ and KHAMIS, Ann. Math. Statist., **29** (1958) 550–557, CHIKKAGOUDAR, Sankhya, **A28** (1966) 93–96). However, one method of this kind will be briefly described in section 3.6 of Chapter VI. The method of this chapter can also be used for estimating the population size; see PATHAK, Sankhya, **A26** (1964) 75–80.

If in Boole's formula:

$$P(B_1 \cup . . . \cup B_n)$$

$$= \sum_{t=1}^{n} P(B_t) - \sum\sum_{t_1<t_2} P(B_{t_1} \cap B_{t_2}) + . . . + (-1)^{n-1}P(B_1 \cap . . . \cap B_n)$$

(with $n = m$) we substitute $B_{t_i} = A - A_{t_i}$, we obtain by using (1)

$$P\{(A - A_1) \cup . . . \cup (A - A_n)\}$$

$$= \sum_{t=1}^{n} P\{A - A_t\} - \sum\sum_{t_1<t_2} P\{A - A_{t_1} - A_{t_2}\}$$

$$+ . . . + (-1)^{n-2} \sum_{t_1<\ldots<t_{n-1}} . . . \sum P(A - A_{t_1} - . . . - A_{t_{n-1}}). \quad (2)$$

[The last term is zero, because $A - A_1 - . . . - A_n$ is empty.] But

$$(A - A_1) \cup . . . \cup (A - A_n) = A - A_1 \cap . . . \cap A_n;$$

therefore

$$P\{(A - A_1) \cup . . . \cup (A - A_n)\} = P(A) - P(A_1 \cap . . . \cap A_n).$$

If we substitute the last term in (2) we obtain the required equation.

Suppose we obtain the sample sequence $u_{i_1}, . . . u_{i_n}$ (from the population $u_1, . . ., u_N$). Each i_j is one of the numbers $1, . . . N$. Some of the i_j may coincide; call the set of different elements from $\{1, . . ., N\}$ among the i_j:$\{l_1, . . ., l_m\}$. So $1 \leq m \leq n$. m is called the *effective sample size*.

Case 1. Let A_k be the event: the sample contains u_{l_k} and contains at most the different units $u_{l_1}, . . ., u_{l_m}$.

Equivalently: Let $\{l_1, . . ., l_m\}$ be a given subset of $\{1, . . ., N\}$ and let k be given ($1 \leq k \leq m$). Then A_k is the event that at least one of $i_1, . . ., i_n$ equals l_k, and each of the i_j ($j = 1, . . ., n$) belongs to $\{l_1, . . ., l_m\}$. Then the probability that for m different specified numbers $l_1, . . . l_m$ the sample consists of exactly $u_{l_1}, . . ., u_{l_m}$ is $P\{A_1 \cap . . . \cap A_m\}$.

Let p_h be the probability of drawing u_h in a single draw; suppose it to be the same for each of the draws. Then

$$P(A) = (p_{l_1} + . . . + p_{l_m})^n$$

$$P(A - A_t) = (p_{l_1} + . . . + p_{l_m} - p_{l_t})^n$$

$$P\{(A - A_{t_1}) \cap (A - A_{t_2})\} = (p_{l_1} + . . . + p_{l_m} - p_{l_{t_1}} - p_{l_{t_2}})^n$$

. . . .

Therefore, the probability that the sample consists of exactly $u_{l_1}, . . ., u_{l_m}$ is

$$P\{A_1 \cap . . . \cap A_m\} = (\sum_{j=1}^{m} p_{l_j})^n - \sum_{t=1}^{m} (\sum_{j \neq t} p_{l_j})^n + \sum\sum_{t<t'} (\sum_{j \neq t,t'} p_{l_j})^n - . . . .$$

If $p_h = 1/N$ for $1 \leq h \leq N$, this equals

$$\binom{m}{N}^n - \binom{m}{1}\left(\frac{m-1}{N}\right)^n + \binom{m}{2}\left(\frac{m-2}{N}\right)^n - \binom{m}{3}\left(\frac{m-3}{N}\right)^n + \ldots = \frac{c_m(n)}{N^n},$$

say, and the expected number of distinct units in the sample is

$$N^{-n} \sum_{k=1}^{n} k c_k(n) \binom{N}{k}.$$

REMARK. From the equiprobability case it is evident that $c_m(n)$ equals the number of times that terms of the form $a_{l_1}^{c_1} \ldots a_{l_m}^{c_m}$ appear in the expansion of $(a_1 + \ldots + a_N)^n$, where the c's are positive integers whose sum is n, and where $\{l_1, \ldots, l_m\}$ is any fixed subset of $\{1, \ldots, N\}$. The numbers $c_m(n)/m!$ are called *Stirling numbers of the second kind*, and are tabulated in Table 24.4 of ABRAMOWITZ and STEGUN (editors), Handbook of Mathematical Functions (U.S. National Bureau of Standards, 1964). By substituting in

$$\Delta^m f(x) = \sum_{j=0}^{m} (-1)^j \binom{m}{j} f(x + (m-j)),$$

we get $c_m(n) = \Delta^m x^n$ evaluated at $x = 0$ and similarly,

$$\Delta^m (N-x)^n \big|_{x=0} = (N-m)^n - \binom{m}{1}(N-m+1)^n + \ldots .$$

Case 2. Let $\{l_1, \ldots, l_m\}$ be a subset of $\{1, \ldots, N\}$ and let k and h be given with $1 \leq k \leq m$, $1 \leq h \leq m$.

Let A_k be the event that the sample contains u_{l_k}, contains at most $u_{l_1}, \ldots, u_{l_m}$, and that the first draw consists of u_{l_h}. This event is equivalent with the event that at least one of $i_1, \ldots, i_n$ equals the given number l_k, that each i_j belongs to $\{l_1, \ldots, l_m\}$ and that i_1 equals l_h.

Let p_r be the probability that the first draw consists of u_r. We suppose that it is the same for each of the draws. Then $A - A_h$ is empty and for $t \neq h$; $t_1, t_2 \neq h; \ldots$:

$$\begin{aligned}
\mathrm{P}\{A\} &= p_{l_h}(p_{l_1} + \ldots + p_{l_m})^{n-1} \\
\mathrm{P}\{A - A_t\} &= p_{l_h}(p_{l_1} + \ldots + p_{l_m} - p_{l_t})^{n-1} \\
\mathrm{P}\{A - A_{t_1} - A_{t_2}\} &= p_{l_h}(p_{l_1} + \ldots + p_{l_m} - p_{l_{t_1}} - p_{l_{t_2}})^{n-1} \\
&\ldots\ldots\ldots\ldots\ldots\ldots
\end{aligned}$$

Therefore the probability that the sample consists of exactly $u_{l_1}, \ldots, u_{l_m}$ and that the first element drawn is u_{l_h} is

$$\mathrm{P}\{A_1 \cap \ldots \cap A_m\}$$

$$= p_{l_h}\{(\sum_{j=1}^{m} p_{l_j})^{n-1} - \sum_{t \neq h} (\sum_{j \neq t} p_{l_j})^{n-1} + \sum_{\substack{t<t' \\ t,t' \neq h}}\sum (\sum_{j \neq t,t'} p_{l_j})^{n-1} + \ldots\}.$$

For $p_r = 1/N$ $(1 \leq r \leq N)$ it equals

$$\frac{1}{N}\left\{\left(\frac{m}{N}\right)^{n-1} - \binom{m-1}{1}\left(\frac{m-1}{N}\right)^{n-1} + \binom{m-1}{2}\left(\frac{m-2}{N}\right)^{n-1} - \binom{m-1}{3}\left(\frac{m-3}{N}\right)^{n-1} + \ldots\right\}$$

$$= \frac{1}{N^n}\left\{m^n - (m-1)^n + \frac{m-1}{2}(m-2)^n - \frac{(m-1)(m-2)}{(2)(3)}(m-3)^n + \ldots\right\} = \frac{1}{m} c_m(n)/N^n.$$

Let $\mathrm{Q}\{l_h | l_1 \ldots l_m\}$ be the conditional probability that the first draw consists of u_{l_h} given that the sample consists of $u_{l_1}, \ldots, u_{l_m}$. It equals the ratio of the last expression to the corresponding one in Case 1. So, for $p_r = 1/N$ and all n,

$$\mathrm{Q}\{l_h | l_1 \ldots l_m\} = 1/m.$$

Case 3. For $m > 1$, and $k \neq k'$, and k, k' belonging to $\{1, \ldots, m\}$, define $\mathrm{Q}\{l_k, l_{k'} | l_1, \ldots, l_m\}$ as the conditional probability that in two given draws we draw given units u_{l_k} and $u_{l_{k'}}$, respectively.

Under simple random sampling it equals

$$\frac{1}{c_m(n)}\left\{m^{n-2} - \binom{m-2}{1}(m-1)^{n-2} + \binom{m-2}{2}(m-2)^{n-2} - \ldots\right\} = \frac{c_m(n) - c_m(n-1)}{m(m-1)c_m(n)}.$$

2. Estimator for Y

Let $a_{\alpha,k} = 1$ if u_k is the αth unit drawn and 0 otherwise. It is evident that

$$\sum_k a_{\alpha,k} = 1.$$

Let $\mathscr{E}a_{\alpha,k} = p_k$. Consider

$$\sum_{i=1}^{N} y_i a_{\alpha,i}/p_i,$$

the observation actually obtained divided by its probability. The mean of the expression is

$$\sum_{i=1}^{N} (y_i p_i/p_i) = \sum_{i=1}^{N} y_i = Y.$$

Note that we assume p_i to be the same for each draw. So

$$\frac{1}{n}\sum_{\alpha=1}^{n}\sum_{i=1}^{N}(y_i a_{\alpha,i}/p_i)$$

has mean

$$\frac{1}{n}\sum_{\alpha=1}^{n} Y = \frac{n}{n} Y = Y.$$

Let

$$t_i = \sum_{\alpha=1}^{n} a_{\alpha,i} = \text{the number of times that } u_i \text{ is in the sample.}$$

Then

$$\frac{1}{n}\sum_{\alpha=1}^{n}\sum_{i=1}^{N}(y_i a_{\alpha,i}/p_i) = \frac{1}{n}\sum_{i=1}^{N}(y_i t_i/p_i).$$

The conditional expected value of

$$\sum_{i=1}^{N}(y_i a_{\alpha,i}/p_i)$$

given that the sample consists of the different units $u_{l_1}, \ldots, u_{l_m}$ is

$$\mathscr{E}\{\sum_{i=1}^{N} y_i a_{\alpha,i}/p_i | l_1, \ldots, l_m\} = \sum_{k=1}^{m}(y_{l_k}/p_{l_k})\mathscr{E}\{a_{\alpha,l_k}|u_{l_1} \ldots u_{l_m}\}$$

$$= \sum_{k=1}^{m} \frac{y_{l_k}}{p_{l_k}} Q(l_k|l_1 \ldots l_m).$$

Its unconditional mean is

$$\sum_{i=1}^{N} y_i \mathscr{E}a_{\alpha,i}/p_i = \sum_{i=1}^{N} y_i = Y.$$

So if $a_i = 1$ when u_i is in the sample, and 0 otherwise,

$$\hat{Y}_m = \sum_{i=1}^{N} a_i y_i Q\{i|l_1 \ldots l_m\}/p_i$$

is an unbiased estimator of Y, and $\hat{\bar{Y}}_m = N^{-1}\hat{Y}_m$ of $\bar{Y}$.

From section 8 of Chapter I it follows that this estimator should have a variance smaller than that of

$$\frac{1}{n}\sum_{i=1}^{N} t_i y_i/p_i$$

or at most equal to this.

As we saw, for simple random sampling $p_i = 1/N$ and, when $i \in \{l_1, \ldots, l_n\}$,

$$Q\{i|l_1 \ldots l_m\} = \frac{1}{m};$$

so

$$\hat{Y}_m = (N/m)\sum_{i=1}^{N} a_i y_i,$$

i.e., N times the average of the $\mathscr{Y}$-values of distinct units in the sample.

We note that for $m = 1$,

$$Q\{l_1|l_1\} = 1;$$

and if $n = 2$ and $m = 2$,

$$Q\{l_1|l_1, l_2\} = \tfrac{1}{2} = Q\{l_2|l_1, l_2\}.$$

Therefore, if $n = 2$, $\hat{Y}_m$ equals

$$\tfrac{1}{2}\sum_{i=1}^{N} t_i y_i/p_i$$

(in particular, under simple sampling, $N\bar{y}$).

However, if $m = 3$,

$$Q\{l_1|l_1, l_2\} = \frac{p_{l_1}\{(p_{l_1} + p_{l_2})^2 - p_{l_1}^2\}}{(p_{l_1} + p_{l_2})^3 - p_{l_1}^3 - p_{l_2}^3} = \frac{2p_{l_1} + p_{l_2}}{3(p_{l_1} + p_{l_2})},$$

which equals $Q\{l_2|l_1, l_2\}$ only if $p_{l_1} = p_{l_2}$.

3. The variance of this estimator

The variance of $\hat{Y}_m$ equals the mean of $V(\hat{Y}_m|m)$, the conditional variance, given that there are m distinct units in the sample (since the conditional mean equals Y and so has 0 variance). We shall compute $V(\hat{Y}_m)$ *just* for the case of simple random sampling. We note that then $V(\hat{Y}_m|m)/N^2$ is the variance of a mean of a simple random sample of m drawn without replacement, and so equals $N^2((1/m) - (1/N))S^2$. So

$$V(\hat{Y}_m) = \mathscr{E}V(\hat{Y}_m|m) = S^2\mathscr{E}((1/m) - (1/N))N^2.$$

We show in section 6 that

$$\mathscr{E}\frac{N}{m} - 1 = \sum_{i=1}^{N-1}\left(\frac{N-i}{N}\right)^{n-1} \quad \text{equals} \quad \frac{N}{n} - \frac{1}{2} + \frac{n-1}{12N} + \cdots,$$

where the dots indicate terms of smaller order in N;† so

$$V(\hat{Y}_m) \approx N^2\left(\frac{1}{n} - \frac{1}{2N} + \frac{n-1}{12N^2}\right)S^2.$$

The variance of $N\bar{y}$ for a sample of n drawn with replacement is $(N^2/n)(1 - (1/N))S^2$ and without replacement is $N^2((1/n) - (1/N))S^2$. So, for $n > 2$, when the smaller order terms are negligible, $\hat{Y}_m$ has a variance smaller than that of the ordinary sample mean of n for sampling with replacement. The variance of $\hat{Y}_m$ is larger than the variance of $\bar{y}_n$, the sample mean of n drawn without replacement (and also when $n = 2$). But in many cases, the cost of sampling is more nearly proportional to the number m of distinct members drawn rather than to the number n of drawings; so that, notwithstanding the fact that the variance of $\hat{Y}_m$ is bigger than the variance of $\bar{y}_n$, $\hat{Y}_m$ may be a more efficient estimator.‡

NOTE. Similarly, in the approximate expressions for the variance and bias for the ratio estimator that we obtained for sampling without replacement, substitution of $\mathscr{E}((1/m) - (1/N))$ for $(1/n) - (1/N)$ yields analogous expressions for sampling with replacement but without repetition of units, provided N is large enough to make $\mathscr{E}m^{-2}$ close to n^{-2}.

† The assertion is correct for $n > 2$. For $n = 1$ the result is, instead, $N - 1$; and for $n = 2$: $\frac{1}{2}(N - 1)$.

‡ For simple random sampling, when $\mathscr{E}m$ is an integer n', one might compare $V(\hat{Y}_m)$ with $V(\bar{y}_{n'})$; since $\mathscr{E}m^{-1} \geq (\mathscr{E}m)^{-1}$, $V(\hat{Y}_m) \geq V(\bar{y}_{n'})$. The relative difference in variance when $\mathscr{E}m$ is within 10% of an integer is less than 5% and usually much less as seen from a table in RAO, Rev. Intern, Statist. Inst., **34** (1966) 124–138. Rao also obtains a similar comparison for method (vi) of section 4 in Chapter VI.

4. Estimator of σ^2

Consider the conditional expected value (for $n > 1$) of

$$s^2 = \frac{1}{n-1} \sum_{i=1}^{N} (y_i - \bar{y})^2 t_i = \frac{1}{n(n-1)} \sum\sum_{i<i'}^{N} (y_i - y_{i'})^2 t_i t_{i'}$$
$$= \frac{1}{n(n-1)} \sum\sum_{\alpha<\alpha'}^{n} \sum\sum_{i\neq i'}^{N} (y_i - y_{i'})^2 a_{\alpha,i} a_{\alpha',i'}$$

given the sample consists of $u_{l_1}, \ldots, u_{l_m}$. Since s^2 is an unbiased estimator of σ^2, this conditional expected value is also unbiased for σ^2 and has a variance less than or at most equal to that of s^2,

$$\mathscr{E}(s^2|u_{l_1}, \ldots, u_{l_m}) = \begin{cases} 0 & \text{if } m = 1 \\ \dfrac{\frac{1}{2}n(n-1)}{n(n-1)} \sum\sum_{k\neq k'}^{m} (y_{l_k} - y_{l_{k'}})^2 \\ \qquad \times\, Q(l_k, l_{k'}|l_1, \ldots, l_m) & \text{if } m > 1. \end{cases}$$

In the case of simple random sampling this equals (by Case 3 above) for $m > 1$:

$$\frac{1}{m(m-1)} \frac{c_m(n) - c_m(n-1)}{c_m(n)} \tfrac{1}{2} \sum\sum_{k\neq k'}^{m} (y_{l_k} - y_{l_{k'}})^2$$
$$= \frac{1}{m-1} \sum_{k=1}^{m} \left(y_{l_k} - \frac{\hat{Y}_m}{N}\right)^2 (1 - d_m(n)),$$

where

$$d_m(n) = c_m(n-1)/c_m(n),$$

which has been tabulated by PATHAK, Sankhya, **A24** (1962) 287–302.

NOTE. For $n < 4$, this expression is identical to s^2.

5. Estimator of $V(\hat{Y}_m)$

$$\sigma^2 = \frac{1}{N} \sum_{i=1}^{N} y_i^2 - \bar{Y}^2 \quad \text{so} \quad \bar{Y}^2 = \frac{1}{N} \sum_{i=1}^{N} y_i^2 - \sigma^2$$

and therefore, if $m > 1$,

$$\frac{\sum_{k=1}^{m} y_{l_k}^2}{m} - \frac{1}{m-1} \sum_{k=1}^{m} \left(y_{l_k} - \frac{\hat{Y}_m}{N}\right)^2 (1 - d_m(n))$$

has mean $\bar{Y}^2$.

Also

$$V(\hat{Y}_m) = \mathscr{E}\hat{Y}_m^2 - N^2\bar{Y}^2,$$

so the following expression has mean $V(\hat{Y}_m)$:

$$\hat{Y}_m^2 - N^2\left\{\frac{\sum_{k=1}^{m} y_{l_k}^2}{m} - \frac{\sum_{k=1}^{m} (y_{l_k} - (\hat{Y}_m/N))^2(1 - d_m(n))}{m-1}\right\},$$

which equals

$$\begin{cases} N^2\left(\dfrac{1}{m} - d_m(n)\right)\dfrac{\sum_{k=1}^{m} (y_{l_k} - (\hat{Y}_m/N))^2}{m-1} & \text{if } m > 1, \\ 0 & \text{if } m = 1. \end{cases}$$

For the case of unequal probabilities see Chapter VI, section 3.3.

6. Appendix: Moments of negative order of the effective sample size

It will be shown here that

$$\mathscr{E}\frac{N}{m} - 1 = \sum_{l=1}^{N-1}\left(\frac{N-l}{N}\right)^{n-1},$$

and that for $n > 2$ this equals

$$N\left\{\frac{1}{n} - \frac{1}{2N} + \frac{n-1}{12N^2} + \mathrm{O}\left(\frac{1}{N^4}\right)\right\}.$$

In the Remark after Case 1 (section 1), put $b_i = 1 - a_i$; and let $q_{i_1,\dots,i}$ be the probability of not including $\{i_1, \dots, i_l\}$ $(l < N)$, and

$$q(l) = \sum_{\{i_1\dots i_l\} \subset \{1,\dots,N\}} q_{i_1,\dots,i_l}.$$

Write for $|x| < 1$

$$(1 - x)^{-d} = 1 + \sum_{r=1}^{\infty} A_r x^r.$$

Now

$$m = \sum_{i=1}^{N} a_i = N - \sum_{i=1}^{N} b_i.$$

Since

$$0 < \sum_{i=1}^{N} b_i \leq N - 1,$$

we can in the following expand for positive integers d (actually the result holds also if d is a negative integer, in which case the following expansion is finite):

$$\mathscr{E}\,\frac{1}{m^d} = \mathscr{E}\left(\frac{1}{N - \Sigma b_i}\right)^d = \frac{1}{N^d}\,\mathscr{E}\left\{1 + A_1\frac{\Sigma b_i}{N} + A_2\frac{(\Sigma b_i)^2}{N^2} + \ldots\right\}$$

$$= \frac{1}{N^d}\left\{1 + \sum_{t=1}^{\infty} A_t\frac{\mathscr{E}\left(\sum_{i=1}^{N} b_i\right)^t}{N^t}\right\}.$$

Now,

$$\mathscr{E}\left(\sum_{i=1}^{N} b_i\right)^t = \sum_{l=1}^{N-1}\ \sum_{\{i_1\ldots i_l\}\subset\{1,\ldots,N\}}\ \sum_{\substack{c_i>0\,(i=1,\ldots,l)\\ \sum_{i=1}^{l} c_i = t}} \mathscr{E} b_{i_1}^{c_1}\ldots b_{i_l}^{c_l}\binom{t}{c_1\ldots c_l}$$

$$= \sum_{l=1}^{N-1} q(l)\Delta^l x^t\big|_{x=0},$$

so

$$\mathscr{E}\left(\frac{1}{m^d}\right) = \frac{1}{N^d}\left\{1 + \sum_{t=1}^{\infty}\frac{A_t}{N^t}\sum_{l=1}^{N-1} q(l)\Delta^l x^t\big|_{x=0}\right\}$$

$$= \frac{1}{N^d}\left\{1 + \sum_{l=1}^{N-1} q(l)\Delta^l \sum_{t=1}^{\infty} A_t\left(\frac{x}{N}\right)^t\Big|_{x=0}\right\}$$

$$= \frac{1}{N^d}\left\{1 + \sum_{l=1}^{N-1} q(l)\Delta^l\left(1 - \frac{x}{N}\right)^{-d}\Big|_{x=0}\right\}$$

$$= \frac{1}{N^d} + \sum_{l=1}^{N-1} q(l)\left\{\frac{1}{(N-l)^d} - \frac{\binom{l}{1}}{(N-l+1)^d} + \ldots + (-1)^l\,\frac{\binom{l}{l}}{N^d}\right\}.$$

To get an explicit expression for $d = 1$, we prove now that

$$\sum_{i=0}^{l}\frac{(-1)^{i-l}\binom{l}{i}}{N-i} = \frac{1}{N-l} - \frac{\binom{l}{1}}{N-l+1} + \ldots + (-1)^l\frac{1}{N}$$

$$= \frac{l!}{N(N-1)\ldots(N-l)}.$$

In general, for f a polynomial of degree $\leq l$, if a rational function

$$\frac{f(x)}{\prod_{i=0}^{l}(x-s_i)}$$

with all different s_i equals

$$\sum_{i=0}^{l}\frac{d_i}{x-s_i},$$

then, for $x \neq s_i$

$$\frac{(x-s_i)f(x)}{\prod_{i=0}^{l}(x-s_i)} = \sum_{k\neq i}\frac{x-s_i}{x-s_k}d_k + d_i,$$

so

$$d_i = \lim_{x\to s_i}\frac{(x-s_i)f(x)}{\prod_{k=0}^{l}(x-s_k)}.$$

Here, taking $s_i = i$, $x = N$, $f(x) = l!$, the coefficient of $(N-i)^{-1}$ is

$$\lim_{x\to i}\frac{(x-i)l!}{\prod_{k=0}^{l}(x-k)} = \frac{l!}{x(x-1)\ldots(x-i-1)(x-i+1)\ldots(x-l)}\bigg|_{x=i},$$

which equals

$$\frac{l!}{i(i-1)\ldots 1(-1)(-2)\ldots(i-l)} = \frac{(-1)^{i-l}l!}{i!(l-i)!} = (-1)^{i-l}\binom{l}{i},$$

so

$$\mathscr{E}\left(\frac{1}{m}\right) = \frac{1}{N}\left\{1 + \frac{1}{N-1}q(1) + \frac{(1)(2)}{(N-1)(N-2)}q(2) + \ldots + \frac{(1)(2)\ldots(N-1)}{(N-1)\ldots(2)(1)}q(N-1)\right\}.$$

In simple random sampling

$$q(l) = \binom{N}{l}((N-l)/N)^n$$

and so, for $d = 1$,

$$\mathscr{E}\left(\frac{1}{m}\right) = \frac{1}{N}\left\{1 + \sum_{l=1}^{N-1}\left(\frac{N-l}{N}\right)^{n-1}\right\} = \frac{1}{N^n}\sum_{l=1}^{N} l^{n-1}.$$

But Bernoulli's formula† gives for $n > 1$:

$$\sum_{l=1}^{N-1} l^{n-1} = \frac{N^n}{n} + B_1^* N^{n-1} + \frac{B_2^*}{2!} N^{n-2}(n-1)$$

$$+ \frac{B_4^*}{4!} N^{n-4}(n-1)(n-2)(n-3)$$

$$+ \frac{B_6^*}{6!} N^{n-6}(n-1)(n-2)\ldots(n-5) + \ldots,$$

where the last term in the expansion is that in N^1 (if n is odd or 2) or in N^2 (if n is even and >2), and where the B_i^* are the Bernoulli numbers

$$B_0^* = 1, \quad B_1^* = -\tfrac{1}{2}, \quad B_2^* = \tfrac{1}{6},$$

$$B_4^* = -\tfrac{1}{30}, \quad B_6^* = \tfrac{1}{42}, \quad B_8^* = -\tfrac{1}{30}, \ldots;$$

$$B_{2\nu+1}^* = 0 \quad (\nu = 1, 2, \ldots).$$

So

$$\mathscr{E}\frac{N}{m} - 1 = \sum_{l=1}^{N-1} \left(\frac{N-l}{N}\right)^{n-1},$$

which for $n > 2$ equals

$$N\left\{\frac{1}{n} - \frac{1}{2N} + \frac{n-1}{12N^2} + \mathrm{O}\left(\frac{1}{N^4}\right)\right\}.$$

Note the formula is exact for $n = 3$ and $n = 4$:

$$\sum_{i=1}^{k} i^2 = \tfrac{1}{6}k(k+1)(2k+1), \quad \sum_{i=1}^{k} i^3 = \tfrac{1}{4}k^2(k+1)^2,$$

so

$$\frac{1}{N^2}\sum_{l=1}^{N-1} (N-l)^2 = \frac{1}{6N}(N-1)(2N-1) = \frac{2N^2 - 3N + 1}{6N}$$

$$= N\left\{\frac{1}{3} - \frac{1}{2N} + \frac{2}{12N^2}\right\},$$

$$\frac{1}{N^3}\sum_{l=1}^{N-1} (N-l)^3 = \frac{1}{4N}(N-1)^2 = N\left\{\frac{1}{4} - \frac{1}{2N} + \frac{3}{12N^2}\right\}.$$

For $n = 2$ the approximation formula is not correct, but Bernoulli's formula still holds.

† See, e.g., COURANT and JOHN, Introduction to Calculus and Analysis, I (Interscience, 1965) p. 628.

CHAPTER V

SELECTION OF SAMPLING UNITS

1. Notation; expected mean squares

Let us divide each u_i into M_i subunits, and measure not only the total for $\mathscr{Y}$ for the u_i that we observed, but also for each of its subunits. The purpose of this may be to enable us to see whether in future we would be better off observing a sample of small units rather than a sample of large units. Often, however, there is a choice not just between two possible sizes of units, a "large" unit and a "small" unit, but, among a range of many possible sizes.

Let y_{ij} be the value of $\mathscr{Y}$ for small unit j within large unit i, and let

$$Y_{i.} = \sum_{j=1}^{M_i} y_{ij}, \quad \bar{Y}_{i.} = Y_{i.}/M_i, \quad Y = \sum_{i=1}^{N} Y_{i.}.$$

So far we have considered the

per (large) unit average value of $\mathscr{Y}$:

$$\bar{Y} = Y/N = \sum_{i=1}^{N} Y_{i.}/N,$$

which is the average of the *totals* for the large units. We shall also want to consider the

per small unit average value of $\mathscr{Y}$:

$$\bar{\bar{Y}} = Y/M_{.} = \Sigma\Sigma y_{ij}/M_{.},$$

which is also the weighted average of the per large unit averages:

$$\bar{\bar{Y}} = \Sigma(M_i \bar{Y}_{i.}/\Sigma M_i),$$

and is related to $\bar{Y}$ by

$$\bar{\bar{Y}} = \bar{Y}/\bar{M};$$

here

$$M_{.} = \sum_{i=1}^{N} M_i, \quad \bar{M} = \frac{M_{.}}{N}.$$

WARNING. Some authors use N_i for our M_i, M for our N, $\bar{Y}$ for our $\bar{\bar{Y}}$, and vice versa.

Denote by $S_w^2(i)$, the (adjusted)† variance of the $\mathcal{Y}$-values of small units *within* the large unit u_i:

$$S_w^2(i) = \frac{1}{M_i - 1} \sum_{j=1}^{M_i} (y_{ij} - \bar{Y}_{i.})^2.$$

We shall, for the time being, consider the case in which all the M_i are equal:

$$M_i = \bar{M} \quad (i = 1, \ldots, N).$$

Then we define the average of the $S_w^2(i)$ by

$$S_w^2 = \sum_{i=1}^{N} S_w^2(i)/N.$$

For a simple random sample, without replacement of n large units we can similarly define

$$s_w^2 = \sum_{i=1}^{N} s_w^2(i) a_i/n,$$

where $a_i = 1$ if u_i is in the sample and 0 otherwise, and

$$n = \sum_{i=1}^{N} a_i.$$

Consider also‡ the adjusted variance between the *large* units of the averages $\bar{Y}_{i.}$ per small unit (with $N - 1$ degrees of freedom):

$$S_l^2 = \sum_{i=1}^{N} (\bar{Y}_{i.} - \bar{\bar{Y}})^2/(N - 1).$$

† From the equation

$$\sum_{j=1}^{M_i} (y_{ij} - \bar{Y}_{i.}) = 0$$

it follows that if, for fixed i, we choose $M_i - 1$ of the quantities $y_{ij} - \bar{Y}_{i.}$, the remaining one is determined; we express this by saying that $S_w^2(i)$ is based on $M_i - 1$ *degrees of freedom* (DF). The adjusted variance is then the variance "adjusted for degrees of freedom", i.e., multiplied by $M_i/(M_i - 1)$.

‡ Cochran, Sampling Techniques (Wiley, 2nd ed., 1963), uses the (adjusted) variance S_u^2 between the large unit totals Y_i; he speaks about putting these on a small unit basis by division by $\bar{M}$, thus implicitly defining

$$S_u'^2 = S_C^2 = \frac{1}{\bar{M}} \sum_{i=1}^{N} (Y_{i.} - \bar{Y})^2/(N - 1).$$

We shall call this S_C^2, because Cochran's notation: S_b^2 – b for *between* – might cause confusion in what follows. (Later on, beginning with section 6 of Chapter 10, he uses S_u^2 in a different sense; for our S_l^2, he, like many other authors, uses S_1^2, then denoting S_w^2 by S_2^2.) Because

$$\sum_{i=1}^{N} (Y_{i.} - \bar{Y})^2 = \bar{M}^2 \sum_{i=1}^{N} (\bar{Y}_{i.} - \bar{\bar{Y}})^2,$$

we have

$$S_C^2 = S_l^2 \bar{M} = \sum_{i=1}^{N} \sum_{j=1}^{M_i} (\bar{Y}_{i.} - \bar{\bar{Y}})^2/(N - 1).$$

Evidently,

$$\sum_{i=1}^{N}\sum_{j=1}^{M_i}(y_{ij} - \bar{\bar{Y}})^2 = \sum_{i=1}^{N}\sum_{j=1}^{M_i}(y_{ij} - \bar{Y}_{i.})^2 + \bar{M}\sum_{i=1}^{N}(\bar{Y}_{i.} - \bar{\bar{Y}})^2$$

$$= N(\bar{M} - 1)S_{\text{w}}^2 + (N - 1)\{\bar{M}S_l^2\}. \qquad (*)$$

Therefore the adjusted variance among the $\mathscr{Y}$-values of the small units (per small unit) is†

$$S^2 = \frac{1}{M_. - 1}\{N(\bar{M} - 1)S_{\text{w}}^2 + \bar{M}(N - 1)S_l^2\}$$

$$\approx \frac{\bar{M} - 1}{\bar{M}} S_{\text{w}}^2 + \frac{N - 1}{N} S_l^2,$$

with $M_. - 1$ degrees of freedom. In terms of unadjusted variances‡

$$\sigma^2 = \frac{1}{N\bar{M}}\{N\bar{M}\sigma_{\text{w}}^2 + N\bar{M}\sigma_l^2\} = \sigma_{\text{w}}^2 + \sigma_l^2.$$

Note that $M. - 1$ is equal to $N(\bar{M} - 1) + N - 1$ – compare (*).

REMARK. The variance of the average of the values of $\mathscr{Y}$ in a simple random sample of $n\bar{M}$ small units (drawn without replacement) is

$$\{(N\bar{M} - n\bar{M})/N\bar{M}\}S^2/n\bar{M},$$

which is larger than the variance $\{(N - n)/N\}S_l^2/n$ of the average of the values of $\mathscr{Y}$ per small unit in a simple random sample of n large units§ if $S_{\text{w}}^2 > S^2$, because the difference between these two sampling variances is equal to

$$\frac{N - n}{N}\frac{S^2}{n\bar{M}} - \frac{N - n}{N}\frac{1}{n}\frac{(M_. - 1)S^2 - N(\bar{M} - 1)S_{\text{w}}^2}{(N - 1)\bar{M}}$$

$$= \frac{N - n}{N - 1}\frac{1}{n}\frac{\bar{M} - 1}{\bar{M}}(S_{\text{w}}^2 - S^2).$$

† An approximate equality sign indicates that a factor $1/M_.$ has been neglected.

‡ Utilizing the remark contained in the Warning of section 12 of Chapter II, we note that this result is in accordance with the decomposition given in section 8 of Chapter I.

§ The variance of the average of the $Y_{i.}$ for the n large units of the sample is

$$\frac{N - n}{N}\frac{1}{n}\sum_{i=1}^{N}(Y_{i.} - \bar{Y})^2/(N - 1),$$

so the variance of the estimator of $\bar{\bar{Y}}$ is $\bar{M}^{-2}$ times this, giving

$$\frac{N - n}{N}\frac{1}{n}S_l^2.$$

Therefore, in order that sampling of large units give a more precise estimator of the average value of $\mathscr{Y}$ than does sampling of small units, the large units should be heterogeneous with respect to the values of $\mathscr{Y}$ of the small units out of which they are built up. The ratio of variances under two methods, here $S^2/\bar{M}S_l^2$, is often referred to as the *efficiency* of the second relative to the first, even though it takes no account of cost differences (see section 2 below).

Let us define the two sample means

$$\bar{y} = \sum_{i=1}^{N} a_i y_{i.}/n, \quad \bar{\bar{y}} = \frac{\sum_{i=1}^{N} a_i m_i \bar{y}_{i.}}{\sum_{i=1}^{N} a_i m_i}.$$

For $m_i = \bar{m}$, $\bar{\bar{y}}$ is also equal to

$$\sum_{i=1}^{N} a_i \bar{y}_{i.}/n$$

and to $\bar{y}/\bar{m}$.

REMARK. Here we write $y_{i.}$ and $\bar{y}_{i.}$ instead of $Y_{i.}$ and $\bar{Y}_{i.}$ and m_i instead of M_i, in order to emphasize that we now speak of sample quantities and in order that we can use this notation also in cases in which we *subsample* $m_i < M_i$ small units from the ith large unit. Let us also write

$$m_. = \sum_{i=1}^{N} a_i m_i, \quad \bar{m} = \frac{m_.}{n}.$$

If $\bar{m} = M_i = \bar{M}$ then $m_.$ is also equal to $n\bar{M}$ and $\bar{m} = \bar{M}$.

Let us also define† in the sample the sums of squares

$$\mathrm{SS_w} = \sum_{i=1}^{N} a_i \sum_{j=1}^{m_i} (y_{ij} - \bar{y}_{i.})^2,$$

† Cochran defines, in a way analogous to S_C^2,

$$s_C^2 = \frac{1}{\bar{M}} \sum_{i=1}^{N} a_i (y_{i.} - \bar{y})^2/(n-1)$$

(which he denotes by s_b^2). In our case one may write this in the form

$$s_C^2 = \sum_{i=1}^{N} a_i \sum_{j=1}^{m_i} (\bar{y}_{i.} - \bar{\bar{y}})^2/(n-1) = \mathrm{SS}_l/(n-1) = \mathrm{MS}_l.$$

With subsampling, we do not have $m_i = M_i$ for all i, and the inner sums will be from 1 to M_i, with all squares being multiplied by a_{ij}, which is 1 if the jth small unit of the ith large unit falls in the sample and 0 otherwise. Also, $s_w^2(i)$ is defined by

$$s_w^2(i) = \sum_{j=1}^{M_i} a_{ij}(y_{ij} - \bar{y}_{i.})^2/(m_i - 1).$$

Then the definition of s_w^2 given above for the case $M_i = \bar{M}$ will be maintained, provided also $m_i = \bar{m}$.

$$\mathrm{SS}_l = \sum_{i=1}^{N} a_i \sum_{j=1}^{m_i} (\bar{y}_{i.} - \bar{\bar{y}})^2,$$

$$\mathrm{SS}_t = \sum_{i=1}^{N} a_i \sum_{j=1}^{m_i} (y_{ij} - \bar{\bar{y}})^2.$$

If $m_i = \bar{m}$, SS_l equals

$$\bar{m} \sum_{i=1}^{N} a_i(\bar{y}_{i.} - \bar{\bar{y}})^2.$$

It is clear that

$$\mathrm{SS}_t = \mathrm{SS}_w + \mathrm{SS}_l.$$

It is also clear that the degrees of freedom are

$$\mathrm{DF}_w = \sum_{i-1}^{N} a_i(m_i - 1) = m_. - n = n(\bar{m} - 1),$$

$$\mathrm{DF}_l = n - 1,$$

$$\mathrm{DF}_t = m_. - 1;$$

and that

$$\mathrm{DF}_t = \mathrm{DF}_w + \mathrm{DF}_l.$$

The mean squares are:

$$\mathrm{MS}_w = \mathrm{SS}_w/\mathrm{DF}_w,$$

$$\mathrm{MS}_l = \mathrm{SS}_l/\mathrm{DF}_l,$$

$$\mathrm{MS}_t = \mathrm{SS}_t/\mathrm{DF}_t.$$

Let us summarize this in the so-called analysis of variance table on p. 219. We may obtain $\mathscr{E}$MS by the method of Chapter II. For example, because, as noted in the Remark above,

$$\mathscr{E}(\bar{\bar{y}} - \bar{\bar{Y}})^2 = \frac{N - n}{N} \frac{1}{n} S_l^2$$

($\bar{\bar{y}}$ being the average of the $\bar{y}_{i.}$ for the large units in the sample), we obtain that $\mathscr{E}\bar{\bar{y}}^2 m_.$ is equal to

$$\frac{N - n}{N} \bar{m} S_l^2 + m_. \bar{\bar{Y}}^2.$$

In more general problems of the Analysis of Variance, we take in our sample $\bar{m}$ small units within each large unit in our sample, instead of all the $\bar{M}$ possible

Analysis of Variance Table

Analysis of variance for a simple random sample of n from among N large units (of size $\bar{M}$) without subsampling ($\bar{m} = \bar{M}$) (on a small unit basis)

Source (of Variation)	SS (Sums of Squares)	DF (Degrees of Freedom)	MS (Mean Squares)	$\mathscr{E}$MS (Expected Value of Mean Squares)
Between Large Units	SS_l	$n - 1$	MS_l	$\bar{m}S_l^2 (= S_C^2)$
Between Small Units Within Large Units	SS_w	$n(\bar{m} - 1)$	$MS_w (= s_w^2)$	S_w^2
"Total" (Between Small Units)	SS_t	$m_. - 1$	MS_t	$\{n(\bar{m} - 1)S_w^2 + (n - 1)\bar{m}S_l^2\}/(m_. - 1)$
Overall Average	$\sum_{i=1}^{N} a_i \sum_{j=1}^{m_i} \bar{y}^2$	1	$m_.\bar{y}^2$	$\frac{N - n}{N}\bar{m}S_l^2 + m_.\bar{Y}^2$
Sum	$\sum_{i=1}^{N} a_i \sum_{j=1}^{m_i} y_{ij}^2$	$m_.$		

small units. In this case, as we shall see in Chapter VII, $\mathscr{E}MS_w$ is still S_w^2, but $\mathscr{E}MS_l$ becomes

$$\frac{\bar{M} - \bar{m}}{\bar{M}} S_w^2 + \bar{m}S_l^2;\dagger$$

the case in the table above corresponds to $\bar{M} = \bar{m}$. Also we obtain in general for $\mathscr{E}MS_t$

$$\left(1 - \frac{(n - 1)\bar{m}}{\bar{M}(m_. - 1)}\right) S_w^2 + \frac{(n - 1)\bar{m}}{m_. - 1} S_l^2$$

and for $\mathscr{E}m_.\bar{y}^2$,

$$\frac{\bar{M} - \bar{m}}{\bar{M}} S_w^2 + \frac{N - n}{N} \bar{m}S_l^2 + m_.\bar{Y}^2.$$

From the table above we see that an unbiased estimator of S^2 is‡

$$\hat{S}^2 = \frac{1}{M_. - 1}\{N(\bar{M} - 1)MS_w + (N - 1)MS_l\}.$$

† So $\{MS_l - (1 - (\bar{m}/\bar{M}))MS_w\}/\bar{m}$ is an unbiased estimator of S_l^2.

‡ With subsampling

$$\hat{S}^2 = \frac{1}{M_. - 1}\left[\left\{N(\bar{M} - 1) + (N - 1)\left(1 - \frac{\bar{M}}{\bar{m}}\right)\right\}MS_w + (N - 1)\frac{\bar{M}}{\bar{m}} MS_l\right].$$

This is approximately equal to

$$\bar{m}^{-1}\{(\bar{m}-1)\text{MS}_{\text{w}}+\text{MS}_l\}$$

if N is large;† if n is large the latter expression is also approximately equal to

$$\text{MS}_{\text{t}}=\frac{n(\bar{m}-1)\text{MS}_{\text{w}}+(n-1)\text{MS}_l}{n\bar{m}-1},$$

but in general MS_{t} is not approximately unbiased for S^2. (For more on this see section 4.6.) The formula for $\hat{S}^2$ enables us to estimate what $V(\bar{y})$ would have been, if we would have obtained a simple random sample of n' small units rather than of n large units, namely

$$\{(M_. - n')/M_.\}n'^{-1}\hat{S}^2.$$

If there are also L strata, the analysis of variance table contains in addition a line for: "Between the strata" with sum of squares SS_{st} (with $L-1$ degrees of freedom) and with (if N, n, $\bar{M}$ and $\bar{m}$ are the same for all strata)

$$\mathscr{E}\text{MS}_{\text{st}}=\frac{\bar{M}-\bar{m}}{\bar{M}}S_{\text{w}}^2+\frac{N-n}{N}\bar{m}S_l^2+n\bar{m}S_{\text{st}}^2.$$

In the other lines: ("Between large units within strata", etc.) the sums of squares are the sums over h of the SS previously given, and so are the degrees of freedom; the expected mean squares are not affected. S_l^2 and S_{w}^2 are the averages over the strata of the corresponding expressions defined previously for any one stratum, and S_{st}^2 is the adjusted variance between strata means.

2. Cost and variance comparisons

Say that the cost of assessing the total for $\mathscr{Y}$ for a large unit is $C = C'\bar{m}$, and for a small unit is c. In the display on the top of the next page take $\bar{M}=\bar{m}$.

Note that $\hat{S}^2/\text{MS}_l$ (see the last line) is an estimator of the relative efficiency mentioned in the previous section.

EXAMPLE. Cochran's book reproduces some data from an unpublished study by F. A. Johnson concerning ways in which a bed of white pine seedlings can be divided into sampling units. The bed contains six rows, each 434 feet long. Johnson obtained data on four types of units into which a bed may be subdivided:

(i) One foot of a single row.
(ii) Two feet of a single row.
(iii) One foot of the width of the bed.
(iv) Two feet of the width of the bed.

† Also with subsampling.

	Simple random sample of $m_.$ from among the small units†	Simple random sample of n from among the large units
Estimator ($\bar{\bar{y}}$) for $\bar{\bar{Y}}$	$\frac{1}{m_.}\sum_{i=1}^{N} a_i \sum_{j=1}^{\bar{m}} y_{ij}$	$\frac{1}{m_.}\sum_{i=1}^{N} a_i y_{i.} = \frac{1}{n}\sum_{i=1}^{N} a_i \bar{y}_{..}$
Total costs	$cm_. = cn\bar{m}$	$C_n = C'n\bar{m}$
Estimator of $V(\bar{\bar{y}})$	$\frac{M_. - m_.}{M_.}\frac{1}{m_.}\hat{S}^2$	$\frac{1}{\bar{m}^2}\frac{N-n}{N}\frac{1}{n}\Sigma a_i(y_{i.} - \bar{y})^2/(n-1)$
	$= \frac{N-n}{N}\frac{1}{n\bar{m}}\hat{S}^2$	$= \frac{N-n}{N}\frac{1}{n}\Sigma a_i(\bar{y}_{i.} - \bar{\bar{y}})^2/(n-1)$
		$= \frac{N-n}{N}\frac{1}{n\bar{m}}\mathrm{MS}_l$
Values of n if total outlay allowed is C_0	$C_0/c\bar{m}$	$C_0/C'\bar{m}$
Estimator of $V(\bar{\bar{y}})$ when the budget is C_0	$\frac{N-n}{N}\hat{S}^2c/C_0$	$\frac{N-n}{N}\mathrm{MS}_l C'/C_0$
Therefore we compare	$\hat{S}^2c$	$\mathrm{MS}_l C'$

† We shall see in section 3 that, if the large units are made up of a random collection of small units from the population, $\mathscr{E}\mathrm{MS}_l = S^2$.

Let us designate (i) as the "small unit". Let us measure the cost of locating a large unit and counting seedlings in it in terms of minutes. Since Johnson took a complete census, the variance of the number of seedlings in each type of unit, called S_u^2 in the second footnote in the first section of this chapter, is known exactly.

In the third and fifth line of the following table we find C' and $\mathscr{E}\mathrm{MS}_l$ as functions of the sizes of the large units for each of the four types (these, of course, take on the values c and $\mathscr{E}\hat{S}^2$ respectively, for units of type (i)):

	Types of units			
	(i)	(ii)	(iii)	(iv)
Relative size of the unit	1	2	6	12
Cost per unit: C	$\frac{15}{44}$	$\frac{15}{31}$	$\frac{15}{13}$	$\frac{15}{9}$
Cost per foot of a row: C'	$\frac{15}{44}$	$\frac{15}{62}$	$\frac{15}{78}$	$\frac{15}{108}$
The variance of the number of seedlings in the unit	2.537	6.746	23.094	68.558
$\mathscr{E}\mathrm{MS}_l$	2.537	3.373	3.849	5.713
$C'\mathscr{E}\mathrm{MS}_l$	0.057 7	0.054 4	0.049 4	0.052 9

The cost per foot decreased as the size of unit increased, since less time was lost in locating the units and in moving from unit to unit. From the table we see that type (iii) is most efficient among the four types.

More complicated cost functions may be appropriate and will be mentioned in Chapter VII.

Corresponding results may be obtained for ratio estimators. Very often the most efficient size of unit is larger when ratio estimators rather than ordinary averages are used. The reason is that ρ often increases with size of unit, while the ratio of the coefficients of variation stays nearly constant (see section 6 of Chapter IIA).

If we want to estimate several characteristics, we take into account the variance for the characteristic of greatest interest. See the remarks on this point in section 8 of Chapter III.

3. Variance functions

We do not expect S_{w}^2 to be constant when the units are enlarged, but rather expect S_{w}^2 to increase somewhat when K, the size of the unit, increases.

The function† $S_{\mathrm{w}}^2(K) = AK^g$ often fits the data well (over not too large‡ a range of K); here we may expect, and indeed usually find, g to be a (small) positive fraction if small units that are contiguous tend to be more alike in the characteristic of interest than small units further apart.

The entire population of $M_.$ small units may also be considered as one large unit. The variance which corresponds to this was called S^2; so if the formula fits in this range (which is probably too much to expect – we are resorting here to considerable extrapolation!), we can equate S^2 to AM^g. Together with the equation $S_{\mathrm{w}}^2 = A\bar{M}^g$ (for $K = \bar{M}$), this allows us to estimate A and g when we substitute $\hat{S}^2$ for S^2 and s_{w}^2 for S_{w}^2. It is, however, safer not to rely on this extrapolation, but rather to obtain observations on a number of sizes of units, and then fit A and g (one should also test for goodness of fit).

We may then obtain $S_i^2(K)$ from the formula

$$S_i^2(K) = \frac{(NK-1)S^2 - N(K-1)S_{\mathrm{w}}^2(K)}{(N-1)K},$$

since S^2, the dispersion of the $\mathscr{Y}$-characteristic of the small units, is not affected by their bundling into large units.

† First used by JESSEN, Research Bull. **304** (1942), Agr. Exper. Sta., Iowa State Coll., Ames.

‡ One would not expect S_{w}^2 to keep increasing markedly with K after K has reached a large value.

Observe that if the $M_.$ small units would have been randomly assembled into N large units (with averages for $\mathscr{Y}$ equal to $\bar{Y}'_{i.}$) such that any large unit is a simple random sample of $\bar{M}$ small units out of $\bar{M}N$,

$$\begin{aligned}\mathscr{E}\mathrm{MS}_l &= \bar{M}\mathscr{E}S_l^2\\ &= \bar{M}\{\sum_{i=1}^{N}\mathscr{E}(\bar{Y}'_{i.})^2 - N\bar{\bar{Y}}^2\}/(N-1)\\ &= \bar{M}\left\{\sum_{i=1}^{N}\left(\bar{\bar{Y}}^2 + \frac{S^2}{\bar{M}}\frac{N\bar{M}-\bar{M}}{N\bar{M}}\right) - N\bar{\bar{Y}}^2\right\}\Big/(N-1) = S^2,\end{aligned}$$

$$\begin{aligned}&\mathscr{E}\mathrm{MS}_{\mathrm{w}}\\ &= \frac{1}{N(\bar{M}-1)}\{\sum_{i=1}^{N}\sum_{j=1}^{M_i}\mathscr{E}y_{ij}^2 - \bar{M}\sum_{i=1}^{N}\mathscr{E}(\bar{Y}'_{i.})^2\}\\ &= \frac{1}{N(\bar{M}-1)}\left\{N\bar{M}\left(\bar{\bar{Y}}^2 + \frac{N\bar{M}-1}{N\bar{M}}S^2\right) - \bar{M}N\left(\bar{\bar{Y}}^2 + \frac{N\bar{M}-\bar{M}}{N\bar{M}}\frac{S^2}{\bar{M}}\right)\right\}\\ &= S^2.\end{aligned}$$

Therefore, we would have $S_l^2 = S^2/\bar{M}$. However, the values of $\mathscr{Y}$ for small units within a large unit will frequently be positively correlated, so we may expect $S_l^2(K) = aS^2/K^h$ with $h < 1$ but h close to 1 (a is usually taken to be 1). This relation was proposed by H. F. SMITH, J. Agric. Science, **28** (1938) 1–23. In fact, it fits well to some yield data on uniformity trials for different sized plots.†

This would yield for $S_{\mathrm{w}}^2(K)$

$$\frac{(NK-1)S^2 - a(N-1)S^2K^{1-h}}{N(K-1)},$$

and so (unless $a = 1$ and $h = 1$) $S_{\mathrm{w}}^2(K)$ would depend on N, which seems unreasonable. As N goes to infinity, however, the expression tends to

$$\frac{K}{K-1}S^2\left(1 - \frac{a}{K^h}\right).$$

† Usually the investigator takes a block of land, subdivides it into plots, and grows different quantities on the different plots, or grows a single variety but gives different treatments to the different plots. On the other hand, in uniformity trials he grows the same variety on every plot and gives every plot the same treatment, in order to investigate the differences that arise solely from differences between the plots.

If we assume that $S_w^2(K)$ is equal to this expression, then†

$$S_l^2(K) = \frac{1}{K(N-1)}\left\{(NK-1)S^2 - N(K-1)\frac{K}{K-1}\left(1-\frac{a}{K^h}\right)S^2\right\}$$
$$= \frac{S^2}{K(N-1)}(aNK^{1-h}-1),$$

which for large N is about aS^2K^{-h} and so still satisfies Smith's equation approximately. This equation fits Smith's yield data even better than Smith's original equation (taking $a = 1$ in both cases).

Nevertheless HENDRICKS, J. Amer. Statist. Ass., **39** (1944) 366–376 concluded that various crop average and yield data are better fitted by Jessen's relation than by Smith's; he (and Jessen) applied the relation $A\bar{M}^g$ to determine an optimum number of farms to be used as unit of observation for estimating various farm data.

4. Expression in terms of the intraclass correlation coefficient

4.1. DEFINITION (FOR CONSTANT M_i)

The intraclass correlation coefficient is a measure of average similarity of any two different small units chosen within the same large unit. So its definition presupposes $M_i \geq 2$. For given numbers $y_{ij} - \bar{\bar{Y}}$, consider the average of the products $(y_{ij} - \bar{\bar{Y}})(y_{ij'} - \bar{\bar{Y}})$ over all pairs (j, j') for $j \neq j'$ and over all i in the population. This is by definition the covariance between the $\mathscr{Y}$-values of all pairs of small units in the same large unit. To obtain the correlation, divide by the product of the standard deviation of the first member of each pair and the standard deviation of the second member of each pair. Since these two standard deviations are equal, we get in the denominator the variance of y_{ij} over the whole population:

$$r = \sum_{i=1}^{N}\sum\sum_{\substack{j\neq j'\\1}}^{\bar{M}}(y_{ij}-\bar{\bar{Y}})(y_{ij'}-\bar{\bar{Y}})/\{N\bar{M}(\bar{M}-1)\}$$
$$\times\{\sum_{i=1}^{N}\sum_{j=1}^{\bar{M}}(y_{ij}-\bar{\bar{Y}})^2/N\bar{M}\}^{-1}$$
$$= \sum_{i=1}^{N}\sum\sum_{\substack{j\neq j'\\1}}^{\bar{M}}(y_{ij}-\bar{\bar{Y}})(y_{ij'}-\bar{\bar{Y}})\{(\bar{M}-1)(N\bar{M}-1)S^2\}^{-1}.$$

† If we proceed similarly with the formula $S_w^2 = A\bar{M}^g$, we get

$$S_l^2 = \{(1-(1/N))\bar{M}\}^{-1}\{(\bar{M}-(1/N))S^2 - (\bar{M}-1)S_w^2\}$$
$$\approx S^2 - A\bar{M}^g(1-(1/\bar{M})),$$

so

$$S_w^2 = (\bar{M}-1)^{-1}\{(\bar{M}-(1/N))S^2 - (1-(1/N))\bar{M}S_l^2\}$$
$$\approx (\bar{M}-1)^{-1}\{(\bar{M}-(1/N))S^2 - (1-(1/N))\bar{M}(S^2 - A\bar{M}^g(1-(1/\bar{M})))\}$$
$$= A\bar{M}^g(1-(1/N)) + (1/N)S^2 \approx A\bar{M}^g.$$

4.2. RELATION TO S^2, S_l^2, S_w^2

Since

$$\begin{aligned}(N-1)\bar{M}^2S_l^2 &= \sum_{i=1}^{N}(Y_{i.}-\bar{Y})^2\\ &= \sum_{i=1}^{N}\{\sum_{j=1}^{\bar{M}}(y_{ij}-\bar{\bar{Y}})\}^2\\ &= \sum_{i=1}^{N}\sum_{j=1}^{\bar{M}}(y_{ij}-\bar{\bar{Y}})^2+\sum_{i=1}^{N}\{\sum_{\substack{j\neq j'\\1}}^{\bar{M}}\sum(y_{ij}-\bar{\bar{Y}})(y_{ij'}-\bar{\bar{Y}})\}\\ &= (N\bar{M}-1)S^2\{1+(\bar{M}-1)r\},\end{aligned}$$

we have

$$\begin{aligned}S_l^2 &= \frac{N\bar{M}-1}{(N-1)\bar{M}^2}S^2\{1+(\bar{M}-1)r\}\\ &= \frac{N-(1/\bar{M})}{N-1}\frac{S^2}{\bar{M}}\{1+(\bar{M}-1)r\}\end{aligned}$$

or

$$\sigma_l^2 = \frac{\sigma^2}{\bar{M}}\{1+(\bar{M}-1)r\}.$$

Since $0\leq\sigma_l^2\leq\sigma^2>0$, it follows that

$$-\frac{1}{\bar{M}-1}\leq r\leq 1.$$

The bounds are actually attained, see section 4.3 below.

Also we have

$$r = \left\{\frac{1-(1/N)}{1-(1/\bar{M}N)}\bar{M}S_l^2-S^2\right\}\bigg/\{(\bar{M}-1)S^2\}$$

or

$$r = \frac{\bar{M}\sigma_l^2-\sigma^2}{(\bar{M}-1)\sigma^2}.$$

Substituting in the latter relation

$$\sigma^2 = \sigma_w^2+\sigma_l^2,$$

we obtain†

$$r = \{\sigma_l^2 - (\bar{M} - 1)^{-1}\sigma_w^2\}/\sigma^2$$

$$= \left\{\left(1 - \frac{1}{N}\right) S_l^2 - S_w^2/\bar{M}\right\}\Big/\{(1 - (N\bar{M})^{-1})S^2\}$$

and from this

$$r = 1 - \frac{\bar{M}}{\bar{M} - 1} \frac{\sigma_w^2}{\sigma^2}$$

$$= 1 - \frac{N\bar{M}}{N\bar{M} - 1} \frac{S_w^2}{S^2}$$

$$\approx 1 - \frac{S_w^2}{S^2}.$$

Useful approximations which are valid when N is large (which need not be the case) are

$$r \approx (S_l^2 - \bar{M}^{-1}S_w^2)/S^2,$$

and

$$r = \frac{\bar{M}(1 - (1/N))S_l^2 - \sigma^2}{(\bar{M} - 1)\sigma^2}$$

$$\approx \frac{\bar{M}S_l^2 - S^2}{(\bar{M} - 1)S^2}.$$

† A direct derivation of the first of these equations also leads to a generalization for the case the M_i are not equal:

$$\sum\sum_{j \neq j'}(y_{ij} - \bar{\bar{Y}})(y_{ij'} - \bar{\bar{Y}}) = \sum\sum_{j \neq j'}(y_{ij} - \bar{Y}_{i.})(y_{ij'} - \bar{Y}_{i.}) + M_i(M_i - 1)(\bar{Y}_{i.} - \bar{\bar{Y}})^2,$$

since

$$2\sum\sum_{j \neq j'}(y_{ij} - \bar{Y}_{i.})(\bar{Y}_{i.} - \bar{\bar{Y}})$$

equals zero; the right-hand side equals

$$[\{\sum_j(y_{ij} - \bar{Y}_{i.})\}^2 - \sum_j(y_{ij} - \bar{Y}_{i.})^2] + M_i(M_i - 1)(\bar{Y}_{i.} - \bar{\bar{Y}})^2.$$

Since we defined r in general as the average of terms $(y_{ij} - \bar{\bar{Y}})(y_{i'.} - \bar{\bar{Y}})$ divided by σ^2, we get

$$r = \left\{- \frac{\sum(M_i - 1)S_w^2(i)}{\sum M_i(M_i - 1)} + \frac{\sum M_i(M_i - 1)(\bar{Y}_{i.} - \bar{\bar{Y}})^2}{\sum M_i(M_i - 1)}\right\}\Big/\sigma^2.$$

See further in section 4.7.

We also obtain

$$S_w^2 = \frac{N\bar{M} - 1}{N\bar{M}} (1 - r)S^2$$
$$= (1 - r)\sigma^2,$$

or

$$\sigma_w^2 = \frac{\bar{M} - 1}{\bar{M}} (1 - r)\sigma^2.$$

4.3. SPECIAL CASES

In the case of complete homogeneity of the large units $\sigma_w^2 = 0$, and so

$$r = 1.$$

In the case of complete heterogeneity $\sigma_w^2 = \sigma^2$, so $\sigma_l^2 = 0$ and

$$r = -1/(\bar{M} - 1).$$

In the case of random clustering of the small units in each of the large units (cf. section 3) $S_w^2 = S^2$, and so

$$r = -1/(N\bar{M} - 1),$$

i.e., r is very close to zero. The case r exactly equal to 0 corresponds to

$$S_w^2/\bar{M} = (1 - (1/N))S_l^2.$$

4.4. COMPARISON, USING r, OF VARIANCES FOR SAMPLING DIFFERENT SIZE UNITS

If we sample n large units from among N, the variance of the sample average, $\bar{y}$, per small unit is

$$\frac{N - n}{N} \frac{1}{n} S_l^2 = \frac{N - n}{N} \frac{1}{n} \frac{N - (1/\bar{M})}{N - 1} \frac{S^2}{\bar{M}} \{1 + (\bar{M} - 1)r\}$$

$$= \frac{N - n}{N - 1} \frac{1}{n\bar{M}} \sigma^2\{1 + (\bar{M} - 1)r\}.$$

On the other hand, if we sample $n\bar{M}$ small units from among the $N\bar{M}$ in the population, the variance of the average is

$$\frac{N\bar{M} - n\bar{M}}{N\bar{M}} \frac{1}{n\bar{M}} S^2 = \frac{N - n}{N} \frac{1}{n\bar{M}} S^2;$$

if N is large this is approximately equal to

$$\frac{N - n}{N - 1} \frac{1}{n\bar{M}} \sigma^2.$$

So in the first method the variance of the average is

$$\frac{N - (1/\bar{M})}{N - 1}\{1 + (\bar{M} - 1)r\}$$

times (or, if N is large, approximately

$$1 + (\bar{M} - 1)r$$

times) the variance of the average when sampling according to the second method. This coefficient is equal to one if and only if

$$(\bar{M} - 1)\{1 + (N\bar{M} - 1)r\} = 0,$$

i.e., if and only if† r is equal to $-(N\bar{M} - 1)^{-1}$, the random clustering case already mentioned above. Usually r is positive (in Chapter II, section 13 is discussed a case in which r is negative). Note that even if $|r|$ is small, the factor $1 + (\bar{M} - 1)r$ may be substantially different from 1, and that small errors in r may greatly affect this factor if $\bar{M}$ is large.

The main advantage of comparison in terms of r, rather than in terms of S_l^2 and S_w^2, is that only one quantity is involved, which for certain classes of surveys and a given characteristic may be nearly constant (while in these cases S_l^2 and S_w^2 or S^2 and a parameter of the variance function may not be constant). A difficulty is that, as just mentioned, it may be important to know r accurately, and that, as may be inferred from the formulae for r in terms of the quantities which we are able to estimate reasonably accurately, it may be hard to estimate r accurately from a single survey. Furthermore, r generally varies somewhat with $\bar{M}$, as will now be discussed.

4.5. FURTHER DETAILS ON THIS COMPARISON: DEPENDENCE OF THE ABOVE FACTOR ON $\bar{M}$

Ordinarily r decreases when $\bar{M}$ increases, but decreases less than proportionally, so that usually the factor $1 + (\bar{M} - 1)r$ increases markedly when $\bar{M}$ increases. This property is generally fulfilled for large N if one of the *variance functions* of section 3 holds:

(i) If $S_w^2(\bar{M}) = A\bar{M}^g$,

$$S_l^2 = \frac{(1 - (1/N\bar{M}))(S^2 - A\bar{M}^g) + (1 - (1/N))A\bar{M}^{g-1}}{1 - (1/N)}$$

$$\approx S^2 - A\bar{M}^g\left(1 - \frac{1}{\bar{M}}\right),$$

† Also if $\bar{M} = 1$ the coefficient is 1. However, this is of no interest, since then both methods are identical, and, in fact, r is not defined if all $M_i = 1$.

so

$$r \approx 1 - (A\bar{M}^g/S^2),$$

and therefore, for this approximation,

$$\frac{\mathrm{d}r}{\mathrm{d}\bar{M}} = -gA\bar{M}^{g-1}/S^2 = -g(1-r)/\bar{M} < 0,$$

and

moreover, $\frac{\mathrm{d}}{\mathrm{d}\bar{M}}\{1 + (\bar{M} - 1)r\} = r\left\{1 + g\left(1 - \frac{1}{\bar{M}}\right)\right\} - g\left(1 - \frac{1}{\bar{M}}\right),$

which is larger than 0 if $r > g'/(1 + g')$ with $g' = g(1 - (1/\bar{M}))$; and, moreover, as mentioned in section 3, g is usually a small positive fraction, so that the displayed expression is often much larger than 0.

(ii) If $S_l^2(\bar{M}) = aS^2\bar{M}^{-h}$ (and $\bar{M} > 1$),

$$r \approx \frac{a\bar{M}^{-h+1} - 1}{\bar{M} - 1},$$

and so, for this approximation,

$$\frac{\mathrm{d}r}{\mathrm{d}\bar{M}} = -(\bar{M} - 1)^{-2}[a\bar{M}^{-h}\{h(\bar{M} - 1) + 1\} - 1].$$

Usually $a \approx 1$; if $a = 1$, the maximum of $\mathrm{d}r/\mathrm{d}\bar{M}$ for $0 \leq h < 1$ is attained when $h = 0$, and if $h = 0$ and $a = 1$, $\mathrm{d}r/\mathrm{d}\bar{M} = 0$. Therefore $\mathrm{d}r/\mathrm{d}\bar{M} < 0$ if $S_l^2(\bar{M}) = S^2\bar{M}^{-h}$ with $h > 0$. Also

$$\frac{\mathrm{d}}{\mathrm{d}\bar{M}}\{1 + (\bar{M} - 1)r\} = a\bar{M}^{-h}(1 - h) > 0,$$

which may be rather large if h is not very close to 1.

Note that, if Smith's relation fits (with $a = 1$), $1 - h$ has a clear interpretation, viz. as the logarithm, to the base $\bar{M}$, of the efficiency factor

$$\{(N - \bar{M}^{-1})/(N - 1)\}\{1 + (\bar{M} - 1)r\}.$$

MAHALANOBIS, Phil. Mag., **B231** (1944) 329–451, noted that, in the different regions and years in Bengal he examined, the h, in the variance function for proportion of area under jute, was decreasing with the average proportion for the region and year. He noted that in years or regions for which the proportion of land under jute is relatively small, plots sown with jute are relatively widely scattered, and so the correlation between proportions of land under jute in fields near to each other is relatively low.

4.6. AN APPROXIMATE RELATION AMONG S^2, $\mathscr{E}\text{MS}_t$ AND r

Writing for large N

$$S^2 \approx \frac{\bar{M}-1}{\bar{M}} S_w^2 + S_l^2,$$

since (in general)

$$\mathscr{E}\text{MS}_t = \left(1 - \frac{(n-1)\bar{m}}{\bar{M}(m_{.}-1)}\right) S_w^2 + \frac{(n-1)\bar{m}}{m_{.}-1} S_l^2,$$

we get by subtracting

$$S^2 - \mathscr{E}\text{MS}_t \approx \frac{\bar{m}-1}{m_{.}-1}\left(S_l^2 - \frac{S_w^2}{\bar{M}}\right),$$

which for small $m_{.}^{-1}$ is about

$$\frac{\bar{m}-1}{\bar{m}} \frac{1}{n}\left(S_l^2 - \frac{S_w^2}{\bar{M}}\right),$$

where

$$S_l^2 - \frac{S_w^2}{\bar{M}} \approx rS^2;$$

so

$$\frac{S^2 - \mathscr{E}\text{MS}_t}{S^2}$$

is about

$$\frac{\bar{m}-1}{\bar{m}} \frac{r}{n},$$

an approximate relation among S^2, $\mathscr{E}\text{MS}_t$ and r which is sometimes useful.

4.7. GENERALIZATION OF r TO THE CASE OF UNEQUAL M_i

In this case, since each member of the set $\{u_{i1}, \ldots, u_{iM_i}\}$ appears $M_i - 1$ times together with some different member of the set in the enumeration of all pairs of members, the appropriate generalizations of $\bar{\bar{Y}}$ and σ^2 as they appear in the formula for r are

$$\bar{\bar{Y}}' = \Sigma(M_i - 1)\sum_{j=1}^{M_i} y_{ij}/\Sigma(M_i - 1)M_i = \Sigma(M_i - 1)M_i\bar{Y}_{i.}/\Sigma(M_i - 1)M_i$$

and

$$\sigma'^2 = \Sigma(M_i - 1)\sum_{j=1}^{M_i}(y_{ij} - \bar{\bar{Y}}')^2/\Sigma(M_i - 1)M_i,$$

respectively. σ_l^2 may then be taken as

$$\sigma_l'^2 = \Sigma(M_i - 1)M_i(\bar{Y}_{i.} - \bar{\bar{Y}}')^2/\Sigma(M_i - 1)M_i,$$

which appears in the expression for r given in the footnote to section 4.2 above. Then

$$r = \left\{\sigma_l'^2 - \frac{\sigma_w^2}{\bar{M} - 1 + \sigma^2(\mathscr{M})/\bar{M}}\right\}\Big/\sigma'^2,$$

where σ_w^2 is the M_i-weighted average of the $\sigma_w^2(i)$. If the M_i are strongly correlated with the $\sigma_w^2(i)$ this may differ much from the unweighted average. Similarly, if the M_i are strongly correlated with the $\bar{Y}_{i.}$, as often happens, there may be substantial differences between $\sigma_l'^2$ and a corresponding unweighted expression.

One can also obtain this r in a form analogous to the formula

$$r = \frac{\bar{M}}{\bar{M} - 1}(\sigma_l^2/\sigma^2) - \frac{1}{\bar{M} - 1},$$

but it is even more difficult to interpret. For this, note that

$$\sum_{j' \neq j} (y_{ij'} - \bar{\bar{Y}}') = M_i(\bar{Y}_{i.} - \bar{\bar{Y}}') - (y_{ij} - \bar{\bar{Y}}'),$$

so that

$$r = \frac{\Sigma M_i^2(\bar{Y}_{i.} - \bar{\bar{Y}}')^2 - \Sigma\Sigma(y_{ij} - \bar{\bar{Y}}')^2}{\Sigma\{(M_i - 1)\Sigma(y_{ij} - \bar{\bar{Y}}')^2\}}.$$

To avoid the complications inherent in r in this case, a slightly different coefficient has been considered; see Chapter VII, section 10.2.

5. The analysis of variance table when the M_i or m_i are not constant

In this case the expected mean squares are different and questions arise as to the definition of S_l^2 and S_w^2.

5.1. WITHOUT SUBSAMPLING

$$\mathscr{E}\text{MS}_w = \mathscr{E}\frac{1}{n}\Sigma a_i S_w^2(i)\frac{m_i - 1}{\bar{m} - 1}$$

will be equal to $N^{-1}\Sigma S_w^2(i)$ if the ratios $(m_i - 1)/(\bar{m} - 1)$ are uncorrelated with the $S_w^2(i)$; in general $\mathscr{E}\text{MS}_w$ will be a weighted average of $S_w^2(1), \ldots, S_w^2(N)$, which will be the closer to the unweighted average, the closer the sizes of the large units are to constancy. If n is not small, $\mathscr{E}\text{MS}_w$ will be close to the size-weighted average of the $S_w^2(i)$. $\mathscr{E}\text{MS}_l$ is close to $\bar{M}$ times the weighted expression

$$\Sigma(M_i/\bar{M})(\bar{Y}_{i.} - \bar{\bar{Y}})^2/(N - 1).$$

It is sometimes stated that $\mathscr{E}\text{MS}_l$ equals $\tilde{m}S_l^2$, where

$$\tilde{m} = \frac{1}{n - 1}\Sigma a_i m_i\left(1 - \frac{m_i}{n\bar{m}}\right).$$

This can be true only in an approximate sense† since $\tilde{m}$ is a random variable. If we suppose that $\bar{Y}_{1.}, \ldots, \bar{Y}_{N.}$ are themselves a random sample of numbers

† In the case of subsampling $\tilde{m}$ is nonrandom only if the subsampling numbers are defined not in relation to the units in the population that may be included, but in relation to the order in which they are to be included. In surveys the latter is rarely the case.

$\xi_1, \ldots, \xi_N$ drawn at random *without regard to the sizes of the large units* from an indefinitely large population (so called superpopulation; see Chapter VIII) with mean μ and variance τ^2, then

$$\mathrm{MS}_l = \frac{1}{n-1}\,[\Sigma a_i m_i \xi_i^2 - (n\bar{m})^{-1}\{\Sigma a_i m_i^2 \xi_i^2 + \sum_{i \neq i'}\sum a_i a_{i'} m_i m_{i'} \xi_i \xi_{i'}\}]$$

has conditional expectation (given $m_1, \ldots, m_N$) of

$$\frac{1}{n-1}\,[\Sigma a_i m_i(\mu^2 + \tau^2) - (n\bar{m})^{-1}\{\Sigma a_i m_i^2(\mu^2 + \tau^2) + \sum_{i \neq i'}\sum a_i a_{i'} m_i m_{i'} \mu^2\}]$$

$$= \frac{1}{n-1}\,\Sigma a_i m_i\left(1 - \frac{m_i}{n\bar{m}}\right)\tau^2 = \tilde{m}\tau^2.$$

Note that

$$\tilde{m} = \bar{m}(1 - n^{-1}c^2(\mathscr{M}))$$

with $c(\mathscr{M})$ the (adjusted) sample coefficient of variation of the sizes, and so $\tilde{m} < \bar{m}$ if the m's are not constant. For surveys the italicized assumption does not appear reasonable in the light of the discussion of section 3. However, we shall below discuss a situation in which a modified assumption of this sort is reasonable.

5.2. WITH SUBSAMPLING

Upon some calculation we obtain

$$\mathscr{E}\mathrm{MS}_{\mathrm{w}} = \mathscr{E}\,\frac{1}{n}\,\Sigma a_i S_{\mathrm{w}}^2(i)\,\frac{m_i - 1}{\bar{m} - 1}$$

$$\mathscr{E}\mathrm{MS}_l = \mathscr{E}\,\frac{1}{n}\,\Sigma a_i\,\frac{M_i - m_i}{M_i}\,S_{\mathrm{w}}^2(i)\,\frac{1 - n^{-1}(m_i/\bar{m})}{1 - n^{-1}}$$

$$+\,\mathscr{E}\,\frac{1}{n-1}\,\Sigma a_i m_i\left(\bar{Y}_{i.} - \frac{\Sigma a_i m_i \bar{Y}_{i.}}{n\bar{m}}\right)^2.$$

If the subsampling ratios are constant or, more generally, uncorrelated with the within-variances, the first component of $\mathscr{E}\mathrm{MS}_l$ will usually be close to $(1 - \mathscr{E}\bar{m}/\bar{M})\Sigma S_{\mathrm{w}}^2(i)/N$. For a constant subsampling ratio, the second component of $\mathscr{E}\mathrm{MS}_l$ will be close to the product of this ratio and $\bar{M}$ times the weighted expression given under 5.1.

A special case of interest about which a general statement is possible is that in which the M_i and the $S^2_w(i)$ do not vary a great deal,† and the subsampling numbers are intended to be (nearly) constant but vary‡ through nonresponse or other practical difficulties *unconnected* with the variation in the sizes of variances of the large units. Then the first component of $\mathscr{E}\text{MS}_l$ will be approximately S^2_w, the average of the $S^2_w(i)$, times 1 minus

$$\mathscr{E}\,\frac{1}{n}\,\Sigma a_i\,\frac{m_i}{\bar{M}}\,\frac{1-n^{-1}(m_i/\bar{m})}{1-n^{-1}}=\frac{1}{\bar{M}}\,\mathscr{E}\tilde{m},$$

while $\mathscr{E}\text{MS}_w$ will be about the average of the $S^2_w(i)$. In the present case the italicized assumption above, but replacing the large-unit sizes by the subsampling numbers, may be reasonable. We have then the approximations

$$\mathscr{E}\text{MS}_l=\left(1-\frac{\mathscr{E}\tilde{m}}{\bar{M}}\right)\Sigma S^2_w(i)/N+(\mathscr{E}\tilde{m})S^2_l,$$

$$\mathscr{E}\text{MS}_w=\Sigma S^2_w(i)/N,$$

from which, by estimating $\mathscr{E}\tilde{m}$ by $\tilde{m}$,§ we may derive the estimator

$$\left\{\text{MS}_l-\left(1-\frac{\tilde{m}}{\bar{M}}\right)\text{MS}_w\right\}\Big/\tilde{m}$$

of S^2_l. This estimator, with $\bar{m}$ instead of $\tilde{m}$, has already been mentioned in section 1 (in connection with constant m_i).

† As will be seen in Chapter VII, in large scale surveys it is often worthwhile to stratify in such a way that within strata this holds.

‡ If they are actually constant, $\mathscr{E}\text{MS}_w = S^2_w$, and, if the M_i and $S^2_w(i)$ are uncorrelated, $\mathscr{E}\text{MS}_l$ is

$$\left(1-\frac{\bar{m}}{\bar{M}^*}\right)S^2_w+\bar{m}S^2_l,$$

where $\bar{M}^*$ is the harmonic mean of the M_i.

§ In section 5 of Chapter VII we give unbiased estimators of S^2_l and the $S^2_w(i)$ which do not depend on special conditions. But those estimators may themselves have a large variance when the m_i are small. Therefore under the present conditions the estimators of S^2_l and $\Sigma S^2_w(i)/N$ implied by the discussion of this section are likely to be better when the m_i are small and n is not very small (so that $\tilde{m}$ is likely to be close to $\mathscr{E}\tilde{m}$).

CHAPTER VI

SAMPLING WITH UNEQUAL PROBABILITIES

1. General

Even without stratification it is not always desirable to give each sample point equal probability. We shall here deal with the theory of sampling with unequal probabilities in a general fashion, and go into applications to specific situations later. We shall use the notation:

t_i = number of times u_i is included in the sample,
a_i = 1 if $t_i > 0$, and 0 if $t_i = 0$,
π_i = probability that u_i is in the sample (so $\pi_i = \mathscr{E}a_i$),
π_{ij} = probability that both u_i and u_j are in the sample ($i \neq j$) (so $\pi_{ij} = \mathscr{E}a_ia_j$),
$\delta_i(s)$ = probability of obtaining u_i in draw number s,
$\delta_{ij}(s, t)$ = probability of obtaining u_i in draw number s and u_j in draw number t.

Evidently

$$\sum_{i=1}^{N} t_i = n,$$

and in sampling of n without replacement $a_i = t_i$ (see footnote to section 2 of Chapter II). Clearly for each i, s, $t(s \neq t)$:

$$\sum_{j=1}^{N} \delta_{ij}(s, t) = \delta_i(s),$$

and for all s:

$$\sum_{=1}^{N} \delta_i(s) = 1.$$

Let us consider unbiased† estimation of Y by

$$\sum_{i=1}^{N} y_i t_i / \mathscr{E} t_i,$$

often referred to as the *Horvitz-Thompson* estimators.‡

For the case of sampling without replacement this becomes

$$\hat{Y} = \sum_{i=1}^{N} y_i a_i / \mathscr{E} a_i;$$

the a_i are random variables with means π_i, so that they have variances $\pi_i(1 - \pi_i)$; and the covariance of a_i and a_j is (for $i \neq j$) $\pi_{ij} - \pi_i \pi_j$. Therefore the variance of the above estimator is

$$V(\hat{Y}) = \sum_{i=1}^{N} \left(\frac{y_i}{\pi_i}\right)^2 \pi_i(1 - \pi_i) + \sum_{i \neq j}\sum \frac{y_i}{\pi_i} \frac{y_j}{\pi_j} (\pi_{ij} - \pi_i \pi_j),$$

which, if all π_{ij} are positive, has an unbiased estimator

$$\sum_{i=1}^{N} a_i \left(\frac{y_i}{\pi_i}\right)^2 (1 - \pi_i) + \sum_{i \neq j}\sum a_i a_j \frac{y_i}{\pi_i} \frac{y_j}{\pi_j} \frac{\pi_{ij} - \pi_i \pi_j}{\pi_{ij}}.$$

Another way of arriving at this estimator of $V(\hat{Y})$ is to write $V(\hat{Y})$ in the form

$$V(\hat{Y}) = \mathscr{E} \hat{Y}^2 - Y^2 = \mathscr{E} \hat{Y}^2 - \sum_{i=1}^{N} y_i^2 - \sum_{i \neq j}\sum y_i y_j,$$

which leads to the estimator

$$\hat{Y}^2 - \sum_{i=1}^{N} \frac{y_i^2}{\pi_i} a_i - \sum_{i \neq j}\sum \frac{y_i y_j}{\pi_{ij}} a_i a_j,$$

which is seen to be identical to the previously given expression.

Sampling with unequal probabilities generally concerns sampling of clusters, and often takes place within a stratum; as a result the case of very small n is of special importance (and will often be mentioned separately), and n/N is often not small.

We shall discuss separately sampling with and without replacement. However, in schemes of sampling in which a fixed number of different units is

† Note that no estimator of Y can be unbiased if $\mathscr{E} t_i = 0$ for some i.

‡ These authors formulated three classes of estimators; for these and a more general classification of estimators under a variety of sampling methods, one may refer to, among others, KOOP, Metrika, 7 (1963) 81–114, 165–204.

assured, such as that of section 3.5, the distinction loses practical relevance.† We shall also make some comparisons between with and without replacement methods.

2. Sampling without replacement

For this case

$$\sum_{i=1}^{N} a_i = n,$$

so

$$\sum_{i=1}^{N} \pi_i = \mathscr{E} \sum_{i=1}^{N} a_i = n,$$

and, since $a_i^2 = a_i$,

$$\sum_{j \neq i} \pi_{ij} = \mathscr{E} \sum_{j \neq i} a_i a_j = \mathscr{E} a_i(n - a_i) = (n-1)\pi_i.$$

Also

$$\sum_{j \neq i} \pi_i \pi_j = \pi_i \Big(\sum_{j=1}^{N} \pi_j - \pi_i \Big) = \pi_i (n - \pi_i),$$

so

$$\sum_{j \neq i} (\pi_i \pi_j - \pi_{ij}) = \pi_i (1 - \pi_i) \quad (i = 1, \ldots, N).$$

Now we note that if we substitute this in the expression for $V(\hat{Y})$ found in section 1, we get

$$\begin{aligned} V(\hat{Y}) &= \sum_i \left(\frac{y_i}{\pi_i}\right)^2 \pi_i(1-\pi_i) - \sum\sum_{i \neq j} (\pi_i \pi_j - \pi_{ij}) \frac{y_i y_j}{\pi_i \pi_j} \\ &= \sum_i \left(\frac{y_i}{\pi_i}\right)^2 \sum_{j \neq i} (\pi_i \pi_j - \pi_{ij}) - \tfrac{1}{2} \sum\sum_{i \neq j} (\pi_i \pi_j - \pi_{ij}) \frac{2 y_i y_j}{\pi_i \pi_j} \\ &= \tfrac{1}{2} \sum\sum_{i \neq j} \left\{ \left(\frac{y_i}{\pi_i}\right)^2 + \left(\frac{y_j}{\pi_j}\right)^2 - \frac{2 y_i y_j}{\pi_i \pi_j} \right\} (\pi_i \pi_j - \pi_{ij}), \end{aligned}$$

† Another interesting plan of this sort is *rejective sampling* in which one makes n independent draws of one unit each with fixed probabilities, but accepts the sample only if the units drawn are all different. See HÁJEK, Ann. Math. Statist., **35** (1964) 1491–1523, and SAMPFORD, Biometrika, **54** (1967) 499–513. The method of the latter paper is given in section 6.4 below. The scheme described in section 3.5 has rejective features.

(where we used symmetry) so that

$$V(\hat{Y}) = \sum_{i<j}\sum (\pi_i\pi_j - \pi_{ij})\left(\frac{y_i}{\pi_i} - \frac{y_j}{\pi_j}\right)^2;$$

and if we take $\pi_i = ny_i/Y$, $V(\hat{Y})$ becomes zero.† From this we learn the crucial fact that, if there is some way of choosing the π_i so that they are not very different from ny_i/Y, $\hat{Y}$ will have a very small variance. Thus, there may be available rough estimates of the y_i or of a fixed but unknown multiple of the y_i; the latter is the case if the y_i are themselves sums of quantities defined for subunits and the average $\mathscr{Y}$-values per subunit fluctuate from unit to unit much less than the number of subunits contained in the different units.‡

The values of the estimator for $V(\hat{Y})$ given in the previous section often come out negative. If all π_{ij}'s are positive, the so-called *Yates and Grundy* estimator

$$\sum_{i<j}\sum a_i a_j \frac{\pi_i\pi_j - \pi_{ij}}{\pi_{ij}}\left(\frac{y_i}{\pi_i} - \frac{y_j}{\pi_j}\right)^2$$

is defined and is positive (when the π_i are not proportional to the y_i) if, as is often the case, $\pi_i\pi_j > \pi_{ij}$ for all i and j.§ It is easily seen, that this estimator is unbiased.

REMARK. Stratified sampling may be looked upon as a particular case of sampling with unequal probabilities without replacement. We use double indices,

$$\hat{Y} = \sum_{h=1}^{L}\sum_{j=1}^{N_h} a_{hj}y_{hj}/\pi_{hj},$$

with

$$\mathscr{E}a_{kj} = \pi_{hj} = \frac{n_h}{N_h},$$

† Since the π_i are probabilities, this is not possible if, for some i, $y_i/Y > 1/n$. It is advisable to put such "large" units in a separate stratum or subdivide them, if possible. Since the π_i must be positive, the method is not feasible if some $y_i \leq 0$.

‡ In that case we are sampling with *probabilities proportionate* to some (exact or approximate) measure of the *size* of the units. In general one speaks of (with or without replacement) sampling with probabilities proportionate to some characteristic $\mathscr{X}$ of the units (with $x_i > 0$) or estimates of these x_i; see further on this in the latter part of section 2 of Chapter VII. Again, in order to be able to make $\pi_i = nx_i/X$, each x_i/X must be less than $1/n$. This can be a rather strong restriction on the choice of $\mathscr{X}$ if n is not small compared with N.

§ Even if all π_{ij} are positive, if some of the ratios $\pi_{ij}/(\pi_i\pi_j)$ are very close to 0, the estimator would be a poor one to use, since it could have a large variance. DES RAJ, J. Amer. Statist. Assoc., **51** (1956) 269–284, gives a simple example in which the Yates-Grundy estimator is negative whenever the estimator given in the previous section is positive! However the π_{ij} of his example do not satisfy the general conditions of section 4.

so that

$$\hat{Y} = \sum_{h=1}^{L} \frac{N_h}{n_h} \sum_{j=1}^{N_h} a_{hj} y_{hj} = \sum_{h=1}^{L} N_h \bar{y}_{h.}.$$

Here, for $j \neq k$,

$$\pi_{hj,hk} = \mathrm{P}\{u_{hj} \in s,\, u_{hk} \in s\} = \frac{n_h(n_h - 1)}{N_h(N_h - 1)},$$

and for $h' \neq h$

$$\pi_{hj,h'k} = \mathrm{P}\{u_{hj} \in s,\, u_{h'k} \in s\} = \pi_{hj}\pi_{h'k}.$$

So

$$V(\hat{Y}) = \sum\sum \frac{y_{hj}^2}{\pi_{hj}} - \sum\sum y_{hj}^2 + \sum_h \sum_{j \neq k}\sum \frac{y_{hj}\, y_{hk}}{\pi_{hj}\pi_{hk}} \pi_{hj,hk}$$

$$- \sum_h \sum_{j \neq k}\sum y_{hj} y_{hk} + \sum_{h \neq h'}\sum \sum_j \sum_k \frac{y_{hj}}{\pi_{hj}} \frac{y_{h'k}}{\pi_{h'k}} (\pi_{hj,h'k} - \pi_{hj}\pi_{h'k})$$

$$= \sum \frac{N_h}{n_h} \sum y_{hj}^2 - \sum\sum y_{hj}^2 + \sum_h \frac{(n_h - 1)/n_h}{(N_h - 1)/N_h} \sum_{j \neq k}\sum y_{hj} y_{hk}$$

$$- \sum_h \{(\sum_j y_{hj})^2 - \sum_j y_{hj}^2\} + 0.$$

Since the third term is equal to

$$\sum \frac{N_h}{n_h} \frac{n_h - 1}{N_h - 1} (Y_{h.}^2 - \sum y_{hj}^2),$$

we get

$$V(\hat{Y}) = \sum \frac{N_h}{n_h} \frac{N_h - n_h}{N_h - 1} \sum y_{hj}^2 - \sum \frac{1}{n_h} \frac{N_h - n_h}{N_h - 1} Y_{h.}^2$$

$$= \sum N_h (N_h - n_h) \frac{1}{n_h} \frac{1}{N_h - 1} \sum (y_{hj} - \bar{Y}_{h.})^2,$$

like in Chapter III.

From the alternative formula for $V(\hat{Y})$

$$V(\hat{Y}) = \tfrac{1}{2} \sum_{i \neq i'}\sum (\pi_i \pi_{i'} - \pi_{ii'}) \left(\frac{y_i}{\pi_i} - \frac{y_{i'}}{\pi_{i'}} \right)^2$$

$$= \tfrac{1}{2} \sum_h \omega_h \left(\frac{N_h}{n_h} \right)^2 \sum_{j \neq k}\sum (y_{hj} - y_{hk})^2 + \tfrac{1}{2} \sum_{h \neq h'}\sum \omega_{hh'} \sum_j \sum_k (y'_{hj} - y'_{h'k})^2,$$

with

$$\omega_h = (n_h/N_h)^2 - \pi_{hj'hk} \qquad (j \neq k),$$

$$\omega_{hh'} = (n_h/N_h)(n_{h'}/N_{h'}) - \pi_{hj,h'k} \quad (h \neq h'),$$

$$y'_{hj} = y_{hj}/\pi_{hj},$$

we see that stratification reduces the variance of $\hat{Y}$ much if it is such that (i) units u_i and $u_{i'}$ for which y'_i and $y'_{i'}$ differ much in absolute value appear in different strata (since $\omega_{hh'} = 0$), and (ii) units u_i and $u_{i'}$ for which y_i and $y_{i'}$ differ little appear in the same strata (since $\omega_h = (n_h/N_h)(N_h - n_h)/(N_h - 1)$).

Note that in stratified sampling ω_h and $\omega_{hh'}$ are *nonnegative*. In general, for any method of sampling, $V(\hat{Y})$ will be reduced if for pairs of units this method makes $\pi_i\pi_{i'} - \pi_{ii'}$ *negative* (and relatively large in absolute value) for any u_i and $u_{i'}$ for which $|y'_i - y'_{i'}|$ is relatively large; this means that the probability of them appearing in the sample together is relatively large. This implies, e.g., that in cluster sampling heterogeneity of the clusters should be striven for, as already remarked in Chapter V (with $\mathscr{Y}$ corresponding to $\mathscr{Y}'$ here).

3. Sampling with replacement

3.1. COMPARISONS WITH SAMPLING WITHOUT REPLACEMENT

We already know that, in sampling with equal probabilities, the variance of the sample mean, for a sample of n drawings with replacement exceeds that for a sample of n drawn without replacement.† We may inquire whether, given a scheme of sampling with replacement, with given probabilities p_i of including u_i in any *one* draw, we can achieve a smaller variance for the corresponding unbiased estimator of $\bar{Y}$, when we sample without replacement and with probabilities $\pi_i = np_i$ of including unit i in the sample. We shall see that for some of the schemes given in this chapter we can answer this question affirmatively, but this does not hold for all cases.

Incidentally, it is clear that the relations at the beginning of section 2 do not in general hold in this case.‡ Thus, $1 - \pi_i$ is the probability that u_i does not

† In Chapter IV, section 3, we obtain the same result for the average $\mathscr{Y}$-value of the different units in the sample, but note some limitation to the significance of this comparison.

‡ In their Chapter 39, KENDALL and STUART, The Advanced Theory of Statistics (Griffin, 1966) Vol. 3, obtain the same relations for sampling with replacement, but their definitions of the π_{ij} are different from ours. (In fact, their relations are identical to those given at the end of page 599 of the paper quoted in the second footnote of the next section, after dividing the first relations by $n\bar{p}_v$ and transferring to the left-hand sides $n\bar{p}_v$ and $n(n-1)\bar{p}_{vv}$, respectively.)

occur in the sample and, because of the independence of the drawings, equals $(1 - p_i)^n$, and so

$$\pi_i = 1 - (1 - p_i)^n = np_i - \binom{n}{2} p_i^2 + \binom{n}{3} p_i^3 + \ldots (-1)^{n-1} \binom{n}{n} p_i^n.$$

E.g., if $N = 5$, $n = 3$, $p_1 = p_2 = p_3 = 0.2$, $p_4 = 0.3$, $p_5 = 0.1$, $\Sigma\pi_i = 2.392$ instead of 3. For $i \neq j$,

$$\pi_{ij} = 6p_i p_j - 3p_i^2 p_j - 3p_i p_j^2,$$

the subtraction being for the cases in which two of the units drawn are u_i and u_j, respectively. So

$$\sum_{j \neq 1} \pi_{1j} = 0.756$$

instead of $2\pi_1 = 0.976$.

3.2. THE MOST USUAL ESTIMATOR

For sampling with replacement we shall first consider the frequently used unbiased estimator

$$\hat{Y} = \sum_{i=1}^{N} y_i t_i / \mathscr{E} t_i.$$

Let

$$\delta_i(s) = p_i \quad (s = 1, \ldots, n).$$

The joint frequency function of $t_1, \ldots, t_N$ is multinomial, hence

$$\mathscr{E} t_i = np_i, \quad V(t_i) = np_i(1 - p_i), \quad \operatorname{Cov}(t_i, t_j) = -np_i p_j \; (i \neq j),$$

and so

$$\begin{aligned}
V(\hat{Y}) &= \sum_{i=1}^{N} \left(\frac{y_i}{\mathscr{E} t_i}\right)^2 V(t_i) + \sum_{i \neq j}\sum \frac{y_i}{\mathscr{E} t_i} \frac{y_j}{\mathscr{E} t_j} \operatorname{Cov}(t_i, t_j) \\
&= \sum_{i=1}^{N} \left(\frac{y_i}{np_i}\right)^2 np_i(1 - p_i) - \sum_{i \neq j}\sum \frac{y_i}{np_i} \frac{y_j}{np_j} np_i p_j \\
&= \frac{1}{n} \sum_{i=1}^{N} \frac{y_i^2}{p_i} - \frac{1}{n} \sum_{i=1}^{N} y_i^2 - \frac{1}{n} \left\{ \left(\sum_{i=1}^{N} \frac{y_i}{p_i} p_i \right)^2 - \sum_{i=1}^{N} \left(\frac{y_i}{p_i}\right)^2 p_i^2 \right\} \\
&= \frac{1}{n} \left\{ \sum_{i=1}^{N} \frac{y_i^2}{p_i} - Y^2 \right\} = \frac{1}{n} \sum_{i=1}^{N} \left(\frac{y_i}{p_i} - Y\right)^2 p_i = \frac{1}{n} \sum_{i < i'}\sum \left(\frac{y_i}{p_i} - \frac{y_{i'}}{p_{i'}}\right)^2 p_i p_{i'}.
\end{aligned}$$

Note that, if $p_i = 1/N$, this becomes $N^2\sigma^2/n$, as it should.

Comparing the final expression for $V(\hat{\hat{Y}})$ with the corresponding expression for $V(\hat{Y})$ (given in section 2) for the case that we take for the p_i in sampling with replacement $1/n$ times the π_i for sampling without replacement, we see that $V(\hat{Y}) < V(\hat{\hat{Y}})$ if

$$\pi_{ij} > \{(n-1)/n\}\pi_i\pi_j \quad \text{for all } i \text{ and } j \neq i$$

holds for sampling without replacement.†

It is important to note‡ that, if $p_i = y_i/Y$, $V(\hat{\hat{Y}}) = 0$; this implies that, if there are ways of choosing the p_i to be close to y_i/Y, the variance of $N\bar{y}$ is very small (see the corresponding discussion for sampling without replacement in section 2).

An unbiased estimator of $V(\hat{\hat{Y}})$ is

$$\frac{1}{n}\sum_{i=1}^{N} t_i\left(\frac{y_i}{p_i} - \hat{\hat{Y}}\right)^2 \bigg/ (n-1) = \frac{1}{n^2}\sum_{i<i'}\sum t_i t_{i'}\left(\frac{y_i}{p_i} - \frac{y_{i'}}{p_{i'}}\right)^2 \bigg/ (n-1),$$

or, in terms of the sample sequence $u_{i_1}, \ldots, u_{i_n}$,

$$\frac{1}{n}\sum_{j=1}^{n}(y_{i_j}p_{i_j}^{-1} - \hat{\hat{Y}})^2/(n-1) = \frac{1}{n}\left\{\frac{1}{n}\sum_{j=1}^{n} y_{i_j}^2 p_{i_j}^{-2} - \frac{1}{n(n-1)}\sum_{j\neq k}\sum y_{i_j}y_{i_k}p_{i_j}^{-1}p_{i_k}^{-1}\right\}$$

$$= \frac{1}{n^2}\sum_{j<k}\sum (y_{i_j}p_{i_j}^{-1} - y_{i_k}p_{i_k}^{-1})^2/(n-1).$$

† On the average over $j \neq i$ this always holds, as for the left-hand side the average is equal to π_i, and for the right-hand side is equal to $\pi_i[1 - (\pi_i/n)]$. It follows from the Remark below that in this case $V(\hat{Y}) < V(\hat{\hat{Y}})$ for all y's if and only if for all y's

$$\sum_{i\neq j}\sum y_i y_j \pi_{ij}/(\pi_i\pi_j) < n^{-1}(n-1)\{\sum_{i\neq j}\sum y_i y_j + \sum y_i^2\},$$

that is, if and only if the quadratic form

$$\sum y_i^2 + \sum_{i\neq j}\sum \lambda_{ij}y_i y_j > 0 \quad \text{for all } y\text{'s},$$

where $\lambda_{ij} = 1 - n\pi_{ij}/(\pi_i\pi_j(n-1))$. A necessary and sufficient condition for this is that all principal minors of the matrix of the form be positive. The smallest order minors are 1 and

$$\det\begin{pmatrix} 1 & \lambda_{ij} \\ \lambda_{ij} & 1 \end{pmatrix} = 1 - \lambda_{ij}^2,$$

so that necessary conditions are $\lambda_{ij}^2 < 1$, that is,

$$\pi_{ij} < 2n^{-1}(n-1)\pi_i\pi_j.$$

‡ The case in which the $\delta_i(s)$ and $\delta_{ij}(s, t)$ vary from draw to draw is discussed in KONIJN, J. Amer. Statist. Ass., **57** (1962) 590–606. If we denote their averages by $\bar{p}_i$ and $\bar{p}_{ij}$, respectively, we may, by applying (6.5) to (6.6) and (6.7) of that paper obtain as a generalization of the above statement, that, if $\hat{\hat{Y}} = \sum y_i\bar{p}_i^{-1}/n$, $V(\hat{\hat{Y}})$ vanishes if the $\bar{p}_i$ are proportional to the y_i. An interesting special case is the one in which successive draws are independent (*Poisson binomial sampling*). In this case $V(\hat{\hat{Y}})$ equals the expression given in the text (replacing p_i by $\bar{p}_i$) *minus* $\sum_i\sum_j y_i y_j \omega_{ij}(n\bar{p}_i\bar{p}_j)^{-1}$, where

$$\omega_{ij} = \sum\{\delta_i(s) - \bar{p}_i\}\{\delta_j(s) - \bar{p}_j\}/n.$$

To show unbiasedness, first show that

$$\frac{1}{n}\sum_{i=1}^{N} t_i\left(\frac{y_i}{p_i} - Y\right)^2\Big/ n,$$

has expected value $V(\hat{Y})$, and so is an unbiased estimator of $V(\hat{Y})$ in the case in which Y is known:

$$\frac{1}{n}\mathscr{E}\sum_{i=1}^{N} t_i\left(\frac{y_i}{p_i} - Y\right)^2\Big/ n = \frac{1}{n}\sum_{i=1}^{N}\left(\frac{y_i}{p_i} - Y\right)^2 \mathscr{E}t_i/n$$

$$= \frac{1}{n}\sum_{i=1}^{N}\left(\frac{y_i}{p_i} - Y\right)^2 p_i = V(\hat{Y}).$$

Now

$$\frac{1}{n}\sum_{i=1}^{N} t_i\left(\frac{y_i}{p_i} - \hat{Y}\right)^2 = \frac{1}{n}\sum_{i=1}^{N} t_i\left(\frac{y_i}{p_i} - Y\right)^2 - (\hat{Y} - Y)^2;$$

so

$$\mathscr{E}\frac{1}{n}\sum_{i=1}^{N} t_i\left(\frac{y_i}{p_i} - \hat{Y}\right)^2 = nV(\hat{Y}) - V(\hat{Y}) = (n-1)V(\hat{Y}).$$

Another proof of unbiasedness operates on the alternative expression given for the estimator; since

$$\mathscr{E}t_i t_{i'} = n(n-1)p_i p_{i'}$$

when $i \neq i'$, the expected value of the estimator equals the last of the string of alternative expressions given for $V(\hat{Y})$. A proof based on the form of estimator in terms of the y_{i_j} is also easy to give, either by observing that for two different draws j and k

$$\mathscr{E}y_{i_j}y_{i_k}p_{i_j}^{-1}p_{i_k}^{-1} = \sum_{i}\sum_{i'} y_i y_{i'} = Y^2$$

and

$$\mathscr{E}y_{i_j}^2 p_{i_j}^{-2} = \sum_{i=1}^{N} y_i^2 p_i^{-1};$$

or that

$$y_{i_j}p_{i_j}^{-1} = \sum_{\alpha=1}^{n} a_{\alpha i} y_i p_i^{-1},$$

where $a_{\alpha i} = 1$ if on the αth drawing we obtain u_i and 0 otherwise.

The latter form also yields a very simple derivation of $V(\hat{Y})$:

$$V(\hat{Y}) = \sum_{\alpha=1}^{n} V(n^{-1}\sum_{i=1}^{N} a_{\alpha i}y_i p_i^{-1}) = n^{-1}V(\sum_{i=1}^{N} a_{\alpha i}y_i p_i^{-1}) = n^{-1}\sum p_i(y_i p_i^{-1} - Y)^2.$$

REMARK.† If we take for the p_i in sampling with replacement $1/n$ times the π_i for sampling without replacement,

$$V(\hat{Y}) - V(\hat{\hat{Y}}) = \frac{n-1}{n} Y^2 - \sum_{i \neq j}\sum \frac{y_i y_j}{\pi_i \pi_j} \pi_{ij}.$$

Now consider the case in which sampling is without replacement, but (in order to avoid requiring the knowledge of the π_{ij}, which are often hard to compute) we put the variance estimator appropriate to sampling with replacement:

$$\frac{n}{n-1} \sum t_i \left(\frac{y_i}{\pi_i} - \frac{\hat{\hat{Y}}}{n}\right)^2$$

in the form

$$\frac{1}{n-1}\left[n \sum t_i \left(\frac{y_i}{\pi_i}\right)^2 - \left\{\sum_{i \neq j}\sum t_i t_j \frac{y_i}{\pi_i}\frac{y_j}{\pi_j} + \sum t_i^2 \left(\frac{y_i}{\pi_i}\right)^2\right\}\right].$$

Since sampling is without replacement, actually $t_i = a_i$, and, since $a_i = a_i^2$, we have

$$n \sum t_i \left(\frac{y_i}{\pi_i}\right)^2 - \sum t_i^2 \left(\frac{y_i}{\pi_i}\right)^2$$

equals $(n-1)\Sigma t_i (y_i/\pi_i)^2$, so that the expectation of the above estimator is

$$\mathscr{E} \sum t_i \left(\frac{y_i}{\pi_i}\right)^2 - \frac{1}{n-1} \mathscr{E} \sum_{i \neq j}\sum t_i t_j \frac{y_i}{\pi_i}\frac{y_j}{\pi_j} = \sum \frac{y_i^2}{\pi_i} - \frac{1}{n-1} \sum_{i \neq j}\sum \pi_{ij} \frac{y_i}{\pi_i}\frac{y_j}{\pi_j}$$

instead of the variance of $\hat{Y}$. So this estimator, as an estimator of $V(\hat{Y})$, has a bias of

$$Y^2 - \frac{n}{n-1} \sum_{i \neq j}\sum \frac{y_i}{\pi_i}\frac{y_j}{\pi_j} \pi_{ij} = \frac{n}{n-1}\{V(\hat{\hat{Y}}) - V(\hat{Y})\}.$$

Therefore, the inappropriate variance estimator has a bias which is $n/(n-1)$ times the reduction in variance due to sampling without replacement. So if (as is often the case) this is positive, the variance may be substantially overestimated (especially if $n = 2$), and if it is negative, the variance may be substantially underestimated by this method.

† The Remark is due to DURBIN, J. Roy. Statist. Soc., **B15** (1953) 262–269. From Chapter VII it follows that the formula for $V(\hat{\hat{Y}}) - V(\hat{Y})$ and the bias of the variance estimator given in the text are not affected by subsampling, since the extra term due to subsampling is, for both methods of sampling primary units, of the form $\Sigma\sigma_i^2/\pi_i$ (where the σ_i^2 are the within-cluster variances).

3.3. THE ESTIMATOR OF CHAPTER IV

In Chapter IV we found† an estimator of Y which in general is different from $\bar{\bar{Y}}$ or $\hat{Y}$, namely $\hat{Y}_m$:

$$\hat{Y}_m = \sum_{k=1}^{m} y_{l_k} \mathrm{Q}\{l_k | l_1, \ldots, l_m\} / p_{l_k},$$

where $\{u_{l_1}, \ldots, u_{l_m}\}$ is the set of (different) units included in the sample and $\mathrm{Q}\{l_k | l_1, \ldots, l_m\}$ is the conditional probability that in one given draw we obtain u_{l_k} given that the sample includes precisely $u_{l_1}, \ldots, u_{l_m}$; and p_i is the probability that u_i is obtained in the first (or any other) draw. We there noted that for $n > 2$ (and when m does not happen to equal n), $\hat{Y}_m$ differs from $\hat{\hat{Y}}$, and that in these cases the variance of $\hat{Y}_m$ is smaller than that of $\hat{\hat{Y}}$.

Since $V(\hat{Y}_m) \leq V(\hat{\hat{Y}})$, a simple overestimate of $V(\hat{Y}_m)$ is $v(\hat{\hat{Y}})$. One can give a more efficient unbiased estimator. As we saw,

$$\frac{1}{n(n-1)} \sum_{j \neq k} \sum y_{i_j} y_{i_k} p_{i_j}^{-1} p_{i_k}^{-1}$$

is an unbiased estimator of Y^2. Its conditional expectation (given that the sample contains precisely $u_{l_1}, \ldots, u_{l_m}$) is also unbiased for Y^2 but has smaller variance (cf. section 8 of Chapter I); if we subtract it from $\hat{Y}_m^2$, we get an improved estimator of $V(\hat{Y}_m)$. Let $m > 1$ and let $\mathrm{Q}\{i, j | l_1, \ldots, l_m\}$ be the probability that in two given draws we obtain, respectively, u_i and u_j, given that the sample contains the units $u_{l_1}, \ldots, u_{l_m}$; and let $\mathrm{P}(l_1, \ldots, l_m)$ be the probability of drawing the different units $u_{l_1}, \ldots, u_{l_m}$. Reasoning exactly like in the derivation of $\hat{Y}_m$, we have for h and $k \in \{1, \ldots, m\}$ with $h \neq k$:

$$\mathrm{Q}\{l_k, l_k | l_1, \ldots, l_m\}$$
$$= p_{l_k}^2 \{(\sum_{j=1}^{m} p_{l_j})^{n-2} - \sum_{\substack{t=1 \\ t \neq k}}^{m} (\sum_{j \neq t} p_{l_j})^{n-2} + \ldots\} \mathrm{P}^{-1}(l_1, \ldots, l_m),$$

$$\mathrm{Q}\{l_h, l_k | l_1, \ldots, l_m\}$$
$$= p_{l_h} p_{l_k} \{(\sum_{j=1}^{m} p_{l_j})^{n-2} - \sum_{t \neq h,k}^{m} (\sum_{j \neq t} p_{l_j})^{n-2} + \ldots\} \mathrm{P}^{-1}(l_1, \ldots, l_m);$$

and the conditional expectation of the estimator

$$\frac{1}{n(n-1)} \sum_{\alpha \neq \beta} \sum_{i} \sum_{j} \sum a_{\alpha i} a_{\beta j} \frac{y_i}{p_i} \frac{y_j}{p_j}$$

† If $m = n - 1$, $\mathrm{Q}\{l_k | l_1, \ldots, l_m\}$ simply becomes

$$\{1 + p_{l_k} (\sum_{h=1}^{m} p_{l_h})^{-1}\} / n$$

for $k = 1, \ldots, m$.

of Y^2 mentioned above is

$$\sum_{k=1}^{m} Q\{l_k, l_k|l_1, \ldots, l_m\}\left(\frac{y_{l_k}}{p_{l_k}}\right)^2 + \sum_{h\neq k}\sum Q\{l_h, l_k|l_1, \ldots, l_m\}\frac{y_{l_h}y_{l_k}}{p_{l_h}p_{l_k}}.$$

The resulting estimator of $V(\hat{Y}_m)$ is

$$\sum_{h<k}\sum [Q\{l_h, l_k|l_1, \ldots, l_m\} - Q\{l_h|l_1, \ldots, l_m\}\, Q\{l_k|l_1, \ldots, l_m\}]\left(\frac{y_{l_h}}{p_{l_h}} - \frac{y_{l_k}}{p_{l_k}}\right)^2$$

$$= \sum_{h}\sum_{k} [Q\{l_h|l_1, \ldots, l_m\}\, Q\{l_k|l_1, \ldots, l_m\} - Q\{l_h, l_k|l_1, \ldots, l_m\}]\frac{y_{l_h}}{p_{l_h}}\frac{y_{l_k}}{p_l}.$$

3.4. PLANS WITH A CONSTANT EFFECTIVE SAMPLE SIZE

The expressions for

$$\sum \pi_i \quad \text{and} \quad \sum_{j\neq i}\pi_{ij}$$

are not the same for sampling with replacement as for sampling without replacement. Since, for given effective sample size m, sampling of n with replacement is *equivalent* to sampling of m without replacement,†

$$\sum_{i=1}^{N}\pi_i = \mathscr{E}\sum_{i=1}^{N}\pi_i(m) = \mathscr{E}m,$$

and

$$\sum_{i=1}^{N}\sum_{j\neq i}\pi_{ij} = \mathscr{E}\sum_{i=1}^{N}(m-1)\pi_i(m) = \mathscr{E}(m-1)m$$

$$= \mathscr{E}m^2 - \mathscr{E}m = \mathscr{E}m(\mathscr{E}m - 1) + V(m).$$

Substantial fluctuations in the value of m would appear to inflate the variance of estimators of Y; because of this, and because the cost‡ of a survey generally varies more with m than with n, there is some interest in considering sampling designs for which m is the same for all samples that have a positive probability of occurrence. For such sampling designs

$$\sum_{j\neq i}\pi_{ij} = (m-1)\pi_i \quad (i = 1, \ldots, N).$$

The analogy of these relations to those in section 2 leads us to consider again the estimator for the variance of $\hat{Y}$ considered there, which was unbiased and

† In the derivation that follows $\pi_i(m)$ stands for the probability that u_i is in a sample of m without replacement, or the probability that u_i is in a sample of n drawings with replacement, given that m of the units drawn are different.

‡ In practice it is often desirable that the cost be pretty closely predictable.

nonnegative if the π_{ij} are positive and $\pi_{ij} \leq \pi_i \pi_j$, and useful if, moreover, none of the ratios $\pi_{ij}/(\pi_i \pi_j)$ are small. (Again we should aim at making the π_i proportional to the y_i.)

3.5. HANURAV'S SCHEME

Hanurav† has recently found a sampling scheme which ensures these conditions.‡ Let us number the π_i such that

$$0 < \pi_1 \leq \pi_2 \leq \ldots \leq \pi_{N-1} \leq \pi_N \leq 1;$$

we shall actually assume that $\pi_N < 1$. Define $p_i = \pi_i/m$; he has described the scheme only for the case $m = 2$, but has stated that he has extended it to all larger m.

Let

$$\delta = \{2(1 - p_N)(p_N - p_{N-1})\}/(1 - p_N - p_{N-1}),$$

then $0 \leq \delta \leq 1$. Conduct a Bernoullian trial with success probability δ. If success results, select one (say u_j) among $u_1, \ldots, u_{N-1}$ with probability for u_k proportional to p_k ($k = 1, \ldots, N - 1$), and use the sample $\{u_N, u_j\}$; if failure results, draw two units from $u_1, \ldots, u_N$ with replacement and with probability for choosing u_k equal to

$$p_k^* = p_k/(1 - p_N - p_{N-1}) \quad (k = 1, \ldots, N - 1),$$

and the probability for choosing u_N equal to

$$p_N^* = 1 - \sum_{k=1}^{N-1} p_k^* \left(= \frac{p_N - \frac{1}{2}\delta}{1 - \delta} = p_{N-1}^* \right).$$

If the two selections coincide, draw two units with replacement from $u_1, \ldots, u_N$ with probability for choosing u_k proportional to p_k^{*2} for $k = 1, \ldots, N$. If the two units coincide again, draw two units with replacement from $u_1, \ldots, u_N$ with probability of obtaining u_k proportional to p_k^* to the power 2^2; and so forth (the next powers being $2^3, \ldots$).

3.6. SAMPFORD'S INVERSE SAMPLING SCHEME

For a method which is related to that of 3.3 in that estimators of Y are based on observations on the different units in the sample, but weighted by the number of their repetitions,§ and are thus easier to compute, we refer to SAMPFORD, Biometrika, **49** (1962) 27–40. Sampling is continued until $n + 1$

† HANURAV, J. Roy. Statist. Soc., **B29** (1967) 374–391.

‡ It would be of interest to relax the condition of constancy of m to one of smallness of $V(m)$ and see whether one can improve on Hanurav's scheme under the relaxed conditions.

§ That is, if the probabilities are proportional to the number of subunits in the units (see section 2); otherwise there has to be an additional weighting.

different units are obtained (hence the name "inverse sampling")† and then the last one is rejected (this only so that the estimator will be unbiased). Also in the unbiased estimator of variance the squares are weighted by the number of repetitions. For approximately equal probabilities the variance of the estimator of Y by this method appears to be about halfway between that of $\hat{Y}$ and $\hat{\hat{Y}}$; and the method is conjectured to have an even greater relative advantage over that of 3.2 when the probabilities are very unequal. PATHAK, Biometrika, **51** (1964) 185–193 gives simpler derivations‡ of some of Sampford's results. He compares Sampford's method with the method of section 5.1, and shows that, when, for all pairs $(u_i, u_{i'})$ of different units, $p_i + p_{i'} < \frac{1}{2}$, the latter gives a smaller variance for $n = 2$. An estimator of smaller variance, based on the $\mathscr{Y}$-values of the n distinct units only, and not on the number of their repetitions, is derived in section 5.2.

4. Schemes of implementation of sampling without replacement§

The question is (like in section 3.5) to specify sampling schemes which lead to specified π_i's and which allow finding unbiased, positive estimators of these variances, which themselves do not have unduly inflated standard errors.

Generally one draws one of $u_1, \ldots, u_N$ such that the probability $\delta_k(1)$ of drawing in the first draw u_k is an assigned number§§ p_k^1; then select one among the remaining set of u's with probabilities proportional to the assigned numbers $p_1^2, \ldots, p_N^2$ (with $\Sigma p_i^2 = 1$), excluding p_k^2, such that the conditional probability of drawing u_l (different from the one drawn, say u_k, in the first draw) is $p_l^2/(1 - p_k^2)$; that is, so that the probability of obtaining u_l in the second drawing, regardless of which u_k $(k \neq l)$ was drawn in the first drawing, is

$$\delta_l(2) = \sum_{k \neq l} p_k^1 \frac{p_l^2}{1 - p_k^2} = p_l^2 \left(A - \frac{p_l^1}{1 - p_l^2}\right),$$

where

$$A = \sum_{k=1}^{N} \frac{p_k^1}{1 - p_k^2}.$$

† Compare the slightly different inverse sampling plans referred to in section 14 of Chapter II.

‡ In its appendix, an expression which is not a conditional probability is designated as such. Note also that the e_i in Corollary 1.1 do not satisfy the conditions of the Theorem.

§ According to what is remarked at the beginning of the last paragraph of section 1, schemes of sections 3.5 and 3.6 are also relevant to sampling without replacement.

§§ Note that in this section the superscripts of the p_k are not powers.

So the probability of including u_i in a sample of 2 is

$$\delta_i(1) + \delta_i(2) = p_i^1 + p_i^2 \left(A - \frac{p_i^1}{1 - p_i^2} \right).$$

Clearly, if $n = 2$, π_{ij} is

$$p_i^1 \frac{p_j^2}{1 - p_i^2} + p_j^1 \frac{p_i^2}{1 - p_j^2}.$$

In general for any n

$$\pi_i = \sum_{k=1}^{n} \delta_i(k),$$

where (denoting the sum over the permutations of $i_1, \ldots, i_{k-1}$ by Σ')

$$\delta_i(k) = \Sigma' \sum_{\{i_1,\ldots,i_{k-1}\} \subset \{1,\ldots,i-1,i+1,\ldots,N\}} \left(p_{i_1}^1 \frac{p_{i_2}^2}{1 - p_{i_1}^2} \cdots \right.$$
$$\left. \times \frac{p_{i_{k-1}}^{k-1}}{1 - p_{i_1}^{k-1} \ldots - p_{i_{k-2}}^{k-1}} \frac{p_i^k}{1 - p_{i_1}^k \ldots - p_{i_{k-1}}^k} \right),$$

and for $n \geq 2$ and $i \neq j$

$$\pi_{ij} = \sum_{k=2}^{n} \delta_{ij}(k),$$

$$\delta_{ij}(k) = \sum_{m<k} \delta_{ij}(m, k) + \sum_{m<k} \delta_{ji}(m, k),$$

where $\delta_{ij}(m, k)$ is the probability that u_i and u_j are obtained in the mth and kth draws respectively, $(1 \leq m < k \leq n)$:

$$\delta_{ij}(m, k) = \Sigma' \sum_{\{i_1,\ldots,i_{k-2}\} \subset \{1,\ldots,N\} - \{i,j\}} \left(p_{i_1}^1 \frac{p_{i_2}^2}{1 - p_{i_1}^2} \cdots \right.$$
$$\times \frac{p_i^m}{1 - p_{i_1}^m - \ldots - p_{i_{m-1}}^m} \frac{p_i^{m+1}}{1 - p_{i_1}^{m+1} - \ldots - p_{i_{m-1}}^{m+1} - p_i^{m+1}}$$
$$\left. \times \ldots \frac{p_j^k}{1 - p_{i_1}^k - \ldots - p_i^k - p_{i_m}^k - \ldots - p_{i_{k-2}}^k} \right).$$

In all this the numbers p_i^s are taken to satisfy $0 < p_i^s < 1$,

$$\sum_{i=1}^{N} p_i^s = 1.$$

Equations of the kind given above are often hard to solve, for p_i^s when $n = 2$† and more so when $n > 2$ (see, however, section 6). If we use the

† For $n = 2$ and $p_i^2 = p_i^1$ an efficient method of solution has been given by BREWER and UNDY, Austral. J. Statist., **4** (1962) 89–100. That paper also shows superiority for $n = 2$ for sampling without over sampling with replacement in the sense referred to in section 3.1, and gives a useful discussion of variance estimation.

estimation methods of section 5, it is, however, unnecessary to solve these equations.

In the appendix to this section we show that, if $n = 2$, $\pi_i\pi_j - \pi_{ij} > 0$, so the estimator for $V(\hat{Y})$ given in section 2 always has positive values.

Let us now consider some *specific* cases:

(i) One plan is to use, for all s, the same sequence of numbers $p_1^s, \ldots, p_N^s$, i.e., to make $p_i^s = p_i^1 (= p_i$, say). For small n the resulting equations are hard to solve; for large n a method is given in section 6.1.

(ii) Another plan, associated with the names Midzuno and Sen, is to draw one of $u_1, \ldots, u_N$ such that the probability of drawing u_k is p_k^1 (call it p_k), then select the other $n - 1$ such that each sequence of $n - 1$ from the remaining u's have equal probability. A sample sequence containing u_i has either (a) u_i first and then any one of the $(N - 1)^{(n-1)}$ possible sequences that do not include u_i; or (b) has some other u_j first, then u_i in the 2nd, 3rd, . . ., or nth draw, and then any of the remaining $N - 2$ u's in the other $n - 2$ draws. So we have

$$\pi_i = p_i \frac{1}{(N-1)\ldots(N-n+1)}(N-1)^{(n-1)}$$

$$+ \sum_{j \neq i} p_j(n-1)\frac{1}{(N-1)\ldots(N-n+1)}(N-2)^{(n-2)}$$

$$= \frac{N-n}{N-1}p_i + \frac{n-1}{N-1}.$$

Similarly

$$\pi_{ij} = (p_i + p_j)(n-1)\frac{1}{(N-1)\ldots(N-n+1)}(N-2)^{(n-2)}$$

$$+ \sum_{k \neq i,j} p_k(n-1)^{(2)}\frac{1}{(N-1)\ldots(N-n+1)}(N-3)^{(n-3)}$$

$$= \frac{n-1}{N-1}\left\{\frac{N-n}{N-2}(p_i + p_j) + \frac{n-2}{N-2}\right\},$$

$$\pi_{ijk} = \frac{(n-1)(n-2)}{(N-1)(N-2)}\left\{\frac{N-n}{N-3}(p_i + p_j + p_k) + \frac{n-3}{N-3}\right\}.$$

The main advantage of this plan is the easy solvability of the p_i in terms of the desired π_i:

$$p_i = \frac{N-1}{N-n}\pi_i - \frac{n-1}{N-n}.$$

But, as this equation also shows, all p_i are positive only if all

$$\pi_i > (n-1)/(N-1),$$

a rather severe restriction.

From the above formulas it follows at once that $\pi_i\pi_j - \pi_{ij}$ is positive for this scheme.

(iii) A combination of the above systems; the first two selections as in (i) and the $(n-2)$ following ones as in (ii). Reasoning like in (ii), we get, if $\pi_i(2)$ and $\pi_{ij}(2)$ refer to the results given for $n = 2$ in (i):

$$\pi_i = \frac{N-n}{N-2}\pi_i(2) + \frac{n-2}{N-2},$$

$$\pi_{ij} = \pi_{ij}(2) + \{\pi_i(2) + \pi_j(2) - 2\pi_{ij}(2)\}\frac{n-2}{N-2}$$

$$+ \{1 - \pi_i(2) - \pi_j(2) + \pi_{ij}(2)\}\frac{n-2}{N-2}\,\frac{n-3}{N-3}.$$

Here positive $\pi_i(2)$ exist only if all

$$\pi_i > (n-2)/(N-2),$$

a rather severe restriction.

From the above formulas it follows that $\pi_i\pi_j - \pi_{ij} > 0$ for this plan as well.

(iv) In method (ii) the probability that $u_{i_1}, \ldots, u_{i_n}$ are included is equal to

$$\sum_{j=1}^{n} p_{i_j} \Big/ \binom{N-1}{n-1},$$

since it is the sum of the probabilities of getting any one among $u_{i_1}, \ldots, u_{i_n}$ in the first draw and the others in the remaining draws. Therefore, if $p_i = x_i/X$,

$$\sum_{j=1}^{n} p_{i_j} \Big/ \binom{N-1}{n-1} = \frac{x}{X} \Big/ \binom{N-1}{n-1} = \frac{\bar{x}}{\bar{X}} \Big/ \binom{N}{n},$$

and the *ratio estimator* $(y/x)X$ *is unbiased* for Y. A better method of sampling with this probability of obtaining $\{u_{i_1}, \ldots, u_{i_n}\}$ was given by LAHIRI, Bull. Intern. Statist. Inst., **33**, Pt. II (Intern. Statist. Confer. India, 1951) 133–140. In this method let C be a known number such that for any subset $\{u_{t_1}, \ldots, u_{t_n}\}$ from $\{u_1, \ldots, u_N\}$,

$$\sum_{j=1}^{n} x_{t_j} \le C,$$

where the x_i are natural numbers.† Let us choose at random an integer c between 1 and C. Then choose n random numbers $i_1, \ldots, i_n$ between 1 and N without replacement (or with replacement, if it is desired to sample with replacement). If

$$c \leq \sum_{j=1}^{n} x_{i_j},$$

$\{u_{i_1}, \ldots, u_{i_n}\}$ is taken to constitute the sample; otherwise select once more n random numbers – and so further, until the sum of the $\mathscr{X}$-values of the units which have these subscripts in the population is larger than equal to c.

To prove that this method gives the desired selection probabilities, it suffices‡ to consider the case $n = 1$ and show that the probability of ultimately choosing some given u_i is x_i/X. The probability of completing the first stage with the choice of u_i is $N^{-1}(x_i/C)$, and so the probability of having to go to the next stage

$$1 - N^{-1}(\Sigma x_j/C) = 1 - (\bar{X}/C).$$

The desired probability is therefore

$$\frac{1}{N}\frac{x_i}{C} + \left(1 - \frac{\bar{X}}{C}\right)\frac{1}{N}\frac{x_i}{C} + \left(1 - \frac{\bar{X}}{C}\right)^2 \frac{1}{N}\frac{x_i}{C} + \ldots = \frac{1}{N}\frac{x_i}{C}\frac{C}{\bar{X}} = \frac{x_i}{N\bar{X}} = \frac{x_i}{X}.$$

† Note that, whereas for (ii) it was required that

$$\pi_i > \frac{n-1}{N-1}$$

for all i, here it needs to be satisfied only as an average over the n units of any sample that may occur, i.e. over the units with the n smallest $\mathscr{X}$-values.

‡ In fact, the case of general n is equivalent to the case of selecting a single combination among the set of all possible combinations of n single units with probability proportionate to their (combined $\mathscr{X}$-) sizes. So $N^{-1}x_i/C$ is replaced by

$$\binom{N}{n}^{-1}(x_{i_1} + \ldots x_{i_n})/C,$$

$1 - N^{-1}X/C$ by

$$1 - \binom{N}{n}^{-1} nX/C,$$

and the probability of ultimately choosing $u_{i_1}, \ldots, u_{i_n}$ by

$$\binom{N-1}{n-1}^{-1}(x_{i_1} + \ldots x_{i_n})/X.$$

If some units have a larger x_i, we may split them into subunits with smaller $\mathscr{X}$-values and consider a unit selected whenever any of its parts is selected. This will reduce the number of steps.†

An unbiased estimator of

$$V((y/x)X) = Y^2 - \mathscr{E}((y/x)X)^2$$

is evidently

$$\left(\frac{y}{x}X\right)^2 - \frac{1}{x}X\binom{N-1}{n-1}\left\{a\Big/\binom{N-1}{n-1} + 2b\Big/\binom{N-2}{n-2}\right\}$$

$$= \left(\frac{y}{x}X\right)^2 - \frac{1}{x}X\left(a + 2\frac{N-1}{n-1}b\right) = N^2\frac{\bar{X}}{\bar{x}}\left\{\left(\frac{\bar{X}}{\bar{x}} - 1\right)\bar{y}^2 + \frac{N-n}{N}\frac{1}{n}s^2(\mathscr{Y})\right\}$$

where a is the sample sum of squares for $\mathscr{Y}$ and b the sum of products y_iy_j for the different pairs in the sample.

(v) Some very simply executed methods for $n = 2$ are given by DURBIN, Appl. Statist., **16** (1967) 152–164. In his main method the p_i^2 are chosen such that $\pi_i = 2p_i$, where $p_i = p_i^1$ (like in the method of section 6.2). In particular, when we first draw u_k, the probability of drawing next u_l different from u_k is taken to be (conditionally)

$$\lambda_k p_l\left(\frac{1}{1-2p_k} + \frac{1}{1-2p_l}\right),$$

where λ_k is a factor of proportionality, which, since the sum over $l \neq k$ equals 1, equals

$$\left\{1 + \sum\frac{p_i}{1-2p_i}\right\}^{-1} = \lambda \quad \text{(say)},$$

independent of k. The probability of getting first u_k and then u_l is thus symmetric in k and l. Therefore, the total probability of getting u_i second equals the probability of getting u_i first – which is p_i – so that $\pi_i = 2p_i$, while

$$\pi_{ij} = 2\lambda p_i p_j\left(\frac{1}{1-2p_i} + \frac{1}{1-2p_j}\right) \quad (i \neq j).$$

Comparing $V(\hat{Y})$ with the variance of $\hat{\hat{Y}}$ of section 2 (with replacement and with probabilities p_i), we get for the expression

$$\tfrac{1}{2}\{\pi_{ij} - \tfrac{1}{2}\pi_i\pi_j\}\{(1-2p_i)(1-2p_j)\}^{-1}:$$

† This method can be extended for use with strata. A sample is selected from the entire population such that, if $\bar{x}_h$ is the average of the $\mathscr{X}$-values of units in the sample that belong to stratum h, the probability of drawing any given sample is proportional to $\sum N_h\bar{x}_h$ for this sample.

$$2\lambda(1 - p_i - p_j) - (1 - 2p_i)(1 - 2p_j)$$

$$\geq \frac{N-2}{N-1}(1 - p_i - p_j) - (1 - 2p_i)(1 - 2p_j) \geq \left(\frac{N-2}{N}\right)^2 \Big/ (N-1) > 0,$$

since the minimum of λ is $\frac{1}{2}(N-2)/(N-1)$; and so $V(\hat{Y}) > V(\hat{Y})$. A similar method was given by BREWER, Austral. J. Statist., **5** (1963) 5–13. There

$$p_k^1 = 2\lambda p_k(1 - p_k)/(1 - 2p_k), \qquad p_l^2 = p_l,$$

where the positive numbers $p_1, \ldots, p_N$ (adding up to 1 and each less than n^{-1}) are given in advance; this also gives π_i and π_{ij} as above. For an extension by Sampford to general n see section 6.3. Durbin also gives a simplified version of his method.

(vi) A still simpler method for $n = 2$ given in the above mentioned paper is applicable when units can be paired off into pairs to which we wish to assign approximately equal π_i's, except possibly for one triple with approximately equal π_i's. For this see the paper quoted. STEVENS, J. Roy. Statist. Soc., **B20** (1958) 393–397 has considered more generally grouping of units with approximately equal π_i. (Stevens' unbiased variance estimator is shown to be positive by RAO, Rev. Intern. Statist. Inst., **34** (1966) 125–138.)

APPENDIX. Proof that

$$\pi_i \pi_j - \pi_{ij} > 0$$

if $n = 2$ (and 0 if $N = 2$, of course), using the general conditions stated at the beginning of this section.

For simplicity of notation we prove only $\pi_1\pi_2 - \pi_{12} > 0$ (for $n = 2$):

$$\pi_1\pi_2 - \pi_{12} = (\pi_{12} + \sum_{j>2} \pi_{1j})(\pi_{12} + \sum_{j>2} \pi_{2j}) - \pi_{12}$$

$$= \pi_{12}(1 - \sum_{i>2} \sum_{j>i} \pi_{ij}) + \sum_{j>2} \pi_{1j} \sum_{j>2} \pi_{2j} - \pi_{12}$$

$$= \sum_{j>2} \pi_{1j} \sum_{j>2} \pi_{2j} - \pi_{12} \sum_{i>2} \sum_{j>i} \pi_{ij}.$$

The product of the

$$\sum_{j>2} \pi_{ij} = \frac{p_i^1}{1 - p_i^2} \sum_{j>2} p_j^2 + p_i^2 \sum_{j>2} \frac{p_j^1}{1 - p_j^2}$$

for $i = 1$ and $i = 2$ is

$$p_1^1 p_2^1 \frac{1}{1 - p_1^2} \frac{1}{1 - p_2^2} (\sum_{j>2} p_j^2)^2 + p_1^2 p_2^2 \left(\sum_{j>2} \frac{p_j^1}{1 - p_j^2}\right)^2 + \pi_{12}(\sum_{k>2} p_k^2)\left(\sum_{j>2} \frac{p_j^1}{1 - p_j^2}\right),$$

where

$$\pi_{12} = \frac{p_1^1 p_2^2}{1 - p_1^2} + \frac{p_2^1 p_1^2}{1 - p_2^2}.$$

On the other hand,

$$\sum_{i>2} \sum_{j>1} \pi_{ij} = \sum_{j>2} \left\{ (\sum_{k>2} p_k^2 - p_j^2) \frac{p_j^1}{1 - p_j^2} \right\},$$

which, if multiplied by π_{12} and subtracted from the last member of the expression above leaves the positive term

$$\pi_{12} \sum_{j>2} p_j^2 \frac{p_j^1}{1 - p_j^2}.$$

5. Estimators for sampling without replacement which do not require calculating inclusion probabilities

5.1. DES RAJ'S ESTIMATOR

DES RAJ, J. Amer. Statist. Ass., **51** (1956) 269–284, gave an estimator designed for plan (i) of section 4 which do not require calculation of the π_i and π_{ij}. If the sample sequence is $u_{i_1}, \ldots, u_{i_n}$, define

$$z_1 = y_{i_1}/p_{i_1},$$

$$z_j = \sum_{k=1}^{j-1} y_{i_k} + (y_{i_j}/p_{i_j})(1 - \sum_{k=1}^{j-1} p_{i_k}) \quad (j = 2, \ldots, n).$$

Writing $\mathscr{E}(.|i, \ldots, i_k)$ for the expected value given that the first k units in the sample sequence are $u_{i_1}, \ldots, u_{i_k}$, we have $\mathscr{E}z_1 = Y$, and for $j > 1$

$$\mathscr{E}\{z_j|i_1, \ldots, i_{j-1}\} = \sum_{k=1}^{j-1} y_{i_k} + \sum_{h \neq i_1, \ldots, i_{j-1}} \left\{ (y_h/p_h)(1 - \sum_{k=1}^{j-1} p_{i_k}) \frac{p_h}{1 - \sum_{k=1}^{j-1} p_{i_k}} \right\}$$

$$= \sum_{k=1}^{j-1} y_{i_k} + \sum_{h \neq i_1, \ldots, i_{j-1}} y_h = Y,$$

so $\mathscr{E}z_j = Y$ for all j. Also for $j < k = 2, \ldots, n$

$$\mathscr{E}\{z_j z_k|i_1, \ldots, i_{k-1}\} = \mathscr{E}z_j \, \mathscr{E}\{z_k|i_1, \ldots, i_{k-1}\} = \mathscr{E}z_k Y = Y^2,$$

so that any two z's are uncorrelated. Des Raj considered as an estimator of Y the average, $\bar{z}$, of the z's. From the above

$$\mathscr{E}\bar{z}^2 - \mathscr{E}\frac{2}{n(n-1)} \sum\sum_{j<k} z_j z_k = \mathscr{E}\bar{z}^2 - Y^2 = V(\bar{z}),$$

so that an unbiased estimator of the variance of $\bar{z}$ is given by

$$\frac{1}{n(n-1)}\sum_{j=1}^{n}(z_j-\bar{z})^2=\bar{z}^2-\frac{2}{n(n-1)}\sum_{j<k}\sum z_j z_k.$$

For $n = 2$ this becomes

$$\tfrac{1}{4}(1-p_{i_1})^2\left(\frac{y_{i_1}}{p_{i_1}}-\frac{y_{i_2}}{p_{i_2}}\right)^2.$$

By the method of section 5.2, a better unbiased estimator is found to be

$$\tfrac{1}{4}(1-p_{i_1})(1-p_{i_2})\left(\frac{y_{i_1}}{p_{i_1}}-\frac{y_{i_2}}{p_{i_2}}\right)^2.$$

For $n = 2$ the variance of $\bar{z}$ is

$$\frac{1}{n^2}\sum_{i>j}\sum p_i p_j\left(\frac{y_i}{p_i}-\frac{y_j}{p_j}\right)^2(2-p_i-p_j),$$

since, by section 3.2,

$$V(z_1)=\sum_{i>j}\sum p_i p_j\left(\frac{y_i}{p_i}-\frac{y_j}{p_j}\right)^2,$$

and, similarly,

$$V(z_2)=\sum_{i>j}\sum p_i p_j\left(\frac{y_i}{p_i}-\frac{y_j}{p_j}\right)^2(1-p_i-p_j).$$

For general n the variance equals

$$\frac{1}{n^2}\sum_{i>j}\sum p_i p_j\left(\frac{y_i}{p_i}-\frac{y_j}{p_j}\right)^2\{1+r_{ij}(1)+\ldots+r_{ij}(n-1)\},$$

where $r_{ij}(k)$ is the probability that u_i and u_j are not included in the sequence $u_{i_1}, \ldots, u_{i_k}$. It can be shown easily that this variance is less than that of

$$n^{-1}\sum_{j=1}^{n} y_{i_j}/p_{i_j},$$

where sampling with replacement is used with the same p_{i_j} as in the above scheme. For that it suffices to note that $V(z_{k+1}) < V(z_k)$, as illustrated by our computation of $V(z_1)$ and $V(z_2)$ above.

The estimator $\bar{z}$ is simple, has a positive estimator of variance, and for large N and not too unequal probabilities has been shown by PATHAK, Sankhya, **29A** (1967) 283–298, to often have a smaller variance than other estimators (e.g., than that of the estimator of section 7 if the coefficient of variation of the p_i is well in excess of 1). Nonetheless, we shall now see that we can improve the estimator.

5.2. AN IMPROVED ESTIMATOR

In this section, in contrast to the previous one, we shall denote by

$$\mathscr{E}\{\cdot|l_1, \ldots, l_n\}$$

an expected value given that the sample sequence $u_{i_1}, \ldots, u_{i_n}$ contains the (different) units $u_{l_1}, \ldots, u_{l_n}$ (without regard to order of appearance). From section 8 of Chapter I we know that the variance of $\mathscr{E}\{\bar{z}|l_1, \ldots, l_n\}$ will never exceed that of $\bar{z}$. Now

$$z_{j+1} - z_j = \left(\frac{y_{i_{j+1}}}{p_{i_{j+1}}} - \frac{y_{i_j}}{p_i}\right)(1 - \sum_{k=1}^{j} p_{i_k}),$$

so

$$\mathscr{E}\{z_{j+1} - z_j|l_1, \ldots, l_n\}$$

$$= \sum\sum \left(\frac{y_{r_k}}{p_{r_k}} - \frac{y_{r_h}}{p_{r_h}}\right)(1 - \sum_{k=1}^{j-1} p_{i_k} - p_{r_h}) \frac{p_{r_h}}{1 - \sum_{t=1}^{j-1} p_{i_t}} \frac{p_{r_k}}{1 - \sum_{t=1}^{j-1} p_{i_t} - p_{r_h}},$$

where the inner summation is over all $\{r_h, r_k\}$ in $\{l_1, \ldots, l_n\} - \{i_1, \ldots, i_{j-1}\}$ and where the outer summation is over all $\{i_1, \ldots, i_{j-1}\}$ in $\{l_1, \ldots, l_n\}$. This expression equals

$$\sum \frac{1}{1 - \sum_{t=1}^{j-1} p_{i_t}} \sum (y_{r_k} p_{r_h} - y_{r_h} p_{r_k}),$$

which equals 0, since, in the inner summation, for each time that $r_h = \alpha$ and $r_k = \beta$, we also have that $r_h = \beta$, $r_k = \alpha$. Therefore $\mathscr{E}\{z_j|l_1, \ldots, l_n\}$ is the same for each j, and so

$$\mathscr{E}\{\bar{z}|l_1, \ldots, l_n\} = \mathscr{E}\{z_1|l_1, \ldots, l_n\}.$$

Let $Q_1\{i|l_1, \ldots, l_n\}$ be the conditional probability of drawing first u_i given that the sample sequence forms the set $\{u_{l_1}, \ldots, u_{l_n}\}$; the corresponding unconditional probability is $\delta_i(1) = p_i$, and the probability of the condition $P(l_1, \ldots, l_n)$. We obtain as conditional expectation of $\bar{z}$ the unbiased estimator

$$\sum_{i=1}^{N} a_i y_i Q_1\{i|l_1, \ldots, l_n\}/p_i.$$

For $n = 2$ this becomes

$$\frac{1}{2 - p_{l_1} - p_{l_2}} \left\{(1 - p_{l_2})\frac{y_{l_1}}{p_{l_1}} + (1 - p_{l_1})\frac{y_{l_2}}{p_{l_2}}\right\},$$

since

$$\mathrm{P}(l_1, l_2) = \pi_{l_1 l_2} = p_{l_1} \frac{p_{l_2}}{1 - p_{l_1}} + p_{l_2} \frac{p_{l_1}}{1 - p_{l_2}}, \ \delta_{l_1 l_2}(1, 2) = \frac{p_{l_1} p_{l_2}}{1 - p_l}.$$

So

$$\mathrm{Q}_1\{l_1 | l_1, l_2\} = \frac{\delta_{l_1 l_2}(1, 2)}{\mathrm{P}(l_1, l_2)} = \frac{1 - p_{l_2}}{2 - p_{l_1} - p_{l_2}}.$$

We can similarly find the variance of $\bar{z}$ (see also below); for $n = 2$ it becomes

$$\sum_{j>i}\sum p_i p_j \left(\frac{y_i}{p_i} - \frac{y_j}{p_j}\right)^2 \frac{1 - p_i - p_j}{2 - p_i - p_j}.$$

In general the last fraction is 1 minus the sum over all $\{l_1, \ldots, l_n\}$ that contain u_i and u_j of

$$\mathrm{Q}_1\{i | l_1, \ldots, l_n\}\, \mathrm{Q}_1\{j | l_1, \ldots, l_n\} \mathrm{P}(l_1, \ldots, l_n)/(p_i p_j).$$

For general n the variance is minimal when the p_i are proportional to the y_i. Let

$$\mathrm{Q}_{(12)}\{i, j | l_1, \ldots, l_n\}$$

be the conditional probability that the first two units drawn are (in either order) u_i and u_j, given that the sample consists of $u_{l_1}, \ldots, u_{l_n}$; the corresponding unconditional probability is

$$\delta_{ij}(1, 2) + \delta_{ij}(2, 1) = p_i \frac{p_j}{1 - p_i} + p_j \frac{p_i}{1 - p_j}.$$

An unbiased estimator of the variance is found to be

$$\sum_{j>i}\sum a_i a_j \left(\frac{y_i}{p_i} - \frac{y_j}{p_j}\right)^2 p_i p_j$$

$$\times \left[\frac{\mathrm{Q}_{(12)}\{i, j | l_1, \ldots, l_n\}}{\delta_{ij}(1, 2) + \delta_{ij}(2, 1)} - \frac{\mathrm{Q}_1\{i | l_1, \ldots, l_n\} \mathrm{Q}_1\{j | l_1, \ldots, l_n\}}{p_i p_j}\right].$$

We first show that, for $j \neq k$,

$$\mathscr{E}\{z_j z_k | l_1, \ldots, l_n\} = \mathscr{E}\{z_1 z_2 | l_1, \ldots, l_n\};$$

for that it suffices to prove, by the same method as used above to show that $\mathscr{E}\{z_j | l_1, \ldots, l_n\}$ does not depend on j, that

$$\mathscr{E}\{z_{j+1} z_k - z_j z_k | l_1, \ldots, l_n\} = 0$$

for $j + 1 = 2, \ldots, k - 1$ and

$$\mathscr{E}\{z_1 z_{k+1} - z_1 z_k | l_1, \ldots, l_n\} = 0$$

for $k = 2, \ldots, n$, since from these equalities follows, respectively, that, in the part of the matrix of the $\mathscr{E}\{z_j z_k | l_1, \ldots, l_n\}$ above the diagonal, the elements of each column are equal to the first element in that column, and that the elements in the first row are equal. Now, like in 5.1, an unbiased estimator of the variance of $\mathscr{E}\{\bar{z} | l_1, \ldots, l_n\}$ is given by

$$\mathscr{E}^2\{\bar{z}|l_1, \ldots, l_n\} - \frac{2}{n(n-1)} \sum_{j>k}\sum \mathscr{E}\{z_j z_k | l_1, \ldots, l_n\}$$
$$= \mathscr{E}^2\{z_1 | l_1, \ldots, l_n\} - \mathscr{E}\{z_1 z_2 | l_1, \ldots, l_n\},$$

where

$$\begin{aligned}\mathscr{E}\{z_1 z_2 | l_1, \ldots, l_n\} \\ &= \mathscr{E}\{y_{i_1} p_{i_1}^{-1}(y_{i_1} + y_{i_1} p_{i_2}^{-1}(1 - p_{i_1})) | l_1, \ldots, l_n\} \\ &= \sum a_i y_i^2 p_i^{-1} \mathrm{Q}_1\{i | l_1, \ldots, l_n\} \\ &\quad + \sum_{i \neq i'}\sum a_i a_{i'} y_i y_{i'} \{\delta_{ii'}(1, 2) + \delta_{ii'}(2, 1)\}^{-1} \mathrm{Q}_{(12)}\{i, i' | l_1, \ldots, l_n\}.\end{aligned}$$

Using the fact that (in obvious notation)

$$\mathrm{P}_1\{l_1, \ldots, l_n | i\} = \sum_{i' \neq i} \frac{p_{i'}}{1 - p_i} \mathrm{P}_{(12)}\{l_1, \ldots, l_n | i, i'\},$$

we obtain the desired expression. The formula given for the variance of $\mathscr{E}\{\bar{z} | l_1, \ldots, l_n\}$ may be obtained almost at once by taking first the conditional expectation of the bracket for fixed (i, j), i.e., by multiplying by $\mathrm{P}(l_1, \ldots, l_n)$ divided by $\mathscr{E} a_i a_j$, the probability of including u_i and u_j, and summing over all those $\{l_1, \ldots, l_n\}$ that contain u_i and u_j and then taking the expectation.

PATHAK and SHUKLA, Sankhya, **A28** (1966) 41–46, showed that the square bracket in our variance estimator is always positive. For $n = 2$ the expression reduces to

$$\frac{(1 - p_{l_1})(1 - p_{l_2})(1 - p_{l_1} - p_{l_2})}{(2 - p_{l_1} - p_{l_2})^2} \left(\frac{y_{l_1}}{p_{l_1}} - \frac{y_{l_2}}{p_{l_2}}\right)^2,$$

since

$$\mathrm{Q}_{(12)}\{l_1, l_2 | l_1, l_2\} = 1.$$

Formulae for the case $n = 3$ are given in the Pathak and Shukla paper [in (21) the factor $p_{(2)} p_{(3)}$ has been omitted].

6. Calculation of the inclusion probabilities for sampling without replacement when n is not necessarily very small

6.1. VAN BEEK AND VERMETTEN'S METHOD FOR LARGE n

Using method (i) of section 4, VAN BEEK and VERMETTEN, Appl. Statist., **15** (1966) 74–95, obtained an approximate value for π_i. The probability of obtaining $u_{i_1}, \ldots, u_{i_n}$ successively is

$$p_{i_1} \frac{p_{i_2}}{1 - p_{i_1}} \frac{p_{i_3}}{1 - p_{i_1} - p_{i_2}} \cdots \frac{p_{i_n}}{1 - p_{i_1} - \cdots - p_{i_{n-1}}},$$

which for n not large equals approximately

$$\prod_{j=1}^{n} p_{i_j}\{1 + (n-1)p_{i_1} + (n-2)p_{i_2} + \cdots + 1 \cdot p_{i_{n-1}}\}.$$

For large n this is a serious underestimate, but, since we shall need ratios of such expressions, this does not matter much.

Consider the sum of these approximate probabilities over all permutations of $i_1, \ldots, i_n$:

$$n! \prod_{j=1}^{n} p_{i_j}\{1 + \tfrac{1}{2}(n-1)\sum_{j=1}^{n} p_{i_j}\} = \tfrac{1}{2}n![2\prod_{j=1}^{n} p_{i_j} + (n-1)\prod_{j=1}^{n} p_{i_j}\sum_{j=1}^{n} p_{i_j}].$$

Let

$$K_l = \Sigma p_{i_1} \cdots p_{i_l},$$

where the sum is taken over all sets

$$\{i_1, \ldots, i_l\} \subset \{i, \ldots, N\};$$

and

$$K_l' = \Sigma p_{i_1} \cdots p_{i_l},$$

where the sum is taken over all sets

$$\{i_1, \ldots, i_l\} \subset \{1, \ldots, N\} - \{i\}.$$

The product $K_1 K_n$ consists of terms containing squares, and of terms (each a product of $n+1$ p's) without squares; each of the latter terms occurs $n+1$ times in $K_1 K_n$. Therefore, the sum over all

$$\{i_1, \ldots, i_n\} \subset \{1, \ldots, N\}$$

of

$$\prod_{j=1}^{n} p_{i_j} \sum_{j=1}^{n} p_i$$

equals $K_1K_n - (n+1)K_{n+1}$. Thus the sum of the approximate probabilities for all such choices is

$$\tfrac{1}{2}n![2K_n + (n-1)\{K_1K_n - (n+1)K_{n+1}\}].$$

Similarly the sum of approximate probabilities for samples of size n containing a given u_i is

$$\tfrac{1}{2}n![2p_iK'_{n-1} + (n-1)\{p_i^2K'_{n-1} + p_i((\sum_{j=1}^{N} p_j - p_i)K'_{n-1} - nK'_n)\}].$$

Consequently, the ratio of the last two expressions is the approximate value for π_i:

$$p_i\left[K'_{n-1} - \frac{n(n-1)}{n+1}K'_n\right][K_n - (n-1)K_{n+1}]^{-1}$$

$$= p_i\frac{1 - \{n(n-1)/(n+1)\}(K'_n/K'_{n-1})}{(K'_n/K'_{n-1})[1-(n-1)(K'_{n+1}/K'_n)] + p_i[1-(n-1)(K'_n/K'_{n-1})]},$$

since $K_1 = 1$ and $K_n = K'_n + p_iK'_{n-1}$.

We obtain a further approximation by replacing K_n by

$$\binom{N}{n}p^n,$$

where p is some average of all the p_i (which lies between their geometric mean ($n = N$) and their arithmetic mean ($n = 1$)), and similarly replacing K'_n by

$$\binom{N-1}{n}(p')^n.$$

This means we write (neglecting the difference between the approximation p' appropriate to K'_n and that to K'_{n-1}):

$$\binom{N}{n}p^n = \binom{N-1}{n}(p')^n + p_i\binom{N-1}{n-1}(p')^{n-1},$$

or

$$p = \left\{1 - \frac{n}{N}\left(1 - \frac{p_i}{p'}\right)\right\}^{1/n}p'$$

$$= \left\{1 - \frac{1}{N}\left(1 - \frac{p_i}{p'}\right)\right\}p' - \frac{n-1}{2N}\frac{1}{N}\left(1 - \frac{p_i}{p'}\right)^2p' + \dots .$$

Taking only the first term we get

$$p' \approx \frac{Np - p_i}{N-1},$$

which gives for π_i the approximation

$$\frac{np_i\{(N-1)/N\}D}{(N-n)p+\{n((N-1)/N)(C/B)-(N-n)/N\}p_i},$$

where

$$A = 1 - \frac{(n-1)(N-n)}{n+1}\frac{Np-p_i}{N-1},$$

$$B = 1 - \frac{(n-1)(N-n-1)}{n+1}\frac{Np-p_i}{N-1},$$

$$C = 1 - \frac{(n-1)(N-n)}{n}\frac{Np-p_i}{N-1},$$

$$D = A/B.$$

We can expand C/B in powers of p_i, taking for p here $1/N$. This gives up to first order terms

$$1 - \frac{(n-1)N}{n(2N+n^2-n-2)} + p_i\frac{(n-1)(n+1)N(N-1)}{n(2N+n^2-n-2)^2};$$

it can be used when $p_i < 2/(n-1)+(n+2)/N$. This gives as approximation for π_i

$$nDp_i\left[(N-n)p+\alpha(n-1)p_i+(1-\alpha)^2(n^2-1)\frac{N}{N-1}p_i^2\right]^{-1}$$

with

$$\alpha = 1 - \frac{N-1}{2(N-1)+n^2-n}.$$

D is less than 1; for $p = 1/N$ it exceeds $1-(1-p_i)/p_i(N-1)$.

As a final correction of the approximations for π_i, we should divide them by n^{-1} times their sum over i.

To solve the p_i from the π_i, solve the quadratic equation in the p_i, taking for np some value near the arithmetic mean of the π_i; then adjust them to make their sum add to 1. A check is obtained by substitution in the exact formula for the π's.

For an application of this method, see the reference quoted above.

6.2. FELLEGI'S METHOD

FELLEGI, J. Amer. Statist. Ass., **58** (1963) 183–201, gave a system of solving for† the p_i^k from the π_i. Let $p_i^1 = \pi_i/n$, so that their sum is unity. Fellegi's

† Note that the superscripts of the p_i in this subsection are not powers.

method consists of setting for each $k = 1, \ldots, n$

$$\delta_i(k) = p_i^1 = \pi_i/n \quad (i = 1, \ldots, N)$$

and solving these equations, which, since $\delta_i(k)$ is a fraction of the p_i^l for $l \leq k$ (given in section 4), are (nonlinear) equations for the p_i^l ($1 < l \leq k$; $i = 1, \ldots, N$).

Since $\delta_i(1) = p_i^1$, we have at once

$$p_i^1 = \delta_i(1) = \pi_i/n.$$

Now, by induction, if for each i we have found $p_i^1, \ldots, p_i^{k-1}$, we obtain (Σ' denoting summation over the permutations of $i_1, \ldots, i_{k-1}$)

$$\mathrm{P}(i_1, \ldots, i_{k-1}) = \sum{}' p_{i_1}^1 \frac{p_{i_2}^2}{1 - p_{i_1}^2} \cdots \frac{p_{i_{k-1}}^{k-1}}{1 - p_{i_1}^{k-1} - \cdots - p_{i_{k-2}}^{k-1}},$$

and obtain for p_i^k ($i = 1, \ldots, N$) the successive approximations ${}_rp_i^k$ ($r = 1, 2, \ldots$) as follows. Defining functions Q_i of $N - 1$ positive arguments whose sum is less than 1 (the functions having a separate definition for each k):

$$\mathrm{Q}_i(a_1, \ldots, a_{i-1}, a_{i+1}, \ldots, a_N) = \sum_{\{i_1, \ldots, i_{k-1}\} \subset \{1, \ldots, i-1, i+1, \ldots, N\}} (1 - a_{i_1} - \cdots - a_{i_{k-1}})^{-1} \mathrm{P}(i_1, \ldots, i_{k-1}),\dagger$$

we obtain successively

$$\begin{aligned}
{}_1p_i^k &= p_i^{k-1}, \quad (i = 1, \ldots, N) \\
{}_2p_1^k &= \delta_1(k)\, \mathrm{Q}_1^{-1}({}_1p_2^k, {}_1p_3^k, \ldots, {}_1p_N^k), \\
{}_2p_2^k &= \delta_2(k)\, \mathrm{Q}_2^{-1}({}_2p_1^k, {}_1p_3^k, \ldots, {}_1p_N^k), \\
{}_2p_3^k &= \delta_3(k)\, \mathrm{Q}_3^{-1}({}_2p_1^k, {}_2p_2^k, {}_1p_4^k, \ldots, {}_1p_N^k), \\
&\cdots\cdots\cdots\cdots \\
{}_3p_1^k &= \delta_1(k)\, \mathrm{Q}_1^{-1}({}_2p_2^k, {}_2p_3^k, \ldots, {}_2p_N^k), \\
{}_3p_2^k &= \delta_2(k)\, \mathrm{Q}_2^{-1}({}_3p_1^k, {}_2p_3^k, \ldots, {}_2p_N^k), \\
{}_3p_3^k &= \delta_3(k)\, \mathrm{Q}_3^{-1}({}_3p_1^k, {}_3p_2^k, {}_2p_4^k, \ldots, {}_2p_N^k), \\
&\cdots\cdots\cdots\cdots \\
&\cdots\cdots\cdots\cdots
\end{aligned}$$

continuing until, for all i, ${}_rp_i^k$ is as close to ${}_{(r-1)}p_i^k$ as desired. The procedure is quite easily programmed for an electronic computer.‡ For $n = 2$, and all

† Therefore $\delta_i(k)\mathrm{Q}_i^{-1}(p_1^k, \ldots, p_{i-1}^k, p_{i+1}^k, \ldots, p_N^k) = p_i^k$.

‡ However, for large n the computation proved to be *prohibitive*, even on a high-speed electronic computer. Even for small n the scheme is rather laborious. Fellegi, while recognizing this, mentions that he devised his scheme so as to be applicable also to rotating samples (see Chapter XI and his paper).

$p_i^1 < \frac{1}{2}$, a proof of convergence of the procedure has been given by BREWER, J. Amer. Statist. Soc., **62** (1967) 79–85.

If $n = 2$, and

$$A = \sum_{i=1}^{N} \frac{p_i^1}{1 - p_i^2},$$

then

$$\delta_i(2) = p_i^1 = p_i^2 \left(A - \frac{p_i^1}{1 - p_i^2}\right) \geq p_i^2(A - p_i^1),$$

so

$$Ap_i^2(1 - p_i^2) = p_i^1 \tag{*}$$

with

$$\sum_{i=1}^{N} p_i^2 = 1;$$

therefore

$$A = \{1 - \sum_{i=1}^{N} (p_i^2)^2\}^{-1}.$$

From given A we find

$$p_i^2 = \tfrac{1}{2} \pm (\tfrac{1}{4} - (p_i^1/A))^{\frac{1}{2}}.$$

Since $p_i^1 \geq p_i^2(1 - p_i^1)$, $p_i^2 \leq \frac{1}{2}$ if $p_i^1 \leq \frac{1}{3}$. Assuming this we get

$$\begin{aligned} p_i^2 &= \tfrac{1}{2} - (\tfrac{1}{4} - (p_i^1/A))^{\frac{1}{2}} \\ &= \tfrac{1}{2} - (\tfrac{1}{4})^{\frac{1}{2}} + \tfrac{1}{2}(\tfrac{1}{4})^{-\frac{1}{2}}(p_i^1/A) + \tfrac{1}{8}(\tfrac{1}{4})^{-1\frac{1}{2}}(p_i^1/A)^2 + \tfrac{1}{16}(\tfrac{1}{4})^{-2\frac{1}{2}}(p_i^1/A)^3 + \ldots \\ &= \frac{p_i^1}{A} + \left(\frac{p_i^1}{A}\right)^2 + 2\left(\frac{p_i^1}{A}\right)^3 + \ldots . \end{aligned}$$

Since† $\pi_{ij} = 2Ap_i^2 p_j^2$ and $\pi_i = 2p_i^1$, the variance of the estimator of Y becomes approximately

$$\frac{1}{2} \sum_{j>i}\sum p_i^1 p_j^1 \left\{2 - \frac{1}{A}\left(1 + \frac{p_i^1 + p_j^1}{A}\right)\right\} \left(\frac{y_i}{p_i^1} - \frac{y_j}{p_j^1}\right)^2 .$$

If the p_i^2 do not vary a great deal, then

$$A \approx 1 + \sum_{i=1}^{N} (p_i^2)^2 = 1 + O\left(\frac{1}{N}\right).$$

† Obtained by substituting (*) in $\pi_{ij} = \{p_i^1 p_j^2/(1 - p_i^2)\} + \{p_j^1 p_i^2/(1 - p_j^2)\}$.

When comparing $V(\hat{Y})$ with the variance of $\hat{Y}$ of section 2 (with replacement with probabilities p_i of drawing u_i at any one draw equal to π_i/n), we obtain, when $n = 2$, the expression

$$V(\hat{\hat{Y}}) - V(\hat{Y}) = \sum_{i \neq j}\sum (Ap_i^2 p_j^2 - p_i^1 p_j^1)(m_i - m_j)^2,$$

with $m_i = y_i/\pi_i$; this is equal to

$$\Sigma(Ap_i^2 - p_i^1)m_i^2 \Sigma p_j^2 + \Sigma(Ap_j^2 - p_j^1)m_j^2 \Sigma p_i^2$$
$$- 2A\Sigma p_i^2 m_i \Sigma p_j^2 m_j + 2\Sigma p_i^1 m_i \Sigma p_j^1 m_j$$
$$= 2A\Sigma p_i^2(m_i - \bar{m}^2) - 2\Sigma p_i^1 (m_i - \bar{m}^1)^2,$$

with $\bar{m}^1 = \Sigma p_i^1 m_i$, $\bar{m}^2 = \Sigma p_i^2 m_i$. Since $Ap_i^2 \geq p_i^1$,

$$V(\hat{\hat{Y}}) - V(\hat{Y}) \geq 2\Sigma p_i^1[(m_i - \bar{m}^2)^2 - (m_i - \bar{m}^1)^2] = 2(\bar{m}^1 - \bar{m}^2)^2;$$

which is positive (unless $\bar{m}^1 = \bar{m}^2$).

6.3. SAMPFORD'S METHOD OF SAMPLING WITHOUT REPLACEMENT

SAMPFORD, Biometrika, **54** (1967) 499–513 extends the last method mentioned in (v) of section 4.

Given are N positive numbers $p_1, \ldots, p_N$ adding to 1, and each less than $1/n$. Compute (different from in the quoted section) $\lambda_i = p_i(1 - np_i)^{-1}$ for each N, and $R_r = \Sigma \lambda_i^r$ for $r = 1, \ldots, n$. Define $L_0 = L_0(\bar{\imath}) = 1$ and compute for each $m = 1, \ldots, n$ and $i = 1, \ldots, N$

$$L_m = m^{-1} \sum_{r=1}^{m} (-1)^{r-1} R_r L_{m-r},$$
$$L_m(\bar{\imath}) = L_m - \lambda_i L_{m-1}(\bar{\imath}),$$
$$q_i = c_1 \lambda_i \sum_{t=2}^{n} (t - np_i) L_{n-t}(\bar{\imath})/n^{t-1}$$

with

$$c_1 = (n-1)^{-1} K_n, \qquad K_n = \sum_{t=1}^{n} t L_{n-t}/n^t.$$

Here the q_i are the p_i^1. For $n = 2$, $p_i^2 = p_i$. So, if on the first draw we select u_i, then on the second draw we select u_l with conditional probability

$$q_{j \cdot i} = p_j/(1 - p_i).$$

Now consider the case $n > 2$. In that case, if we select u_i on the first draw, we compute for all $j \neq i$ and all $m \leq n$

$$L_m(\bar{\imath}\bar{\jmath}) = L_m(\bar{\imath}) - \lambda_j L_{m-1}(\bar{\imath}\bar{\jmath})$$

and

$$q_{j\cdot i} = c_2(i)\lambda_j \sum_{t=3}^{n} \{t - n(p_i + p_j)\}L_{n-t}(\bar{i}\bar{j})/n^{t-2},$$

with

$$c_2(i) = \frac{1}{n-2}\, c_1\lambda_i/q_i,$$

so that the sum of the $q_{j\cdot i}$ over $j \neq i$ is 1. We then draw the second unit such that the conditional probability of drawing u_j is $q_{j\cdot i}$. If $n = 3$, $p_k^3 = p_k$, so that the conditional probability of obtaining u_k on the third draw is $p_k/(1 - p_i - p_j)$.

Suppose $n > 3$. Having drawn u_i and u_j, we now compute for all $m \leq n$ and for all k different from i and j

$$L_m(\bar{i}\bar{j}\bar{k}) = L_m(\bar{i}\bar{j}) - \lambda_k L_{m-1}(\bar{i}\bar{j}\bar{k}),$$
$$q_{k\cdot ij} = c_3(ij)\lambda_k \Sigma\{t - n(p_i + p_j + p_k)\}L_{n-t}(\bar{i}\bar{j}\bar{k})/n^{t-3},$$

with

$$c_3(ij) = \frac{1}{n-3}\, c_2(i)\lambda_j/q_{j\cdot i}.$$

The process may be continued in a straightforward manner. The resulting π_i equals np_i and the formula for π_{ij} is

$$\pi_{ij} = K_n\lambda_i\lambda_j \sum_{t=2}^{n} \{t - n(p_i + p_j)\}L_{n-t}(\bar{i}\bar{j})/n^{t-2}$$

(for a proof see Sampford's paper).

The calculations proceed *very rapidly*. We programmed the procedure to compute the q's and the π_{ij}, and to draw the serial numbers of the units to be included in the sample. On the CDC 6600, with a population of 32 units this took (exclusive of the time required for compiling the program) 0.9 seconds for a sample of 24 and 0.65 seconds for a sample of 15. The method of the following section involves far fewer computations, but the number of drawings required grows rapidly with sample size.

6.4. SAMPFORD'S REJECTIVE METHOD

Sampford also proposes a rejective method (see the last footnote of section 1) which yields the same π_i and π_{ij} and so the same variances and for which the procedure is shorter if $n > 2$ but small. The first drawing is made with probabilities p_i and the unit drawn is replaced, and subsequent drawings (with replacement) are made with probabilities $\lambda_i/\Sigma\lambda_i$, but only accepted if each time we draw a unit different from the units drawn previously (that is, if a unit occurs more than once in the sample, a new sample is drawn).

From the definitions in the previous section we find that the probability there of obtaining a sample consisting of the set $\{u_{i_1}, \ldots, u_{i_n}\}$ of units is

$$K_n\lambda_{i_1} \ldots \lambda_{i_n}(1 - p_{i_1} \ldots - p_{i_n}), \tag{*}$$

which equals $(n-1)!^{-1}K_n(\Sigma\lambda_i)^{n-1}$ times the coefficient of $t_{i_1} \ldots t_{i_n}$ in

$$W(t) = (\Sigma p_i t_i)(\Sigma\lambda_i t_i/\Sigma\lambda_i)^{n-1}.$$

But $W(t)$ is the probability generating function for with-replacement sampling in which the first drawing is made with probabilities p_i and the others with probabilities $\lambda_i/\Sigma\lambda_i$. The probability of a sample in such with-replacement sampling in which u_i occurs r_i times is the coefficient of

$$t_1^{r_1} \ldots t_N^{r_N}$$

in $W(t)$. Since in sampling with replacement the probabilities of all samples in which no unit occurs more than once are proportional to (*), the probabilities of the without-replacement scheme may also be achieved with the rejective scheme. The constant $(n-1)!^{-1}K_n(\Sigma\lambda_i)^{n-1}$ is the expected number of samples that must be drawn to obtain a sample without duplicates; in practice a sample may be discarded as soon as a duplicate unit is drawn. In our example this constant was quite prohibitive (for sampling 15 units it was of the order of 50).

7. A random split method

This method, originated by RAO, HARTLEY and COCHRAN in J. Roy. Statist. Soc., **B24** (1962) 482–491, gives estimators of the same form as those discussed above. Let p_i $(i = 1, \ldots, N)$ be given positive numbers adding to one. We split the population at random into n mutually exclusive sets of sizes $N_1, \ldots, N_n$. Let $\mathscr{G}$ denote the collection of possible splits of the population into n mutually exclusive sets of these sizes, and g an element of $\mathscr{G}$. Evidently there are

$$N!/\prod_{s=1}^{n}(N_s!)$$

different splits possible. For given $g \in \mathscr{G}$ let $\{u(g, s, 1), \ldots, u(g, s, N_s)\}$ be the sth set produced by the particular g. We now select, independently in each set, a unit such that the probability of drawing $u(g, s, k)$ (given we are in g and its sth set) equals

$$P_{gsk} = p(g, s, k)/\sum_{h=1}^{N_s} p(g, s, h).$$

Let $y(g, s, k)$ be the $\mathscr{Y}$-value of $u(g, s, k)$. We estimate Y by

$$\hat{Y} = \sum_{g\in\mathscr{G}} a_g \sum_{s=1}^{n} \sum_{k=1}^{N_s} a_{gsk}\, y(g, s, k)/P_{gsk},$$

where, for given g, $a_{gsk} = 1$ if $u(g, s, k)$ is drawn in the sample, and zero otherwise; and where $a_g = 1$ if g is the selected split, and 0 otherwise.

Since $\mathscr{E}\{\hat{Y}|g\} = Y$, $\mathscr{E}\hat{Y} = Y$. For fixed g and s the a_{gsk} have means P_{gsk}, and variances $P_{gsk}(1 - P_{gsk})$, and the covariance between a_{gsh} and a_{gsk} $(h \neq k)$ is equal to $-P_{gsh}P_{gsk}$; hence the conditional variance of

$$\sum_{k=1}^{N_s} a_{gsk}y(g, s, k)/P_{gsk}$$

is

$$\begin{aligned}
V\{\sum_{k=1}^{N_s} a_{gsk}y(g, s, k)/P_{gsk}|g, s\} \\
&= \sum_{k=1}^{N_s} y^2(g, s, k)/P_{gsk} - \sum_{k=1}^{N_s} y^2(g, s, k) - \sum_{h \neq k}\sum y(g, s, h)y(g, s, k) \\
&= \sum_{k=1}^{N_s} P_{gsk}\{y(g, s, k)P_{gsk}^{-1} - \sum_{h=1}^{N_s} y(g, s, h)\}^2 \\
&= \sum_{\substack{h<k \\ 1}}^{N_s}\sum p(g, s, h)\, p(g, s, k)\left\{\frac{y(g, s, h)}{p(g, s, h)} - \frac{y(g, s, k)}{p(g, s, k)}\right\}^2.
\end{aligned}$$

Since, for given g, a_{gsh} and a_{gtk} are independent for $s \neq t$, we have

$$V\{\hat{Y}|g\} = \sum_{s=1}^{n} \sum_{g \in \mathscr{G}} a_g V\{\sum_{k=1}^{N_s} a_{gsk}y(g, s, k)/P_{gsk}|g, s\}.$$

The probability that a pair of observations falls in the sth set of a random split is $N_s(N_s - 1)/N(N - 1)$, so

$$\begin{aligned}
V(\hat{Y}) &= \mathscr{E}V\{\hat{Y}|g\} \\
&= \mathscr{E}\sum_{s=1}^{n} \sum_{g \in \mathscr{G}} a_g \sum_{\substack{h<k \\ 1}}^{N_s}\sum p(g, s, h)\, p(g, s, k)\left\{\frac{y(g, s, h)}{p(g, s, h)} - \frac{y(g, s, k)}{p(g, s, k)}\right\}^2 \\
&= \sum_{s=1}^{n} \frac{N_s(N_s - 1)}{N(N - 1)} \sum_{\substack{i<j \\ 1}}^{N}\sum p_i p_j \left\{\frac{y_i}{p_i} - \frac{y_j}{p_j}\right\}^2 \\
&= \frac{\sum_{s=1}^{n} N_s(N_s - 1)}{N(N - 1)}\left(\sum_{i=1}^{N} \frac{y_i^2}{p_i} - Y^2\right) \\
&= \frac{\sum_{s=1}^{n} N_s(N_s - 1)}{N(N - 1)} \sum_{i=1}^{N} p_i\left\{\frac{y_i}{p_i} - Y\right\}^2.
\end{aligned}$$

So we should strive for the p_i to be proportional to the y_i.

We remark that, for the estimator $\hat{\hat{Y}}$ for sampling with replacement of section 3, we have

$$V(\hat{\hat{Y}}) = \frac{1}{n} \sum_{i=1}^{N} p_i \left(\frac{y_i}{p_i} - Y\right)^2,$$

so that

$$V(\hat{Y}) = n \sum_{s=1}^{n} \frac{N_s(N_s - 1)}{N(N-1)} V(\hat{\hat{Y}}).$$

This is minimal for $N_1 = \ldots = N_n = N/n$, giving for the right-hand side of the equation $(1 - (n-1)/(N-1))\, V(\hat{\hat{Y}})$, which is less than $V(\hat{\hat{Y}})$. Note that with equal probabilities the variance of the usual estimator of Y in sampling without replacement is also $(1 - (n-1)/(N-1))V(\hat{\hat{Y}})$! If N/n is an integer, we take all N_s equal; if $N = nK + k$, with $K > 0$ and $0 < k < n$, let us take $N_1 = \ldots = N_k = K + 1$, $N_{k+1} = \ldots = N_n = K$, so that

$$V(\hat{Y}) = \left\{1 - \frac{n-1}{N-1} + \frac{k(n-k)}{N(N-1)}\right\} V(\hat{\hat{Y}}).$$

Noting that

$$\mathscr{E}\left\{\sum_{k=1}^{N_s} a_{gsk} \frac{y^2(g, s, k)}{p^2(g, s, k)} \sum_{h=1}^{N_s} p(g, s, h) | g, s\right\} = \sum_{k=1}^{N_s} \frac{y^2(g, s, k)}{p(g, s, k)},$$

we find, since the probability that an observation falls in set s of a split is N_s/N,

$$\mathscr{E}\left\{\sum_{s=1}^{n} \sum_{g \in \mathscr{G}} a_g \sum_{k=1}^{N_s} a_{gsk} \left(\frac{y(g, s, k)}{p(g, s, k)}\right)^2 \sum_{h=1}^{N_s} p(g, s, h)\right\} = \sum_{s=1}^{n} \frac{N_s}{N} \sum_{i=1}^{N} \frac{y_i^2}{p_i} = \sum_{i=1}^{N} \frac{y_i^2}{p_i}.$$

Since, for $v(\hat{Y})$ an unbiased estimator of $V(\hat{Y})$, we have

$$\mathscr{E}\{v(\hat{Y}) - \hat{Y}^2\} = -Y^2,$$

we obtain by adding this to the previous equation

$$\mathscr{E}\left\{v(\hat{Y}) + \sum_{s=1}^{n} \sum_{g \in \mathscr{G}} a_g \sum_{k=1}^{N_s} a_{gsk} \left(\frac{y(g, s, k)}{p(g, s, k)} - \hat{Y}\right)^2 \sum_{h=1}^{N_s} p(g, s, h)\right\}$$
$$= \sum_{i=1}^{N} p_i \left\{\frac{y_i}{p_i} - Y\right\}^2.$$

The right-hand side is the mean of $\{N(N-1)/\Sigma N_s(N_s - 1)\}v(\hat{Y})$, so $(1 - (N(N-1)/\Sigma N_s(N_s - 1)))$ times $v(\hat{Y})$ plus the second term in the sum above has zero expectation, giving for an unbiased estimator of $V(\hat{Y})$

$$v(\hat{Y}) = \frac{\sum_{s=1}^{n} N_s^2 - N}{N^2 - \sum_{s=1}^{n} N_s^2} \sum_{g \in \mathscr{G}} a_g \sum_{s=1}^{n} \sum_{k=1}^{N_s} a_{gsk} \left(\frac{y(g, s, k)}{p(g, s, k)} - \hat{Y}\right)^2 \sum_{h=1}^{N_s} p(g, s, h).$$

If $N_1 = \ldots = N_k = K + 1$, $N_{k+1} = \ldots = N_n = K$, the factor in front of the summation sign becomes

$$\frac{N^2 + k(n-k) - Nn}{N^2(n-1) - k(n-k)}.$$

Let $\mathscr{U}$ be the set of $u(g, s, k)$ in the sample; denote by $\mathscr{Z}$ the set of couples

$$(u, y(u)), \quad u \in \mathscr{U}.$$

One can show that $\mathscr{Z}$ is sufficient for $\mathscr{Y}$ and compute $\mathscr{E}\{\hat{Y}|\mathscr{Z}\}$ (see Chapter I, section 8), leading to an estimator of smaller variance. HÁJEK, Ann. Math. Statist., **35** (1964) 1491–1523, remarks that $\mathscr{E}\{\hat{Y}|\mathscr{Z}\}$ is unfortunately very difficult to compute.

The random split method may also be carried out with double sampling, so that first a random sample of n' is chosen, and probabilities p_i determined on the basis of an auxiliary characteristic $\mathscr{X}$ observed on the units of that sample. The sample of n' being random, *any* splitting of it into n sets (of sizes $N_1, \ldots, N_n$, with $\Sigma N_s = n'$), when not based on its observed characteristics, is also random. To estimate Y, we, of course, have to multiply the expression given above by the ratio to n' of the population size, N. The variance of the estimator will be the variance of its conditional expectation, $N(N - n')(n')^{-1}S^2$, plus the mean of the conditional variance, which is

$$\frac{N}{N-1} \frac{\Sigma N_s(N_s - 1)}{n'(n'-1)} \Sigma p_i \left(\frac{y_i}{p_i} - Y\right)^2,$$

where the factor $N/(N - 1)$ is due to the fact that Y appearing in the variance formula for the single sampling case is, in the conditional variance, replaced by a random variable.

8. Estimation of the variance of an estimator if the method of sampling had been different — sampling proportional to the values of $\mathscr{X}$ compared with simple random sampling using a ratio estimator based on $\overline{\mathscr{X}}$, and with stratification according to the values of $\mathscr{X}$†

In Chapter III, section 10 we studied the case in which we obtained a simple random sample and on the basis of it desired to estimate the variance of our

† For the estimation of the effect of stratification on a cluster sample see SUKHATME, Sampling Theory of Surveys with Applications (Indian Society of Agric. Statist. and Iowa State University Press, 1954) sections 7.17 and 8.11. For ratio estimators see MOKASHI, J. Ind. Soc. Agr. Statist., **6** (1954) 77–82, and RAO and CHAWLA, J. Ind. Soc. Agr. Statist., **8** (1956) 91–101 (the latter for the case of unequal probabilities). Most of the section given here is based on a communication of J. N. K. Rao (1962).

estimator of Y if we had used stratified sampling instead, and vice versa. Now we shall make similar comparisons with sampling proportional to the values of $\mathscr{X}$.

(a) The sample has been obtained by sampling proportionate to the value of $\mathscr{X}$ with replacement. In this case, according to section 3.2, using $p_i = x_i/X$:

$$\hat{Y} = \frac{1}{n} X \sum_{i=1}^{N} t_i \frac{y_i}{x_i},$$

$$v(\hat{Y}) = \frac{X^2}{n(n-1)} \sum_{i=1}^{N} t_i \left(\frac{y_i}{x_i} - \frac{\hat{Y}}{X}\right)^2 = \frac{X^2}{n-1}\left\{\sum_{i=1}^{N} t_i \frac{y_i^2}{nx_i^2} - \left(\frac{1}{n}\sum_{i=1}^{N} t_i \frac{y_i}{x_i}\right)^2\right\},$$

where t_i is the number of times u_i is included, so that $\mathscr{E} t_i = np_i$.

(i) Suppose first we wish to estimate the variance of the appropriate estimator under simple random sampling:

$$V(N\bar{y}) = \frac{N(N-n)}{N-1} \frac{1}{n}(\sum_{i=1}^{N} y_i^2 - Y^2/N) = \frac{N-n}{N-1}\frac{1}{n}\sum_{i<i'}\sum (y_i - y_{i'})^2,$$

$$V(\hat{Y}_R) \approx \frac{N(N-n)}{N-1}\frac{1}{n}\left(\sum_{i=1}^{N} y_i^2 - 2\frac{Y}{X}\sum_{i=1}^{N} y_i x_i + \frac{Y^2}{X^2}\sum_{i=1}^{N} x_i^2\right).$$

Since

$$\mathscr{E} X \frac{1}{n} \sum_{i=1}^{N} t_i \frac{y_i^2}{x_i} = \sum_{i=1}^{N} y_i^2, \quad V(\hat{Y}) = \mathscr{E}\hat{Y}^2 - Y^2,$$

an unbiased estimator of $V(N\bar{y})$ will be

$$\frac{N(N-n)}{N-1}\frac{1}{n}\left[X\frac{1}{n}\sum_{i=1}^{N} t_i \frac{y_i^2}{x_i} - \{\hat{Y}^2 - v(\hat{Y})\}/N\right],$$

and another the nonnegative estimator

$$\frac{N-n}{N-1}\frac{1}{n-1} X^2 \sum_{i<i'}\sum t_i t_{i'} (y_i - y_{i'})^2/(n^2 x_i x_{i'}),$$

since

$$\mathscr{E} t_i t_{i'} = n(n-1)(x_i/X)(x_{i'}/X)$$

if $i \neq i'$. To obtain an estimator of $V(\hat{Y}_R)$, replace, respectively:

$\sum y_i^2$ by its unbiased estimator

$$(X/n)\sum_{i=1}^{N} t_i y_i^2/x_i,$$

$YX^{-1}\sum y_i x_i$ by its estimator

$$(\hat{Y}X^{-1}/n)\sum_{i=1}^{N} t_i y_i x_i X/x_i = (\hat{Y}/n)\sum_{i=1}^{N} t_i y_i,$$

$Y^2X^{-2}\sum x_i^2$ by its estimator

$$\{\hat{Y}^2 - v(\hat{Y})\}X^{-2}(X/n)\sum_{i=1}^{N} t_i x_i^2/x_i = \{\hat{Y}^2 - v(\hat{Y})\}(X^{-1}/n)\sum_{i=1}^{N} t_i x_i;$$

thus obtaining

$$\frac{N(N-n)}{N-1}\frac{1}{n}\left\{X\frac{1}{n}\sum_{i=1}^{N} t_i\frac{y_i^2}{x_i} - 2\hat{Y}\frac{1}{n}\sum_{i=1}^{N} t_i y_i + X^{-1}(\hat{Y}^2 - v(\hat{Y}))\frac{1}{n}\sum_{i=1}^{N} t_i x_i\right\}.$$

An estimator like the second estimator of $V(N\bar{y})$, obtained by replacing $\mathscr{Y}$ by $\hat{\mathscr{D}}$ with $\hat{d}_i = y_i - \hat{Y}X^{-1}x_i$, is

$$\frac{N-n}{N-1}\frac{1}{n-1}X^2\sum_{i<i'}\sum t_i t_{i'}(\hat{d}_i - \hat{d}_{i'})/(n^2 x_i x_{i'}).$$

(ii) Now suppose we wish to estimate the variance of the estimator of Y under stratification by $\mathscr{X}$-values. We shall use the $\mathscr{Y}'$ symbol of Chapter II, section 11.

We wish to estimate

$$\sum_{h=1}^{L} N_h(N_h - n_h)\frac{1}{n_h}S_h^2,$$

where, for example,

$$S_1^2 = \{\sum_{i=1}^{N_1} y_i^2 - N_1\bar{Y}_1^2\}/(N_1 - 1).$$

When

$$\sum_{i=1}^{N_1} t_i > 1,$$

we may estimate S_1^2 by the immediately derived expression

$$X^2\left\{\frac{1}{n}\sum_{i=1}^{N_1} t_i\frac{y_i^2}{x_i^2} - \left(\frac{1}{n}\sum_{i=1}^{N} t_i\frac{y_i}{x_i}\right)^2\Big/\sum_{i=1}^{N} t_i\right\}\Big/\left(\sum_{i=1}^{N} t_i - 1\right),$$

which is approximately unbiased if we may neglect

$$\mathrm{P}\{\sum_{i=1}^{N_1} t_i \leq 1\}.$$

(b) The sample has been obtained by simple random sampling or stratified sampling.

(i) Estimation from a simple random sample. Since:

$$V(\hat{Y}) = X\frac{1}{n}\sum_{i=1}^{N}\frac{y_i^2}{x_i} - \frac{1}{n}Y^2 = \frac{1}{n}\sum_{i<i'}\sum x_i x_{i'}\left(\frac{y_i}{x_i} - \frac{y_{i'}}{x_{i'}}\right)^2,$$

we may estimate $V(\hat{Y})$ unbiasedly by

$$\frac{N(N-1)}{n(n-1)}\frac{1}{n}\sum_{i<i'}\sum a_i a_{i'} x_i x_{i'}\left(\frac{y_i}{x_i} - \frac{y_{i'}}{x_{i'}}\right)^2 = \frac{N(N-1)}{n-1}\left\{\bar{x}\frac{1}{n}\Sigma a_i\frac{y_i^2}{x_i} - \bar{y}^2\right\},$$

since the probability that u_i and $u_{i'}$ $(i' \neq i)$ both appear in a simple random sample of n is $n(n-1)/N(N-1)$.

An unbiased estimator which is not necessarily positive is obtained by substituting in

$$V(\hat{Y}) = n^{-1}(X\Sigma y_i^2/x_i - Y^2)$$

the simple random sampling estimators $(N/n)\Sigma a_i y_i^2/x_i$ for $\Sigma y_i^2/x$ and $N^2\{\bar{y}^2 - ((N-n)/Nn)s^2\}$ for Y^2, giving

$$\frac{N}{n^2}X\,\Sigma a_i\frac{y_i^2}{x_i} - \frac{N^2}{n}\bar{y}^2 + N^2\frac{N-n}{N}\frac{1}{n^2}s^2,$$

which equals

$$\frac{N^2}{n}\left\{\bar{X}\frac{1}{n}\Sigma a_i\frac{y_i^2}{x_i} + \frac{N-n}{N}\frac{1}{n(n-1)}\Sigma a_i y_i^2 - \frac{N-1}{N}\frac{n}{n-1}\bar{y}^2\right\}.$$

A comparison with the previous estimator shows that for large n the difference between the estimators of variance lies in the middle term (which, if $\Sigma a_i y_i^2$ and $\bar{X}\Sigma a_i y_i^2/x_i$ do not differ much, is much smaller than the first term), and in

$$(N^2/n)(\bar{X} - \bar{x})n^{-1}\Sigma a_i y_i^2/x_i.$$

(ii) Estimation from a stratified sample:

$$\mathscr{E}X\frac{1}{n}\sum_{h=1}^{L}\frac{N_h}{n_h}\sum_{i=1}^{N_h} a_{hi}\frac{y_{hi}^2}{x_{hi}} = X\frac{1}{n}\sum_{h=1}^{L}\sum_{i=1}^{N_h}\frac{y_{hi}^2}{x_{hi}} = X\frac{1}{n}\sum_{i=1}^{N}\frac{y_i^2}{x_i};$$

and so, since

$$V(\hat{Y}_{\text{st}}) = \mathscr{E}(\sum_{h=1}^{L} N_h\bar{y}_h)^2 - Y^2,$$

an unbiased estimator of

$$X\frac{1}{n}\sum_{i=1}^{N} t_i\frac{y_i^2}{x_i} - \frac{1}{n}Y^2$$

is

$$X\frac{1}{n}\sum_{h=1}^{L}\frac{N_h}{n_h}\sum_{i=1}^{N_h} a_{hi}\frac{y_{hi}^2}{x_{hi}} - \frac{1}{n}\{(\sum_{h=1}^{L} N_h\bar{y}_h)^2 - v(\hat{Y}_{st})\}.$$

9. Double sampling for determining probabilities of including units in the sample

(a) Suppose we wish to draw units with probabilities proportional to the values of a characteristic $\mathscr{X}$ for which we do not have advance information. For example, if the units are composed of subunits, x_i may be the number of subunits contained in u_i. One method may be to draw without replacement a simple random sample s' of n' units, measure $\mathscr{X}$ on them, and then draw (with replacement) a subsample of n units for measuring $\mathscr{Y}$ with conditional probabilities p_i proportional to the $\mathscr{X}$ values in s'.

Let a_i' be one if u_i appears in s' and 0 otherwise, and let $x' = \Sigma a_i' x_i$ and $y' = \Sigma a_i' y_i$; so x' and y' denote the total $\mathscr{X}$ and $\mathscr{Y}$ values, respectively, in s', and $n' = \Sigma a_i'$. Let t_i be the number of times u_i appears in the subsample. Then $p_i = a_i' x_i/x'$, and we may estimate Y by

$$\hat{Y}_{n',n} = \frac{N}{n'}\sum_{i=1}^{N}\frac{t_i}{np_i}y_i.$$

Now

$$\mathscr{E}\left\{\sum\frac{t_i}{np_i}y_i|s'\right\} = \sum a_i' y_i,$$

and, since $\mathscr{E}a_i' = n'/N$, $\mathscr{E}\hat{Y}_{n',n} = Y$.

From Chapter II it follows that

$$V(\Sigma a_i' y_i) = \{(N-n')/N\}n'S^2.$$

From section 3.2 we have

$$V\left\{\sum\frac{t_i}{np_i}y_i|s'\right\} = \frac{1}{n}\Sigma a_i' p_i\left(\frac{y_i}{p_i} - y'\right)^2 = \frac{1}{n}\sum_{i<j}\sum a_i' a_j'\left(\frac{y_i}{p_i} - \frac{y_j}{p_j}\right)^2 p_i p_j.$$

The expected value of this under simple random sampling without replacement is, since

$$\mathscr{E}a_i' a_j' = n'(n'-1)/\{N(N-1)\}$$

when $i \neq j$:

$$\frac{1}{n}\frac{n'(n'-1)}{N(N-1)}\sum_{i<j}\sum\left(\frac{y_i}{p_i} - \frac{y_j}{p_j}\right)^2 p_i p_j = \frac{n'(n'-1)}{N(N-1)}V\{\hat{Y}_n((x_i/X))\},$$

where $V\{\hat{Y}_n((x_i/X))\}$ is the variance of the estimator of section 3.2, with given probabilities x_i/X, based on n observations (clearly this variance is independent of X). Thus

$$V(\hat{Y}_{n',n}) = \frac{N}{n'}\left[(N-n')S^2 + \frac{n'-1}{N-1}V\{\hat{Y}_n((x_i/X))\}\right]$$

$$= \frac{N}{N-1}\,\frac{n'-1}{n'}\,V\{\hat{Y}_n((x_i/X))\} + N^2\frac{N-n'}{Nn'}\,S^2.$$

Since n' is to be rather large, we see that $V(\hat{Y}_{n',n})$ equals approximately the variance of the estimator based on n observations with known probabilities x_i/X, plus the variance of the estimator of Y based on a simple random sample of n' without replacement.

Since

$$\mathscr{E}\left\{\sum \frac{t_i}{np_i} y_i^2|\mathscr{s}'\right\} = \sum a_i' y_i^2,$$

and

$$\mathscr{E}\left\{\frac{1}{n(n-1)}\sum_{i\neq j}\sum t_i t_j y_i y_j/(p_i p_j)|\mathscr{s}'\right\} = y'^2$$

(the latter due to the (conditional) independence of t_i and t_j),

$$S^2 = \mathscr{E}\frac{1}{n'-1}\,\Sigma a_i'(y_i - \bar{y}')^2$$

may be unbiasedly estimated by

$$\hat{S}^2{}_{n',n} = \frac{1}{n(n'-1)}\left\{\Sigma t_i y_i^2/p_i - \frac{1}{n'(n-1)}\sum_{i\neq j}\sum t_i t_j y_i y_j/p_i p_j\right\}.$$

Also $\mathscr{E}V\{(N/n')\Sigma(t_i/np_i)y_i|\mathscr{s}'\}$ may be unbiasedly estimated by

$$\frac{(N/n')^2}{n(n-1)}\Sigma t_i\left\{y_i/p_i - \frac{1}{n}(\Sigma t_j y_j/p_j)\right\}^2,$$

since in section 3.2 we showed that $(n'/N)^2$ times this expression is an unbiased estimator of $V\{\hat{Y}_n((p_i))|\mathscr{s}'\}$. Therefore

$$\hat{V}(\hat{Y}_{n',n}) = \left(\frac{N}{n'}\right)^2\left[\frac{1}{n(n-1)}\Sigma t_i\left\{y_i/p_i - \frac{1}{n}(\Sigma t_j y_j/p_j)\right\}^2\right.$$

$$\left. + \frac{N-n'}{n}\,\frac{n'}{n'-1}\,\frac{1}{n}\left\{\Sigma t_i y_i^2/p_i - \frac{1}{n'(n-1)}\sum_{i\neq j}\sum t_i t_j y_i y_j/p_i p_j\right\}\right].$$

(b) If we do not know the total X for the population, we may use an advance sample $\mathscr{s}'$ of size n' to estimate X (as $\hat{X}_{n'} = N\bar{x}'$) and esimate an upper bound for the x_i; and then take a sample independent of $\mathscr{s}'$ of size n to compute $\hat{Y} = \hat{X}_{n'}\Sigma(t_i/nx_i)y_i$ by the method of Lahiri (given in section 4 (iv)), for which a knowledge of X is not required and the x_i need only be ascertained for certain of the units.

Since $\hat{X}_{n'}$ and $X\Sigma(t_i/nx_i)y_i$ are independent with

$$\mathscr{E}\hat{X}_{n'}/X = 1, \qquad V(\hat{X}_{n'}/X) = N^2\frac{N-n'}{N}\frac{1}{n'}C(\mathscr{X}\mathscr{X}),$$

$$\mathscr{E}\left(X\sum\frac{t_i}{nx_i}y_i\right) = Y, \qquad V\left(X\sum\frac{t_i}{nx_i}y_i\right) = V\{\hat{Y}_n((x_i/X))\},$$

we have

$$V\left(\hat{X}_{n'}\sum\frac{t_i}{nx_i}y_i\right)$$

$$= \mathscr{E}\left[\left(\sum\frac{t_i}{nx_i}y_i\right)^2 V\left\{\hat{X}_{n'}|\sum\frac{t_i}{nx_i}y_i\right\}\right] + V\left[\sum\frac{t_i}{nx_i}y_i\,\mathscr{E}\left\{\hat{X}_{n'}|\sum\frac{t_i}{nx_i}y_i\right\}\right]$$

$$= \left[\left\{V\left(\sum\frac{t_i}{nx_i}y_i\right)^2 V(\hat{X}_{n'}) + (Y/X)^2\right\}V(\hat{X}_{n'})\right] + V\left(\sum\frac{t_i}{nx_i}y_i\right)X^2$$

$$= V\{\hat{Y}_n((x_i/X))\} + C(\mathscr{X}\mathscr{X})\frac{N-n'}{N}\frac{1}{n'}[Y^2 + V\{\hat{Y}_n((x_i/X))\}],$$

where the first term on the right-hand side is the variance we would obtain in single-stage sampling if X were known. Compare this with the expression under (a) above

$$V(\hat{Y}_{n',n}) = \frac{N}{N-1}\frac{n'-1}{n'}V\{\hat{Y}_n((x_i/X))\} + C(\mathscr{Y}\mathscr{Y})\frac{N-n'}{N}\frac{1}{n'}Y^2.$$

The present method is evidently more advantageous only if $C(\mathscr{X}\mathscr{X})$ is much smaller than $C(\mathscr{Y}\mathscr{Y})$.

An estimator is obtained by noting that

$$\left\{V\left(\sum\frac{t_i}{nx_i}y_i\right) + (Y/X)^2\right\}V(\hat{X}_{n'})$$

is estimated unbiasedly by $(\Sigma(t_i/nx_i)y_i)^2v(\hat{X}_{n'})$, and

$$V\left(\sum\frac{t_i}{nx_i}y_i\right)X^2 = V\left(\sum\frac{t_i}{nx_i}y_i\right)[\{X^2 + V(\hat{X}_{n'})\} - V(\hat{X}_{n'})]$$

by

$$\frac{1}{n(n-1)} \Sigma t_i \left(\frac{y_i}{x_i} - \frac{1}{n} \Sigma \frac{t_j}{x_j} y_j\right)^2 \{\hat{X}^2_{n'} - v(\hat{X}_{n'})\}.$$

REMARK. RAJ, Ann. Math. Statist., **35** (1964) 900–902, to whom these results are due, also gives the additional terms to the variance if the units are subsampled.

(c) Results for the estimator in section 4 (iv) using two-phase sampling, with and without subsampling, are given by CHIKKAGOUDAR, Ann. Instit. Statist. Math., **19** (1967) 131–142 (the second sample is a subsample of the first). He discusses sampling without replacement in the second phase only. He also considers the case in which the initial sample of n' is drawn without replacement and classified according to the strata, and then n_h units are subsampled from the units of $\mathscr{s}'$ falling in stratum h with probability proportional to their total $\mathscr{X}$-value, without or with replacement.

REMARK. Double sampling for ratio estimation with subsampling is discussed by SUKHATME and KOSHAL, J. Ind. Soc. Agr. Statist., **11** (1959) 128–144.

10. Difference estimation with unequal probabilities and double sampling

In Chapter IIA, section 13.2 we considered the use of double sampling with difference estimators. We now suppose that there is information available, prior to the time of first sampling, which allows us to draw the first sample $\mathscr{s}'$ with certain probabilities p_i (with replacement). Let t'_i be the number of times u_i is included in $\mathscr{s}'$. In the first sample we only observe $\mathscr{X}$.

(a) First we investigate the case in which $\mathscr{s}'$ is subsampled by taking a simple random sample without replacement of size n from it. Let $a_i = 1$ if u_i is in this sample and 0 otherwise. Our estimator will be

$$\sum \frac{t'_i}{n'p_i} \left[\frac{a_i}{n/n'} y_i + k \left\{x_i - \frac{a_i}{n/n'} x_i\right\}\right],$$

which has conditional mean given $\mathscr{s}'$ of $\Sigma(t'_i/n'p_i)y_i$; the mean and variance of the latter expression are Y and $V\{\hat{Y}_{n'}((p_i))\}$, in the notation of the previous section.

The conditional variance given $\mathscr{s}'$ of our estimator is, if

$$z_i = (y_i - kx_i)/p_i \quad \text{and} \quad \bar{z}' = \Sigma t'_i x_i/n',$$

$$\frac{n'-n}{n'} \frac{1}{n} \Sigma t'(z_i - \bar{z}')^2/(n'-1) = \frac{n'-n}{n'} \frac{1}{n} \frac{1}{n'} \sum_{i>j}\sum t'_i t'_j (z_i - z_j)^2/(n'-1),$$

since our estimator equals the average over the subsample of the $\mathscr{Z}$-values plus an expression which for given s' is constant. The expected value of this conditional variance is, since

$$\mathscr{E}t_i't_j' = n'(n'-1)p_ip_j$$

when $i \neq j$:

$$\frac{n'-n}{n'}\frac{1}{n}\sum_{i>j}\sum p_ip_j(z_i - z_j)^2 = \frac{n'-n}{n'}\frac{1}{n}\Sigma p_i(z_i - \Sigma p_iz_i)^2$$

$$= \frac{n'-n}{n'}\frac{1}{n}\Sigma p_i\left(\frac{y_i - kz_i}{p_i} - Y + kX\right)^2 = \frac{n'-n}{n'}V\{(\hat{Y} - k\hat{X})_n((p_i))\}.$$

So the variance of our estimator is

$$V\{\hat{Y}_{n'}((p_i))\} + \frac{n'-n}{n'}V\{(\hat{Y} - k\hat{X})_n((p_i))\}$$

$$= V\{\hat{Y}_n((p_i))\} + \frac{n'-n}{n'}[V\{k\hat{X}_n((p_i))\} - 2\operatorname{Cov}\{\hat{Y}_n((p_i)), k\hat{X}_n((p_i))\}].$$

The first term on the right-hand side shows the variance of the ordinary estimator (with given p_i but without prior sampling or the use of an auxiliary characteristic $\mathscr{X}$).

Now, if $\bar{z} = \Sigma t_i'a_iz_i/n$,

$$\Sigma t_i'a_i(z_i - \bar{z})^2/(n-1)$$

has conditional expected value

$$\Sigma t_i'(z_i - \bar{z}')^2/(n'-1) = \frac{1}{n'}\sum_{i<j}\sum t_i't_j'(z_i - z_j)^2/(n'-1)$$

and expected value $\Sigma p_i(z_i - Y + kZ)^2$, which is $n'V\{\hat{Y}_{n'}((p_i))\}$ if $k = 0$. So the variance may be estimated by

$$\frac{1}{n'}\Sigma t_i'a_i\left(\frac{y_i}{p_i} - \frac{1}{n}\Sigma t_j'a_j\frac{y_j}{p_j}\right)^2 \Big/ (n-1) + \frac{n'-n}{n'}\frac{1}{n}\Sigma t_i'a_i(z_i - \bar{z})^2/(n-1).$$

(b) After s' is drawn, a sample of size n is obtained independent of s' by the same method as s', and both $\mathscr{Y}$ and $\mathscr{X}$ observed. The estimator of Y can be written

$$\Sigma\frac{t_i}{np_i}y_i + k\left\{\Sigma\frac{t_i'}{n'p_i}x_i - \Sigma\frac{t_i}{np_i}x_i\right\} = \Sigma\frac{t_i}{np_i}(y_i - kx_i) + k\Sigma\frac{t_i'}{n'p_i}x_i.$$

Therefore its variance is

$$V\{(\hat{Y} - k\hat{X})_n((p_i))\} + k^2V\{\hat{X}_{n'}((p_i))\},$$

which is estimated by

$$\frac{1}{n}\Sigma t_i(z_i - \bar{z})^2/(n-1) + \frac{k^2}{n'}\Sigma t_i'\left(\frac{x_i}{p_i} - \frac{1}{n'}\Sigma t_j'\frac{x_j}{p_j}\right)^2 \Big/ (n'-1).$$

REMARK to (a) and (b). The approximate variances for ratio estimators may be again obtained by substituting the ratio for k.

CHAPTER VII

CLUSTER SAMPLING

1. Definition and uses

In a formal sense cluster sampling means that the population is divided into subpopulations, and that some but not all of these are represented in the sample; those represented may be included in the sample in their entirety, or they may be subsampled. In stratified sampling we also subdivide the population into subpopulations, but we include *all* of the subpopulations and we *always* subsample them.

Subdivision into strata is usually planned so as to reduce the variances of our estimators. On the other hand, what compels us to use clusters (even though this will generally lead to larger variances) is the fact that for many surveys the cost of an unclustered survey of adequate size would be unduly high.

Some examples:

(a) High cost of travel between units; in this case clusters will be made compact (in a transportation-economic sense).

(b) There exist lists of clusters of units (or we may be able to compile them at moderate expense), but not lists of the smallest sampling units which we wish to sample ("listing units"); and the cost of compiling lists of all these small units in the population is large compared with the cost of measuring $\mathscr{Y}$ for the ultimate listing units. E.g., there may be lists or maps of city blocks, but not of individual dwelling units; or there may be lists of households, but not of individuals, etc.

(c) For small-sized units it may be difficult to fix accurate boundaries (e.g., in small fields or parts of buildings). In this case we may decide to divide the population into somewhat larger units, and to sample among these (without subsampling).

(d) In addition, some of the administrative considerations we noted in the chapter on stratified sampling may play a role in the decision to use clustering, though these play a lesser role in deciding to use clusters than in the decision to use strata.

It should not be forgotten that we can and often do use both clustering and stratification.

The difference in aim between clustering and stratification leads to different criteria for setting up clusters than for setting up strata. In contrast to stratified sampling, in cluster sampling the variance for estimators of Y is made small by making each cluster as much as possible representative of the diversity of the entire population, because then r is small (see Chapter V, section 4). However, insofar as cost considerations lead to clusters for which r is not small, the disadvantageous effect of this on the variance of the estimators may be reduced by subsampling.

In Chapter II, section 13, we already discussed a case of cluster sampling in which $\mathscr{Y}$ is such that its possible values for a subunit are 1 or 0 only. In this chapter $\mathscr{Y}$ will be a general property.

If we are interested in the total yield or the per acre yield of wheat, what should be the units? Clearly, there is no natural unit in this and many other cases. One investigator may take as unit an acre, another a quarter-acre (and then we can consider each acre as a cluster of 4 subunits, each consisting of 4 one quarter-acres), etc. Evidently, the chapter on choice of units is also relevant to cluster sampling.

As we shall see, cluster sampling plans sometimes lead to complex formulae and operations. Deming has developed some plans which avoid most of these complexities; we postpone their discussion to the end of Chapter IX.

2. Notation

Because of what was said at the end of the previous section, we shall use the same notation as in Chapter V. Corresponding to the N large units of Chapter V we have here N "primary sampling units" (psu's). Among them we choose n (which are called clusters). Corresponding to the small units of Chapter V we have in this chapter subunits; the ith unit contains M_i subunits.

Moreover, let us denote by $\mathscr{s} = (u_{j_1}, \ldots, u_{j_n})$ the sampling sequence of primary sampling units (i.e., the sequence of clusters). u_{j_k} contains M_{j_k} subunits. We obtain observations on each one of them; or we obtain (independently from unit to unit) a sample $\mathscr{s}_{j_k}$ of size m_{j_k} from u_{j_k}. The sample $\mathscr{s}_{j_k}$ is called an *ultimate* cluster if we do not subsample it any further. Let us define

$\bar{M}$ as the average of the M_i in the population,
$\bar{m}$ as the average of the m_{j_k} in the sample.

REMARK. Note that even if, for every k, $m_{j_k} = M_{j_k}$, $\bar{m} = \bar{M}$ does *not* necessarily hold.

Let us also define

$$y_{j_k\cdot} = \sum_{h \in \mathcal{O}_{j_k}} y_{j_k h},$$

$$\bar{y}_{j_k\cdot} = y_{j_k\cdot}/m_{j_k},$$

$$\mathrm{SS_w}(j_k) = \sum_{h \in \mathcal{O}_{j_k}} (y_{j_k h} - \bar{y}_{j_k\cdot})^2,$$

$$s^2_{\mathrm{w}}(j_k) = \mathrm{SS_w}(j_k)/(m_{j_k} - 1).$$

3. Cluster sampling without subsampling

3.1. ESTIMATORS OF Y AND $\bar{\bar{Y}}$ WITH THEIR VARIANCES AND ESTIMATORS OF THE LATTER

Here $m_i = M_i$, $y_{i.} = Y_{i.}$ and $\bar{y}_{i.} = \bar{Y}_{i.}$. We distinguish the following cases:

(a) All M_i equal (to $\bar{M}$). The usual estimator for $\bar{\bar{Y}}$ is

$$\bar{\bar{y}} = \sum_{k=1}^{n} \bar{y}_{j_k\cdot}/n,$$

and has variance

$$\frac{N-n}{N}\,\frac{1}{n}\,\frac{1}{\bar{M}^2}\sum_{i=1}^{N}(Y_{i.} - \bar{Y})^2/(N-1) = \frac{N-n}{N}\,\frac{1}{n}\sum_{i=1}^{N}(\bar{Y}_{i.} - \bar{\bar{Y}})^2/(N-1),$$

which is estimated by

$$\frac{N-n}{N}\,\frac{1}{n}\sum_{k=1}^{n}(\bar{y}_{j_k\cdot} - \bar{\bar{y}})^2/(n-1).$$

Y is estimated by $N\bar{M}\bar{\bar{y}}$.

If we have a property $\mathscr{X}$ related to $\mathscr{Y}$ for which $\bar{\bar{X}}$ is known, we may use a ratio estimator of the form

$$\frac{\Sigma a_i \bar{y}_{i.}}{\Sigma a_i \bar{x}_{i.}}\,\bar{\bar{X}}.$$

(b) Not all M_i are equal. In this case the bias of $N\bar{M}\bar{\bar{y}}$ as estimator of Y may be large. An unbiased estimator for $Y = N\bar{Y}$ is

$$N\bar{y} = N\sum_{k=1}^{n} y_{j_k\cdot}/n.$$

The fact that it is unbiased, and that its variance is

$$\frac{1}{n}\,\frac{N-n}{N}\,N^2\sum_{i=1}^{N}(Y_{i.} - \bar{Y})^2/(N-1)$$

is due to $\bar{y}$ being the average of the $\mathscr{Y}$-totals for the simple random sample of n primary units from among $\{u_1, \ldots, u_N\}$. Since

$$\bar{\bar{Y}} = Y/\sum_{i=1}^{N} M_i = \bar{Y}/\bar{M},$$

the estimator for $\bar{\bar{Y}}$ corresponding to the one given for Y is $\bar{y}/\bar{M}$.

If the M_i vary much compared with the $\bar{Y}_{i.}/\bar{Y}$ and are not negatively correlated with the latter, we are better off using the ratio estimators†

$$\hat{Y}_R = N\bar{M}\frac{\bar{y}}{\bar{m}} = N\bar{M}\bar{\bar{y}}, \quad \hat{\bar{\bar{Y}}}_R = \frac{\bar{y}}{\bar{m}} = \bar{\bar{y}},$$

since their variances depend on the variation among the $\bar{Y}_{i.}$ rather than among the $Y_{i.} = M_i\bar{Y}_{i.}$:

$$V(\hat{Y}_R) \approx \frac{N-n}{N}\frac{1}{n}N^2\sum_{i=1}^{N}(Y_{i.} - M_i\bar{\bar{Y}})^2/(N-1)$$

$$= \frac{N-n}{N}\frac{1}{n}N^2\sum_{i=1}^{N}M_i^2(\bar{Y}_{i.} - \bar{\bar{Y}})^2/(N-1),$$

$$V(\hat{\bar{\bar{Y}}}_R) = \frac{1}{(N\bar{M})^2}V(\hat{Y}_R);$$

whereas

$$V(N\bar{y}) = \frac{N-n}{N}\frac{1}{n}N^2\sum_{i=1}^{N}(M_i\bar{Y}_{i.} - \bar{Y})^2/(N-1)$$

$$= \frac{N-n}{N}\frac{1}{n}N^2\sum_{i=1}^{N}\{(\bar{Y}_{i.}M_i - \bar{Y}M_i/\bar{M}) + \bar{Y}(M_i - \bar{M})/\bar{M}\}^2/(N-1)$$

$$= V(\hat{Y}_R) + \frac{N-n}{N}\frac{1}{n}N^2\{2\bar{\bar{Y}}\sum_{i=1}^{N}M_i(\bar{Y}_{i.} - \bar{\bar{Y}})(M_i - \bar{M})/(N-1)$$

$$+ \bar{\bar{Y}}^2\sum_{i=1}^{N}(M_i - \bar{M})^2/(N-1)\}$$

$$= V(\hat{Y}_R) + \frac{N-n}{N}\frac{1}{n}Y^2\{2C(\mathscr{M}\mathscr{M}\mathscr{Y}) + 2C(\mathscr{M}\mathscr{Y}) + C(\mathscr{M}\mathscr{M})\}$$

(assuming $\bar{y}$ is positive). Perhaps it is intuitively clearer to make the comparison in terms of size and total $\mathscr{Y}$-values of the primary units. Since

$$\hat{\bar{\bar{Y}}}_R = y/\Sigma M_{j_k},$$

† In general, even though at present we do not consider subsampling, $\bar{m}$ does not have to equal $\bar{M}$ (see the Remark in section 2). In order to remain valid in the case of subsampling, it may be preferable to write $\hat{\bar{M}}$ instead of $\bar{m}$, where $\hat{\bar{M}} = \Sigma a_i M_i/n$.

we have from section 6 of Chapter IIA the following approximate condition for its variance to be less than that of $\bar{y}/\bar{M}$: the correlation between $\mathscr{Y}^*$ and $\mathscr{M}$ exceeds half $(C(\mathscr{M}\mathscr{M})/C(\mathscr{Y}^*\mathscr{Y}^*))^{\frac{1}{2}}$, where the $\mathscr{Y}^*$-characteristic of a primary unit is the *total* of the $\mathscr{Y}$-values of its subunits and is assumed positive.

We estimate $V(\hat{\bar{Y}}_R)$ by

$$\frac{N-n}{N}\,\frac{1}{n}\,N^2\sum_{k=1}^{n} M_{j_k}^2(\bar{y}_{j_k\cdot}-\bar{\bar{y}})^2/(n-1),$$

and $V(N\bar{y})$ by

$$\frac{N-n}{N}\,\frac{1}{n}\,N^2\sum_{k=1}^{n}(M_{j_k}\bar{y}_{j_k\cdot}-\bar{y})^2/(n-1).$$

Note that, while $\bar{y}/\bar{M}$ requires knowledge of $\bar{M}$, $\hat{\bar{Y}}_R$ does not; on the other hand, for estimating the variance of the estimators of $\bar{\bar{Y}}$ as indicated above, the knowledge of $\bar{M}$ is required in both cases.† The situation is reversed for estimation of Y.‡

Sometimes we estimate $\bar{\bar{Y}}$ by

$$(1/n)\sum_{k=1}^{n}\bar{y}_{j_k\cdot}.$$

Its bias equals

$$N^{-1}\{\Sigma\bar{Y}_{i\cdot}-\Sigma M_i\bar{Y}_{i\cdot}/\bar{M}\}=-\bar{M}^{-1}\Sigma(\bar{Y}_{i\cdot}-\bar{\bar{Y}})(M_i-\bar{M})/N,$$

so that its relative bias (for positive Y) is, but for the factor $N/(N-1)$, equal to $-C(\mathscr{M}\mathscr{Y})$. Therefore its square bias makes only a small contribution to the mean square error if $C^2(\mathscr{M}\mathscr{Y})$ is small compared with $(1/n)C(\mathscr{Y}\mathscr{Y})$. But

$$V(\bar{y}/\bar{M})=\frac{N-n}{N}\,\frac{1}{n}\,\sigma^2(\mathscr{Y}^*/\bar{M})$$

exceeds

$$V\left(\frac{1}{n}\sum_{k=1}^{n}\bar{y}_{j_k\cdot}\right)=\frac{N-n}{N}\,\frac{1}{n}\,\Sigma\left(\bar{Y}_{i\cdot}-\frac{\Sigma\bar{Y}_{i\cdot}}{N}\right)^2\Big/(N-1)$$

$$=\frac{N-n}{N}\,\frac{1}{n}\,\sigma^2(\mathscr{Y}^*/\mathscr{M})$$

if $C(\mathscr{M}\mathscr{Y})$ positive, and could occasionally exceed it quite a bit.

† One may, however, use $1/\hat{\bar{M}}^2$ instead of $1/\bar{M}^2$ in these estimators; as we observed previously, in the case of the ratio estimator, this may even be preferable.

‡ Sometimes one may wish to resort to double sampling in order to first obtain an estimate of $\bar{M}$.

There are additional ways of handling the case in which there are large differences among the M_i. If we know the sizes M_i or if we have ways of estimating them we can:

(i) Divide the population into several *strata* according to *size* groups.

(ii) Sample psu's with probabilities proportional to the sizes of these units (pps) or to estimates† of the sizes (ppes). The ppes estimator of $\bar{\bar{Y}}$ is $1/N\bar{M}$ times $\hat{Y}$, where

$$\hat{Y} = \sum_{i=1}^{N} t_i Y_{i.}/(np_i)$$

and its variance is discussed in section 3.2 of Chapter VI. The estimator

$$(1/n)\sum_{k=1}^{n} \bar{y}_{j_k.}$$

is appropriate for pps sampling and is obtained from the ppes estimator by putting $p_i = M_i/(N\bar{M})$. Its variance is

$$\frac{1}{n}\sum_{i=1}^{N}\frac{M_i}{\bar{M}}(\bar{Y}_{i.} - \bar{\bar{Y}})^2/N.$$

[We may compare this with sampling with equal probabilities without replacement, in which case the formula for the estimator is $\bar{y}/\bar{m}$ and the variance of the estimator is

$$\frac{N-n}{N}\,\frac{1}{n}\sum_{i=1}^{N}\left(\frac{M_i}{\bar{M}}\right)^2(\bar{Y}_{i.} - \bar{\bar{Y}})^2/(N-1),$$

which one usually would expect to be larger than the former variance.]

An estimator for the variance is given in Chapter VI:

$$\frac{1}{n}\sum_{k=1}^{n}(\bar{y}_{j_k.} - \bar{\bar{y}})^2/(n-1).$$

† Sometimes these estimates are obtained from a large sample, and the values of the $\mathscr{Y}$-characteristic from a subsample. In section 6.3 (ii) of Chapter I is given a method of pps or ppes sampling, applicable if the sizes or their estimates are known in advance. (Actually, that method can be used under less stringent conditions, see Example 1 of Chapter IX, section 11.) If these are not known in advance, we may use the method explained in section 4 (iv) of Chapter VI for drawing a sample of 1, by setting the n of that discussion equal to 1; to obtain a sample of n with replacement, repeat the procedure n times. It is also possible to sample *without* replacement. – For comparison with stratification and ratio (and related) estimation, see also the first footnote of section 1 of Chapter III. In this comparison one should also take into account any variation of cost with size of clusters, since under pps or ppes sampling the larger units have a larger probability of being included than the smaller ones.

Actually, as we saw in Chapter VI, (ii) is preferable if the probabilities are approximately proportional to the $Y_{i.} = M_i \bar{Y}_{i.}$ rather than to the M_i.† Sometimes we use instead of the M_i a property which is correlated with the M_i. For the ratio estimator of $\bar{\bar{Y}}$

$$\frac{\Sigma t_i Y_{i.}/p_i}{\Sigma t_i M_i/p_i}$$

we at once find the approximate variance

$$(N\bar{M})^{-2}\frac{1}{n}\Sigma p_i\left(\frac{Y_{i.}}{p_i} - \frac{M_i}{p_i}\bar{\bar{Y}}\right)^2 = N^{-2}\frac{1}{n}\Sigma\frac{1}{p_i}\left(\frac{M_i}{\bar{M}}\right)^2(\bar{Y}_{i.} - \bar{\bar{Y}})^2,$$

with estimator

$$(N\hat{\bar{M}})^{-2}\frac{1}{n}\Sigma t_i\left(\frac{Y_{i.}}{p_i} - \frac{M_i}{p_i}\hat{\bar{\bar{Y}}}_R\right)^2\Big/(n-1)$$

$$= N^{-2}\frac{1}{n}\Sigma t_i\left(\frac{M_i/\hat{\bar{M}}}{p_i}\right)^2(\bar{Y}_{i.} - \hat{\bar{\bar{Y}}}_R)^2(n-1)$$

where

$$N\hat{\bar{M}} = \frac{1}{n}\,\Sigma t_i M_i/p_i.$$

Note that for the pps estimator $p_i = M_i/(N\bar{M})$, so that $\bar{M}$ must be known; indeed, in that case the ratio estimator is identical with the pps estimator

$$\frac{1}{n}\sum_{k=1}^{n}\bar{y}_{j_k.}$$

given above.

Like in case (a), if there is a property $\mathscr{X}$, related to $\mathscr{Y}$, for which $\bar{\bar{X}}$ (or X) is known, we can use a ratio estimator, a difference estimator or a regression estimator for ($\bar{\bar{Y}}$ or Y) instead of the estimators given above.

3.2 CONFIDENCE INTERVALS AND BOUNDS FOR $\bar{\bar{Y}}$

One usually gives upper confidence bounds for $\bar{\bar{Y}}$ of the form: $\bar{\bar{Y}}_e + tv^{\frac{1}{2}}(\bar{\bar{Y}}_e)$ (e = estim.) with t obtained from a table of the standard normal distribution, and similarly for lower confidence bounds and intervals. The purpose of the following is to present some empirical data on the use of such a procedure (with $t = 2$). Clearly, there is a need for far more investigation.

† Therefore $\hat{Y}$ is often appropriate, even if the variation of the M_i is relatively small (e.g. due to prior stratification). For example, if Y_i is the expenditure on entertainment of family i and the p_i are proportional to rough and ready assessments of family income, $n^{-1}\Sigma t_i Y_{i.}/p_i$ may be a very good estimator of Y when, in the population or subpopulation under consideration, such expenditures are nearly proportional to family income – as may often be the case.

The presentation is based on the population of cities discussed in section 4 of Chapter IIA. This population was divided into potential clusters in two very different ways. One partition was an essentially random one, accomplished by dividing the cities into 32 classes according to the first or first two letters of their names, but combining some classes in order to reduce the variation in the sizes of the classes. (The measure of homogeneity, as defined in section 10.2 below, should therefore be of the order of $-\frac{1}{790}$; for the $\mathscr{Y}$-values it was -0.004 and for the $\mathscr{X}$-values -0.010.) The second partition was obtained by dividing the states in which the cities are situated into 20 geographically contiguous and socio-economically rather homogeneous areas,† with not too radically varying numbers of cities.‡ (One would therefore expect a high measure of homogeneity; for the $\mathscr{Y}$-values it was 0.152, for the $\mathscr{X}$-values 0.223.) Since in practice the populations are partitioned into potential clusters for cost and administrative reasons, it is unfortunately not usual to obtain clusters as heterogeneous as may be obtained by a random partitioning; neither is it common to get very homogeneous clusters.

In either case we sampled with probabilities proportional to size, without replacement, according to the method of section 6.3 of Chapter VI. The measure of size used was the sum of the populations of the cities included in a potential cluster (in the first partition it was slightly modified by assigning to a city with a population over 150 000 the same weight as a city with 150 000 inhabitants). The first partition was defined so as to make these measures of size vary rather little, namely from 1 196 000 to 1 613 000 (the number of cities per class varied from 18 to 34).§ In the second partition the measures of size varied substantially, from 873 000 to 5 677 000. Moreover, the high variances made it necessary to include (at least) 6 of the 20 areas in the sample with probability 1. For the remaining 14 areas the measures of size varied from 873 000 to 2 351 000 (containing 15 to 52 cities).

For the estimator of $\overline{\overline{Y}}$ (divided by $\overline{\overline{X}}$ to maintain comparability with the data presented in section 4 of Chapter IIA) we obtained on the basis of 20 000 replications ($\mathscr{E}m$ = expected number of cities in the sample):

† The areas were: California (including Hawaii); other West Coast; Northern Mountain; Arizona, New Mexico, Texas; the Dakotas, Nebraska, Kansas; Minnesota, Iowa; Wisconsin, Michigan; Illinois; Indiana; Missouri, Arkansas, Oklahoma; Ohio; West Virginia, Kentucky, Tennessee; Louisiana, Mississippi, Alabama; Georgia, Florida; Virginia, the Carolinas; Pennsylvania, Maryland (including the District of Columbia); New Jersey, Delaware; Connecticut, Rhode Island; other New England; New York.

‡ However, the first region has 101 cities with a population of 5 677 000, very much more than any of the other regions.

§ To investigate the effect of greater inequality in the measures of size, we also took for these the numbers of cities in the classes.

	$\mathscr{E}m$	l	r	$\bar{w}$
first partition:				
samples of 15 clusters	381	1.1	6.9	0.13
samples of 24 clusters	611	1.0	5.5	0.08
second partition:				
samples of 15 clusters	663	1.5	9.3	0.14

It is seen that the first partition yields about the same $\bar{w}$ as a simple random sample of size $\mathscr{E}m$, but l and r are larger.† In the second partition $\bar{w}$ is quite substantially larger than for a simple random sample of size $\mathscr{E}m$, and so are r and l. Even in comparison with a simple random sample of a size that yields about the same $\bar{w}$, both l and r are very much larger. (The differences between the values of $\bar{w}$ given above and those for the corresponding ratio estimators are very small.)

For comparison with the data in the last footnote of section 15.1 of Chapter II, we give the percentages of 20 000 repetitions in which the confidence intervals did not include $\bar{\bar{X}}$: 6.7 for samples of 15 clusters and 5.7 for samples of 24 clusters in the first partition (with $l = 1.9$ and 1.5, respectively),‡ and 7.2 for samples of 15 clusters in the second partition (with $l = 0$). In percentage of $\bar{\bar{X}}$ the average half-widths were 7 and 4 in the first, and 7 in the second partition.

4. Some general results on subsampling

With subsampling the estimator of Y or $\bar{\bar{Y}}$ is often a linear function of quantities computed from the subsamples taken from $\mathscr{s}$. From the subsample from the kth selected primary unit compute a statistic z_{j_k}, and define $z_i = z_{j_k}$ if $u_{j_k} = u_i$. Let the conditional mean of z_{j_k}, given that z_{j_k} was computed from unit u_{j_k}, be μ_{j_k}, the conditional variance $\sigma^2_{j_k}$. The conditional covariance is zero by independence of sampling. (In the special case of *single stage* sampling, when we observe all elements of each included primary unit, $\mu_{j_k} = z_{j_k}$ and $\sigma_{j_k} = 0$.) Suppose we wish to estimate

$$T = T(\mu) = T(\mu_1, \ldots, \mu_N) = \sum_{i=1}^{N} \mu_i$$

by an unbiased estimator of the form

$$t = t(z) = t(z_{j_1}, \ldots, z_{j_n}) = \sum_{i=1}^{N} b_i(\mathscr{s}) z_i,$$

where, for given $\mathscr{s} \in \mathscr{S}$, the $b_i(\mathscr{s})$ are numbers to be fixed in advance. Clearly $b_i(\mathscr{s}) = 0$ for $u_i \notin \mathscr{s}$.

† With sampling probabilities proportional to numbers of cities, $\bar{w}$ was larger (0.17 $\bar{\bar{X}}$ and 0.10 $\bar{\bar{X}}$, respectively); and so was r (10.5 and 9.7), but l was very small (0.4 and 0.3).

‡ Using number of cities, $\bar{w}$ did not change appreciably; r was somewhat larger (6.0 and 5.1), and l somewhat smaller (1.4 and 1.2).

By Chapter I, section 8(b), the variance of t is

$$V(\sum_{i=1}^{N} b_i(s)\mu_i) + \sum_{i=1}^{N} \mathscr{E} b_i^2(s)\sigma_i^2$$

with

$$V(\sum_{i=1}^{N} b_i(s)\mu_i) = \sum_{i=1}^{N} \mu_i^2 V(b_i(s)) + \sum\sum_{\substack{i \neq j \\ 1}}^{N} \mu_i\mu_j \operatorname{Cov}(b_i(s), b_j(s)),$$

the variance in *single* stage sampling.

Let

$$Q(\mu_{j_1}, \ldots, \mu_{j_n}) = \sum_{i=1}^{N} \mu_i^2 c_i(s) + \sum\sum_{\substack{i \neq j \\ 1}}^{N} \mu_i\mu_j d_{ij}(s),$$

where $c_i(s) = 0$ if $u_i \notin s$, $d_{ij}(s) = 0$ if $u_i \notin s$ or if $u_j \notin s$, be an unbiased estimator of

$$V(\sum_{i=1}^{N} b_i(s)\mu_i),$$

where for given $s \in \mathscr{S}$, the $c_i(s)$ and the $d_{ij}(s)$ are given numbers. Thus $Q(\mu_{j_1}, \ldots, \mu_{j_n})$ would be an unbiased estimator of $V(t)$ *if there were no subsampling.*

THEOREM. Let $\hat{\sigma}_{j_k}^2$ be a conditionally unbiased estimator of $\hat{\sigma}_{j_k}^2$ for $k = 1, \ldots, n$; and suppose that Q is defined as above with, for $i \neq j \in \{1, \ldots, N\}$,

$$\mathscr{E} c_i(s) = V(b_i(s)), \quad \mathscr{E} d_{ij}(s) = \operatorname{Cov}(b_i(s), b_j(s)).$$

Then, if t is unbiased for T,

$$Q(z_{j_1}, \ldots, z_{j_n}) + \sum_{k=1}^{n} e_{j_k}(s)\hat{\sigma}_{j_k}^2$$

is an unbiased estimator of $V(t)$; here the $e_i(s)$ vanish for $u_i \notin s$ and, for given $s \in \mathscr{S}$, are any fixed numbers which satisfy

$$\mathscr{E} e_i(s) = \mathscr{E} b_i(s).$$

Proof.

$$\mathscr{E}\{z_{j_k}^2 | u_{j_k} = u_i\} = \sigma_i^2 + \mu_i^2,$$

and, since the z_{j_k} are conditionally independent, for $k \neq l$

$$\mathscr{E}\{z_{j_k} z_{j_l} | u_{j_k} = u_i, u_{j_l} = u_j\} = \mu_i\mu_j,$$

so

$$\mathscr{E}\{Q(z_{j_1}, \ldots, z_{j_n}) | s\} = Q(\mu_{j_1}, \ldots, \mu_{j_n}) + \sum_{i=1}^{N} c_i(s)\sigma_i^2,$$

so, taking expectations on both sides,

$$\mathscr{E}Q(z_{j_1}, \ldots, z_{j_n}) = V(\sum_{i=1}^{N} b_i(\mathscr{s})\mu_i) + \mathscr{E}\sum_{i=1}^{N} c_i(\mathscr{s})\sigma_i^2.$$

Now

$$\mathscr{E}c_i(\mathscr{s}) = V(b_i(\mathscr{s})) = \mathscr{E}b_i^2(\mathscr{s}) - \mathscr{E}^2 b_i(\mathscr{s});$$

and, if $\mathscr{E}t = T$ identically in the value of one or more characteristics of $u_{11}, \ldots, u_{NM_N}$, then $\mathscr{E}b_i(\mathscr{s}) = 1$ for all i, and so

$$\mathscr{E}^2 b_i(\mathscr{s}) = \mathscr{E}b_i(\mathscr{s}) \quad \text{and} \quad \mathscr{E}c_i(\mathscr{s}) = \mathscr{E}b_i^2(\mathscr{s}) - \mathscr{E}b_i(\mathscr{s}).$$

Hence, since $\mathscr{E}e_i(\mathscr{s}) = \mathscr{E}b_i(\mathscr{s})$,

$$\mathscr{E}\{Q(z_{j_1}, \ldots, z_{j_n}) + \sum_{i=1}^{N} e_i(\mathscr{s})\hat{\sigma}_i^2\}$$

$$= V(\sum_{i=1}^{N} b_i(\mathscr{s})\mu_i) + \mathscr{E}\sum_{i=1}^{N} b_i^2(\mathscr{s})\sigma_i^2 - \mathscr{E}\sum_{i=1}^{N} b_i(\mathscr{s})\sigma_i^2 + \mathscr{E}\sum_{i=1}^{N} e_i(\mathscr{s})\sigma_i^2$$

$$= V(\sum_{i=1}^{N} b_i(\mathscr{s})\mu_i) + \mathscr{E}\sum_{i=1}^{N} b_i^2(\mathscr{s})\sigma_i^2 = V(t).$$

In the next section we give simple applications of the above results.

NOTE. The usual estimators of variance in the absence of subsampling are such that *automatically*

$$\mathscr{E}c_i(\mathscr{s}) = V(b_i(\mathscr{s})) \quad \text{and} \quad \mathscr{E}d_{ij}(\mathscr{s}) = \operatorname{Cov}(b_i(\mathscr{s}), b_j(\mathscr{s}))$$

for $i \neq j$. That is, if we write $V(\Sigma b_i(\mathscr{s}))$ in the form

$$\Sigma f(\mathscr{E}t_i)\mu_i^2 + \sum_{i \neq j}\sum g(\mathscr{E}t_i t_j, \mathscr{E}t_i, \mathscr{E}t_j)\mu_i\mu_j,$$

its usual estimate is

$$\Sigma f(\mathscr{E}t_i)\mu_i^2 t_i/\mathscr{E}t_i + \sum_{i \neq j}\sum g(\mathscr{E}t_i t_j, \mathscr{E}t_i, \mathscr{E}t_j)\mu_i\mu_j t_i t_j/\mathscr{E}t_i t_j.$$

In sampling without replacement we can write

$$b_i(\mathscr{s}) = a_i(\mathscr{s})/\pi_i,$$

where $a_i(\mathscr{s}) = 1$ if $u_i \in \mathscr{s}$ and 0 otherwise; and $\pi_i = \mathscr{E}a_i(\mathscr{s})$, the probability that $u_i \in \mathscr{s}$. If $z_i' = z_i/\pi_i$, we have

$$t = \sum_{k=1}^{n} z_{j_k}', \quad \sigma_i'^2 = \sigma_i^2/\pi_i^2 \quad \text{and} \quad \hat{\sigma}_i'^2 = \hat{\sigma}_i^2/\pi_i^2.$$

The estimator

$$Q(\pi_{j_1}z'_{j_1}, \ldots, \pi_{j_n}z'_{j_n}) + \sum_{k=1}^{n} \pi_{i_k}\hat{\sigma}'^2_{j_k}$$

is unbiased for $V(t)$, since

$$\sum_{i=1}^{N} b_i(\mathscr{s})\hat{\sigma}_i^2 = \sum_{i=1}^{N} a_i(\mathscr{s})\,\pi_i^{-1}\pi_i^2\hat{\sigma}'^2_i = \sum_{i=1}^{N} a_i(\mathscr{s})\pi_i\hat{\sigma}'^2_i.$$

For $V(t)$ we get

$$V(t) = V\{t(\mu_{j_1}, \ldots, \mu_{j_n})\} + \sum_{i=1}^{N} \pi_i\sigma'^2_i = V\{t(\mu_{j_1}, \ldots, \mu_{j_n})\} + \sum_{i=1}^{N} \sigma_i^2/\pi_i,$$

as

$$\Sigma\mathscr{E}b_i^2(\mathscr{s})\sigma_i^2 = \Sigma\mathscr{E}a_i^2(\mathscr{s})\pi_i^{-2}\pi_i^2\sigma'^2_i = \Sigma\mathscr{E}a_i(\mathscr{s})\sigma'^2_i = \Sigma\pi_i\sigma'^2_i = \Sigma\sigma_i^2/\pi_i.$$

The theorem given above is found in somewhat more restricted form in DES RAJ, J. Amer. Statist. Ass., **6** (1966) 391–396.

REMARK. If we obtain l independent estimators $t_1(z), \ldots, t_l(z)$ by obtaining l times a sample of n clusters (replacing it before each new drawing), we obtain the estimator $\bar{t}(z) = \Sigma t_\alpha(z)/l$, with variance equal to $1/l$ times the variance of any one $t_\alpha(z)$. A simple estimator of that variance is

$$l^{-1}\Sigma\{t_\alpha(z) - \bar{t}(z)\}^2/(l-1).$$

$t_\alpha(z)$ can also be built up as a sum of similar expressions $t_{h\alpha}(z)$ from each of a number of strata. Compare section 14.2 of Chapter III. In that case the above formula for an estimator of variance remains valid; an estimator with smaller standard deviation is

$$l^{-1}\sum_h\sum_\alpha\{t_{h\alpha}(z) - \bar{t}_h(z)\}^2/(l-1).$$

5. Subsampling when primary units have been obtained by sampling without replacement and the estimator is linear

5.1. EQUAL PROBABILITY SAMPLING OF PRIMARY UNITS

Let us first consider estimation of

$$\bar{\bar{Y}} = \sum_{i=1}^{N} (M_i/\bar{M})\,\bar{Y}_{i.}/N.$$

One way of making t an unbiased estimator for $\bar{\bar{Y}}$ is to take for μ_{j_k} the expression $(1/N\bar{M})M_{j_k}\bar{Y}_{j_k.}$ and for z_{j_k} the expression $(1/N\bar{M})M_{j_k}\bar{y}_{j_k.}$; for *simple* sampling without replacement from among the u_i,

$$b_i(\mathscr{s}) = a_i(\mathscr{s})(n/N)^{-1},$$

and so

$$t = \sum_{k=1}^{n} (M_{j_k}/\bar{M})\bar{y}_{j_k\cdot}/n.$$

Thus

$$V\left(\frac{N}{n}\sum_{i=1}^{N} a_i(\mathscr{s})\mu_i\right) = V\left(\frac{1}{n}\sum_{k=1}^{n} \frac{M_{j_k}}{\bar{M}}\bar{Y}_{j_k\cdot}\right)$$

$$= \frac{N-n}{N}\frac{1}{n}\sum_{i=1}^{N}\left(\frac{M_i\bar{Y}_{i.}}{\bar{M}} - \bar{\bar{Y}}\right)^2\Big/(N-1).$$

Let σ_i^2 be the conditional variance of $(1/N\bar{M})M_i\bar{y}_{i\cdot}$, which is

$$\left(\frac{1}{N}\frac{M_i}{\bar{M}}\right)^2 \frac{M_i - m_i}{M_i}\frac{1}{m_i} S_{\mathrm{w}}^2(i)$$

for simple random sampling (without replacement) within u_i and is then estimated by

$$\hat{\sigma}_i^2 = \left(\frac{1}{N}\frac{M_i}{\bar{M}}\right)^2 \frac{M_i - m_i}{M_i}\frac{1}{m_i} s_{\mathrm{w}}^2(i).$$

Now

$$V(t) = V\left(\frac{N}{n}\sum_{k=1}^{n}\mu_{j_k}\right) + \mathscr{E}\frac{N^2}{n^2}\sum_{k=1}^{n}\sigma_{j_k}^2$$

$$= \frac{1}{n}\left\{\frac{N-n}{N}S_l^2 + \sum_{i=1}^{N}(N\sigma_i)^2/N\right\},$$

where

$$S_l^2 = \frac{1}{\bar{M}^2}\sum_{i=1}^{N}(Y_{i.} - \bar{Y})^2/(N-1).$$

In order to estimate this variance, we first obtain the estimator for

$$V((N/n)\sum_{k=1}^{n}\mu_{j_k})$$

in the absence of subsampling:

$$\frac{N-n}{N}\frac{1}{n}\sum_{k=1}^{n}\left(\frac{M_{j_k}}{\bar{M}}\bar{Y}_{j_k\cdot} - t(\mu)\right)^2\Big/(n-1).$$

Application of the theorem above gives as estimator for $V(t)$

$$v(t) = \frac{N-n}{N}\frac{1}{n}s_l^2 + \frac{N}{n}\sum_{k=1}^{n}\hat{\sigma}_{j_k}^2,$$

where

$$s_t^2 = \sum_{k=1}^{n} \left(\frac{M_{j_k}}{\bar{M}} \bar{y}_{j_k \cdot} - t \right)^2 \Big/ (n-1).$$

Remark (iv) in section 5 of Chapter II yields another derivation of $v(t)$. Note that for *any method of sampling whatsoever* within the u_{j_k} that yields conditionally unbiased estimators $\bar{y}_{j_k \cdot}$ and $\hat{\sigma}^2_{j_k}$ of μ_{j_k} and $\sigma^2_{j_k}$, respectively,

$$\mathscr{E} s_t^2/n = \frac{1}{n} S_t^2 + \frac{N}{n} \sum_{i=1}^{N} \sigma_i^2 = V(t) + \frac{1}{N} S_t^2,$$

which follows from $\mathscr{E} v(t) = V(t)$, and so

$$\frac{N-n}{N} \frac{1}{n} \mathscr{E} s_t^2 = \frac{N-n}{N} \frac{1}{n} S_t^2 + \frac{N}{n} \sum_{i=1}^{N} \sigma_i^2 - \mathscr{E} \frac{N}{n} \sum_{k=1}^{n} \hat{\sigma}^2_{j_k}$$

$$= \frac{N-n}{N} \left\{ \frac{1}{n} S_t^2 + \frac{N}{n} \sum_{i=1}^{N} \sigma_i^2 \right\}.$$

For n/N small, $V(t)$ equals approximately (relatively speaking) $\mathscr{E}(1/n)s_t^2$. Offhand one might have thought that $(1/n)s_t^2$ is an unbiased estimator of $(1/n)S_t^2$, so that to estimate $V(t)$ a *correction* is needed to account for the variation within the primary units. From the above it is clear that *hardly any such correction is needed.* Moreover, as

$$\mathscr{E} \frac{N-n}{N} \frac{1}{n} s_t^2 < V(t) < \mathscr{E} \frac{1}{n} s_t^2,$$

$(1/n)s_t^2$ is a conservative estimator of $V(t)$ and $((N-n)/Nn)s_t^2$ underestimates $V(t)$.

Since the simple formula $(1/n)s_t^2$ is often a good estimator of $V(t)$, its use is widespread. This approximation is particularly useful if subsampling is systematic (see Chapter IX), for in that case there is no obvious way of estimating the $\sigma^2_{j_k}$.

From the above formula it also follows that an unbiased estimator† of S_t^2 is

$$s_t^2 - N^2 \sum_{k=1}^{n} \hat{\sigma}^2_{j_k}/n.$$

† Like in Chapter V, an estimator of S_t^2 may be used to judge alternative sampling plans. We may also consider as an alternative plan simple random sampling of k subunits. For this we have to estimate $(N\bar{M})^{-1}(N\bar{M} - k)k^{-1}(N\bar{M} - 1)\{\Sigma\Sigma y_{ir}^2 - (N\bar{M})^{-1}(\Sigma\Sigma y_{ir})^2\}$. The term $\Sigma\Sigma y_{ir}^2$ in this expression may be estimated by $(N/n)\Sigma M_{j_k}$ Av $(y^2_{j_k r})$, where Av = sample average within the cluster, and $(\Sigma\Sigma y_{ir})^2$ by $(N\bar{M})^2\{t^2 - v(t)\}$.

WARNING. This estimator may have a large sampling variance even if that of s_l^2 or of

$$N^2 \sum_{k=1}^{n} \hat{\sigma}_{j_k}^2 / n$$

is moderate. Its value may sometimes turn out to be negative. In certain cases other estimators may be preferable – see Chapter V, section 5. Also, for small n, $(1/n)s_l^2$ may have a large sampling variance.

The estimator of Y corresponding to t above is $N\bar{M}$ times the above expression; neither it nor its estimated variance requires knowledge of $\bar{M}$. If $\bar{M}$ is unknown, we may use the estimator for $\bar{\bar{Y}}$ of Example 1 in the next section.

REMARK. Here

$$t = \frac{1}{n} \sum_{k=1}^{n} \frac{M_{j_k}}{\bar{M}} \bar{y}_{j_k\cdot}.$$

If all the M_i are equal, we sample a fixed number within each primary unit, and t is equal to the sample average, $\bar{\bar{y}}$. Consider now the case in which the M are not equal, but the *subsampling fraction* $f_2 = m_i/M_i$ is (as nearly as possible) *constant* for all i, and so

$$f_2 = \frac{1}{\bar{M}} \sum_{i=1}^{N} m_i/N.$$

Then

$$t = \frac{1}{n\bar{M}} \sum_{k=1}^{n} \frac{M_{j_k}}{m_{j_k}} y_{j_k\cdot} = \frac{1}{nf_2\bar{M}} \sum_{k=1}^{n} y_{j_k\cdot}$$

$$= \frac{1}{nf_2\bar{M}} y = y \left\{ \frac{n}{N} \sum_{i=1}^{N} m_i \right\}^{-1},$$

$$V(t) = \frac{N-n}{N} \frac{1}{n} S_l^2 + \frac{1-f_2}{N\bar{M}f_2} \frac{1}{n} \sum_{i=1}^{N} \frac{M_i}{\bar{M}} S_{\mathrm{w}}^2(i),$$

$$v(t) = \frac{N-n}{N} \frac{1}{n} s_l^2 + \frac{1-f_2}{N\bar{M}f_2} \frac{1}{n} \sum_{k=1}^{n} \frac{M_{j_k}}{\bar{M}} s_{\mathrm{w}}^2(j_k).$$

t is (approximately) *self-weighting*;

$$nf_2\bar{M} = \frac{n}{N} \sum_{i=1}^{N} m_i$$

is the *expected sample size*

$$\mathscr{E} \sum_{k=1}^{n} m_{j_k}$$

and is often denoted by $n\bar{m}$, despite the confusion that might result. On the other hand, the sample average is

$$\sum_{k=1}^{n} y_{j_k \cdot} / \sum_{k=1}^{n} m_{j_k} = y/(n\bar{m}) = \bar{\bar{y}},$$

which has a random denominator. This estimator is a special case of the first example in the following section.

For the variance of

$$(1/n) \sum_{k=1}^{n} \bar{y}_{j_k \cdot},$$

the *biased estimator* discussed in section 3, we obtain the same formula as for t above, except that S_l^2 is replaced by

$$\Sigma(\bar{Y}_{i \cdot} - (\Sigma \bar{Y}_{i \cdot}/N))^2/(N-1)$$

and s_l^2 by the corresponding sample formula; in simple random subsampling σ_i^2 is the conditional variance of $\bar{y}_{i \cdot}/N$.

5.2. UNEQUAL PROBABILITY SAMPLING OF PRIMARY UNITS

We have to use $b_i(s) = a_i(s)/\pi_i$, where $\pi_i = \mathscr{E} a_i(s)$ is not a constant; the corresponding modifications are easily made in the above formulae. For example, $((N-n)/Nn)S_l^2$ is replaced by

$$V\left(\sum \frac{a_i(s)}{\pi_i} \frac{M_i}{N\bar{M}} \bar{Y}_{i \cdot}\right) = \frac{1}{(N\bar{M})^2} \sum_{i<j}\sum (\pi_i \pi_j - \pi_{ij}) \left(\frac{M_i \bar{Y}_{i \cdot}}{\pi_i} - \frac{M_j \bar{Y}_{j \cdot}}{\pi_j}\right)^2.$$

For the estimators of section 5 in Chapter VI with subsampling, we find estimators of variance at once. E.g., for that of section 5.2, if $n = 2$ and if $\hat{Y}_{l_1 \cdot}$ and $\hat{Y}_{l_2 \cdot}$ are conditionally unbiased estimators of $Y_{l_1 \cdot}$ and $Y_{l_2 \cdot}$ with estimators of their conditional variances equal to $\hat{\sigma}_{l_1}^2$ and $\hat{\sigma}_{l_2}^2$, we estimate the variance of our estimator by

$$(2 - p_{l_1} - p_{l_2})^{-2}(1 - p_{l_1})(1 - p_{l_2})(1 - p_{l_1} - p_{l_2})\left(\frac{\hat{Y}_{l_1 \cdot}}{p_{l_1}} - \frac{\hat{Y}_{l_2 \cdot}}{p_{l_2}}\right)^2$$
$$+ (2 - p_{l_1} - p_{l_2})^{-1}\left(\frac{1 - p_{l_2}}{p_{l_1}} \hat{\sigma}_{l_1}^2 + \frac{1 - p_{l_1}}{p_{l_2}} \hat{\sigma}_{l_2}^2\right).$$

For the method of Durbin given in (v) of section 4 of Chapter VI we get as an estimator of the variance of $\frac{1}{2}((\hat{Y}_{l_1 \cdot}/p_{l_1}) + (\hat{Y}_{l_2 \cdot}/p_{l_2}))$:

$$\frac{1}{4}\left[\left(\frac{\hat{Y}_{l_1 \cdot}}{p_{l_1}} - \frac{\hat{Y}_{l_2 \cdot}}{p_{l_2}}\right)^2 \left(\frac{\pi_{l_1}\pi_{l_2}}{\pi_{l_1 l_2}} - 1\right) + \pi_{l_1} v\left(\frac{\hat{Y}_{l_1 \cdot}}{p_{l_1}}\right) + \pi_{l_2} v\left(\frac{\hat{Y}_{l_2 \cdot}}{p_{l_2}}\right)\right].$$

Another method of sampling with unequal probabilities is due to Wilks and is discussed in section 8.

6. Subsampling when primary units have been obtained by sampling without replacement and a ratio estimator is used

Instead of considering the statistics z_{j_k}, let us consider the pairs (v_{j_k}, w_{j_k}) with conditional means (ϕ_{j_k}, ψ_{j_k}). We wish to estimate the variances of the ratio estimator

$$\hat{R} = \frac{t(v)}{t(w)}$$

for $R = T(\phi)/T(\psi)$, where $T(\psi)$ is known.

Let

$$x_i = (v_i - Rw_i)/T(\psi), \quad R^0 = \frac{t(\phi)}{t(\psi)}, \quad \mathscr{E}x_i = \xi_i.$$

Let

$$\xi_i^0 = \xi_i|_{R=R^0}$$

and note that

$$\sum_{i=1}^{N} \xi_i = 0, \quad \sum_{i=1}^{N} b_i(\mathscr{s})\xi_i^0 = 0.$$

By analogy to the previous discussion of ratio estimators we use the approximation

$$\hat{R} - R = \frac{T(\psi)}{t(w)} \frac{t(v) - Rt(w)}{T(\psi)} \approx \frac{t(v) - Rt(w)}{T(\psi)} = \sum_{i=1}^{N} b_i(\mathscr{s})x_i$$

in order to find the approximate variance of $\hat{R}$ by the method of section 4. Let us first estimate the variance of

$$\sum_{i=1}^{N} b_i(\mathscr{s})\xi_i$$

in the case in which there is no subsampling. Since the ξ_{j_k} depend on R, which is unknown, let us estimate this variance on the basis of the $\xi_{j_k}^0$ which are known in this case,† and let us call this estimator (which is a quadratic form in $\xi_{j_1}^0, \ldots, \xi_{j_n}^0$) as a function of $\mu_{j_1}, \ldots, \mu_{j_n}$

$$F(\mu_{j_1}, \ldots, \mu_{j_n}),$$

where μ_{j_k} equals the pair (ϕ_{j_k}, ψ_{j_k}). Then, according to the proposition of section 4, we obtain the following approximately unbiased estimator for $V(\hat{R})$ with subsampling:

$$F(z_{j_1}, \ldots, z_{j_n}) + \sum_{k=1}^{n} b_{j_k}\hat{\sigma}_{j_k}^2,$$

where z_{j_k} is the pair (v_{j_k}, w_{j_k}).

† Similar to substitution of $\hat{\mathscr{D}}$ for $\mathscr{D}$ in Chapter II, section 9.

In the case of sampling without replacement, note that, if the π_i are equal,

$$\sum_{k=1}^{n} \xi_{j_k}^{0} = 0,$$

and so, under simple random sampling among the u_i and in the absence of subsampling,

$$\sum_{k=1}^{n} \xi_{j_k}^{02}/(n-1)$$

is the usual estimator of

$$\sum_{i=1}^{N} \xi_i^2/(N-1),$$

corresponding to

$$\sum_{k=1}^{n} \hat{d}_{j_k}^2/(n-1)$$

estimating

$$\sum_{i=1}^{N} D_i^2/(N-1)$$

in ratio estimation.

EXAMPLE 1. Let us estimate

$$R = \sum_{i=1}^{N} M_i \bar{Y}_{i.} / \sum_{i=1}^{N} M_i = \bar{\bar{Y}}$$

by

$$\hat{R} = \sum_{k=1}^{n} M_{j_k} \bar{y}_{j_k.} / \sum_{k=1}^{n} M_{j_k}$$

when sampling among the u_i in simple random sampling without replacement. Let

$$v_i = M_i \bar{y}_{i.}, \quad w_i = M_i;$$

therefore, since

$$\begin{aligned} b_i(\mathscr{s}) &= (n/N)^{-1} a_i(\mathscr{s}), \\ \phi_i &= M_i \bar{Y}_{i.}, \\ \psi_i &= M_i, \\ t(w) &= t(\psi) = (N/n) \sum_{k=1}^{n} M_{j_k}, \\ T(\psi) &= N\bar{M}, \\ x_i &= \frac{1}{N} \frac{M_i}{\bar{M}} (\bar{y}_{i.} - \bar{\bar{Y}}), \end{aligned}$$

$$\xi_i = \frac{1}{N}\frac{M_i}{\bar{M}}(\bar{Y}_{i.} - \bar{\bar{Y}}),$$

$$V\left(\frac{N}{n}\sum_{k=1}^{n}\xi_{j_k}\right) = \frac{N-n}{N}\frac{N^2}{n}\sum_{i=1}^{N}(\xi_i - \sum_{i=1}^{N}\xi_i/N)^2/(N-1)$$

$$= \frac{N-n}{N}\frac{1}{n}\sum_{i=1}^{N}\left(\frac{M_i}{\bar{M}}\right)^2(\bar{Y}_i - \bar{\bar{Y}})^2/(N-1).$$

If the m_{j_k} had been equal to the M_{j_k}, we would have known $\xi^0_{j_1}, \ldots, \xi^0_{j_n}$:

$$\xi^0_{j_k} = \frac{1}{N}\frac{M_{j_k}}{\bar{M}}(\bar{Y}_{j_k.} - R^0),$$

where under that hypothesis

$$R^0 = \frac{t(\phi)}{t(\psi)} = \sum_{k=1}^{n} M_{j_k}\bar{Y}_{j_k.} \Big/ \sum_{k=1}^{n} M_{j_k};$$

so we could have estimated

$$V((N/n)\sum_{k=1}^{n}\xi_{j_k})$$

by

$$F(\mu_{j_1}, \ldots, \mu_{j_n}) = \frac{N-n}{N}\frac{1}{n}N^2\sum_{k=1}^{n}\left(\xi^0_{j_k} - \frac{\sum_{k=1}^{n}\xi^0_{j_k}}{n}\right)^2 \Big/ (n-1)$$

$$= \frac{N-n}{N}\frac{1}{n}\sum_{k=1}^{n}\left(\frac{M_{j_k}}{\bar{M}}\right)^2(\bar{Y}_{j_k.} - R^0)^2/(n-1),$$

since here $\Sigma\xi^0_{j_k} = 0$. Therefore the proposition of section 4 gives for $V(\hat{R})$ in case there is subsampling the estimator

$$\frac{N-n}{N}\frac{1}{n}\sum_{k=1}^{n}\left(\frac{M_{j_k}}{\bar{M}}\right)^2(\bar{y}_{j_k.} - \hat{R})^2/(n-1) + \frac{N}{n}\sum_{k=1}^{n}\hat{\sigma}^2_{j_k}$$

or replacing $\bar{M}$ by $(1/n)\Sigma M_{j_k}$. In case we have simple random sampling (without replacement) within all the u_{j_k},

$$\hat{\sigma}^2_{j_k} = \frac{1}{N^2}\left(\frac{M_{j_k}}{\bar{M}}\right)^2\frac{M_{j_k} - m_{j_k}}{M_{j_k}}\frac{1}{m_{j_k}}s^2_{j_k}.$$

Of course,

$$V(\hat{R}) \approx V\left(\frac{N}{n}\sum_{k=1}^{n}\xi_{j_k}\right) + \frac{N}{n}\sum_{i=1}^{N}\sigma^2_i.$$

EXAMPLE 2. Let us estimate

$$R = \sum_{i=1}^{N} M_i \bar{Y}_{i.} \Big/ \sum_{i=1}^{N} M_i \bar{X}_{i.} = Y/X$$

by

$$\hat{R} = \sum_{k=1}^{n} M_{j_k} \bar{y}_{j_k.} \Big/ \sum_{k=1}^{n} M_{j_k} \bar{x}_{j_k.}$$

when sampling among the u_i is simple without replacement. Let $b_i(\mathscr{s})$ and v_i be like in Example 1 and let $w_i = M_i \bar{x}_{i.}$. Then

$$x_i = \frac{1}{N} \frac{M_i}{\bar{M}} (\bar{y}_{i.} - R\bar{x}_i)/\bar{\bar{X}},$$

$$\xi_i = \frac{1}{N} \frac{M_i}{\bar{M}} (\bar{Y}_{i.} - R\bar{X}_{i.})/\bar{\bar{X}},$$

$$V\left(\frac{N}{n} \sum_{k=1}^{n} \xi_{j_k}\right) = \frac{N-n}{N} \frac{1}{n} \frac{1}{\bar{\bar{X}}^2} \sum_{i=1}^{N} \left(\frac{M_i}{\bar{M}}\right)^2 (\bar{Y}_{i.} - R\bar{X}_{i.})^2/(N-1),$$

and

$$V(\hat{R}) \approx V\left(\frac{N}{n} \sum_{k=1}^{n} \xi_{j_k}\right) + \frac{N}{n} \sum_{i=1}^{N} \sigma_i^2.$$

In simple sampling within clusters, σ_i^2 is equal to

$$\frac{1}{N^2} \left(\frac{M_i}{\bar{M}}\right)^2 \frac{M_i - m_i}{M_i} \frac{1}{m_i} S_{\text{w}}^2(\mathscr{D},i)/\bar{\bar{X}}^2,$$

where

$$S_{\text{w}}^2(\mathscr{D},i) = \sum_{j=1}^{M_i} \{(y_{ij} - Rx_{ij}) - (\bar{Y}_{i.} - R\bar{X}_{i.})\}^2/(M_i - 1).$$

We may estimate

$$V((N/n) \sum_{k=1}^{n} \xi_{j_k})$$

if the μ_{j_k} are known by

$$F(\mu_{j_1}, \ldots, \mu_{j_n}) = \frac{N-n}{N} \frac{1}{n} \frac{1}{\bar{\bar{X}}^2} \sum_{k=1}^{n} \left(\frac{M_i}{\bar{M}}\right)^2 (\bar{Y}_{j_k.} - R^0 \bar{X}_{j_k.})^2/(n-1).$$

Therefore we estimate $V(\hat{R})$ in case there is subsampling by

$$\frac{N-n}{N} \frac{1}{n} \frac{1}{\bar{\bar{X}}^2} \sum_{k=1}^{n} \left(\frac{M_{j_k}}{\bar{M}}\right)^2 (\bar{y}_{j_k.} - \hat{R}\bar{x}_{j_k.})^2/(n-1) + \frac{N}{n} \sum_{k=1}^{n} \hat{\sigma}_{j_k}^2$$

or replacing $\bar{M}\bar{\bar{X}}$ by $(1/nN)\Sigma M_{j_k} \bar{X}_{j_k}$.

NOTE. Again it is possible to use unequal probabilities for selecting the primary units.

REMARK. RAO, J. Ind. Soc. Agric. Statist., **16** (1964) 175–188, has considered extension of some of the methods mentioned in section 7.2 of Chapter IIA to the present case, and refers to further literature.

7. Subsampling when primary units have been obtained by sampling with replacement

In section 3 we discussed some estimators of $\overline{\overline{Y}}$ in cluster sampling without subsampling. Generalizations to the case of subsampling can be of different kinds with respect to any unit u_i for which $t_i(\mathscr{s})$, the number of times u_i appears in $\mathscr{s}$, exceeds 1:

7.1. EACH TIME A PRIMARY UNIT IS DRAWN, A SUBSAMPLE OF PREDETERMINED SIZE IS TAKEN FROM IT

We draw a sample of m_i from each such u_i, replace it, draw again, independently, a sample of m_i from u_i, etc., until we have drawn $t_i(\mathscr{s})$ such samples. [In this case we cannot write our estimators in the form

$$\sum_{i=1}^{N} b_i(\mathscr{s})z_i,$$

since the subsample we obtain from unit u_i (and therefore the *random* quantity computed from it†) depends on $\mathscr{s}$ not only through the fact that $u_i \in \mathscr{s}$. Therefore the theorem of section 4 does not apply; however, in this case estimation of the variance of estimators is even simpler due to the independence of the n drawings.] Suppose in the kth drawing we obtain u_{j_k} and compute $z(j_k, r_{j_k k})$, a conditionally unbiased estimator of μ_{j_k} with conditional variance $\sigma^2_{j_k}$; here $r_{j_k k}$ indicates that in the kth drawing we obtained u_{j_k} for $r_{j_k k}$th time $(1 \leq r_{j_k k} \leq t_{j_k}(\mathscr{s}))$.

(i) Consider first the unbiased estimator

$$\bar{g} = (1/n)\sum_{k=1}^{n} g_k,$$

with each

$$g_k = \sum_{i=1}^{N} a_{ik}z(i, r_{ik})/p_i$$

an independent unbiased estimator of $\overline{\overline{Y}}$ and all g_k having the same distribution. Here $a_{ik} = 1$ if on the kth occasion we draw u_i and 0 otherwise, and

† Recall that $b_i(\mathscr{s})$ is *fixed* for given $\mathscr{s}$.

$p_i = \mathscr{E} a_{ik}$, assumed independent of k. For $z(i, r_{ik})$ we take $M_i \bar{y}_{i.}(r_{ik})/(N\bar{M})$, where $\bar{y}_{i.}(r_{ik})$ is the average observation of the r_{ik}th sample from u_i. So, as

$$\sum_k a_{ik} \bar{y}_{i.}(r_{ik})$$

is t_i times the average, $\bar{y}_{i.}$, over all samples from u_i,

$$\bar{g} = \frac{1}{n} \sum t_i \frac{M_i}{N\bar{M}} \bar{y}_{i.}/p_i.$$

The conditional mean of g_k is then

$$\sum_{i=1}^{N} a_{ik} \mu_i / p_i.$$

In section 3.2 of Chapter VI we had a similar expression (for estimating Σy_i instead of $\Sigma \mu_i = \bar{\bar{Y}}$) with y_i in place of μ_i, and t_i and $\mathscr{E} t_i = np_i$ in place of a_{ik} and $\mathscr{E} a_{ik} = p_i$, respectively; so we get

$$V(\sum_{i=1}^{N} a_{ik} \mu_i / p_i) = \sum_{i=1}^{N} p_i((\mu_i/p_i) - \bar{\bar{Y}})^2.$$

Since the conditional variance of g_k is

$$\sum_{i=1}^{N} a_{ik}^2 \sigma_i^2 / p_i^2,$$

which has mean

$$\sum_{i=1}^{N} \sigma_i^2 / p_i,$$

we obtain

$$\begin{aligned} V(\bar{g}) &= \frac{1}{n} V(g_k) \\ &= \frac{1}{n} \sum_{i=1}^{N} p_i \left(\frac{\mu_i}{p_i} - \bar{\bar{Y}} \right)^2 + \frac{1}{n} \sum_{i=1}^{N} \sigma_i^2 / p_i \\ &= \frac{1}{n} \sum_{i=1}^{N} p_i \left(\frac{Y_{i.}}{p_i N \bar{M}} - \bar{\bar{Y}} \right)^2 + \frac{1}{n} \sum \frac{\sigma_i^2}{p}, \end{aligned}$$

where if subsampling is simple without replacement

$$\sigma_i^2 = \left(\frac{M_i}{N\bar{M}} \right)^2 \frac{M_i - m_i}{M_i} \frac{1}{m_i} S_{\text{w}}^2(i).$$

An unbiased estimator of $V(\bar{g})$ is

$$\frac{1}{n} \sum_{k=1}^{n} (g_k - \bar{g})^2/(n-1),$$

for *any subsampling* method yielding conditionally unbiased estimators of the μ_{j_k}, which follows from the independence of the g_k and also from the fact that its mean is

$$\frac{1}{n(n-1)}\left(\sum_{k=1}^{n} \mathscr{E}g_k^2 - n\mathscr{E}\bar{g}^2\right) = \frac{1}{n(n-1)}\{nV(g_k) + n\bar{\bar{Y}}^2 - nV(\bar{g}) - n\bar{\bar{Y}}^2\}$$

$$= \frac{1}{n-1}\{V(g_k) - V(\bar{g})\} = V(\bar{g}).$$

The component $(1/n)\Sigma\sigma_i^2/p_i$ may be estimated by

$$\frac{1}{n}\sum_{i=1}^{N} \frac{t_i}{np_i}(\hat{\sigma}_i^2/p_i) = \frac{1}{n^2}\sum_{i=1}^{N} t_i\hat{\sigma}_i^2/p_i^2.$$

If $p_i = M_i/(N\bar{M})$, the estimator becomes simply $\bar{g} = (1/n)\Sigma t_i\bar{y}_{i\cdot}$.

(ii) For cases in which pps or ppes sampling would be an efficient method, but the sizes of primary units are only known approximately and $\bar{M}$ is not known, the ratio estimator

$$\hat{\bar{Y}}_{R\,\text{ppes}} = \frac{\sum_{k=1}^{n} g_k}{(N\bar{M})^{-1}\sum_{k=1}^{n} M_{j_k}/p_{j_k}} = \frac{\Sigma t_i M_i \bar{y}_{i\cdot}/p_i}{\Sigma t_i M_i/p_i}$$

has only a small bias. Moreover, its variance may well be smaller than that of $\bar{g}$.

More generally, we consider the estimator

$$\hat{R} = \sum_{k=1}^{n} g_k \Big/ \sum_{k=1}^{n} h_k.$$

Then we replace z by (z, w) and g by (g, h), where w is the same function of the $\mathscr{X}$-observations as z is of the $\mathscr{Y}$-observations, and h_k the same function of the w's as g_k is of the z's. Like in the last section (Example 2), replace $\mathscr{Y}$ by $(\mathscr{Y} - R\mathscr{X})/\bar{\bar{X}}$. We then obtain, since $\Sigma(Y_i - RX_i) = 0$,

$$V(\hat{R}) \approx \frac{1}{n}\sum_{i=1}^{N} p_i\left(\frac{Y_{i\cdot} - RX_{i\cdot}}{p_i X}\right)^2 + \frac{1}{n}\sum_{i=1}^{N}\frac{M_i^2}{p_i X^2}\,\frac{M_i - m_i}{M_i}\,\frac{1}{m_i}\,S_{\mathrm{w}}^2(i).$$

Similarly we obtain

$$v(\hat{R}) = \frac{1}{n(n-1)}\,\frac{1}{\bar{\bar{X}}^2}\sum_{k=1}^{n}(g_k - \hat{R}h_k)^2$$

$$= \frac{1}{n(n-1)}\,\frac{1}{X^2}\,\Sigma t_i M_i^2\left(\frac{\bar{y}_{i\cdot} - \hat{R}\bar{x}_{i\cdot}}{p_i}\right)^2$$

or with in place of X an estimator thereof: $\Sigma t_i M_i \bar{x}_{i\cdot}/p_i$.

For the special case above

$$\bar{X}_{i.} = 1, \quad R = \bar{\bar{Y}}, \quad \sum_{k=1}^{n} h_k = (N\bar{M})^{-1} \sum_{k=1}^{n} M_{j_k}/p_{j_k}, \quad w_i = M_i;$$

the variance is approximately

$$V(\hat{\bar{\bar{Y}}}_{R\,\text{ppes}}) \approx \frac{1}{n} \sum_{i=1}^{N} p_i \left(\frac{Y_{i.} - \bar{\bar{Y}} M_i}{p_i N\bar{M}} \right)^2 + \frac{1}{n} \sum_{i=1}^{N} \left(\frac{M_i}{N\bar{M}} \right)^2 \frac{1}{p_i} \frac{M_i - m_i}{M_i} \frac{1}{m_i} S_{\text{w}}^2(i)$$

$$= \frac{1}{n} \sum_{i=1}^{N} \frac{1}{p_i} \left(\frac{M_i}{N\bar{M}} \right)^2 \left\{ (\bar{Y}_{i.} - \bar{\bar{Y}})^2 + \frac{M_i - m_i}{M_i m_i} S_{\text{w}}^2(i) \right\}$$

with estimator

$$v(\hat{\bar{\bar{Y}}}_{R\,\text{ppes}}) = \frac{1}{n(n-1)} \sum_{k=1}^{n} \left(\frac{M_{j_k}}{p_{j_k} N\hat{\bar{M}}} \right)^2 (\bar{y}_{j_k.}(r_{j_k k}) - \hat{\bar{\bar{Y}}}_{R\,\text{ppes}})^2,$$

where

$$N\hat{\bar{M}} = \frac{1}{n} \Sigma M_{j_k}/p_{j_k}.$$

REMARK. If we sample a total of

$$\sum_{k=1}^{n} m_{j_k} = \sum_{i=1}^{N} t_i(\mathscr{s}) m_i$$

units among the $N\bar{M}$ possible ones, the expected number of units sampled is

$$n \sum_{i=1}^{N} p_i m_i.$$

Now our estimators of $\bar{\bar{Y}}$ become (approximately) *self-weighting* if $M_{j_k}/(p_{j_k} m_{j_k})$ is (as closely as possible) a constant c, so that

$$\sum_{i=1}^{N} p_i m_i$$

is then equal to $N\bar{M}/c$. Therefore, f_0, the *expected overall sampling fraction*, equals

$$f_0 = \mathscr{E} \sum_{k=1}^{n} m_{j_k}/(N\bar{M}) = n \sum_{i=1}^{N} p_i m_i/(N\bar{M}) = n/c.$$

Consequently, if we choose f_0 in advance (in a self-weighting sample), the sampling fraction from $u_{j_k} = u_i$ is, in each drawing from u_i, taken to be as close as possible to

$$\frac{m_i}{M_i} = \frac{1}{cp_i} = \frac{f_0}{np_i}.$$

If the p_i are proportional to the M_i, m_i is a constant, which often is very advantageous, since it leads to *equal work loads* in each cluster.

7.2. FROM EACH PRIMARY UNIT DRAWN ONE SUBSAMPLE OF PREDETERMINED SIZE IS OBTAINED

From each such u_i we draw a simple sample of size m_i. For each j_k that equals i, we take the same value for z_{j_k}; that is, we give this value a weight equal to the number of times $t_i(\mathscr{s})$ it is obtained in the sample. In the notation of the previous section, we take

$$b_i(\mathscr{s}) = t_i(\mathscr{s})/\mathscr{E}t_i(\mathscr{s}), \quad \text{and} \quad \hat{\bar{Y}} = (1/n)\sum_{i=1}^{N} t_i(\mathscr{s})z_i/p_i$$

(with $z_i = M_i\bar{y}_{i.}/N\bar{M}$). From section 3.2 of Chapter VI we obtain

$$V(\sum_{i=1}^{N} b_i(\mathscr{s})\mu_i) = \frac{1}{n}\sum_{i=1}^{N} p_i\left(\frac{\mu_i}{p_i} - \bar{\bar{Y}}\right)^2.$$

The expected value of the conditional variance of

$$\sum_{i=1}^{N} b_i(\mathscr{s})z_i$$

is

$$\sum_{i=1}^{N} \mathscr{E}b_i^2(\mathscr{s})\sigma_i^2 = \frac{1}{n}\sum_{i=1}^{N} \sigma_i^2/p_i + \frac{n-1}{n}\sum_{i=1}^{N} \sigma_i^2,$$

as

$$\mathscr{E}t_i^2(\mathscr{s}) = \mathscr{E}(\sum_k a_{ik})^2 = \sum_k \mathscr{E}a_{ik}^2 + \sum\sum_{k \neq k'} \mathscr{E}a_{ik}a_{ik'}$$
$$= np_i + n(n-1)p_i^2;$$

and so the unconditional variance of our estimator is

$$V(\hat{\bar{Y}}) = \left\{\frac{1}{n}\sum_{i=1}^{N} p_i\left(\frac{\mu_i}{p_i} - \bar{\bar{Y}}\right)^2 + \frac{1}{n}\sum_{i=1}^{N} \sigma_i^2/p_i\right\} + \frac{n-1}{n}\sum_{i=1}^{N} \sigma_i^2,$$

where the term in brackets on the right-hand side equals the variance of the estimator under 7.1 above.

The increase in the variance is only partly compensated by the fact that the expected number of different subunits observed will be smaller in this case than in 7.1; this has been shown by RAO, J. Ind. Soc. Agric. Statist., **13** (1961) 211–217.

To estimate the variance we apply the theorem of section 4: In section 3.2 of Chapter VI with μ_i in place of y_i,

$$F(\mu_{j_1}, \ldots, \mu_{j_n}) = \frac{1}{n(n-1)}\sum_{i=1}^{N} t_i(\mathscr{s})\left(\frac{\mu_i}{p_i} - n^{-1}\sum_{i=1}^{N} \frac{t_i(\mathscr{s})\mu_i}{p_i}\right)^2;$$

so

$$F(z_{j_1}, \ldots, z_{j_n}) = \frac{1}{n(n-1)} \sum_{i=1}^{N} t_i(\mathscr{s}) \left(\frac{z_i}{p_i} - \hat{\hat{Y}}\right)^2,$$

and

$$v(\hat{\hat{Y}}) = \frac{1}{n(n-1)} \sum_{i=1}^{N} t_i(\mathscr{s}) \left(\frac{z_i}{p_i} - \hat{\hat{Y}}\right)^2 + \frac{1}{n} \sum_{i=1}^{N} t_i(\mathscr{s}) \hat{\sigma}_i^2 / p_i.$$

One can similarly consider estimators of ratios. Instead of defining $b_i(\mathscr{s})$ as above, we may define it as

$$b_i(\mathscr{s}) = a_i(\mathscr{s}) \mathrm{Q}(i \mid \mathscr{s}) / p_i;$$

see section 3.3 of Chapter VI.

7.3. FROM EACH PRIMARY UNIT DRAWN ONE SUBSAMPLE IS OBTAINED OF A SIZE WHICH DEPENDS ON THE NUMBER OF TIMES THE PRIMARY UNIT IS DRAWN

From such a u_i we draw $t_i(\mathscr{s})\, m_i$ observations instead of m_i observations. (If subsampling is without replacement we can adopt this plan only if the probability of $t_i(\mathscr{s})\, m_i$ exceeding M_i is negligible [which may not be feasible for large $\mathscr{N}$]; and if, when such an event does occur, we take M_i observations from u_i. A modification which avoids this difficulty is discussed in the next section.) For the same reason as mentioned under 7.1, we cannot write our estimator in the form

$$\sum_{i=1}^{N} b_i(\mathscr{s}) z_i.$$

We shall instead write the estimator as

$$\bar{l}(\mathscr{s}) = \frac{1}{n} \sum_{i=1}^{N} t_i(\mathscr{s}) l_i(\mathscr{s})$$

with

$$l_i(\mathscr{s}) = z_i(\mathscr{s}) / p_i \quad \text{and} \quad n = \sum_{i=1}^{N} t_i(\mathscr{s}).$$

The variance of the conditional expectation of the estimator is again

$$(1/n) \sum_{i=1}^{N} p_i((\mu_i / p_i) - \bar{\bar{Y}})^2.$$

Here

$$\mu_i = (N\bar{M})^{-1} \bar{Y}_{i.} M_i.$$

In subsampling without replacement, with $z_i(\mathscr{s})$ equal to $M_i/N\bar{M}$ times the average of $\mathscr{Y}$-values in the sample of $t_i(\mathscr{s})\, m_i$ from unit u_i, the mean of the conditional variance of our estimator is

$$\frac{1}{n^2N^2\bar{M}^2}\sum_{i=1}^{N}\mathscr{E}t_i^2(\mathscr{s})\frac{1}{p_i^2}M_i^2\left(\frac{1}{t_i(\mathscr{s})m_i}-\frac{1}{M_i}\right)S_{\mathrm{w}}^2(i)$$

$$=\frac{1}{n(N\bar{M})^2}\sum_{i=1}^{N}\frac{M_i^2S_{\mathrm{w}}^2(i)}{m_ip_i}-\frac{1}{n(N\bar{M})^2}\sum_{i=1}^{N}\frac{M_iS_{\mathrm{w}}^2(i)}{p_i}-\frac{n-1}{n(N\bar{M})^2}\sum_{i=1}^{N}M_iS_{\mathrm{w}}^2(i)$$

$$=\frac{1}{n(N\bar{M})^2}\sum_{i=1}^{N}\frac{1}{p_i}M_i^2\frac{M_i-m_i}{M_i}\frac{1}{m_i}S_{\mathrm{w}}^2(i)-\frac{n-1}{n(N\bar{M})^2}\sum_{i=1}^{N}M_iS_{\mathrm{w}}^2(i),$$

where the last term represents the gain over the method discussed in 7.1. However, by the present method more units have to be observed. Nonetheless, as shown by RAO, J. Ind. Soc. Agric. Statist., **13** (1961) 211–217, even with the expected numbers of different subunits observed in each u_i equal for both methods, there remains an advantage with the present method.

SUKHATME, Sample Theory of Surveys with Applications (Ind. Soc. for Agric. Statist. and Iowa State University Press, 1954) Chapter 8, section 10, derives an estimator of the variance of $\bar{l}(\mathscr{s})$. He shows that if

$$s_l^2=\sum_{i=1}^{N}t_i(\mathscr{s})\{l_i(\mathscr{s})-\bar{l}(\mathscr{s})\}^2/(n-1)$$

and if $s_{\mathrm{w}}^2(i)$ denotes $(t_i(\mathscr{s})\, m_i - 1)^{-1}$ times the sum of squares of deviations of the observations from their average in the subsamples from u_i,

$$v(\bar{l}(\mathscr{s}))=\frac{s_l^2}{n}+\frac{1}{n(n-1)}\frac{1}{(N\bar{M})^2}$$

$$\times\sum_{i=1}^{N}t_i(\mathscr{s})\left\{\left(1-\frac{1}{t_i(\mathscr{s})}\right)\frac{1}{m_i}\left(\frac{M_i}{p_i}\right)^2-(n-1)\frac{M_i}{p_i}\right\}s_{\mathrm{w}}^2(i).$$

Sukhatme mentions the wide use of this scheme in India for estimating acreage under crops, the u_i being villages, divided into subunits of 8 consecutive survey numbers (usually each survey number constitutes a *different* field) in the village, and 4 subunits are selected (giving an equal chance to all subunits in the village). The probability of including a village is chosen to be proportional to the total number of survey numbers in the village. Use of natural units such as these is very economical; and the variance of the estimates is relatively small because of the high correlation between the number of fields and the total crop acreage (due to the fact that all uncultivated land in the village is usually given one single survey number).

One can similarly consider estimators of ratios.

8. A method of two-stage sampling, without replacement at both stages, in which primary units are included with unequal probabilities

The $t_i(\mathscr{s})\, m_i$ of section 7.3 may exceed the M_i, especially if, due to N being large, many M_i are small. WILKS, Bull. Intern. Statist. Inst., **37**, Pt. II (1960) 241–248 gives a method for which this is not possible. He achieves this by making the $t_i(\mathscr{s})$ joint hypergeometric, instead of multinomial, variables with expected values proportional to the M_i; we shall call them m_i'. For simplicity, we consider only the case in which m_i (as defined previously) is a constant and call it t. Suppose that each M_i is an entire multiple of t: $M_i = M_i't$. We shall take a fixed total number, m, of subunits in the sample; denote m/t by m'. Then

$$\mathscr{E}m_i' = m'M_i/M,$$

where

$$M = \sum_{i=1}^{N} M_i.$$

In the present section it will also be convenient to denote $m_i't$ by m_i (in a sense different from that in section 7.3). If $m_i > 0$ we include subpopulation i and subsample m_i subunits from it; if $m_i = 0$ we do not include it in our sample. Clearly, if we denote

$$\sum_{i=1}^{N} M_i'$$

by M',

$$V(m_i') = m' \frac{M' - m'}{M' - 1} \frac{M_i'}{M'} \frac{M' - M_i'}{M'},$$

$$\operatorname{Cov}(m_i', m_j') = -m' \frac{M' - m'}{M' - 1} \frac{M_i'}{M'} \frac{M_j'}{M'} \quad (i \neq j).$$

Let $\bar{y}$ be the average of the $\mathscr{Y}$-values for the sample. $\mathscr{E}\bar{y} = \bar{\bar{Y}}$ and, we shall show,

$$V(\bar{y}) = \frac{1}{m'} \frac{M' - m'}{M' - 1} \left\{ \frac{1}{t} \frac{1}{M'} \Sigma(M_i' - 1) S_w^2(i) + \Sigma \frac{M_i'}{M'} (\bar{Y}_{i.} - \bar{\bar{Y}})^2 \right\}$$

$$= \frac{1}{m'} \frac{M' - m'}{M' - 1}$$

$$\times \left\{ \frac{1}{t} \frac{1}{M} \Sigma\Sigma(y_{ij} - \bar{Y}_{i.})^2 + \frac{1}{M} \Sigma M_i(\bar{Y}_{i.} - \bar{\bar{Y}})^2 - \frac{t-1}{t} \frac{1}{M} \Sigma S_w^2(i) \right\}$$

$$= \frac{1}{m'} \frac{M' - m'}{M' - 1} \left\{ \frac{1}{t} \frac{1}{M} \Sigma M_i S_w^2(i) + \frac{1}{M} \Sigma M_i(\bar{Y}_{i.} - \bar{\bar{Y}})^2 - \frac{1}{M} \Sigma S_w^2(i) \right\}.$$

(In simple random sampling we have $t = 1$, and we have

$$M_i, \quad M = \sum_{i=1}^{N} M_i, \quad \text{and} \quad m = \sum_{i=1}^{N} m_i,$$

respectively, in place of

$$M'_i, \quad M' = \sum_{i=1}^{N} M'_i, \quad \text{and} \quad m' = \sum_{i=1}^{N} m'_i;$$

respectively.) The last term subtracted in the bracket,

$$\bar{M}^{-1} \sum_{i=1}^{N} S_{\mathrm{w}}^2(i)/N,$$

is small compared with

$$(1/t) \sum_{i=1}^{N} (M_i/M) S_{\mathrm{w}}^2(i)$$

if, as is often the case, $\bar{M}$ is much larger than t and there is no very large correlation between the M_i and the $S_{\mathrm{w}}^2(i)$.

Now let us derive the formula given for $V(\bar{\bar{y}})$, i.e., let us find

$$\mathscr{E}\left[\sum\left\{\frac{m_i}{m}(\bar{y}_{i.} - \bar{Y}_{i.}) + \left(\frac{m_i}{m} - \frac{M_i}{M}\right)\bar{Y}_{i.}\right\}\right]^2.$$

Given the primary units selected, the conditional covariance of $\bar{y}_{i.}$ with $\bar{y}_{j.}$ $(j \neq i)$ and with the second term in the curly brackets is zero, and the conditional variance of $\bar{y}_{i.}$ is $(m_i^{-1} - M_i^{-1}) S_{\mathrm{w}}^2(i)$, so that $V(\bar{\bar{y}})$ is equal to

$$\begin{aligned}
\Sigma \mathscr{E} \frac{1}{m}\left(\frac{m'_i}{m'} - \frac{m'_i}{M'_i}\frac{m'_i}{m'_i}\right) S_{\mathrm{w}}^2(i) &+ \Sigma \bar{Y}_{i.}^2 . V\left(\frac{m'_i}{m'}\right) + \sum_{i \neq j}\sum \bar{Y}_{i.} \bar{Y}_{j.} \operatorname{Cov}\left(\frac{m'_i}{m'}, \frac{m'_j}{m'}\right) \\
&= \frac{1}{m} \Sigma \left\{\frac{M'_i}{M'} - \frac{1}{M'}\left(\frac{M' - m'}{M' - 1} - \frac{M' - m'}{M' - 1}\frac{M'_i}{M'} + \frac{M'_i}{M'} m'\right)\right\} S_{\mathrm{w}}^2(i) \\
&\quad + \frac{1}{m'}\frac{M' - m'}{M' - 1}\left\{\Sigma \bar{Y}_{i.}^2 . \frac{M'_i}{M'}\frac{M' - M'_i}{M'} - \sum_{i \neq j}\sum \bar{Y}_{i.} \bar{Y}_{j.} \frac{M'_i}{M'}\frac{M'_j}{M'}\right\} \\
&= \frac{M' - m'}{M' - 1}\left\{\frac{1}{m}\frac{1}{M'} \Sigma(M'_i - 1) S_{\mathrm{w}}^2(i) + \frac{1}{m'} \Sigma \frac{M'_i}{M'}(\bar{Y}_{i.} - \bar{\bar{Y}})^2\right\}.
\end{aligned}$$

To estimate this variance, Wilks considers the following quantities

$$\begin{aligned}
G_1 &= \Sigma m'_i s_i^2, \\
G_2 &= \Sigma m'_i s_i^2 / M_i, \\
G_3 &= \Sigma m'_i (\bar{y}_{i.} - \bar{\bar{y}})^2,
\end{aligned}$$

which have expectations

$$\mathscr{E}G_1 = m'\sum \frac{M_i}{M} S_w^2(i),$$

$$\mathscr{E}G_2 = \frac{m'}{M} \Sigma S_w^2(i),$$

$$\mathscr{E}G_3 = \frac{M' - m'}{M' - 1}\frac{1}{t}\left\{\Sigma S_w^2(i) - \sum \frac{M_i}{M} S_w^2(i)\right\}$$

$$+ (m' - 1)\frac{M'}{M' - 1}\sum \frac{M_i}{M}(\bar{Y}_{i.} - \bar{\bar{Y}})^2.$$

The latter may be obtained as follows:

$$G_3 = \Sigma m_i' \bar{y}_{i.}^2 - m'\bar{\bar{y}}^2$$
$$= \Sigma m_i'(\bar{y}_{i.} - \bar{Y}_{i.})^2 + 2\Sigma m_i' \bar{y}_{i.}\bar{Y}_{i.} - \Sigma m_i' \bar{Y}_{i.}^2 - m'\bar{\bar{y}}^2$$

has mean

$$\mathscr{E}\Sigma m_i'\left(\frac{1}{m_i} - \frac{1}{M_i}\right)S_w^2(i) + 2m'\sum \frac{M_i}{M}\bar{Y}_{i.}^2 - m'\sum \frac{M_i}{M}\bar{Y}_{i.}^2 - m'\{V(\bar{\bar{y}}) + \bar{\bar{Y}}^2\}$$

$$= \frac{1}{t}\left(1 - \frac{m'}{M'}\right)\Sigma S_w^2(i) + m'\sum \frac{M_i}{M}(\bar{Y}_{i.} - \bar{\bar{Y}})^2 - m'V(\bar{\bar{y}})$$

$$= \frac{1}{t}\frac{M' - m'}{M' - 1}\left\{\left(\frac{M' - 1}{M'} + \frac{1}{M'}\right)\Sigma S_w^2(i) - \sum \frac{M_i}{M} S_w^2(i)\right\}$$

$$+ \left(m' - \frac{M' - m'}{M' - 1}\right)\sum \frac{M_i}{M}(\bar{Y}_{i.} - \bar{\bar{Y}})^2.$$

Therefore

$$\sum \frac{M_i}{M}(\bar{Y}_{i.} - \bar{\bar{Y}})^2 = \mathscr{E}\frac{1}{m' - 1}\left\{\frac{M' - 1}{M'}G_3 - \frac{M - m}{M}\frac{1}{m}(MG_2 - G_1)\right\}$$

and we may estimate $V(\bar{\bar{y}})$ by

$$v(\bar{\bar{y}}) = \frac{1}{m'}\frac{M' - m'}{M' - 1}\left\{\left(\frac{1}{tm'} + \frac{1}{m' - 1}\frac{M - m}{M}\frac{1}{m}\right)G_1\right.$$

$$\left. - \left(\frac{1}{m'} + \frac{1}{m' - 1}\frac{M - m}{m}\right)G_2 + \frac{1}{m' - 1}\frac{M' - 1}{M'}G_3\right\}$$

$$= \frac{1}{m' - 1}\frac{M - m}{M}\frac{1}{m}\left\{G_1 - \frac{M}{m'}G_2 + tG_3\right\}$$

$$= \frac{1}{m' - 1}\frac{M' - m'}{M'}\left\{\frac{1}{t}\sum \frac{m_i}{m}s_i^2 + \sum \frac{m_i}{m}(\bar{y}_{i.} - \bar{\bar{y}})^2 - \frac{1}{m}\sum\left(\frac{m_i}{m}s_i^2\Big/\frac{M_i}{M}\right)\right\}.$$

This estimator is different from the one given by Wilks; also the expression is simpler than his, and its form is more similar to the expression for $V(\bar{y})$. It appears that Wilks's claim that the estimator given by him for

$$\sum \frac{M_i}{M} (\bar{Y}_i - \bar{Y})^2$$

is unbiased is incorrect.

Wilks also considers minimizing $V(\bar{y})$ subject to a given bound on the total cost. Unfortunately he uses as cost function $c_1 m' + c_2 m$. It is evident, however, that such a function cannot be appropriate; a more appropriate one would be $c_0 + c_1 n + c_2 m$, where n is the number of primary units included in the sample. Since n is a random variable, we may try to minimize the variance subject to a bound on

$$c_0 + c_1 \mathscr{E} n + c_2 m.$$

We may find $\mathscr{E} n$ from the equation

$$\mathscr{E} n = \Sigma \mathrm{P}\{m_i' > 0\},$$

where

$$\mathrm{P}\{m_i' > 0\} = 1 - \left\{\binom{M_i'}{0}\binom{M' - M_i'}{m'} \Big/ \binom{M'}{m'}\right\}$$

$$= 1 - \frac{(M' - m')^{(M_i')}}{(M')^{(M_i')}}.$$

9. Subsampling when using the method of random splits

In Chapter VI, section 7, we gave a random split method. In this section we apply the results of section 4 to this method when subsampling is used. We replace the population total for $u(g, s, k)$ by an estimator whose conditional variance is $\hat{\sigma}^2(g, s, k)$, say. The extra term in the estimated variance obtained by the method of section 4 is then

$$\sum_{g \in \mathscr{G}} a_g \sum_{s=1}^{n} \sum_{k=1}^{N_s} a_{gsk} \hat{\sigma}^2(g, s, k) / P_{gsk}.$$

For example, we take for the estimator of the total for $u(g, s, k)$:

$$M_{gsk} \bar{y}(g, s, k),$$

so that, with s_{gsk}^2 the sample mean square within a primary unit,

$$\hat{\sigma}^2(g, s, k) = M_{gsk}^2 \left(\frac{1}{m_{gsk}} - \frac{1}{M_{gsk}}\right) s_{gsk}^2.$$

REMARK. STUART, Rev. Intern. Statist. Inst., **32** (1964) 193–201 gives conditions under which, when subsampling is used, the random split method improves results as compared with sampling without replacement of the psu's. He shows that, if n/N is small, advantages, if any, will be negligible. He also generalizes the random split method. Other generalizations are given by CHIKKAGOUDAR, Austral. J. Statist., **9** (1967) 57–70.

10. Comparison of the variance of an estimator under two-stage sampling with that under single stage sampling

10.1. SIMPLEST CASE

Let us consider the simplest case: simple random sampling, without replacement, of n clusters and simple subsampling, without replacement, of $\bar{m}$ within each selected cluster, when all potential clusters are of equal size. Then

$$V(\bar{\bar{y}}) = \frac{N-n}{N}\frac{1}{n}S_i^2 + \frac{\bar{M}-\bar{m}}{\bar{M}}\frac{1}{n\bar{m}}S_{\mathrm{w}}^2.$$

Substituting the formulas of Chapter V, section 4, we obtain

$$V(\bar{\bar{y}}) = \frac{N\bar{M}-1}{N\bar{M}}\frac{1}{n\bar{m}}S^2[1-k+\{(\bar{m}-1)-k(\bar{M}-1)\}r]$$

$$= \frac{N\bar{M}-1}{N\bar{M}}\frac{1}{n\bar{m}}S^2[1+(\bar{m}-1)r-k\{1+(\bar{M}-1)r\}],$$

where

$$k = \frac{n-1}{N-1}\frac{\bar{m}}{\bar{M}},$$

which is less than the overall sampling fraction f, and so is usually small. We may compare this with the variance of $\bar{y}$ when drawing a simple random sample of $n\bar{m}$ small units: $(1-f)S^2/n\bar{m}$. For small f the variance of the cluster sample is therefore about $\{1+(\bar{m}-1)r\}$ times that of the simple random sample.

Some special cases are of interest:

The case $n = N$ is that of stratified sampling, giving

$$\frac{N\bar{M}-1}{N\bar{M}}(1-f)\frac{S^2}{n\bar{m}}(1-r),$$

where $f = \bar{m}/\bar{M}$. This shows that in this case the variance in stratified sampling is always less than in simple random sampling when $r > 0$ (we shall generalize this result below). As $r > -(1/(\bar{M}-1))$, the maximum relative increase in variance due to stratification is $1/(\bar{M} - 1)$. If we neglect $1/(N\bar{M})$, r represents the relative gain due to stratification in this case.†

The case $\bar{m} = 1$ is also interesting; since $r > -(1/(\bar{M} - 1))$, the variance in cluster sampling is always less than S^2/n, the approximate variance of $\bar{y}$ in simple random sampling. For $\bar{m} = 2$, this is true for $(n - 1)/(N - 1) \leq \frac{1}{2}$ as long as r is less than $k/\{1 - k(\bar{M} - 1)\}$, because, for $(n - 1)/(N - 1) \leq \frac{1}{2}$, the denominator in the above expression is positive. This explains why often *small* $\bar{m}$ is aimed at.

The case $\bar{m} = \bar{M}$ leads back to the formula

$$V(\bar{y}) = (\sigma^2/n\bar{M})\{(N - n)/(N - 1)\}\{1 + (\bar{M} - 1)r\}$$

for single stage sampling of whole clusters.

One might also consider sampling $n\bar{m}/\bar{M}$ clusters without subsampling, for which the variance of $\bar{y}$ is

$$\left(\frac{\bar{M}}{n\bar{m}} - \frac{1}{N}\right)S_l^2,$$

which exceeds the variance for the subsampling design by

$$\frac{1}{n}\left(\frac{\bar{M}}{\bar{m}} - 1\right)\left\{S_l^2 - \frac{1}{\bar{M}}S_{\mathrm{w}}^2\right\},$$

where the expression in the curly brackets is approximately rS^2. This indicates that r is the main factor favoring subsampling when the only variable cost consists in obtaining information on the small units (see further in the next section); this is evident, since, if the clusters are homogeneous, it is wasteful to include all its elements in the sample, unless the cost of including all differs little from that of including only some.

The above results are, of course, (approximately) applicable to ratio estimators as well, if we attach $(\mathscr{D})$ to the symbols appearing in the formulas.

† The variance of the mean of a stratified sample, when all strata are of size $\bar{N}$, and $\bar{n}$ are taken for each stratum, is $(1 - (\bar{n}/\bar{N}))(1/L\bar{n})\Sigma S_h^2/L$. For an unstratified sample of this size the variance of the sample mean is $(1 - (\bar{n}/\bar{N}))(1/L\bar{n})S^2$, where (see Chapter III, Remark (i) at end of section 8):

$$S^2 = \Sigma S_h^2/L + \frac{N}{N-1}\left\{\Sigma(\bar{Y}_h - \bar{Y})^2/L - \frac{1}{\bar{N}}\frac{L-1}{L}\Sigma S_h^2/L\right\}.$$

Therefore the relative variance gain from stratification is

$$\left\{\Sigma(\bar{Y}_h - \bar{Y})^2/L - \frac{L-1}{L}\frac{1}{\bar{N}}\Sigma S_h^2/L\right\}\Big/\sigma^2 = \left\{\sigma_l^2 - \frac{L-1}{L}\frac{1}{\bar{N}}S_{\mathrm{w}}^2\right\}\Big/\sigma^2.$$

As $\sigma_l^2/\sigma^2 = \{1 + (\bar{N} - 1)r\}/\bar{N}$ and $S_{\mathrm{w}}^2/\sigma^2 = (1 - r)$, this is $r + (1/L\bar{N})(1 - r)$. When $1/L\bar{N} = 1/N$ is small compared with r, we can neglect it, and the relative variance gain from stratification is approximated by r.

10.2. CASE OF CONSTANT SUBSAMPLING FRACTION

For making comparisons by similar methods in more general cases, the intraclass correlation coefficient is not convenient; some difficulties with its generalization have already been indicated in section 4.7 of Chapter V. HANSEN, HURWITZ and MADOW, Sample Survey Methods and Theory (Wiley, 1953) have for this purpose introduced *a measure of homogeneity* of the clusters. This measure is suggested by analogy with the intraclass correlation coefficient, which, as we saw, is also a measure of homogeneity (similarity).

In Chapter V, section 4, we discussed the case of equal M_i and found

$$\sigma^2 = \frac{N\bar{M}-1}{N\bar{M}} S^2 = \frac{N-1}{N} S_l^2 + \frac{\bar{M}-1}{\bar{M}} S_{\mathrm{w}}^2,$$

$$r = \left\{\frac{N-1}{N} S_l^2 - \frac{1}{\bar{M}} S_{\mathrm{w}}^2\right\}\Big/\sigma^2$$

$$= \left\{\frac{N-1}{N}\,\frac{\bar{M}}{\bar{M}-1} S_l^2 - \frac{1}{\bar{M}-1}\sigma^2\right\}\Big/\sigma^2.$$

In the present chapter we obtained, under simple cluster sampling in a single stage, for the variance of the estimator of $\bar{\bar{Y}}$ in the case of the unbiased estimator and the ratio estimator (where $X_{i.} = M_i$), respectively, $(1/n)(N-n)/N$ times

$$\frac{1}{\bar{M}^2}\Sigma(Y_{i.} - \bar{Y})^2/(N-1) = \sum\left(\frac{M_i}{\bar{M}}\bar{Y}_{i.} - \bar{\bar{Y}}\right)^2\Big/(N-1),$$

and (approximately)

$$\frac{1}{\bar{M}^2}\sum\left(Y_{i.} - \frac{M_i}{\bar{M}}\bar{Y}\right)^2\Big/(N-1) = \sum\left(\frac{M_i}{\bar{M}}\right)^2(\bar{Y}_{i.} - \bar{\bar{Y}})^2/(N-1);$$

so that we have two different generalizations of S_l^2 of importance. It will often be convenient to denote the second by S_l^{*2}. If the M_i differ more than a little, we shall (in the case of equal probability sampling) usually use the ratio estimator, so that, in cases it matters, the second generalization is the more important one. This is one reason why, for purposes of defining a measure of homogeneity, we shall use the second generalization of S_l^2. Another reason is that often we also consider ratio estimators using for $\mathscr{X}$ a characteristic different from that used above. In that case we get for our estimator of $\bar{\bar{Y}}$ an approximate variance of $(1/n)(N-n)/N$ times

$$\sum\left(\frac{M_i}{\bar{M}}\right)^2\bar{D}_{i.}^2/(N-1) = \frac{1}{\bar{M}^2}\Sigma D_{i.}^2/(N-1),$$

where

$$D_{i.} = Y_{i.} - \frac{\bar{Y}}{\bar{\bar{X}}} X_{i.}$$

[so that, if $X_{i.} = M_i$, this gives

$$\left(\frac{1}{\bar{M}^2}\right) \Sigma \left(Y_{i.} - \frac{\bar{Y}}{\bar{M}} M_i\right)^2 \Big/ (N-1)].$$

We shall denote this by $S_l^2(\mathscr{D})$; in case $x_{ij} \equiv 1$, we may also write S_l^{*2}.

With simple subsampling we have for the variance of the estimator of $\bar{\bar{Y}}$ the additional term

$$\frac{1}{nN} \Sigma \left(\frac{M_i}{\bar{M}}\right)^2 \left(\frac{1}{m_i} - \frac{1}{M_i}\right) S_{\mathrm{w}}^2(i)$$

when we use the unbiased estimator, and the approximation to the variance in the case of the ratio estimator gives the same additional term;† while for general $\mathscr{X}$ the term is only modified by replacing $S_{\mathrm{w}}^2(i)$ by $S_{\mathrm{w}}^2(\mathscr{D},i)$ [so that, if $x_{ij} \equiv 1$,

$$(M_i - 1)S_{\mathrm{w}}^2(\mathscr{D},i) = \sum_{j=1}^{M_i} \left\{ y_{ij} - \frac{Y}{X} x_{ij} - \left(\bar{Y}_{i.} - \frac{Y}{X} \bar{X}_{i.}\right)\right\}^2$$

equals

$$\Sigma\{y_{ij} - \bar{\bar{Y}} - (\bar{Y}_{i.} - \bar{\bar{Y}})\}^2 = \Sigma(y_{ij} - \bar{Y}_{i.})^2 = (M_i - 1)S_{\mathrm{w}}^2(i)].$$

For our measure of homogeneity we shall consider the important case of an (as closely as possible) *constant subsampling fraction* f_2 and define $\bar{m} = f_2 \bar{M}$ (see Remark in section 5.1); then the additional term is

$$\frac{1}{n\bar{m}} \frac{\bar{M} - \bar{m}}{\bar{M}} \bar{S}_{\mathrm{w}}^2,$$

where

$$\bar{S}_{\mathrm{w}}^2 = \Sigma \frac{M_i}{\bar{M}} S_{\mathrm{w}}^2(i)/N,$$

the weighted average of the $S_{\mathrm{w}}^2(i)$. (For general ratio estimators we replace $\bar{S}_{\mathrm{w}}^2$ by $\bar{S}_{\mathrm{w}}^2(\mathscr{D})$.)

We then define the measure of homogeneity by

$$r^* = \left\{\frac{N-1}{N} S_l^{*2} - \frac{1}{\bar{M}} \bar{S}_{\mathrm{w}}^2\right\} \Big/ \sigma^{*2},$$

† The number $X_{i.} = M_i$ is known for each cluster included in the sample, in contrast to $Y_{i.}$ which is estimated by subsampling; therefore $\mathscr{X}$ does not contribute to $S_{\mathrm{w}}^2(i)$.

with

$$\sigma^{*2} = \frac{N-1}{N} S_i^{*2} + \frac{\bar{M}-1}{\bar{M}} \bar{S}_w^2;$$

so that also

$$r^* = \left\{\frac{N-1}{N} \frac{\bar{M}}{\bar{M}-1} S_i^{*2} - \frac{1}{\bar{M}-1} \sigma^{*2}\right\} \Big/ \sigma^{*2}$$

$$= 1 - \frac{\bar{S}_w^2}{\sigma^{*2}},$$

and

$$\frac{1}{n\bar{m}} \{1 + r^*(\bar{m}-1)\}\sigma^{*2} = \frac{N-1}{N} \frac{1}{n} S_i^{*2} + \frac{\bar{M}-\bar{m}}{\bar{M}} \frac{1}{n\bar{m}} \bar{S}_w^2,$$

which is $((n-1)/Nn)S_i^{*2}$ more than the approximate variance of the ratio estimator of $\bar{\bar{Y}}$. From the above equations it is evident that, like r (compare section 4 of Chapter V), r^* equals its minimum $-(1/(\bar{M}-1))$ if $S_i^{*2} = 0$, and its maximum 1 if $\bar{S}_w^2 = 0$; while with random clustering r^* is usually just below 0.† These properties may serve to justify the name measure of homogeneity.

σ^{*2} need not coincide with σ^2, defining the latter as $(N\bar{M})^{-1}\Sigma\Sigma(y_{ij} - \bar{\bar{Y}})^2$; nor r^* with the intraclass correlation coefficient r. We find

$$\sigma^{*2} - \sigma^2 = \sum \left(\frac{M_i}{\bar{M}} - 1\right) \frac{M_i}{\bar{M}} (\bar{Y}_{i.} - \bar{\bar{Y}})^2/N - \frac{1}{\bar{M}} \sum \left(\frac{M_i}{\bar{M}} - 1\right) S_w^2(i)/N$$

$$= \text{Cov}\left\{\left(\frac{\Delta\mathscr{M}}{\bar{M}}\right)^2, (\Delta\mathscr{Y})^2\right\} + \sigma^2\left\{\frac{\mathscr{M}}{\bar{M}}\right\}\sigma^2\{\mathscr{Y}\} + \text{Cov}\left\{\frac{\mathscr{M}}{\bar{M}}, (\Delta\mathscr{Y})^2 - \frac{\mathscr{W}}{\bar{M}}\right\},$$

where $\mathscr{M}$, $\mathscr{Y}$ and $\mathscr{W}$ are characteristics of the primary units, $\Delta\mathscr{M}$ and $\Delta\mathscr{Y}$ deviations from their averages $\bar{M}$ and $\bar{\bar{Y}}$, and $w(u_i) = S_w^2(i)$. Therefore, in cases in which this difference is numerically large, it is likely to be positive, as usually the $S_w^2(i)$ would be small compared with the expressions $\bar{M}(\bar{Y}_{i.} - \bar{\bar{Y}})^2$, and the correlation of $\mathscr{M}$ or $(\Delta\mathscr{M})^2$ with $(\Delta\mathscr{Y})^2$ would be positive rather than negative. In practice σ^{*2} generally exceeds σ^2 if the M_i vary much while the totals $Y_{i.}$ are far from proportional to them; but otherwise σ^{*2} and σ^2 differ little.

Similar quantities may, of course, be obtained for ratio estimators, with

$$d_{ij} = y_{ij} - (Y/X)x_{ij}$$

† In Chapter V, section 4 we obtained in this case $r = -1/(N\bar{M} - 1)$; this continues to hold approximately with any set of given $M_1, \ldots, M_N$ which are not all equal, if their coefficient of variation is small compared to that of $\mathscr{Y}$.

in place of y_{ij}. If the $Y_{i.}$ are roughly proportional to the $X_{i.}$, $\sigma^{*2}(\mathscr{D})$ will be close to $\sigma^2(\mathscr{D})$. This happens often in practice; for example, the $Y_{i.}$ and the $X_{i.}$ may be roughly proportional to the M_i.

We saw that, when the M_i vary much, we may resort to stratification by size or sampling proportional to size (or estimated size) to reduce variances. In those cases one may again obtain quantities similar to the above ones and rather similar results. In particular, if these devices are successful in their aim, σ^{*2} will be close to σ^2.

The properties discussed at the end of section 4 of Chapter V, especially in connection with two variance functions, also appear to hold for r^* in empirical studies reported on in the book by HANSEN *et al.*; see their Vol. I, Chapter 6: Table 3 in section 8; and Table 9 and Fig. 6 in section 27.

It follows from the above that all the results of section 1 are applicable to the case of constant subsampling fractions, provided asterisks are attached to some of the symbols. This includes the discussion of stratified versus unstratified sampling when the stratification is proportional.

Finally we compare, without using measures of homogeneity, the variance of the self-weighted estimator of the per unit average under different circumstances:

(i) The sample contains elements from all the L strata with proportional allocation, and the total sample size is a.

(ii) The sample contains elements from only l of the L subpopulations, sampled with equal probabilities without replacement, a constant subsampling fraction $\bar{m}/M$ is applied, and a ratio estimator is computed.

(iii) Only l of the L subpopulations are included in the sample with probabilities proportional to size, and a constant overall expected sampling fraction is applied, giving $\bar{m}$ observations from each subpopulation included.

In the first case the variance is

$$\left(\frac{1}{a} - \frac{1}{L\bar{M}}\right)\bar{S}_{\mathrm{w}}^2,$$

in the second case about

$$\left(\frac{1}{l} - \frac{1}{L}\right)S_l^{*2} + \left(\frac{1}{l\bar{m}} - \frac{1}{l\bar{M}}\right)\bar{S}_{\mathrm{w}}^2,$$

and in the third case

$$\frac{1}{l}\sum\frac{M_i}{\bar{M}}(\bar{Y}_{i.} - \bar{\bar{Y}})^2/N + \frac{1}{l\bar{m}}\bar{S}_{\mathrm{w}}^2 - \frac{1}{l\bar{M}}\sum S_{\mathrm{w}}^2(i)/N.$$

Clearly, if $a = l\bar{m}$, the first variance will nearly always be smaller than the other two when $(1/l)S_l^{*2}$, or, respectively, the usually smaller

$$\frac{1}{l}\sum \frac{M_i}{\bar{M}}(\bar{Y}_{i.} - \bar{\bar{Y}})^2/N,$$

is not relatively negligible.

10.3. GAINS DUE TO STRATIFICATION COMPARED FOR TWO-STAGE AND SINGLE STAGE SAMPLING

We only consider the simplest case and first assume that each stratum contains the same number, $\bar{N}$, of clusters. We shall estimate

$$\bar{\bar{Y}} = \frac{1}{L\bar{N}}\sum_{h=1}^{L}\sum_{i=1}^{\bar{N}} Y_{hi.}/\bar{M} = \frac{1}{L\bar{N}}\Sigma\Sigma\bar{Y}_{hi.},$$

where

$$\bar{Y}_{hi.} = Y_{hi.}/\bar{M} = \sum_{j=1}^{\bar{M}} y_{hij}/\bar{M}$$

has average $\bar{\bar{Y}}_h$.

If we take a stratified sample of clusters, the variance of the average

$$(1/L\bar{n})\sum_{h=1}^{L}\sum_{i=1}^{\bar{N}} a_{hi}\bar{Y}_{hi.}$$

per small unit is

$$\left(1 - \frac{\bar{n}}{\bar{N}}\right)\frac{1}{L\bar{n}}\Sigma S_h^2/L$$

with

$$S_h^2 = \frac{1}{\bar{N}-1}\sum_{i=1}^{\bar{N}}(\bar{Y}_{hi.} - \bar{\bar{Y}}_h)^2$$

$$= \frac{\bar{N}}{\bar{N}-1}\{\sum_{i=1}^{\bar{N}}(\bar{Y}_{hi.} - \bar{\bar{Y}})^2/\bar{N} - (\bar{\bar{Y}}_h - \bar{\bar{Y}})^2\};$$

so the variance is

$$\left(1 - \frac{\bar{n}}{\bar{N}}\right)\frac{1}{L\bar{n}}\frac{\bar{N}}{\bar{N}-1}\{\Sigma\Sigma(\bar{Y}_{hi.} - \bar{\bar{Y}})^2/(L\bar{N}) - \Sigma(\bar{\bar{Y}}_h - \bar{\bar{Y}})^2/L\}.$$

For a simple random sample of clusters the variance of the average per small unit is

$$\left(1 - \frac{\bar{n}}{\bar{N}}\right)\frac{1}{L\bar{n}}\frac{L\bar{N}}{L\bar{N}-1}\Sigma\Sigma(\bar{Y}_{hi.} - \bar{\bar{Y}})^2/(L\bar{N}).$$

By section 4.2 of Chapter V

$$\Sigma\Sigma(\bar{Y}_{hi.} - \bar{\bar{Y}})^2/(L\bar{N}) = (\sigma^2/\bar{M})\{1 + (\bar{M} - 1)r\},$$

where r is the intraclass correlation coefficient among the small units in the entire population, and

$$\sigma^2 = \Sigma\Sigma\Sigma(y_{hij} - \bar{\bar{Y}})^2/(L\bar{N}\bar{M}).$$

So the relative variance gain from stratification is

$$\frac{\bar{N}}{\bar{N} - 1}\frac{\bar{M}}{\sigma^2\{1 + (\bar{M} - 1)r\}}$$

$$\times \left\{\left(1 - \frac{1}{L\bar{N}}\right)\Sigma(\bar{\bar{Y}}_h - \bar{\bar{Y}})^2/L - \frac{1}{\bar{N}}\left(1 - \frac{1}{L}\right)\Sigma\Sigma(\bar{Y}_{hi.} - \bar{\bar{Y}})^2/(L\bar{N})\right\}$$

$$= \frac{\bar{M}}{\sigma^2\{1 + (\bar{M} - 1)r\}}\left(1 - \frac{1}{L}\right)\left\{\left(1 + \frac{1}{\bar{N} - 1} + \frac{1}{L - 1}\right)\Sigma(\bar{\bar{Y}}_h - \bar{\bar{Y}})^2/L\right.$$

$$\left. - \frac{1}{\bar{N} - 1}\frac{1}{L}\Sigma\Sigma(\bar{Y}_{hi.} - \bar{\bar{Y}}_h)^2/(L\bar{N})\right\}.$$

From the footnote in 10.1 we obtain that the relative variance gain from stratification, if we sample small units rather than clusters, is

$$\sigma^2\left\{\Sigma(\bar{\bar{Y}}_h - \bar{\bar{Y}})^2/L - \left(1 - \frac{1}{L}\right)\frac{1}{\bar{N}\bar{M}}\Sigma\Sigma\Sigma(y_{hij} - \bar{\bar{Y}}_h)^2/(L\bar{N}\bar{M} - L)\right\}$$

$$= \frac{1}{\sigma^2}\left\{\Sigma(\bar{\bar{Y}}_h - \bar{\bar{Y}})^2/L - \left(1 - \frac{1}{L}\right)\frac{1}{\bar{N}\bar{M} - 1}\Sigma\Sigma(\bar{Y}_{hi.} - \bar{\bar{Y}}_h)^2/(L\bar{N})\right.$$

$$\left. - \left(1 - \frac{1}{L}\right)\frac{1}{\bar{N}\bar{M} - 1}\Sigma\Sigma\Sigma(y_{hij} - \bar{Y}_{hi.})^2/(L\bar{N}\bar{M})\right\}.$$

The ratio of these relative gains is at least

$$\frac{\bar{M}}{1 + r(\bar{M} - 1)}\left(1 - \frac{1}{L}\right)$$

times

$$\left\{1 - \frac{1}{\bar{N} - 1}\frac{\Sigma\Sigma(\bar{Y}_{hi.} - \bar{\bar{Y}}_h)^2/(L\bar{N})}{\Sigma(\bar{\bar{Y}}_h - \bar{\bar{Y}})^2/L}\right\} + \frac{1}{\bar{N} - 1} + \frac{1}{L - 1}.$$

It follows from section 4.2 of Chapter V that the term in curly brackets equals

$$\frac{\bar{N}r_l}{1 + (\bar{N} - 1)r},$$

where r_l is the intraclass correlation of large units in the entire population. So the ratio of the relative gains is at least

$$\frac{\bar{M}}{1+r(\bar{M}-1)}\left(1-\frac{1}{L}\right)\left\{\frac{\bar{N}r_l}{1+(\bar{N}-1)r_l}+\frac{1}{\bar{N}+1}+\frac{1}{L-1}\right\}.$$

For small r and r_l large, the relative variance gain from stratification may therefore be *very substantially larger* for cluster sampling than for sampling of small units.

If we allow the N_h to vary among the different strata and make the n_h proportional to the N_h, the relative variance gain from stratification with sampling of small units is

$$\frac{1}{\sigma^2}\left\{\Sigma\frac{N_h}{\bar{N}}(\bar{\bar{Y}}_h-\bar{\bar{Y}})^2/L-\frac{1}{\bar{N}\bar{M}-1}\Sigma\left(1-\frac{1}{L}\frac{N_h}{\bar{N}}\right)\Sigma(\bar{Y}_{hi.}-\bar{\bar{Y}})^2/(\theta_h L\bar{N})\right.$$
$$\left.-\frac{1}{\bar{N}\bar{M}-1}\Sigma\left(1-\frac{1}{L}\frac{N_h}{\bar{N}}\right)\Sigma\Sigma(y_{hij}-\bar{Y}_{hi.})^2/(\theta_h L\bar{N}\bar{M})\right\},$$

where

$$\theta_h=(\bar{N}\bar{M}-1)/(N_h\bar{M}-1).$$

The relative variance gain from stratification with cluster sampling is given by the same expression as above with $\Sigma(\bar{\bar{Y}}_h-\bar{\bar{Y}})^2/L$ replaced by $\Sigma(N_h/\bar{N})(\bar{\bar{Y}}_h-\bar{\bar{Y}})^2/L$, provided the $N_h/(N_h-1)$ are very close to $\bar{N}/(\bar{N}-1)$. Under this assumption we may conclude exactly as above, with r_l replaced by a number which lies between r_l and r_l^*, since r_l^* is based on the definition of S_l^2 with weights $(N_h/\bar{N})^2$ instead of $(N_h/\bar{N})$.

11. Optimum allocation of the observations in a two-stage cluster sample

11.1. CLUSTERS OF EQUAL SIZE

From section 5.1 (see Remark at the end), if we estimate $\bar{\bar{Y}}$ by $\bar{\bar{y}}$, we obtain

$$V(\bar{\bar{y}})=\frac{1}{n}\left(S_l^2-\frac{S_w^2}{\bar{M}}\right)+\frac{1}{n\bar{m}}S_w^2-\frac{1}{N}S_l^2=V'-\frac{1}{N}S_l^2,$$

where the last term is independent of n and $\bar{m}$, and

$$S_w^2=\Sigma S_w^2(i)/N.$$

Suppose the cost is†

$$C = \{c_1 + c_2\bar{m}\}n.$$

We wish to minimize V' subject to fixed C or vice versa; therefore, as in Chapter III, section 8, we first need to find the minimum of

$$V'C = (c_1 + c_2\bar{m})\left(S_l^2 - \frac{S_w^2}{\bar{M}} + \frac{S_w^2}{\bar{m}}\right)$$

$$= c_1S_w^2/\bar{m} + c_2\bar{m}\left(S_l^2 - \frac{S_w^2}{\bar{M}}\right) + c_1\left(S_l^2 - \frac{S_w^2}{\bar{M}}\right) + c_2S_w^2.$$

The form of this is $a/\bar{m} + b\bar{m} + c$, with $a > 0$. As a function of a real positive variable $\bar{m}$ this has a minimum at

$$\bar{m} = (a/b)^{\frac{1}{2}} = (c_1/c_2)^{\frac{1}{2}}S_w/(S_l^2 - (S_w^2/\bar{M}))^{\frac{1}{2}} \approx \left(\frac{c_1}{c_2}\frac{1-r}{r}\right)^{\frac{1}{2}}$$

if $b > 0$,‡ and for $b < 0$ is steadily decreasing.

Consider first the case $b > 0$ and $(a/b)^{\frac{1}{2}} \leq \bar{M}$. The solution in integers must satisfy

$$\frac{a}{\bar{m}} + b\bar{m} \leq \frac{a}{\bar{m}-1} + b(\bar{m}-1)$$

and

$$\frac{a}{\bar{m}} + b\bar{m} \leq \frac{a}{\bar{m}+1} + b(\bar{m}+1).$$

These inequalities may also be expressed as

$$\bar{m}(\bar{m}-1) \leq a/b \leq \bar{m}(\bar{m}+1). \qquad (*)$$

Therefore the solution is as follows: Take the integer l which satisfies

$$l \leq (a/b)^{\frac{1}{2}} \leq l + 1.$$

If $a/b \leq l(l+1)$, take $\bar{m} = l$ [so that $a/b \leq \bar{m}(\bar{m}+1)$, the right-hand side of (*)], otherwise take $\bar{m} = l + 1$ [so that $a/b > (\bar{m}-1)\bar{m}$, the left-hand side of (*)].

† c_1 includes cost (if any) of preparing lists of all subunits contained in the selected primary units (see 11.2). However, if any such unit is included in its entirety, listing its subunits may become unnecessary. Therefore use of this cost function may indicate that subsampling is desirable when a one-stage sample is actually more efficient. So, in case of doubt, one ought to compare the results of the calculations based on this cost function (in case these do not indicate single stage sampling) with the optimum achievable by single stage sampling.

‡ At $\bar{m} = (a/b)^{\frac{1}{2}}$, the function equals $2(ab)^{\frac{1}{2}} + c$, and in general differs from this value by $\{(a/\bar{m})^{\frac{1}{2}} - (b\bar{m})^{\frac{1}{2}}\}^2$, which is positive except at $\bar{m} = (a/b)^{\frac{1}{2}}$.

If $b \leq 0$ or $(a/b)^{\frac{1}{2}} \geq \bar{M}$, we take $\bar{m} = \bar{M}$ (no subsampling).

We then find n from the equation fixing the cost or the variance.

The variance of $\bar{y}$ is not usually sensitive to departures of $\bar{m}$ from the optimum, that is, $V(\bar{y})$ is a function which over a large range about $\bar{m}_{\text{opt.}}$ is very flat. For a given choice of $\bar{m}$, Brooks, J. Amer. Statist. Ass., **50** (1955) 398–415 gives a table which shows, for given c_1/c_2, the range of values of $S_{\text{w}}^2/(S_l^2 - (S_{\text{w}}^2/\bar{M}))$ for which the variance of $\bar{y}$ using a given $\bar{m}$ is within 90% of the variance of $\bar{y}$ using the optimum choice of $\bar{m}$. For almost all the cases examined by Brooks these ranges are very wide. Since $S_{\text{w}}^2/(S_l^2 - (S_{\text{w}}^2/\bar{M}))$ is usually only roughly known, this result is very helpful. If cost of travel between the primary sample units is high, the cost function may contain in addition a factor of the form $c'n^{\frac{1}{2}}$. The book of Hansen, Hurwitz and Madow discusses this case and finds that the factor $(c_1/c_2)^{\frac{1}{2}}$ in $(a/b)^{\frac{1}{2}}$ is now replaced by $\{(c_1 + c'/d)/c_2\}^{\frac{1}{2}}$, where d equals

$$\frac{\{1 + 4(C/c')(c_1 + c_2\bar{m}/c')\}^{\frac{1}{2}} - 1}{(c_1 + c_2\bar{m})/c'}$$

if C is fixed, and

$$2\left\{\left(S_l^2 - \frac{S_{\text{w}}^2}{\bar{M}} + \frac{S_{\text{w}}^2}{\bar{m}}\right)\bigg/ V'\right\}^{\frac{1}{2}}$$

if V' is fixed.

These equations may be solved by first using the solution obtained earlier for the case $c' = 0$, to obtain a first aproximation to d, and then finding a second approximation for $\bar{m}$ using the previously given formula, etc. Then n is found from the equation $n = (\frac{1}{2}d)^2$. According to the book quoted, B. Tepping and B. Skalak showed that the procedure converges.

We now discuss the case in which there is *stratification*. Then

$$V(\bar{\bar{y}}_{\text{st}}) = \frac{1}{(\Sigma N_h\bar{M}_h)^2}\,\Sigma(N_h\bar{M}_h)^2\left\{\frac{1}{n_h}\left(S_{lh}^2 - \frac{S_{\text{w}h}^2}{\bar{M}_h}\right) + \frac{1}{n_h\bar{m}_h}\,S_{\text{w}h}^2 - \frac{1}{N_h}\,S_{lh}^2\right\},$$

$$C = \Sigma c_{1h}n_h + \Sigma c_{2h}n_h\bar{m}_h.$$

If we denote the subsampling fractions in stratum h, $\bar{m}_h/\bar{M}_h$, by f_{2h}, and denote the overall sampling fraction in stratum h, $(n_n/N_h)f_{2h}$, by f_{0h}, we can write the variance as

$$V(\bar{\bar{y}}_{\text{st}}) = \frac{1}{(\Sigma N_h\bar{M}_h)^2}$$
$$\times\left\{\Sigma\frac{(N_h\bar{M}_h)^2}{n_h}\left(S_{lh}^2 - \frac{S_{\text{w}h}^2}{\bar{M}_h}\right) + \Sigma\frac{N_h\bar{M}_h}{f_{0h}}\,S_{\text{w}h}^2 - \Sigma N_h\bar{M}_h^2S_{lh}^2\right\},$$

and C as

$$C = \Sigma c_{1h} n_h + \Sigma c_{2h} N_h \bar{M}_h f_{0h}.$$

If we apply the methods of Chapter III, section 8, to the terms containing f_{0h}, we obtain the condition that the f_{0h} should be proportional to $S_{\text{w}h}/c_{2h}^{\frac{1}{2}}$, and similarly that the n_h should be proportional to

$$N_h \bar{M}_h (S_{lh}^2 - (S_{\text{w}h}^2/\bar{M}_h))^{\frac{1}{2}}/c_{1h}^{\frac{1}{2}},$$

where $S_{lh}^2 > S_{\text{w}h}^2/\bar{M}_h$. These results, together with knowledge of the budget allowance or the prescribed variance of $\bar{\bar{y}}$, may be used as starting points for fixing the $\bar{m}_h$ and n_h. (Since the variance of $\bar{\bar{y}}$ is not usually sensitive to departures from the optimum of the f_{0h}, an overall sampling fraction f_0, the same for all strata, often will give good results.)

11.2. CLUSTERS OF UNEQUAL SIZE†

In this case the cost function has to take into account separately‡ the (variable) cost of preparing lists of all the

$$\sum_{k=1}^{n} M_{j_k}$$

subunits contained in the selected primary units. These costs are a multiple (say c_l) of the number of such subunits. We shall denote the additional cost expended on each primary unit by c_{a}. So our cost function is

$$c_{\text{a}} n + c_l \sum_{k=1}^{n} M_{j_k} + c_2 \sum_{k=1}^{n} m_{j_k}.$$

Since the cost is a random variable, we shall here consider optimizing in terms of expected cost. If, however, the variance of the cost is not small, this approach is not satisfactory. (One may, instead, consider a bound on the cost which may be exceeded with a small probability only.)

11.2.1. *Simple random or stratified sampling of the primary units.* We first examine unstratified simple random sampling of the primary units. Consider the three estimators of $\bar{\bar{Y}}$:

$$\frac{1}{n} \Sigma a_i \frac{M_i}{\bar{M}} \bar{y}_{i.}, \quad \frac{\Sigma a_i M_i \bar{y}_{i.}}{\Sigma a_i M_i}, \quad \bar{\bar{X}} \frac{\Sigma a_i M_i \bar{y}_{i.}}{\Sigma a_i M_i \bar{x}_{i.}} = \bar{\bar{X}} \hat{R},$$

discussed in sections 5 and 6.

† As remarked previously, the difficulties of unequal size clusters can be, at least partly, avoided by stratification of the clusters according to size or by their selection with probabilities proportionate to estimated size. Sometimes it is worthwhile to combine one or more clusters, and/or to subdivide a number of clusters, in order to obtain clusters of more equal sizes.

‡ In the previous section the listing costs, if any, amounted to $c_l \bar{M} n$, and so $c_l \bar{M}$ could be included in c_1.

The variance of the first of these expressions is

$$\frac{N-n}{N}\frac{1}{n}S_l^2+\frac{1}{nN}\Sigma\left(\frac{M_i}{\bar{M}}\right)^2\left(\frac{1}{m_i}-\frac{1}{M_i}\right)S_{\text{w}}^2(i)$$
$$=\frac{1}{n}\left\{S_l^2-\frac{\bar{S}_{\text{w}}^2}{\bar{M}}+\frac{1}{N}\Sigma\left(\frac{M_i}{\bar{M}}\right)^2\frac{S_{\text{w}}^2(i)}{m_i}\right\}-\frac{1}{N}S_l^2,$$

with

$$S_l^2=\Sigma\left(\frac{M_i}{\bar{M}}\bar{Y}_{i.}-\bar{\bar{Y}}\right)^2\Big/(N-1),$$
$$\bar{S}_{\text{w}}^2=\Sigma\frac{M_i}{\bar{M}}S_{\text{w}}^2(i)/N.$$

The approximate variance of the second expression is the same with

$$S_l^2=\Sigma\left(\frac{M_i}{\bar{M}}\right)^2(\bar{Y}_{i.}-\bar{\bar{Y}})^2/(N-1),$$

while the approximate variance of $\bar{\bar{X}}\hat{R}$ has the same formula as that for $\Sigma a_i M_i\bar{y}_{i.}/\Sigma a_i M_i$, except that $\mathscr{Y}$ is replaced by $\mathscr{D}$ [e.g., S_l^2 becomes

$$\Sigma\left(\frac{M_i}{\bar{M}}\right)^2\bar{D}_{i.}^2/(N-1),$$

where $\bar{D}_{i.}=\bar{Y}_{i.}-R\bar{X}_{i.}$].

The expected value of the cost is

$$c_{\text{a}}n+c_l n\bar{M}+c_2 n\sum_{i=1}^{N}m_i/N=n\{c_1+c_2\sum_{i=1}^{N}m_i/N\},$$

where $c_1=c_{\text{a}}+c_l\bar{M}$.

If the problem were a continuous one, the optimum values of m_i would be those that minimize

$$\left\{S_l^2-\frac{\bar{S}_{\text{w}}^2}{\bar{M}}+\frac{1}{N}\Sigma\left(\frac{M_i}{\bar{M}}\right)^2\frac{S_{\text{w}}^2(i)}{m_i}\right\}\{c_1+c_2\Sigma m_i/N\}.$$

Now

$$\frac{c_2}{(N\bar{M})^2}(\Sigma m_i)\{\Sigma M_i^2S_{\text{w}}^2(i)/m_i\}$$
$$=\frac{c_2}{(N\bar{M})^2}\left[\sum_{i>i'}\sum\left\{\frac{m_i}{m_{i'}}M_{i'}^2S_{\text{w}}^2(i')+\frac{m_{i'}}{m_i}M_i^2S_{\text{w}}^2(i)\right\}+\Sigma M_i^2S_{\text{w}}^2(i)\right]$$
$$=\frac{c_2}{N\bar{M}^2}\left[\sum_{i>i'}\sum\left\{\left(\frac{m_i}{m_{i'}}\right)^{\frac{1}{2}}M_{i'}S_{\text{w}}(i')-\left(\frac{m_{i'}}{m_i}\right)^{\frac{1}{2}}M_iS_{\text{w}}(i)\right\}^2\right.$$
$$\left.+\sum_{i>i'}\sum M_iM_{i'}S_{\text{w}}(i)\,S_{\text{w}}(i')\right]$$

and the additional terms in the product are

$$c_1\left(S_l^2 - \frac{\bar{S}_w^2}{\bar{M}}\right) + \frac{1}{N}\Sigma\left\{c_2\left(S_l^2 - \frac{\bar{S}_w^2}{\bar{M}}\right)m_i + c_1\left(\frac{M_i}{\bar{M}}\right)^2 S_w^2(i)/m_i\right\}$$

$$= c_1\left(S_l^2 - \frac{\bar{S}_w^2}{\bar{M}}\right) + \frac{1}{N}\Sigma\left\{c_2^{\frac{1}{2}}\left(S_l^2 - \frac{\bar{S}_w^2}{\bar{M}}\right)^{\frac{1}{2}} m_i^{\frac{1}{2}} - c_1^{\frac{1}{2}}\frac{M_i}{\bar{M}} S_w(i)/m_i^{\frac{1}{2}}\right\}^2$$

$$+ \frac{2}{N\bar{M}}(c_1c_2)^{\frac{1}{2}}\left(S_l^2 - \frac{\bar{S}_w^2}{\bar{M}}\right)^{\frac{1}{2}} \Sigma M_i S_w(i).$$

Both terms that depend on the m_i vanish if and only if

$$m_i = \left(\frac{c_1}{c_2}\right)^{\frac{1}{2}} \frac{M_i}{\bar{M}} S_w(i) \Big/ \left(S_l^2 - \frac{\bar{S}_w^2}{\bar{M}}\right)^{\frac{1}{2}}.$$

(The solution is real only if $S_l^2 > \bar{S}_w^2/\bar{M}$, but even then may exceed M_i.)

Our knowledge of the $S_w^2(i)$ is rarely sufficient to choose m_i as approximations to the above m_i. Then one of the following two cases may apply:†

either the $S_w(i)$ are expected not to vary much;
or it is possible to group the potential clusters into strata according to the different orders of magnitude of the $S_w(i)$.

In the first of the two cases we use a subsampling fraction‡ f_2 which is the same for all clusters; in the second case, a subsampling fraction f_{2h} which is the same for all clusters in stratum h. In the first case the expected value of the cost becomes

$$n\{c_1 + c_2 f_2 \bar{M}\},$$

and the variance

$$\frac{1}{n}\left\{S_l^2 - \frac{\bar{S}_w^2}{\bar{M}} + \frac{1}{\bar{M}f_2}\bar{S}_w^2\right\} - \frac{1}{N}S_l^2;$$

the best choice of f_2 minimizes the product of the bracketed terms. This is again a problem of the form: minimize $a/f_2 + bf_2 + c$; for $S_l^2 - (\bar{S}_w^2/\bar{M}) > 0$; the solution is

$$f_2 = \left(\frac{c_1}{c_2}\right)^{\frac{1}{2}} \frac{1}{\bar{M}} \bar{S}_w \Big/ \left(S_l^2 - \frac{\bar{S}_w^2}{\bar{M}}\right)^{\frac{1}{2}},$$

provided this does not exceed 1. (In other cases the optimum f_2 is 1.) For the second case we can obtain the same expression within each stratum.

† If also these do not apply, theory cannot guide us in the selection of the subsampling numbers.

‡ Some authors write $\bar{m}$ for $f_2\bar{M}$; there is a danger that this will be confused with the random variable which is the (average) number of observations chosen in the selected sequence of clusters.

With stratification we may investigate *best allocation over the strata.* Consider the estimators obtained by substituting in $\Sigma N_h\bar{M}_h\hat{\bar{\bar{Y}}}_h/\Sigma N_h\bar{M}_h$ for $\hat{\bar{\bar{Y}}}_h$ any of the three above estimators of $\bar{\bar{Y}}$ applied to stratum h; the variance formula is

$$\frac{1}{(\Sigma N_h\bar{M}_h)^2}\left\{\Sigma\frac{(N_h\bar{M}_h)^2}{n_h}S_{lh}^2+\Sigma\frac{N_h}{n_h}\Sigma\frac{M_{hi}^2}{m_{hi}}S_{\mathrm{w}h}^2(i)\right.$$
$$\left.-\Sigma\frac{N_h}{n_h}\Sigma M_{hi}S_{\mathrm{w}h}^2(i)-\Sigma N_h\bar{M}_h^2S_{lh}^2\right\}.$$

We shall assume constant overall sampling fractions

$$f_{0h}=(n_h/N_h)f_{2h}$$

in stratum h for each h. Then the variance formula becomes

$$\frac{1}{(\Sigma N_h\bar{M}_h)^2}\left\{\Sigma\frac{(N_h\bar{M}_h)^2}{n_h}\left(S_{lh}^2-\frac{\bar{S}_{\mathrm{w}h}^2}{\bar{M}_h}\right)+\Sigma\frac{N_h\bar{M}_h}{f_{0h}}\bar{S}_{\mathrm{w}h}^2-\Sigma N_h\bar{M}_h^2S_{lh}^2\right\},$$

where

$$\bar{S}_{\mathrm{w}h}^2=\frac{1}{N_h}\sum_{i=1}^{N_h}\frac{M_{hi}}{\bar{M}_h}S_{\mathrm{w}h}^2(i);$$

and the expected cost formula

$$\Sigma c_{1h}n_h+\Sigma c_{2h}f_{0h}N_h\bar{M}_h.$$

If we apply the methods of Chapter III, section 8, to the terms in which the f_{0h} appear, we obtain the condition that the f_{0h} should be proportional to $\bar{S}_{\mathrm{w}h}/c_{2h}^{\frac{1}{2}}$.

Similarly we find the condition that the n_h should be proportional to

$$N_h\bar{M}_h(S_{lh}^2-(\bar{S}_{\mathrm{w}h}^2/\bar{M}_h))^{\frac{1}{2}}/c_{1h}^{\frac{1}{2}}$$

if the bracket is positive. These results may be taken as points of departure for fixing the n_h and f_{0h}. They are immediate generalizations of the case of constant size clusters.

11.2.2. *Unequal probability unstratified and stratified sampling of the primary units.* We first examine sampling of the primary units with unequal probabilities (with replacement) without stratification. Consider the estimators of $\bar{\bar{Y}}$:

$$\frac{1}{n}\sum_{k=1}^{n}\frac{M_{j_k}}{N\bar{M}}\bar{y}_{j_k\cdot}/p_{j_k},\quad\frac{\sum_{k=1}^{n}M_{j_k}\bar{y}_{j_k\cdot}/p_{j_k}}{\sum_{k=1}^{n}M_{j_k}/p_{j_k}},\quad\bar{\bar{X}}\frac{\sum_{k=1}^{n}M_{j_k}\bar{y}_{j_k\cdot}/p_{j_k}}{\sum_{k=1}^{n}M_{j_k}\bar{x}_{j_k\cdot}/p_{j_k}}=\bar{\bar{X}}\hat{R},$$

discussed in section 7.

The variance of the first of these expressions is

$$\frac{1}{n} S_l^2 + \frac{1}{n} \sum \left(\frac{M_i}{N\bar{M}}\right)^2 \frac{1}{p_i} \left(\frac{1}{m_i} - \frac{1}{M_i}\right) S_w^2(i)$$

$$= \frac{1}{n} \left\{ S_l^2 - \frac{\bar{S}_w^2}{\bar{M}} + \frac{1}{N} \sum \left(\frac{M_i}{\bar{M}}\right)^2 \frac{1}{Np_i} \frac{S_w^2(i)}{m_i} \right\},$$

where

$$S_l^2 = \Sigma p_i \left(\frac{M_i}{N\bar{M}} \frac{\bar{Y}_{i.}}{p_i} - \bar{\bar{Y}} \right)^2 = \sum \left(\frac{M_i}{N\bar{M}} \right)^2 \frac{\bar{Y}_{i.}^2}{p_i} - \bar{\bar{Y}}^2,$$

$$\bar{S}_w^2 = \sum \frac{1}{p_i N} \frac{M_i}{\bar{M}} S_w^2(i)/N.$$

The approximate variance of the second expression is the same, except that

$$S_l^2 = \sum \frac{1}{p_i} \left(\frac{M_i}{N\bar{M}} \right)^2 (\bar{Y}_{i.} - \bar{\bar{Y}})^2.$$

The approximate variance of $\bar{\bar{X}}\hat{R}$ is the same as that for the second expression, except that $\mathscr{Y}$ is replaced by $\mathscr{D}$ (e.g., S_l^2 becomes

$$\Sigma(1/p_i)(M_i/N\bar{M})^2 \bar{D}_{i.}^2,$$

where $\bar{D}_{i.} = \bar{Y}_{i.} - R\bar{X}_{i.}$). The expected value of the cost is†

$$n\{c_1' + c_2 \sum_{i=1}^{N} p_i m_i\},$$

where

$$c_1' = c_a' + c_l \sum_{i=1}^{N} p_i M_i.$$

Here

$$c_a' = c_a \frac{1}{n} \sum_{i=1}^{N} \{1 - (1 - p_i)^n\},$$

where c_a/n is multiplied by the probability that u_i is included at least once in the sample. For simplicity we shall assume that the dependence of c_a' on the p_i may be neglected; since

$$\frac{1}{n} \sum_{i=1}^{N} \{1 - (1 - p_i)^n\} = 1 - \frac{n-1}{2} \sum_{i=1}^{N} p_i^2 + \frac{(n-1)(n-2)}{6} \sum_{i=1}^{N} p_i^3 - \ldots,$$

we may do so (and then $c_a' \approx c_a$) if no p_i is much larger than the others.

† The cost of obtaining the p_i is not included in this. If we have to make a preliminary survey (of *all* the large units in the population!) to establish their values, the additional cost may well be prohibitive, unless it can be prorated over several surveys.

The product of expected cost and variance is

$$\left\{S_l^2 - \frac{\bar{S}_w^2}{\bar{M}} + \sum\left(\frac{M_i}{N\bar{M}}\right)^2 \frac{S_w^2(i)}{p_i m_i}\right\}\{c_1' + c_2\Sigma p_i m_i\}.$$

Now

$$\frac{c_2}{(N\bar{M})^2}(\Sigma p_i m_i)\{\Sigma M_i^2 S_w^2(i)/(p_i m_i)\}$$

$$= \frac{c_2}{(N\bar{M})^2}\sum_{i>i'}\sum\left\{\frac{p_i m_i}{p_{i'} m_{i'}} M_{i'}^2 S_w^2(i') + \frac{p_{i'} m_{i'}}{p_i m_i} M_i^2 S_w^2(i)\right\} + \frac{c_2}{(N\bar{M})^2}\Sigma M_i^2 S_w^2(i)$$

$$= \frac{c_2}{(N\bar{M})^2}\sum_{i>i'}\sum\left\{\left(\frac{p_i m_i}{p_{i'} m_{i'}}\right)^{\frac{1}{2}} M_{i'} S_w(i') - \left(\frac{p_{i'} m_{i'}}{p_i m_i}\right)^{\frac{1}{2}} M_i S_w(i)\right\}^2$$

$$+ \frac{c_2}{(N\bar{M})^2}\sum_i\sum_{i'} M_i M_{i'} S_w(i) S_w(i')$$

and the additional terms in the product are

$$c_1'\left(S_l^2 - \frac{\bar{S}_w^2}{\bar{M}}\right) + \sum\left\{c_2^{\frac{1}{2}}\left(S_l^2 - \frac{\bar{S}_w^2}{\bar{M}}\right)^{\frac{1}{2}}(p_i m_i)^{\frac{1}{2}} - (c_1')^{\frac{1}{2}}\left(\frac{M_i}{N\bar{M}}\right) S_w(i)/(p_i m_i)^{\frac{1}{2}}\right\}$$

$$+ 2(c_2 c_1')^{\frac{1}{2}}\left(S_l^2 - \frac{\bar{S}_w^2}{\bar{M}}\right)^{\frac{1}{2}}\sum\frac{M_i}{N\bar{M}} S_w(i).$$

The terms that depend on m_i vanish if and only if

$$m_i = \left(\frac{c_1'}{c_2}\right)^{\frac{1}{2}} \frac{M_i}{\bar{M}} \frac{S_w(i)}{Np_i} \bigg/ \left(S_l^2 - \frac{\bar{S}_w^2}{\bar{M}}\right)^{\frac{1}{2}}.$$

(The solution is real only if $S_l^2 > \bar{S}_w^2/\bar{M}$, and even then may exceed M_i.)

Because of lack of knowledge of the $S_w(i)$, we usually do one of two things: either we assign a single expected overall sampling fraction f_0 to the whole population; or we group the potential clusters into strata according to different orders of magnitudes of the $S_w(i)$, and within each stratum to use a constant expected overall sampling fraction f_{0h}.

Let us discuss the case in which we do not stratify and have an expected overall sampling fraction

$$f_0 = np_i m_i/M_i.$$

The variances now become respectively,

$$\frac{1}{n}\sum\left(\frac{M_i}{\bar{M}}\right)^2 \frac{1}{Np_i}\left(\bar{Y}_{i\cdot}^2 - \frac{S_w^2(i)}{M_i}\right)\bigg/N - \frac{1}{n}\bar{\bar{Y}}^2 + \frac{1}{N\bar{M}}\frac{1}{f_0}\sum\frac{M_i}{\bar{M}} S_w^2(i)/N,$$

$$\frac{1}{n}\sum\left(\frac{M_i}{\bar{M}}\right)^2\frac{1}{Np_i}\left\{(\bar{Y}_{i.}-\bar{Y})^2-\frac{S_{\text{w}}^2(i)}{M_i}\right\}\Big/N+\frac{1}{N\bar{M}}\frac{1}{f_0}\sum\frac{M_i}{\bar{M}}S_{\text{w}}^2(i)/N,$$

$$\frac{1}{n}\sum\left(\frac{M_i}{\bar{M}}\right)^2\frac{1}{Np_i}\left\{\bar{D}_{i.}^2-\frac{S_{\text{w}}^2(\mathscr{D},i)}{M_i}\right\}\Big/N+\frac{1}{N\bar{M}}\frac{1}{f_0}\sum\frac{M_i}{\bar{M}}S_{\text{w}}^2(\mathscr{D},i)/N,$$

and the expected cost

$$n\{c_{\text{a}}'+c_l\Sigma p_iM_i\}+c_2f_0N\bar{M}.$$

We shall look for optimum values of the p_i, f_0 and n, subject to

$$\sum_{i=1}^{N}p_i=1,$$

assuming that $\delta_i^2>0$, where δ_i^2 equals

$$\bar{Y}_{i.}^2-S_{\text{w}}^2(i)/M_i,\quad(\bar{Y}_{i.}-\bar{Y})^2-S_{\text{w}}^2(i)/M_i\quad\text{or}\quad\bar{D}_{i.}^2-S_{\text{w}}^2(\mathscr{D},i)/M_i,$$

respectively. If, for some i, $\delta_i^2<0$, it is generally inadvisable to use two stages, since this is an indication of negative intraclass correlation.

The optimum p_i will be determined only for the two ratio-type estimators. By the usual methods† we find that the p_i are optimal if they are proportional to the $M_i\delta_i/(c_{\text{a}}'+c_lM_i)^{\frac{1}{2}}$. Usually it will not be possible to know these quantities in advance. However, according to Chapter V, δ_i^2 may be expected to have a mean of about rS^2; and we saw there that r will usually be a decreasing function of $\bar{M}$, but decrease so slowly that $(\bar{M}-1)r$ will increase with $\bar{M}$. Roughly speaking, it may therefore be expected that the optimum p_i is proportional to M_i raised to a power between $\frac{1}{2}$ and 1 if c_{a} is much larger than c_lM_i and between 0 and $\frac{1}{2}$ if c_{a} is much smaller than c_lM_i. In the intermediate case we might take p_i proportional to $M_i^{\frac{3}{4}}$.

To obtain the optimum f_0/n for given p_i, we note that the product of the variance and the expected cost is of the form $a/(f_0/n)+b(f_0/n)+c$ with‡

$$a=\{c_{\text{a}}'+c_l\sum_{i=1}^{N}p_iM_i\}\sum_{i=1}^{N}M_iS_{\text{w}}^2(i)/(N\bar{M})^2,$$

$$b=c_2\left(\sum_{i=1}^{N}\frac{M_i^2}{p_i}\delta_i^2-\theta\right)\Big/(N\bar{M}),$$

† The minimum of $\alpha_0+\Sigma(\alpha_i/\pi_i)$ with respect to the $\pi_i=np_i$ with $\beta_0+\Sigma\beta_i\pi_i$ fixed, or vice versa, is attained for positive α_i and β_i, if π_i is proportional to $(\alpha_i/\beta_i)^{\frac{1}{2}}$. For fixed f_0, if we substitute π_i for np_i and $\Sigma\pi_i$ for n, this is our problem with

$$\alpha_i=\{(N\bar{M})^{-1}M_i\delta_i\}^2,\quad\beta_i=c_{\text{a}}'+c_lM_i$$

(both are independent of f_0!).

‡ In the expression for a, replace $S_{\text{w}}(i)$ by $S_{\text{w}}(\mathscr{D},i)$ for the third estimator. These replacements also apply to the expressions in the next paragraph.

where $\theta = 0$ except in the case of the first (nonratio) estimator, where it is Y^2; and the optimum f_0/n is $(a/b)^{\frac{1}{2}}$. Together with a bound for the variance or for the expected cost, this determines f_0 (if this f_0 turns out to exceed 1, take it to be equal to 1) and n.

If we solve, simultaneously for the p_i and f_0/n, the continuous optimization problem for the two ratio-type estimators, we find for f_0/n the ratio of

$$\{\Sigma M_i S_{\mathrm{w}}^2(i)\}^{\frac{1}{2}}/(N\bar{M}) \quad \text{and} \quad c_2^{\frac{1}{2}}\Sigma\{M_i\delta_i/(c_{\mathrm{a}}' + c_l M_i)^{\frac{1}{2}}\}/(N\bar{M})^{\frac{1}{2}},$$

and for $m_i = f_0 M_i/(np_i)$:

$$\left\{\frac{c_{\mathrm{a}}' + c_l M_i}{c_2}\right\}^{\frac{1}{2}} \left\{\frac{1}{N}\Sigma \frac{M_i}{M} S_{\mathrm{w}}^2(i)\right\}^{\frac{1}{2}} \Big/ \delta_i,$$

which is comparable to what was found in 11.2.1.

With stratification we may investigate the best *allocation over the strata*. We shall consider the estimator $\Sigma N_h \bar{M}_h \hat{\bar{Y}}_{\mathrm{ppes},h}/\Sigma N_h \bar{M}_h$. We shall use a constant expected overall sampling fraction f_{0h} within stratum h;

$$f_{0h} = (m_{hi}/M_{hi}) n_h p_{hi}.$$

The variance is then

$$\frac{1}{(\Sigma N_h \bar{M}_h)^2} \sum_{h=1}^{L} \frac{1}{n_h} \left\{\sum_{i=1}^{N_h} \frac{Y_{hi.}^2}{p_{hi}} - Y_h^2\right\} + \frac{1}{(\Sigma N_h \bar{M}_h)^2} \sum_{h=1}^{L} \frac{1}{f_{0h}} \sum_{i=1}^{N_h} M_{hi} S_{\mathrm{w}h}^2(i)$$
$$- \frac{1}{(\Sigma N_h \bar{M}_h)^2} \sum_{h=1}^{L} \frac{1}{n_h} \sum_{i=1}^{N_h} \frac{M_{hi}}{p_{hi}} S_{\mathrm{w}h}^2(i).$$

The expected cost is

$$\Sigma n_h(c_{\mathrm{a}h}' + c_{lh}\Sigma p_{hi} M_{hi}) + \Sigma c_{2h} f_{0h} N_h \bar{M}_h = \Sigma c_{1h} n_h + \Sigma c_{2h} f_{0h} N_h \bar{M}_h.$$

Since the terms depending on f_{0h} are exactly the same as in 11.2.1, we obtain the same proportionality condition as there. The same result holds for ratio type estimators (with extensive stratification they may involve too much risk of bias). The determination of the n_h for fixed p_{hi} is along similar lines as in 11.2.1.

12. Double sampling for assignment to clusters

Sometimes a frame of sampling units is available, but not arranged according to a classification into those N parts which we would – for cost or administrative reasons – like to use as primary sampling units. An example is given in DALENIUS, Sampling in Sweden (Almqvist and Wiksell, 1957) section 6.7. There it was desired, in a survey of motor cars, to use as primary sampling units administrative areas, but it was not feasible to arrange the register of cars by these areas.

A method of handling this situation is the following. Let the population contain M elements. First select a simple random sample $\mathscr{s}'$ of m' elements from the frame, and classify them according to the primary units to which they belong. Suppose the ith primary unit in the population contains M_i units, then

$$M = \sum_{i=1}^{N} M_i.$$

We let a_{ij} be 1 if the jth element in the ith primary unit belongs to $\mathscr{s}'$, and 0 otherwise. Then $\mathscr{s}'$ contains

$$m'_i = \sum_{j=1}^{M_i} a_{ij}$$

elements from the ith primary unit with

$$\sum_{i=1}^{N} m'_i = m'.$$

Some, say N^*, of these may be different from 0.

The next step is to make drawings out of the list of primary units that contain some elements in $\mathscr{s}'$ with certain defined probabilities, and then obtain the value of the characteristic $\mathscr{Y}$ for all members of the selected primary units, or subsample the selected primary units.

REMARK. If m' is fairly large, the probability that some m'_i are zero becomes small, and we can – possibly by a slight increase in m' – enforce the equality of N^* and N (and neglect the effect of this modification on the probabilities to be computed). If the primary units differ widely in size, this may not, however, be possible for the sample size originally contemplated. In the following we shall, therefore, have to take the possibility of N^* being smaller than N into account. Note also that, if N^* is much smaller than N, it may not be possible to draw a *preassigned* number, n, of primary units out of these N^* primary units without replacement; but we can always make n drawings with replacement.

We shall here describe only some possible plans of sampling and estimation. Somewhat more information is contained in a paper by GOSH, Ann. Math. Statist., **34** (1963) 587–597, under the title of *post cluster sampling* (his discussion appears to contain many errors).

Let $b_i = 1$ if the ith primary sampling unit is included in the sample of clusters and 0 otherwise.

(a) Simple sampling of n^* clusters without replacement, no subsampling.

Note that n^* is necessarily random. It is mathematically simpler to consider sampling of a fixed number, n, of primary units out of N (including some not represented in $\mathscr{s}'$) than sampling n^*, out of the N^* represented in $\mathscr{s}'$, also because this makes the b's independent of the a's. For each i, the expected value of the b_i is n/N in the former case, and the conditional expected value is n^*/N^* in the latter case.

A simple unbiased estimator of $\bar{\bar{Y}}$ would be

$$\hat{\bar{\bar{Y}}} = \frac{1}{M}\sum_{i=1}^{N}\frac{b_i}{n/N}\sum_{j=1}^{M_i}\frac{a_{ij}}{m'/M}y_{ij},$$

as against the estimator

$$\frac{1}{M}\sum_{i=1}^{N}\frac{b_i}{n^*/N^*}\sum_{j=1}^{M_i}\frac{a_{ij}}{m'/M}y_{ij},$$

whose variance is harder to obtain. The latter estimator has, for given $\mathscr{s}'$, conditional expected value

$$(1/m')\sum_{i=1}^{N}\sum_{j=1}^{M_i}a_{ij}y_{ij},$$

since

$$\mathscr{E}\{b_i|\mathscr{s}'\} = n^*/N^*,$$

and so is unbiased.

We now derive the variance of $\hat{\bar{\bar{Y}}}$. Note that the M a's are hypergeometric with means m'/M, variances

$$(m'/M)(M-m')/M,$$

and covariances

$$-(m'/M)\{(M-m')/M(M-1)\};$$

and that the N b's are hypergeometric with means n/N, variances

$$(n/N)(N-n)/N,$$

and covariances

$$-(n/N)\{(N-n)/N(N-1)\}.$$

$V(\hat{\bar{\bar{Y}}})$ equals

$$\left(\frac{N}{nm'}\right)^2 \sum_i\sum_j y_{ij}^2 V(b_i a_{ij}) + \sum_i\sum_{j\neq l}\sum y_{ij}y_{il}\operatorname{Cov}(b_i a_{ij}, b_i a_{il})$$
$$+ \sum_{i\neq k}\sum\sum_j\sum_l y_{ij}y_{kl}\operatorname{Cov}(b_i a_{ij}, b_k a_{kl}).$$

Now

$$V(b_ia_{ij}) = \mathscr{E}[V\{b_ia_{ij}|a_{ij}\}] + V[\mathscr{E}\{b_ia_{ij}|a_{ij}\}]$$
$$= \mathscr{E}[a_{ij}^2V\{b_i|a_{ij}\}] + V[a_{ij}\mathscr{E}\{b_i|a_{ij}\}],$$

which, since the b's are independent from the a's, and since $a_{ij}^2 = a_{ij}$, equals

$$\mathscr{E}a_{ij}V(b_i) + Va_{ij}(\mathscr{E}b_i)^2 = \frac{m'}{M}\frac{n}{N}\frac{N-n}{N} + \frac{m'}{M}\frac{M-m'}{M}\left(\frac{n}{N}\right)^2.$$

Similarly,

$$\text{Cov}\,(b_ia_{ij}, b_ka_{kl}) = \mathscr{E}a_{ij}a_{kl}\,\text{Cov}\,(b_i, b_k) + \text{Cov}\,(a_{ij}, a_{kl})\mathscr{E}b_i\mathscr{E}b_k,$$

which, for $k = i$, equals

$$\frac{m'(m'-1)}{M(M-1)}\frac{n}{N}\frac{N-n}{N} - \frac{m'}{M}\frac{M-m'}{M}\frac{1}{M-1}\left(\frac{n}{N}\right)^2$$

and for $k \neq i$ equals

$$-\frac{m'(m'-1)}{M(M-1)}\frac{n}{N}\frac{N-n}{N}\frac{1}{N-1} - \frac{m'}{M}\frac{M-m'}{M}\frac{1}{M-1}\left(\frac{n}{N}\right)^2.$$

Since

$$\sum_i\sum_{j\neq l}\sum y_{ij}y_{il} = \sum_i Y_{i.}^2 - \sum_i\sum_j y_{ij}^2,$$

and

$$\sum_{i\neq k}\sum_j\sum_l\sum y_{ij}y_{kl} = M^2\overline{Y}^2 - \sum_i Y_{i.}^2 = (N-1)(-S_{\text{BT}}^2 + M^2\overline{Y}^2/N),$$

where

$$S_{\text{BT}}^2 = \frac{1}{N-1}\sum_{i=1}^{N}\left(Y_{i.} - \frac{M\overline{Y}}{N}\right)^2$$

is the variance of the totals for the primary units, we have

$$V(\hat{\overline{Y}}) = \frac{1}{m'nM}\Bigg[N(M-m')S^2 + \frac{N-n}{M-1}\{N(m'-1)S_{\text{BT}}^2 + M(M-m')\overline{Y}^2\}\Bigg].$$

We may compare this with obtaining a simple random sample of

$$\mathscr{E}\sum_{i=1}^{N} b_i \sum_{j=1}^{M_i} a_{ij} = m'n/N$$

elements; the amount by which $V(\hat{\bar{Y}})$ exceeds the variance of the mean per element in simple random sampling is

$$\frac{N-n}{nM}\left\{\frac{m'-1}{m'}N\frac{1}{M-1}S^2_{\text{BT}}+\frac{M-m'}{M-1}\frac{1}{m'}M\bar{\bar{Y}}^2-S^2\right\}.$$

We can also compare $V(\hat{\bar{Y}})$ with the variance of the ordinary estimator we would obtain by simple random sampling of n clusters,†

$$\{(N-n)/Nn(N-1)\}\sum_{i=1}^{N}(\bar{Y}_{i.}-\bar{\bar{Y}})^2;$$

for small variations among the M_i, this is approximately equal to

$$\{(N-n)/Nn\bar{M}^2\}S^2_{\text{BT}},$$

where $\bar{M}=M/N$. The coefficient of S^2_{BT} in $V(\hat{\bar{Y}})$ is

$$\frac{N-n}{N}\frac{1}{n}\frac{1}{\bar{M}^2}\frac{m'-1}{m'}\frac{M}{M-1},$$

about the same as the coefficient of S^2_{BT} in the variance of the ordinary estimator. So the increase in variance is about

$$\frac{N}{m'n}\frac{M-m'}{M}\left(S^2+\frac{M}{M-1}\frac{N-n}{N}\bar{\bar{Y}}^2\right).$$

(b) Sampling clusters with probabilities proportionate to their sizes or estimated sizes, no subsampling.

If the M_i vary much, one would often expect to be better off by sampling the clusters with unequal rather than with equal probabilities. We shall discuss here making n drawings with replacement out of the N primary sampling units in the population with probabilities $P_i=M_i/M$ (if these quantities are known) or with probabilities

$$p_i=(1/m')\sum_{j=1}^{M_i}a_{ij}=m'_i/m'$$

(if the M_i/M are unknown).

We shall discuss the estimator

$$\hat{\bar{Y}}_P=\frac{1}{M}\sum_{i=1}^{N}\frac{b_i}{nP_i}\sum_{j=1}^{M_i}\frac{a_{ij}}{m'/M}y_{ij}$$

in the former case, and $\hat{\bar{Y}}_p$ in the latter, where $\hat{\bar{Y}}_p$ is defined like $\hat{\bar{Y}}_P$, but with p_i in place of P_i for each i, and omitting those primary units for which $p_i=0$.

† Actually, a comparison with $\mathscr{E}n^*$ clusters would be more appropriate.

Here $\mathscr{E}b_i = nP_i$ in the first method and $\mathscr{E}\{b_i|\mathscr{s}'\} = np_i$ in the second. Evidently $\hat{\bar{\bar{Y}}}_P$ is unbiased; but also $\hat{\bar{\bar{Y}}}_p$ is unbiased. For

$$\mathscr{E}(a_{ij}) = \mathscr{E}\{a_{ij}|p_i > 0\}\,\mathrm{P}\{p_i > 0\} + \mathscr{E}\{a_{ij}|\sum_{j=1}^{M_i} a_{ij} = 0\}\mathrm{P}(p_i = 0) \qquad (*)$$

$$= \mathscr{E}\{a_{ij}|p_i > 0\}\,\mathrm{P}\{p_i > 0\},$$

so

$$\mathscr{E}\hat{\bar{\bar{Y}}}_p = \frac{1}{m'}\sum_{i=1}^{N}\mathscr{E}\mathscr{E}\{Z_i|I_i\},$$

where

$$Z_i = I_i\frac{b_i}{np_i}\sum_{j=1}^{M_i} a_{ij}y_{ij}$$

and $I_i = 1$ if $p_i > 0$ and 0 otherwise. But

$$\mathscr{E}\{Z_i|I_i = 1\} = \sum_{j=1}^{M_i}\mathscr{E}\{a_{ij}|p_i > 0\}y_{ij}$$

and

$$\mathscr{E}\{Z_i|I_i = 0\} = 0,$$

so

$$\mathscr{E}\mathscr{E}\{Z_i|I_i\} = \sum_{j=1}^{M_i}\mathscr{E}\{a_{ij}|p_i > 0\}y_{ij}\mathrm{P}(I_i = 1) + 0\cdot\mathrm{P}(I_i = 0)$$

$$= \sum_{j=1}^{M_i}\mathscr{E}(a_{ij})y_{ij}$$

using (*), and this is equal to

$$(m'/M)\sum_{j=1}^{M_i} y_{ij};$$

and so $\mathscr{E}\hat{\bar{\bar{Y}}}_p = \bar{\bar{Y}}$.

We now compute

$$V(\hat{\bar{\bar{Y}}}_P) = (m')^2\mathscr{E}A - \bar{\bar{Y}}^2,$$

where

$$A = \mathscr{E}\{(\Sigma A_i)^2|\mathscr{s}'\}, \quad A_i = n^{-1}b_iP_i^{-1}\sum_j a_{ij}y_{ij}.$$

Now

$$\mathscr{E}(\sum_j a_{ij}y_{ij})^2 = \sum_j(\mathscr{E}a_{ij})y_{ij}^2 + \sum\sum_{j\neq k}\mathscr{E}(a_{ij}a_{ik})y_{ij}y_{ik}$$

$$= \frac{m'}{M}\sum_j y_{ij}^2 + \frac{m'(m'-1)}{M(M-1)}(Y_{i\cdot}^2 - \sum_j y_{ij}^2)$$

$$= \frac{m'}{M(M-1)}[(M-m')\sum_j y_{ij}^2 + (m'-1)M_i^2\bar{Y}_{i\cdot}^2],$$

and for $i \neq i'$

$$\mathscr{E}(\sum_j a_{ij}y_{ij})(\sum_j a_{i'j}y_{i'j}) = \frac{m'(m'-1)}{M(M-1)} Y_{i.}Y_{i'.}$$

Therefore

$$A = n^{-2}\sum_i \mathscr{E}\{b_i^2|\mathscr{s}'\}P_i^{-2}(\sum_j a_{ij}y_{ij})^2$$
$$+ n^{-2}\sum\sum_{i\neq i'} \mathscr{E}\{b_ib_{i'}|\mathscr{s}'\}P_i^{-1}P_{i'}^{-1}(\sum_j a_{ij}y_{ij})(\sum_j a_{i'j}y_{i'j}),$$

with

$$\mathscr{E}\mathscr{E}\{b_i^2|\mathscr{s}'\} = \mathscr{E}b_i^2 = nP_i\{1 + (n-1)P_i\},$$
$$\mathscr{E}\mathscr{E}\{b_ib_{i'}|\mathscr{s}'\} = \mathscr{E}[\mathscr{E}\{b_i|\mathscr{s}'\}\mathscr{E}\{b_{i'}|\mathscr{s}'\}] = n^2P_iP_{i'},$$

so

$$\mathscr{E}A = \frac{m'}{n(M-1)}[(M-m')\sum_i\{\sigma_W^2(i) + \bar{Y}_{i.}^2\} + (m'-1)M(\sigma_{BA}^2 + \bar{\bar{Y}}^2)]$$
$$+ \frac{(n-1)m'}{nM(M-1)}$$
$$\times [(M-m')M(\sigma^2 + \bar{\bar{Y}}^2) + (m'-1)(N\sigma_{BT}^2 + M^2\bar{\bar{Y}}^2/N)]$$
$$+ \frac{m'(m'-1)}{M(M-1)}\left[-N\sigma_{BT}^2 + \frac{M^2(N-1)}{N}\bar{\bar{Y}}^2\right],$$

where

$$\sigma_{BA}^2 = \sum_{i=1}^N \frac{M_i}{M}(\bar{Y}_{i.} - \bar{\bar{Y}})^2$$

is the variance of the averages for the primary units, and

$$\sigma_{BT}^2 = \frac{N-1}{N} S_{BT}^2;$$

and

$$V(\hat{\bar{\bar{Y}}}_P) = \frac{1}{nm'(M-1)}\Bigg[(M-m')\Sigma\sigma_W^2(i) + (m'-1)M\sigma_{BA}^2$$
$$+ (M-m')\sum\left(\bar{Y}_{i.}^2 - \frac{\bar{\bar{Y}}^2}{N}\right) - \frac{(m'-1)N}{M}\sigma_{BT}^2$$
$$+ (n-1)(M-m')\sigma^2 + M(m'-1)\left(1-\frac{1}{N}\right)\bar{\bar{Y}}^2\Bigg].$$

This may be compared with $(1/n)\sigma^2_{\text{BA}}$, the variance of the estimator of $\bar{\bar{Y}}$ under ordinary cluster sampling with probabilities proportional to the M_i. The increase will often be very large; we shall therefore not compute $V(\hat{\bar{\bar{Y}}}_p)$, which will be even larger.

13. Domains in cluster sampling

If the cluster sizes are equal (or nearly so) and we do not subsample, we may estimate Y' by

$$N \sum_{j=1}^{n} y'_{i_j}/n.$$

By dividing this by $\Sigma M'_i$ (or, if this is unknown, by its corresponding estimator), we obtain an estimator of $\bar{\bar{Y}}_1$. We similarly estimate Y'/X'. If we subsample, y'_{i_j} has to be replaced by $\bar{y}'_{i_j\cdot}M_{i_j}$.

Now suppose that the cluster sizes are markedly unequal and that we do not subsample. For simple sampling of clusters we generally estimate Y' by $(\Sigma y'_{i_j}/\Sigma M_{i_j})N\bar{M}$, but could use instead $(\Sigma y'_{i_j}/\Sigma M'_{i_j})X'$ if $X' = \Sigma M'_i$ is known (especially if $N\bar{M}$ is not). To estimate $\bar{\bar{Y}}_1$ we may use $\Sigma y'_{i_j}/\Sigma M'_{i_j}$ or divide $\Sigma y'_{i_j}/\Sigma M_{i_j}$ by $\bar{\bar{X}}_1 = X'/\Sigma M_i$ if the latter is known. With unequal probability sampling, Y' may be estimated† by $\Sigma(y'_{i_j}/\pi_{i_j})$; to estimate $\bar{\bar{Y}}_1$, divide this by $\Sigma(M'_{i_j}/\pi_{i_j})$. In all these formulae, if there is subsampling, y'_{i_j} is replaced by $\bar{y}'_{i_j\cdot}M_{i_j}$. One similarly estimates Y'/X' for a general characteristic $\mathscr{X}$, or uses knowledge, if any, of X' in estimating Y' or $\bar{\bar{Y}}_1$.

Formulae for variances and their estimators follow from the theory given previously. Particularly simple results are obtained, in the case of sampling of clusters with probabilities π_i (with replacement), when the design is self-weighting. Let f_0 be the expected overall sampling fraction, then the estimator of Y' is

$$\Sigma \bar{y}'_{i_j\cdot}M_{i_j}/\pi_{i_j} = \frac{1}{f_0}\,\Sigma y'_{i_j\cdot} = \frac{1}{f_0}\,y'$$

(since $f_0 = \pi_i m_i/M_i$) with estimated variance

$$v\left(\frac{1}{f_0}y'\right) = \frac{1}{f_0^2}n\Sigma\left(y'_{i_j\cdot} - \frac{1}{n}y'\right)^2\Big/(n-1);$$

and the estimator of $\bar{\bar{Y}}_1$ if X' is not known is

$$\frac{1}{f_0}y'\Big/\left(\frac{1}{f_0}x'\right) = y'/x'$$

† This estimator fails to exploit knowledge, if any, of the sum of the π_i over the domain.

with estimated variance

$$(x'/n)^{-2}\frac{1}{n}\sum\left(y'_{i_j\cdot}-\frac{y'}{x'}x'_{i_j\cdot}\right)^2\bigg/(n-1)$$

$$=\left(\frac{1}{x'}\right)^2\frac{n}{n-1}\sum\left\{\left(y'_{i_j\cdot}-\frac{y'}{n}\right)-\frac{y'}{x'}\left(x'_{i_j\cdot}-\frac{x'}{n}\right)\right\}^2$$

$$=\left(\frac{1}{x'}\right)^2\left\{v(y')+\left(\frac{y'}{x'}\right)^2v(x')-2\frac{y'}{x'}\operatorname{cov}(y',x')\right\}.$$

14. Differences between ratio estimators for two domains in cluster sampling

If

$$\hat{R}'=\frac{\Sigma t_iM_i\bar{y}'_{i.}/\mathscr{E}t_i}{\Sigma t_iM_i\bar{x}'_{i.}/\mathscr{E}t_i},\quad \hat{R}''=\frac{\Sigma t_iM_i\bar{y}''_{i.}/\mathscr{E}t_i}{\Sigma t_iM_i\bar{x}''_{i.}/\mathscr{E}t_i},$$

where t_i is the number of times the ith potential cluster is included in the sample, and where $\bar{x}'_{i.}$ and $\bar{x}''_{i.}$ may refer to two nonoverlapping domains or to the same domain (with $\bar{x}'_{i.}=\bar{x}''_{i.}=\bar{x}_{i.}$), then

$$\operatorname{Cov}(\hat{R}',\hat{R}'')\approx\frac{1}{X'X''}\mathscr{E}\sum\left(\frac{M_i}{\mathscr{E}t_i}\right)^2(\bar{y}'_{i.}-R'\bar{x}'_{i.})(\bar{x}'_{i.}-R''\bar{x}''_{i.})t_i.$$

Thus in the case of simple cluster sampling without replacement we get without subsampling

$$\operatorname{Cov}(\hat{R}',\hat{R}'')\approx\frac{N-n}{N}\frac{1}{n}\frac{1}{\bar{X}'\bar{X}''}\sum_{i=1}^{N}(y'_i-R'x'_i)(y''_i-R''x''_i)/(N-1).$$

The variance of $\hat{R}'-\hat{R}''$ is then readily computed, and so is its estimator.

In the case of sample with replacement with probabilities p_i, we get as estimator

$$v(\hat{R}'-\hat{R}'')=\frac{1}{n(n-1)}\sum\left(\frac{M_i}{p_i}\right)^2\left\{\frac{\bar{y}'_{i.}-\hat{R}'\bar{x}'_{i.}}{\hat{X}'}-\frac{\bar{y}''_{i.}-\hat{R}''\bar{x}''_{i.}}{\hat{X}''}\right\}^2,$$

which becomes, if $p_i=M_i/\Sigma M_i$,

$$v(\Sigma t_i\bar{y}'_{i.}/\Sigma t_i\bar{x}'_{i.}-\Sigma t_i\bar{y}''_{i.}/\Sigma t_i\bar{x}''_{i.})$$

$$=\frac{1}{n(n-1)}\Sigma t_i\left(\frac{\bar{y}'_{i.}-\hat{R}'\bar{x}'_{i.}}{\bar{\bar{x}}'}-\frac{\bar{y}''_{i.}-\hat{R}''\bar{x}''_{i.}}{\bar{\bar{x}}''}\right)^2.$$

For $x'_{i.} = m'_i$, this becomes

$$\frac{(\Sigma t_i m_i)^2}{n(n-1)} \sum t_i \left(\frac{m'_i}{m_i} \frac{\bar{y}_{i(1).} - \hat{R}'}{m'} - \frac{m''_i}{m_i} \frac{\bar{y}_{i(2).} - \hat{R}''}{m''} \right)^2,$$

which for equal m_i equals

$$v(\bar{y}_{1.} - \bar{y}_{2.}) = \frac{n}{(n-1)} \sum t_i \left\{ \frac{m'_i}{m'} (\bar{y}_{i(1).} - \bar{y}_{1.}) - \frac{m''_i}{m''} (\bar{y}_{i(2).} - \bar{y}_{2.}) \right\}^2.$$

CHAPTER VIII

THE USE OF MODELS

1. Stochastic models

So far we have considered a property $\mathscr{Y}$, defined on a population

$$\mathscr{P} = \{u_1, . . ., u_N\},$$

its distribution function F_N on $\mathscr{P}$, and some of the parameters thereof, like $\bar{Y}$ and $\sigma^2(\mathscr{Y})$. A particular method of drawing samples gives rise to a sample space $\mathscr{S}$ together with a probability function P on $\mathscr{S}$, and hence to sampling distributions of various statistics.

Probabilities came about entirely through the way we decided to sample (*designed*† probabilities). However, it is often desirable to conceive of probabilities coming about in other ways as well, through a sampling process not of the statistician's design. Such a sampling process is usually one whose existence we postulated (without physical evidence that the sampling was actually carried out), resulting in a *postulated* probability distribution. In each case the justification of such a postulation will depend on an interplay of deduction and observation, such as familiarly practiced in scientific activity.‡

We speak about a probability model, or stochastic model, if we specify a reasonable mechanism (model) for generating the phenomena and if this

† Designed probabilities are also very important in computer simulation (Monte Carlo), and in experimentation. For example, if we wish to be able to draw valid conclusions afterwards, it is essential that the experimental treatments are assigned to the units by some randomization procedure. Often, stratified or cluster sampling is used in experimentation, rather than simple random sampling, in order to reduce variances while maintaining applicability of the conclusions to a widely defined population. Incidentally, methods of analysis which rely on these designed probabilities only are called *randomization methods of analysis* (an important subclass being the *permutation methods*); often, one allows as well for postulated probabilities to enter the analysis.

‡ Often a sampling process is postulated without *any* empirical check on its approximate validity; in fact, again and again statisticians are forced to do so by circumstances. It should be noted that, in cases in which checks have been conducted, it has been found frequently that unsuspected factors were present and that, moreover, these would completely invalidate the conclusions that may be drawn from the data with the help of the postulate.

specification implies their probability distribution.† (Usually the specification of the mechanism is not sufficiently specific to lead to one single distribution, but merely leads to a class of distributions.) The word "reasonable" refers to the fact that a mechanism is postulated on the basis of *some* analysis of the situation, either a scientific one (i.e., part of the structure of science), or just a somewhat vague "understanding" of what brings about the phenomena. (Sometimes a mechanism like this is referred to as "explanatory", or – in a certain sense of the word – "analytic".)

If we are unable to specify a reasonable mechanism, we often find it possible to resort to one of a class of simple formulae which experience shows to give workable results in a wide range of applications. We may refer to these as interpolatory‡ stochastic formulae or models.§

Here we shall only be able to discuss the subject of stochastic models in a sketchy manner. §§ However, we cannot avoid the subject altogether, not only since it is necessary in the subsequent chapters, but also since the student should not leave the topic of survey theory without seeing how it is related to the large body of theory that deals with statistical analysis in the presence of postulated probabilities.

We begin by giving an example of an analytic model.

2. Example of an analytic model

Let $\mathscr{P}$ be the (large) set of all door-to-door salesmen of a certain article of which each customer buys at most one. We confine ourselves to a given area and period of time, and denote by $y(u_i)$ the number of these articles sold by the ith salesman in that period. Then $NF_N(v)$ is the number of salesmen in the area who sell v or fewer of the articles during the period.

† A table of random numbers, generally used for obtaining designed probabilities, may itself have been constructed with the help of a (real) physical mechanism which *incorporates* a probability model (see Chapter I, section 6), and which has been most thoroughly tested (see below) against observations. This is also an example of a case in which the specification leads to a *single* distribution. – In contrast to a *stochastic* model, a *deterministic* model is one in which the mechanism generates a unique (completely predictable) outcome.

‡ "Interpolation" is usually taken to involve fitting of a function to a set of observed points such that the fitted function passes through all the observed points; when using here the word "interpolatory", we do not, of course, mean to imply such a requirement.

§ Many would withhold the name "model" from an interpolatory formula. However, in practice it is difficult to say at which point we cease having a reasonable mechanism and have just an interpolatory formula. Indeed, formulas widely useful for interpolation not infrequently also appear in cases in which they possess an analytic basis.

§§ The student might find it helpful at this point to review the preliminary discussion of the topic in section 10 of Chapter I.

It may be reasonable to suppose that F_N belongs to a family of functions $\{F_{N,\alpha}\}$ where $NF_{N,\alpha}(v)$ is the largest integer contained in $NG_\alpha(v)$ with

$$G_\alpha(v) = \sum_{j \leq v} \binom{k}{j} \alpha^j (1 - \alpha)^{k-j},$$

and α is in some range between 0 and 1.

Why is it reasonable to suppose so? Say that, after deducting time for work preparation and for contacting those who turn out not to be potential customers, the work week allows for visits to k potential customers. Also suppose that the customers whom each salesman contacts during the period constitute a simple random sample from a large† population of potential customers, of which an unknown proportion, α, will, if exposed to a salesman, buy the article. Then the probability that $y(u_i) \leq v$ is $G_\alpha(v)$ for each i. Since N is large, we may expect $NF_N(v)$ to equal approximately $NF_{N,\alpha}(v)$ for some α. More precisely, the probability that $L(v)$, the fraction of the N salesmen who sell v or less articles,‡ takes on the value l, is with the above mechanisms equal to

$$\binom{N}{lN} \{G_\alpha(v)\}^{lN}\{1 - G_\alpha(v)\}^{N(1-l)},$$

and therefore

$$\mathscr{E}L(v) = G_\alpha(v), \quad V(L(v)) = G_\alpha(v)\{1 - G_\alpha(v)\}/N;$$

and if N is large, the probability is large that $L(v)$ is in a small interval around its expectation $G_\alpha(v)$, i.e., $F_N(v) \approx G_\alpha(v)$.

Similarly, under the above mechanism, the average $\bar{Y}$ of $y(u_1), \ldots, y(u_N)$, being the average of N independent binomial variables, each with mean $k\alpha$ and variance $k\alpha(1 - \alpha)$, has a mean $k\alpha$ and a variance $k\alpha(1 - \alpha)/N$ (which is small for large N); and

$$\sigma^2(\mathscr{Y}) = \sum_{i=1}^{N} \{y(u_i) - \bar{Y}\}^2/N$$

has mean $k\alpha(1 - \alpha)(N - 1)/N$ and variance (see section 9)

$$\frac{k\alpha(1 - \alpha)}{N}\left[2k\alpha(1 - \alpha)\left(1 - \frac{1}{N}\right) + \{1 - 6\alpha(1 - \alpha)\}\left(1 - \frac{1}{N}\right)^2\right],$$

which is again small if N is large.

† Because the population of potential customers is assumed large, we may as well postulate sampling with replacement (even though this contradicts the assumption of 1 purchase per customer).

‡ Up to now this fraction was looked upon as an unknown constant ($F_N(v)$); with the introduction of the model we look upon it as itself generated by a random process, and so a random variable.

3. General remarks

The following features of our example are characteristic of stochastic models in general:

(a) If the model is valid, the distribution function, F_N, of $\mathscr{Y}$ in $\mathscr{P}$ belongs to a given class, $\mathscr{C}$, of distribution functions, $\mathscr{C}$ being indexed by a real valued parameter, here α. (If also k were unknown, we would have a two-parameter family.) Generally we attempt to keep the number of parameters small, and make the distribution function depend on the parameters in some easily formulated way ("parametric specification"). Contrast this with a situation in which some such model is lacking, so that we can only specify F_N with the help of N parameters. The fact that F_N may be specified in terms of a very limited number of parameters leads to a possibility of estimating and specifying confidence intervals for Y, $\sigma^2(\mathscr{Y})$, etc. more efficiently than in the absence of a parametric specification, particularly if the number of parameters is very much smaller than N. This is one of the main reasons for introducing postulated probability distributions. We shall presently observe the efficiency of a parametric specification in our example.

(b) There can be no certainty that the model is valid; in fact, there is usually some evidence that it is not. (We shall below enumerate various weaknesses of the model postulated for our example.) This does not prevent it from being useful for drawing conclusions which are not *sensitive* to the invalid features of the model. Thus, we saw in Chapter II, section 15, that, for finding useable confidence intervals for $\bar{Y}$, only very general properties of F_N are important (such as absence of widely separated peaks); on the other hand, in Chapter II, section 16, we remarked that, if N is large and if it is true that F_N is approximately normal, we would use a tolerance interval quite different from the one put forth in the absence of such rather precise information. *Sensitivity testing* is an important aspect in the use of models.

(c) Usually a model can be embedded into a wider one; that is, one which contains the given model as a particular case. And there often are ways of testing statistically whether the data are consistent with the given model, Ω_0, assuming the wider model, Ω, to hold. Such "*testing* of a model"† will not be discussed here,‡ but it should be noted that such testing is always relative to

† Essentially such a test rejects Ω_0, if, under each of the distributions of Ω_0, the probability for the event $E(t, \mathscr{s})$ to occur is small. Here t is a certain statistic over $\mathscr{s}$ (called "testing statistic"), $\mathscr{s}$ the sample to be taken, and $E(t, \mathscr{s})$ denotes the event that $t(\mathscr{s})$ falls in some usually preassigned part ("critical region" or "region of rejection") of the range of possible values of t.

‡ We make an exception in section 4 of Chapter XII, where there arises a need to discuss a particular example of such testing.

the wider model, and that the assessment of Ω_0 depends on the entire body of evidence on which it rests, not just on the present data.

(d) The model is specified in precise and relatively simple mathematical terms, even though such precision and simplicity† of the model may be absurd. Without a mathematical formulation, and a simple one at that, we are unable to deduce conclusions from the model (see also section 6 below). But we should keep in mind that this excessive precision and simplicity will also cause some of the possible conclusions to be ridiculous. As one example, often it is mathematically much simpler to let N be infinite, rather than a large number; but we shall see in section 7 that this will have certain absurd consequences.

4. Uses of the model in our example

Suppose we wish to estimate

$$\sigma^2 = \sum_{i=1}^{N} \{y(u_i) - \bar{Y}\}^2/N$$

on the basis of a sample of n. For simplicity we shall consider sampling with replacement (or equivalently: n much smaller than N).‡ In the absence of the postulated model we could use the estimator

$$s^2 = \Sigma t_i\{y(u_i) - \bar{y}\}^2/(n-1).$$

We shall compare this with the case in which we rely on the above model. In that case we saw, that, when N is large, σ^2 is distributed in a small interval around $k\alpha(1-\alpha)$ and $\bar{Y}$ in a small interval around $k\alpha$. This suggests that $\bar{y}(1-(\bar{y}/k))$ would be a good estimator of σ^2.

Now the (conditional) mean of $\bar{y}(1-(\bar{y}/k))$ is

$$\mathscr{E}\bar{y} - \frac{1}{k}\mathscr{E}\bar{y}^2 = \bar{Y} - \frac{1}{k}\{\bar{Y}^2 + \mathscr{E}(\bar{y}-\bar{Y})^2\} = \bar{Y}\left(1-\frac{\bar{Y}}{k}\right) - \frac{1}{kn}\sigma^2,$$

which with our mechanism would be close to§

$$k\alpha(1-\alpha) - \frac{k\alpha(1-\alpha)}{kn} = \alpha(1-\alpha)\left(k-\frac{1}{n}\right);$$

† One aspect of this simplicity is exclusion of a host of factors which are expected to have only minor or incidental effects (collectively they may be included by way of a stochastic error term). In the experimental sciences, a counterpart to this abstraction is the controlled experiment, in which it is attempted to include only those forces and circumstances which are represented in the model.

‡ If we sample without replacement and wish to estimate $S^2 = \sigma^2 N(N-1)^{-1}$, we find that $\mathscr{E}S^2 = k\alpha(1-\alpha) = \mathscr{E}s^2 = \mathscr{E}\hat{\sigma}^2$, where $s^2 = \Sigma a_i\{y(u_i) - \bar{y}\}^2/(n-1)$. The analysis of section 9 relates to sampling with replacement.

§ Actually it would have mean equal to $\alpha(1-\alpha)(k-(1/n))$ plus a term of order $1/N$.

so an improved estimator of $k\alpha(1-\alpha)$ would be

$$\hat{\sigma}^2 = \frac{k}{k-(1/n)}\,\bar{y}\left(1-\frac{\bar{y}}{k}\right).$$

In section 9 we compute the variance of these estimators s^2 and $\hat{\sigma}^2$ under the assumption that our mechanism is as above and find

$$V(s^2) - V(\hat{\sigma}^2) = 2\left(\frac{1}{n-1} - \frac{1}{kn-1}\right)\{k\alpha(1-\alpha)\}^2.$$

5. Limitations of the model of our example

It is not necessarily so that the manner in which the salesmen of the example get in touch with the customers is similar to the selection of a simple random sample; sales to one customer may affect sales to another; k need not be the same for each salesman; α is likely to vary with the persuasiveness of the different salesmen and may increase at the expense of decreasing k.

It is interesting to note that if $\bar{\alpha}$ is the average of the α's for the different salesmen, we may still expect $\bar{Y}$ to equal $k\bar{\alpha}$ approximately, but that σ^2 is now with large probability in a small interval about a number larger than $k\alpha(1-\alpha)(N-1)/N$, and markedly so if the variation of the α's is large.† But then F_N cannot approximately equal $F_{N,\alpha}$ for some α, since, for $\alpha = \bar{\alpha}$, G_α would give the correct mean but not the correct variance.

The question as to whether variation among the α's would seriously invalidate the estimator of σ^2 would need some investigation. Variation among the α's can easily be allowed for at the expense of introducing an extra parameter into the model. To allow for the other possible defects mentioned is more difficult. E.g., in order to get a mathematical model, one would have to specify in which way sales to one customer may affect sales to another; and it is not so obvious how this may be done, since dependence between these

† Write $y_j(u_i) = 1$ if salesman i sells to his jth contact during this period, and 0 otherwise Then

$$y(u_i) = \sum_{j=1}^{k} y_j(u_i),$$

so

$$\mathscr{E}y^2(u_i) = k\alpha_i + k(k-1)\alpha_i^2,$$

hence

$$\mathscr{E}Y^2 = k\{(k-1)\Sigma\alpha_i^2 + N\bar{\alpha} + k\mathop{\Sigma\Sigma}_{i\neq i'}\alpha_i\alpha_{i'}\} = k\{N\bar{\alpha} + k(N\bar{\alpha})^2 - \Sigma\alpha_i^2\},$$

and

$$\mathscr{E}N^{-1}\Sigma\{y(u_i) - \bar{Y}\}^2 = k\left\{\frac{N-1}{N}\,\bar{\alpha}(1-\bar{\alpha}) + \left(k - \frac{N-1}{N}\right)N^{-1}\Sigma(\alpha_i - \bar{\alpha})^2\right\}.$$

sales can take many forms. Allowing for many different modes of dependence (and, more generally, for a large class of these and other defects of our original model) may make the model so complicated that its use becomes of no advantage in any problem.

6. Superpopulations

6.1. CONCEPT, USES, EXAMPLES

A conceptual scheme often employed is one in which the population $\mathscr{P}$ is itself thought of as having been drawn from a larger population Π, called a superpopulation. In our example let $\mathscr{P}$ denote, instead of the population of salesmen, the collection of customers visited by them; and let $z(u_i)$ be defined as 1 if the ith customer in $\mathscr{P}$ buys the article and 0 otherwise. Then Z is the same as Y in the example. Let Π be the population of potential customers (which was assumed large), then $\{u_1, \ldots, u_{kN'}\}$ is conceived of as a simple random sample (with replacement) from Π. Therefore, e.g., Z has a sampling distribution (in particular, Z has a binomial distribution) with mean α and variance $\alpha(1 - \alpha)/(kN')$. [Note: A good salesman will actually attempt to take his customers not at random from Π, but will try to pick out a more promising subpopulation!]

One use to which we already put the superpopulation was for obtaining better estimators. Superpopulations are also used for the purpose of comparing different proposed sampling schemes and estimators. In practice it may often be that a proposed scheme and estimator are going to be used on a number of populations, which are similar in certain respects to the population for which we proposed the scheme and estimator. In that case it is of some interest to know whether the proposal, if adopted for all these uses, may on the average be efficient over such uses.† An example which we shall study in the chapter on systematic sampling is the case of a population in which the items appear in a certain order and we may assume that in the superpopulation the correlation between the $\mathscr{Y}$-value, y_i, of u_i, and its place, i, in the population, satisfies certain properties.

Another example would be if ratio estimators are proposed for a class of problems in which we may suppose that the $\mathscr{Y}$- and $\mathscr{X}$-values of the units are approximately proportional, or, more formally, are proportional on the average in the superpopulation; it is easy to show that in such problems the ratio estimator will be very efficient.

A mathematical specification together with the statement that "it holds approximately" (without specifying precisely what is meant by this) cannot be

† In such cases, only the conclusion about the efficiency of the proposed method would be affected by alterations in the superpopulation model, not its validity.

operated on mathematically. Introduction of a superpopulation may be a way of giving mathematical content to the above statement in quotation marks, while enabling us freely to apply any mathematical operations without fear of logical contradiction (although, as mentioned in section 3(b), there may be no way to prevent contradiction with reality).

The concept of a superpopulation is more or less implicitly involved in many investigations. Thus in Chapter I, section 9, we already commented on the difference between the target population and the sampled one. These differences may sometimes usefully be conceived of as differences between a superpopulation Π and a sample $\mathscr{P}$ drawn from it by simple random sampling. This may be the case with the discrepancies of examples (c) and (d) of section 9, Chapter I, insofar as we have reason to believe they are not of a systematic nature (systematic tendency to deviate in a certain direction). Ultimately the test of validity of such a postulate lies in the empirical investigation of such discrepancies, i.e., of samples (of populations) of such discrepancies from the postulated superpopulation.

6.2. A MODEL FOR MEASUREMENT ERRORS

Let us work this case out in some detail. Thus let us suppose that for each member u_i of $\mathscr{P}$, $y(u_i)$ would equal some unknown number v_i *but for* a fluctuation e_i, which is conceived of having been drawn from a superpopulation. This formalizes the idea that, if we would repeat measuring $\mathscr{Y}$ on u_i by the method employed, errors not of a systematic nature would tend to cancel out, and we would obtain on the average v_i.† Thus, for each i, $\mathscr{E}e_i = 0$.

We write

$$y_i = v_i + e_i, \quad \mathscr{E}e_i e_{i'} = \sigma_{ii'}.$$

Then

$$\bar{y} = \frac{1}{n} \sum_{i=1}^{N} a_i v_i + \bar{e},$$

where

$$\bar{e} = \frac{1}{n} \sum_{i=1}^{N} a_i e_i,$$

and $a_i = 1$ if u_i is in the sample and 0 otherwise. Consider simple random sampling from $\mathscr{P}$ without replacement. Since this process of sampling from $\mathscr{P}$ goes on without regard to the postulated sampling from the superpopulation,

$$\mathscr{E}\bar{e} = \frac{1}{n} \sum_{i=1}^{N} \mathscr{E}a_i \, \mathscr{E}e_i,$$

† One may instead refer to the superpopulation as the set of u_{it}, where u_{it} is the tth observation on the ith unit. This is in closer accord with the use made above of the term superpopulation, but it requires stating that the numbering of the observations is an a priori one and that the t actually associated with a unit u_i in the sample is random.

which equals 0, and so

$$\mathscr{E}\bar{y} = \frac{1}{N}\sum_{i=1}^{N} \nu_i = \bar{\nu} \quad \text{(say).}$$

Let us denote the average of $\sigma_1^2, \ldots, \sigma_N^2$ by

$$\sigma_e^2 = \frac{1}{N}\sum_{i=1}^{N} \sigma_i^2,$$

and the average of the $\sigma_{ii'}$ for $i \neq i' \in \{1, \ldots, N\}$ by

$$r_e\sigma_e^2 = \frac{1}{N(N-1)}\sum\sum_{i\neq i'} \sigma_{ii'}.$$

Also write

$$\sigma_\nu^2 = \frac{1}{N}\sum_{i=1}^{N} (\nu - \bar{\nu})^2 = \frac{N-1}{N} S_\nu^2.$$

The mean of

$$\frac{1}{N-1}\sum_{i=1}^{N} (y_i - \bar{Y})^2$$

$$= \frac{1}{N-1}\left\{\sum_{i=1}^{N} (\nu_i - \bar{\nu})^2 + \sum_{i=1}^{N} e_i^2 - \frac{1}{N}\left(\sum_{i=1}^{N} e_i\right)^2 + 2\sum_{i=1}^{N} e_i (\nu_i - \bar{\nu})\right\}$$

is

$$\frac{1}{N-1}\left\{\sum_{i=1}^{N} (\nu_i - \bar{\nu})^2 + \sum_{i=1}^{N} \sigma_i^2 - \sigma_e^2(1 + (N-1)r_e) + 0\right\} = S_\nu^2 + \sigma_e^2(1 - r_e);$$

and so the mean of the conditional variance of $\bar{y}$ given $e_1, \ldots, e_N$ equals

$$\frac{N-n}{N}\,\frac{1}{n}\,\mathscr{E}\,\frac{1}{N-1}\sum_{i=1}^{N} (y_i - \bar{Y})^2 = \frac{N-n}{N}\,\frac{1}{n}\,\{S_\nu^2 + \sigma_e^2(1 - r_e)\}.$$

On the other hand, the variance of $\bar{y}$ equals the sum of the last mentioned expression and the variance,

$$N^{-2}\{\Sigma\sigma_i^2 + \sum\sum_{i\neq i'}\sigma_{ii'}\},$$

of the conditional expectation of

$$\bar{y} = \bar{\nu} + (e_1 + \ldots + e_N)/N,$$

to wit

$$V(\bar{y}) = \frac{N-n}{N}\frac{1}{n}\{S_v^2 + \sigma_e^2(1-r_e)\} + \frac{1}{N}\sigma_e^2\{1+(N-1)r_e\}$$

$$= \frac{N-n}{N}\frac{1}{n}S_v^2 + \frac{\sigma_e^2}{n}\{1+(n-1)r_e\}.$$

Note that $\sigma_{ii'}$ need not be 0 for all $i \neq i'$. Units which are measured by the same instrument or person, or by a member of the same team of interviewers, or coded by the same coder, or which are measured at about the same time, often have correlated errors.† Moreover, the assumption stated above that sampling from $\mathscr{P}$ is independent of sampling from the superpopulation may not represent the state of affairs adequately. We shall give further attention to these matters in section 3 of Chapter X.

If the σ_i^2 are constant, and if for $i \neq i'$ the $\sigma_{ii'}$ all have the same value, we can also state that the e_i are all drawings from a *single* population (with replacement if $\sigma_{ii'} = 0$).

Sometimes it is supposed that there exists a "true" or "correct" value of $\mathscr{Y}$ for u_i, i.e., a value which corresponds to a method of measuring $\mathscr{Y}$ superior to the one that has been employed. We can then write

$$v_i = \mu_i + \alpha_i,$$

where μ_i is the "correct" value and α_i is called *bias of measurement or systematic error.*

Note that, if y_i and μ_i – the observed and "correct" values – can be only 1 or 0, it is only possible for σ_i^2 to be positive with $\mathscr{E}e_i = 0$ if the model disallows vanishing α_i.‡

A sensible definition of v_i in this case is: the probability that for unit u_i the $\mathscr{Y}$-value will be 1. Then e_i equals $1 - v_i$ with probability v_i, and $-v_i$ with probability $1 - v_i$; so $\mathscr{E}e_i = 0$ and $\sigma_i^2 = v_i(1 - v_i)$. Such a definition

† $-N^{-1}\sigma_e^2\{1+(N-1)r_e\}$ represents the bias of $(1 - nN^{-1})s^2/n$ as an estimator of the variance of $\bar{y}$; even for r_e a very small positive number, this bias may be relative large in absolute value. However, if r_e vanishes, variance of $\bar{y}$ lies between the expectations of $(1 - nN^{-1})s^2/n$ and of s^2/n, so that for small n/N the *usual estimator is a good one* (and s^2/n a somewhat safer one). Therefore, the *most important source of bias* in the usual estimator of the variance of $\bar{y}$ is *positive correlation between the response errors.*

‡ For if $\sigma_i^2 > 0$, e_i must take on at least 2 different values, each with positive probability, and for $\alpha_i = 0$ these can only be 0 or -1 if $\mu_i = 1$, and 1 or 0 if $\mu_i = 0$; but there does not exist $a < 1$ such that $a \times 0 + (1 - a) \times (-1) = 0$, nor does there exist $a > 0$ such that $a \times 1 + (1 - a) \times 0 = 0$.

corresponds, for example, to the many cases in which we observe activities in which individuals engage from time to time, but not in a strictly periodic fashion.†

The bias α_i may be a split into two components:

$$\bar{\alpha} = \sum_{i=1}^{N} \alpha_i/N,$$

called the constant bias; and $\beta_i = \alpha_i - \bar{\alpha}$, the variable bias component. Constant biases are not detectable from the data but may be detected by a more elaborate method of measurement. The α_i may well be related to the μ_i in some systematic way; e.g., the method of measuring may consistently overestimate certain values of μ_i and consistently underestimate certain other values of μ_i.

Sometimes we also suppose that the μ_i and α_i are drawn from a superpopulation with means μ and $\bar{\alpha}$ and variances σ_μ^2 and σ_α^2, and covariance of each pair (α_i, μ_i) equal to $\sigma_{\mu\alpha}$ and for each pair $(\mu_i, \alpha_{i'})$ with $i \neq i'$ equal to 0. Then

$$\mathscr{E} \sum_{i=1}^{N} (\nu_i - \bar{\nu})^2/N = \sigma_\mu^2 + \sigma_\alpha^2 + 2\sigma_{\mu\alpha}.$$

Note that it is not usually necessary to postulate "correct values" μ_i, since we can work entirely in terms of the ν_i; but then we cannot talk about $\sigma_{\mu\alpha}$.

REMARK. In the above-mentioned case in which μ_i is 1 or 0, let f be the fraction of cases in which the μ-value is 1. Since α_i equals $\nu_i - 1$ if $\mu_i = 1$ and ν_i if $\mu_i = 0$, we have that

$$\bar{\alpha} = f(\bar{\nu} - 1) + (1 - f)\bar{\nu} = \bar{\nu} - f,$$

so that, when $\bar{\alpha} = 0$, $f = \bar{\nu}$. Also $\bar{\alpha} = 0$ implies $\bar{\mu} = \bar{\nu}$, and so $\sigma_{\mu\alpha} = \mathscr{E}\mu_i\alpha_i$ equals 1 times the average of $\nu_i - 1$ over cases for which $\mu_i = 1$ plus 0 times the average of ν_i over the other cases, which is *less* than 0.‡ As a result when $\bar{\alpha} = 0$ *the variance of* $\bar{y}$ *does not exceed* σ_μ^2 *if* $r_e = 0$; indeed,

† These belong to the class of cases in which *variability is inherent* in the individual behavior itself, and cannot be looked upon as "error", in the usual sense of the word. We often have some possibility of designing our investigation in a way which leads to less variability in the results. Thus lengthening the period of observation to one in which the activity may take place more than once makes for a more stable determination of its average rate. In the example of section 2, if the time period considered is one-kth of a week, $y(u_i)$ is 1 or 0, but, to estimate $\bar{Y}$ and the inherent variability σ^2, we would usually want to take observations over a much longer period (the two estimators of σ^2 in section 4 coincide for $k = 1$!).

‡ One could only obtain 0 if $\nu_i = 1$ in all the former cases and therefore (since $f = \bar{\nu}$) $\nu_i = 0$ in all the latter cases; that is, if all $\sigma_\alpha^2 = 0$. This, by what was noted earlier, implies all $\sigma_i^2 = 0$.

$$\frac{N-n}{N}\frac{1}{n}S_v^2 + \frac{\sigma_e^2}{n} = \frac{1}{n}\left\{\frac{N-n}{N-1}\frac{\Sigma(v_i-\bar{v})^2}{N} + \frac{\Sigma v_i(1-v_i)}{N}\right\}$$

$$= \frac{1}{n}\left\{\frac{N-n}{N-1}\frac{\Sigma(v_i-\bar{v})^2}{N} + \bar{v}(1-\bar{v}) - \frac{\Sigma(v_i-\bar{v})^2}{N}\right\}$$

$$= \frac{1}{n}\left\{\bar{v}(1-\bar{v}) - \frac{n-1}{N-1}\sigma_v^2\right\} \leq \frac{1}{n}\sigma_\mu^2,$$

with equality only if the v_i are all equal or if $n = 1$.

6.3. NEW INSIGHTS OBTAINABLE FROM A SUPERPOPULATION MODEL

Certain theoretical results may be obtained without the use of the concept of a superpopulation, but the use of such a concept may, nevertheless, facilitate their derivation or throw some new light on their meaning. As an example, let us obtain the formula

$$\left(\frac{1}{n} - \frac{1}{N}\right) S^2$$

for the variance of $\bar{y}$ in a simple random sample taken without replacement from a population $\mathscr{P} = \{u_1, \ldots, u_N\}$. A similar formula appeared in subsampling, and this suggests regarding simple random sampling without replacement as a form of subsampling.

Indeed, we may, without loss of generality, consider $\mathscr{P}$ as having been drawn by simple random sampling with replacement from *any finite* set $\Pi = \{v_1, \ldots, v_A\}$. That is, for $i = 1, \ldots, N$, $u_i = v_{\alpha_i}$, where $\alpha_1, \ldots, \alpha_N$ is a sequence of numbers, each of which is conceived of as having been drawn, independently of the others, from among $\{1, \ldots A\}$. (One may think of the v's as a set of A identical balls marked 1 to A, from which we obtain the N drawings with replacement; those drawn we give a second mark, 1 to N, respectively.)

Moreover, we may assign to the elements $v_1, \ldots, v_A$ $\mathscr{Y}$-values in any way compatible with the set of $\mathscr{Y}$-values on $\mathscr{P}$. Denote the resulting average $\mathscr{Y}$-value on Π by $\bar{X}$ and the variance of these $\mathscr{Y}$-values on Π by $\sigma^2(\mathscr{X})$. Then

$$\mathscr{E}\bar{Y} = \bar{X}, \quad V(\bar{Y}) = \mathscr{E}(\bar{Y} - \bar{X})^2 = \frac{1}{N}\sigma^2(\mathscr{X}).$$

Then a simple random sample of n from $\mathscr{P}$ can also be looked upon as a simple random sample of n taken with replacement from Π, and so

$$\mathscr{E}\bar{y} = \bar{X}, \quad V(\bar{y}) = \mathscr{E}(\bar{y} - \bar{X})^2 = \frac{1}{n}\sigma^2(\mathscr{X}).$$

Moreover, the result we seek represents

$$V\{\bar{y} \mid y(u_1), \ldots, y(u_N)\},$$

the conditional variance of $\bar{y}$, given the $\mathscr{Y}$-values on $\mathscr{P}$. According to section 8 of Chapter I, its expected value equals

$$V(\bar{y}) - V[\mathscr{E}\{\bar{y} \mid y(u_1), \ldots, y(u_N)\}],$$

which, as

$$\mathscr{E}\{\bar{y} \mid y(u_1), \ldots, y(u_N)\} = \bar{Y},$$

is equal to

$$V(\bar{y}) - V(\bar{Y}) = \left(\frac{1}{n} - \frac{1}{N}\right)\sigma^2(\mathscr{X}).$$

Now the formula we seek must be a function of $y(u_1), \ldots, y(u_N)$ which is symmetric in these N arguments. A slight extension of Theorem 3 of HALMOS, Ann. Math. Statist., **17** (1946) 34–43, implies that there is *only* one such function, f, with $\mathscr{E}f\{y(u_1), \ldots, y(u_N)\}$ equal $\sigma^2(\mathscr{X})$ no matter what be the finite population Π or the compatible distribution of $\mathscr{Y}$-values on it (and so, in particular, no matter what be $\sigma^2(\mathscr{X})$.† But $\mathscr{E}S^2 = \sigma^2(\mathscr{X})$). Therefore, the required result is $((1/n) - (1/N))S^2$.

7. Infinite populations

Often it is not evident what we should consider the size of the superpopulation to be. Thus, in considering the effect of a drug on some patients with a certain disease, it is clear that often our real aim is to make statements which apply to all patients who at any time may contact the disease in question;‡ and this is an indefinitely large population.§ In analytic surveys this feature is very common.

Infinite populations have the advantage that they can often be described in mathematically more tractable ways than finite populations. Numerous functions F_N occurring in practice have been described by approximating F_N independently of N, to a degree sufficiently accurate to many of the uses made of them.

To give only a simple illustration, for many practical purposes, heights and weights of people can usually be well described by normal distributions, or more generally by functions which represent distributions of continuous random variables. While actual measurements proceed with some given degree of subdivision of the scale (e.g., to the centimeter, to the millimeter, etc.) and therefore represent a discontinuous variable, in these cases it is almost always more expedient to formulate the model in terms of continuous variables. This involves the danger of drawing mathematical conclusions

† The theorem is based on $\mathscr{E}f\{y(u_1), \ldots, y(u_N)\}$ being the same for *all* (compatible) distributions of $\mathscr{Y}$-values over Π; by making A large enough we can get arbitrarily close to any desired distribution, cf. section 6.3 (ii) of Chapter I.

‡ In fact, our primary aim often is to reveal properties of the mechanism (model) governing the drug and the disease. The population of patients is generated by the continued working of the mechanism. – In practice there will be changes in the course of time (thus the disease may change its nature or die out altogether), but this may not be relevant to our investigations. In order to emphasize that such changes are being abstracted from, a population like this is often called a *hypothetical* one. Thus, one may consider the outcome of coin tosses and, if matters like the inevitable wearing out of the coins are considered irrelevant, they are abstracted from, leading to a *constant probability* (distribution) of heads. (It may, however, be more appropriate to allow the probabilities to change in a prescribed way – one then speaks of a *stochastic process*.)

§ From infinite populations the sampling process is necessarily an entirely imaginary one (to see this one should ask oneself how he could draw a simple random sample from such a population).

which are due merely to the mathematical oversimplification, but this – as we have already noted – is an inevitable feature of the use of models. One, often noted, example is that in such cases any reasonable test of the hypothesis that the observed measurements constitute a simple random sample from an infinite population with a specified continuous distribution function F, when carried out on a sufficiently large sample of observations, must lead to a negative conclusion.

8. Example of an interpolatory formula

Consider two characteristics, $\mathscr{Y}$ and $\mathscr{X}$, defined for members of a population $\mathscr{P}$. We often expect that there is some simple approximate relation between the $\mathscr{Y}$- and $\mathscr{X}$-values for the members of $\mathscr{P}$. For example, between the sales of grocery stores and the numbers of their employees in a given region there is likely to be a relation approximately describable by an increasing function, though we may not be able to devise a theory which leads us to a particular form of increasing function.

In such a case we can attempt to "fit" various simple equations to the data. One of the simplest of these, often successful in problems of this kind, states that y_i is approximately equal to a constant plus a multiple of x_i.† An exact relation $y_i = \alpha + \beta x_i$ will evidently not fit most data. In view of that fact, we may postulate that when the population is divided into strata according to the (say K) different $\mathscr{X}$-values, the average, $\bar{Y}_{(k)}$, of $\mathscr{Y}$-values for the stratum whose $\mathscr{X}$-value is $x^{(k)}$ will be found to be $\alpha + \beta x^{(k)}$, with α and β some (unknown) constants (*linear regression postulate*). Also, that σ_k^2, the mean square, for units with $\mathscr{X}$-value $x^{(k)}$, of deviations of $\mathscr{Y}$-values from $\bar{Y}_{(k)}$, be a positive constant, σ^2, the *same* for each k (*homoskedasticity postulate*). (The latter assumption is often replaced by some specific *heteroskedasticity* assumption, e.g., when all $x^{(k)}$ are positive, that σ_k^2 is proportionate to some power of $x^{(k)}$.)

Such a postulate may, however, lead to some difficulties. In many cases $K = N$, and it is clear that then such constants α and β would usually fail to exist.‡ If K is not small (though smaller than N), it may often be possible to check on the assumption by obtaining $\bar{Y}_{(k)}$ for a few values of k; one would expect to find again that no such α and β exist. Therefore, the use of the postulate would be mostly confined to the case in which the number of different $\mathscr{X}$-values in $\mathscr{P}$ is very small. Even in this case one would not usually

† We exclude the case in which the relation is exact or practically so, which, except for coincidences, would occur only when it is a definitional or an established scientific relation.

‡ Moreover σ^2 would be 0 in this case; and more generally, if for one $x^{(k)}$ the $\mathscr{P}$ has only one member, σ_k^2 for that stratum vanishes.

expect the sort of assertions about the σ_k^2 mentioned above to be valid if some of the strata contain very few units.

It is therefore more appropriate to postulate that $\mathscr{P}$ has *itself been drawn from an infinite* population Π in which we have linear regression and homoskedasticity or some specified type of heteroskedasticity. Alternatively, we may postulate that, for each member u_i of $\mathscr{P}$, $y(u_i)$ would equal exactly $\alpha + \beta x(u_i)$ but for fluctuations, which for all members of $\mathscr{P}$ are conceived of as having been drawn from one *population of fluctuations* with zero mean and some mean square error σ^2 (or which, for each member of $\mathscr{P}$ *with given $\mathscr{X}$-value*, are conceived of as having been drawn from one population of fluctuations with zero mean and some mean square error, possibly dependent in some specified way on that $\mathscr{X}$-value).† Note that when $\beta = 0$ we have as a particular case the model of section 6, where $\mu_i = \bar{\mu}$; with $\beta \neq 0$ and $x(u_i) = \mu_i$ we have the more general case of section 6, with the important exception that there the μ_i are not known.

As an application of the above we show first that, if‡ in the superpopulation

$$\mathscr{E}\{y_i|x_i\} = (Y/X)x_i,$$

then, if the probability for $\bar{x}$ to vanish is 0, the ratio estimator of Y is conditionally unbiased given the sequence $x_{j_1}, \ldots, x_{j_n}$ of $\mathscr{X}$-values observed:

$$\mathscr{E}\left\{\frac{y}{x} X | x_{j_1}, \ldots, x_{j_n}\right\} = Y.$$

Indeed,

$$\frac{X}{x}\mathscr{E}\{y|x_{j_1}, \ldots, x_{j_n}\} = \frac{X}{x}\sum_{k=1}^{n} \mathscr{E}\{y_{j_k}|x_{j_k}\} = \frac{X}{x}\sum_{k=1}^{n} \frac{Y}{X}x_{j_k} = \frac{X}{x}\frac{Y}{X}x = Y.$$

The conditional variance of $(y/x)X$ is therefore the conditional expectation of the sum of the squares of the

$$(X/x)d_{j_k} = \{(X/x)(y_{j_k} - (Y/X)x_{j_k})\}.$$

† A stratified version of these two models has been applied to problems of analytic uses of surveys in KONIJN, J. Amer. Statist. Ass., **57** (1962) 590–606. (An an alternative to the methods of estimating the variances of the estimators of α and β given there, one may consider the methods of section 14.2 of Chapter III.) One should not confuse the estimation of the model parameter β discussed in that paper with the estimation of the population parameter $\Sigma(y_i - \bar{Y})(x_i - \bar{X})/\Sigma(x_i - \bar{X})^2 = \sigma(\mathscr{Y}\mathscr{X})/\sigma^2(\mathscr{X})$, which in the notation for stratified sampling becomes

$$\frac{\Sigma\sigma_h(\mathscr{Y}\mathscr{X}) + N^{-1}\Sigma N_h(\bar{Y}_h - \bar{Y})(\bar{X}_h - \bar{X})}{\Sigma\sigma_h^2(\mathscr{X}) + N^{-1}\Sigma N_h(\bar{X}_h - \bar{X})^2}.$$

‡ If $\mathscr{E}\{y_i|x_i\} = \beta x_i$, then $\mathscr{E}\{Y|x_1, \ldots, x_N\} = \beta X$ and $\beta = \mathscr{E}\{YX^{-1}|x_1, \ldots, x_N\}$; for N large relative to the conditional variance, YX^{-1} should differ only trivially from its conditional expectation.

Hájek† has remarked that under these circumstances we may therefore expect that

$$\frac{N-n}{N}\, n \left(\frac{X}{x}\right)^2 \sum_{k=1}^{n} \left(y_{j_k} - \frac{y}{x} x_{j_k}\right)^2 \Big/ (n-1) = \frac{N-n}{N}\,\frac{1}{n} N^2 \left(\frac{\bar{X}}{\bar{x}}\right)^2 s^2(\hat{\mathscr{D}})$$

should be a good estimator of the conditional variance. This may be compared with the previously given approximate variance estimator

$$\frac{N-n}{N}\,\frac{1}{n} N^2 s^2(\hat{\mathscr{D}}).$$

Since $\mathscr{E}(\bar{X}/\bar{x})^2 > 1$, this comparison shows, incidentally, that on the average, under these circumstances, it is better to use, in the denominator of the estimator of $V(\hat{R})$, $\bar{x}^2$ instead of the known value of $\bar{X}^2$ (cf. the Remark in section 9 of Chapter II).‡

Approximate validity of the conditions given not only typify the case in which to expect small bias, but also, especially, relatively small variance of the ratio estimator (cf. the end of section 9 of Chapter II or the end of section 6 of Chapter IIA). Therefore the latter remark is particularly appropriate whenever the ratio estimator is very good.

As the present edition goes to the press, a paper comes to hand by ROYALL, Biometrika, **57** (1970) 377–387, to which it seems imperative to refer in this book (which otherwise, as mentioned in the Foreword, has to disregard the most recent literature). Under examina tion is the estimation of Y from samples of size n when the values of a characteristic $\mathscr{X}$ are readily available for all members of the population. It is assumed that $y_1, \ldots, y_N$ may be considered independent drawings from a superpopulation Π such that, for units with $\mathscr{X}$-value x, the conditional expected value of its $\mathscr{Y}$-value is a constant multiple of x. Various assumptions are considered for the conditional variance of this $\mathscr{Y}$-value, the typical one being that it is proportional to x. As was briefly mentioned at the end of last footnote in section 6 of Chapter IIA, these conditions are, together with high correlation, particularly favorable to ratio estimation. One of Royall's results is that then, if the criterion of optimality is smallness of the expected value, τ^2, of δ^2, the mean over repeated sampling (from any fixed population in the sample space of Π) of the squared deviations of Y from its estimate, the best sample would *not* be *random*, but would consist of those units which have the highest $\mathscr{X}$-values. (In this case δ^2 is simply the squared difference between Y and its estimate $X(y/x)$.)

† Appl. Mathem., **3** (1958) 384–398.

‡ Nonetheless, we have refrained from advocating the general use of

$$(\bar{X}/\bar{x})^2\{(N-n)/Nn\}s^2(\hat{\mathscr{D}})$$

instead of

$$\{(N-n)/Nn\}s^2(\hat{\mathscr{D}})$$

for estimating $V(\hat{Y}_R)$. Note that the reasoning given is insufficient. Either one has to assume that n is large enough, not only for $s^2(\hat{\mathscr{D}})$ to be virtually unbiased for $S^2(\mathscr{D})$, but also for the expectation of $(\bar{X}/\bar{x})^2 s^2(\hat{\mathscr{D}})$ to virtually equal $\mathscr{E}(\bar{X}/\bar{x})^2$ times $\mathscr{E}s^2(\hat{\mathscr{D}})$; or, as Hájek does, one has to make some normality assumption.

Regarding this result one must note that, under this method of obtaining the observations, nothing is guaranteed about the closeness to Y of the estimator under repeated sampling from the fixed population, $\mathscr{P}$, of interest, Y being an *unknown constant*, about which one can only make probability statements relating to sampling of $\mathscr{P}$ from the superpopulation. (If, on the other hand, the interest is in the mean of Y over the superpopulation, smallness of τ^2 would evidently not be the appropriate criterion.) In all cases, when comparing properties of different procedures, one should consider the sensitivity of these properties to modifications in such of the assumptions as one has rarely an opportunity to submit to close scrutiny. In the present case the stated criterion, or any reasonable alternative criterion, is likely to be very sensitive to deviations from the randomness of the drawing from Π – an essentially unverifiable assumption – and may also be sensitive to deviations from the proportionality assumption.†

9. Appendix: The variance of two estimators of σ^2

It is found in SUKHATME, Sampling Theory of Surveys with Applications (Indian Society of Agricultural Statistics and Iowa State College, 1954) p. 36 (also in KENDALL and STUART, The Advanced Theory of Statistics (Griffin, 1963) Vol. 1, problem 10.13) that, in sampling n with replacement, the variance of the sample variance is

$$\frac{\mu_4 - \mu_2^2}{n} + \frac{2\mu_2^2}{n(n-1)},$$

where μ_2 and μ_4 are the second and fourth central moments of the underlying distribution – in our case, the binomial distribution with parameters α and k:

$$\mu_2 = k\alpha(1-\alpha),$$
$$\mu_4 = k\alpha(1-\alpha)\{3(k-2)\,\alpha(1-\alpha)+1\}.$$

This gives for the variance of the sample variance

$$V(s^2) = \{k\alpha(1-\alpha)/n\}\{2\alpha(1-\alpha)(kn/(n-1)-3)+1\}.$$

† Royall also computes δ^2 for some published populations, and finds that in most cases δ^2 is less if one observes the units with the largest $\mathscr{X}$-values than if one samples randomly, and often much less. However, all but three of these populations (nos. 13, 11 and 2, in decreasing order of size) are so small (less than 45) that one could not draw samples of reasonable size from them without including most of the population (in which case the difference in δ^2 between the two methods gets small). As was already pointed out in section 7 of Chapter I, with small samples purposive selection will often give better results than simple random sampling, but the limited significance of this fact is also brought out there. As to the larger of the populations, for no. 13 with purposive selection δ^2 is already twice that with random sampling when $n = 8$, and already more than 3 times when $n = 12$. For population no. 11, these relations are about $\frac{1}{2}$ when $n = 8$, and already $2\frac{1}{4}$ when $n = 20$, the case considered by the original discussor of that population. For population no. 2 we obtain that already when $n = 20$ with purposive selection δ^2 is $4\frac{1}{2}$ times that with random sampling.

In section 2 we used $\{(N-1)/N\}^2$ times the variance of

$$\sum_{i=1}^{N}\{y(u_i)-\bar{Y}\}^2/(N-1)$$

in sampling N from the same underlying population, which therefore is

$$\{k\alpha(1-\alpha)/N\}\left[2k\alpha(1-\alpha)\frac{N-1}{N}+\{1-6\alpha(1-\alpha)\}\left(\frac{N-1}{N}\right)^2\right].$$

To compute the variance of

$$\hat{\sigma}^2=\frac{k}{k-(1/n)}\bar{y}\left(1-\frac{\bar{y}}{k}\right),$$

we need the variance of $\bar{y}-k^{-1}\bar{y}^2$,

$$\begin{aligned}V(\bar{y}^2)&=\mathscr{E}\bar{y}^4-(\mathscr{E}\bar{y}^2)^2\\&=\mathscr{E}(\bar{y}-\bar{Y})^4+4\bar{Y}\mathscr{E}\bar{y}^3-6\bar{Y}^2\mathscr{E}\bar{y}^2+3\bar{Y}^4-(\mathscr{E}\bar{y}^2)^2\\&=\mathscr{E}(\bar{y}-\bar{Y})^4+4\bar{Y}[\mathscr{E}(\bar{y}-\bar{Y})^3+3\bar{Y}\{\mathscr{E}(\bar{y}-\bar{Y})^2+\bar{Y}^2\}-2\bar{Y}^3]\\&\qquad-6\bar{Y}^2\{\mathscr{E}(\bar{y}-\bar{Y})^2+\bar{Y}^2\}+3\bar{Y}^4-\{\mathscr{E}(\bar{y}-\bar{Y})^2+\bar{Y}^2\}^2\\&=\mu_4(\bar{y})+4\bar{Y}\mu_3(\bar{y})+4\bar{Y}^2\mu_2(\bar{y})-\mu_2^2(\bar{y}).\end{aligned}$$

But, if x_j is the jth observed $\mathscr{Y}$-value *minus* $\bar{Y}$,

$$\mu_4(\bar{y})=\frac{1}{n^4}\mathscr{E}(\sum_{j=1}^{n}x_j)^4=\frac{1}{n^4}\sum_{j=1}^{n}\mathscr{E}x_j^4+\frac{3}{n^4}\sum_{j\neq k}\sum\mathscr{E}x_j^2x_k^2$$

plus sums involving $\mathscr{E}x_jx_kx_l^2$ for different subscripts and $\mathscr{E}x_jx_kx_lx_m$ for different subscripts, each of which vanish; so

$$\mu_4(\bar{y})=\frac{\mu_4}{n^3}+\frac{3(n-1)}{n^3}\mu_2^2.$$

Also

$$\mu_3(\bar{y})=\frac{1}{n^3}\sum_{j=1}^{n}\mathscr{E}x_j^3$$

plus sums involving $\mathscr{E}x_jx_k^2$ for different subscripts and $\mathscr{E}x_jx_kx_l$ for different subscripts, each of which vanish; therefore,

$$\mu_3(\bar{y})=\frac{\mu_3}{n^2}.$$

Finally

$$\mu_2(\bar{y})=\frac{\mu_2}{n}.$$

Hence

$$V(\bar{y}^2) = \frac{\mu_4}{n^3} + 4\bar{Y}\frac{\mu_3}{n^2} + 4\bar{Y}^2\frac{\mu_2}{n} + \frac{2n-3}{n^3}\mu_2^2.$$

Also

$$\begin{aligned}\text{Cov}(\bar{y}, \bar{y}^2) &= \mathscr{E}(\bar{y} - \bar{Y})\bar{y}^2 \\ &= \mathscr{E}(\bar{y} - \bar{Y})\{(\bar{y} - \bar{Y})^2 + 2\bar{y}\bar{Y} - \bar{Y}^2\} \\ &= \mathscr{E}(\bar{y} - \bar{Y})^3 + 2\bar{Y}\mathscr{E}(\bar{y} - \bar{Y})^2,\end{aligned}$$

which equals

$$\frac{\mu_3}{n^2} + 2\bar{Y}\frac{\mu_2}{n}.$$

Therefore,

$$\begin{aligned}V\left(\bar{y} - \frac{1}{k}\bar{y}^2\right) \\ &= \frac{\mu_2}{n} - \frac{4}{k}\bar{Y}\frac{\mu_2}{n} - \frac{2}{k}\frac{\mu_3}{n^2} + \frac{1}{k^2}\left(\frac{\mu_4}{n^3} + 4\bar{Y}\frac{\mu_3}{n^2} + 4\bar{Y}^2\frac{\mu_2}{n} + \frac{2n-3}{n^3}\mu_2^2\right) \\ &= \frac{\mu_2}{n}\left(1 - \frac{4\bar{Y}}{k} + \frac{4\bar{Y}^2}{k^2}\right) - \frac{2\mu_3}{kn^2}\left(1 - \frac{2\bar{Y}}{k}\right) + \frac{1}{k^2n^3}(\mu_4 + (2n-3)\mu_2^2).\end{aligned}$$

Multiplying this by $\{k/(k - (1/n))\}^2$ and substituting for μ_2 and μ_4 the expressions regarded above, and for $\bar{Y}$ and μ_3

$$\bar{Y} = k\alpha, \quad \mu_3 = k\alpha(1-\alpha)(1-2\alpha),$$

we obtain

$$\begin{aligned}V(\hat{\sigma}^2) &= \frac{k\alpha(1-\alpha)}{n}\left(\frac{1}{1-(1/kn)}\right)^2 \\ &\qquad \times \left\{(1-2\alpha)^2\left(1 - \frac{2}{kn}\right) + \frac{2\alpha(1-\alpha)}{kn} + \frac{1}{k^2n^2} - \frac{6\alpha(1-\alpha)}{k^2n^2}\right\} \\ &= \frac{k\alpha(1-\alpha)}{n}\left\{(1-2\alpha)^2 + \frac{2\alpha(1-\alpha)}{kn-1}\right\}.\end{aligned}$$

But

$$V(s^2) = \frac{k\alpha(1-\alpha)}{n}\left\{1 + 2\alpha(1-\alpha)\left(k + \frac{k}{n-1} - 3\right)\right\}.$$

Hence

$$V(s^2) - V(\hat{\sigma}^2) = 2k^2\alpha^2(1-\alpha)^2\left(\frac{1}{n-1} - \frac{1}{kn-1}\right).$$

CHAPTER IX

SYSTEMATIC RANDOM SAMPLING

1. Definitions

Systematic random sampling is also referred to briefly as systematic sampling, although sometimes the latter term is reserved for a selection according to some systematic scheme without introducing an element of randomness.

Systematic random sampling includes a class of methods of sampling. Suppose we wish to draw a sample of n^0 out of a population of N. Let k be a natural number such that $n^0k \leq N \leq n^0(k+1)$. Define n as the largest integer contained in N/k. A common systematic sampling design is as follows: we assume the population is linearly ordered in some way such that we can refer to the units by number without ambiguity. A natural number x is selected between 1 and k, such that each of the numbers 1 to k has the same probability of being selected. If x takes on the value t and $t > N - nk$, the sample, $\mathscr{s}_t$, consists of the n units $u_t, u_{t+k}, \ldots, u_{t+(n-1)k}$; if x takes on the value t and $t \leq N - nk$ (this cannot happen if $N = nk$), the sample, $\mathscr{s}_t$, consists of the $(n+1)$ units $u_t, u_{t+k}, \ldots, u_{t+nk}$.† This type of design is called "*every kth systematic sampling*",‡ and is very widely applied. k is called the *sampling interval.*

EXAMPLE. $N = 149$, and we desire a sample of $n^0 = 60$. $k = 2$ and so $n = 74$, $\mathscr{s}_1 = \{u_1, u_3, \ldots, u_{147}, u_{149}\}$, $\mathscr{s}_2 = \{u_2, u_4, \ldots, u_{148}\}$. The actual sample sizes will be 75 or 74, rather than 60. Some may prefer to use in this case a sampling interval of 3 with $n = 49$, so that $\mathscr{s}_1 = \{u_1, u_4, \ldots, u_{148}\}$, $\mathscr{s}_2 = \{u_2, u_5, \ldots, u_{149}\}$, $\mathscr{s}_3 = \{u_3, u_6, \ldots, u_{147}\}$ of sizes 50, 50 and 49, respectively, both closer to 60 and not exceeding 60.

As emerges in the above example, the method may lead to a sample size quite different from the one aimed at. Moreover, as we shall see later, the

† $\mathscr{s}_1, \ldots, \mathscr{s}_k$ exhausts the population, since, by definition of k and n, $N - nk < k$; and we draw either a sample of size $n + 1$ with probability $(N/k) - n$ or one of size n with probability $n + 1 - (N/k)$.

‡ If units (or objects representing them) are compactly packed one alongside the other and are of uniform size (length, weight, etc.), a very good approximation to this type of sampling can be achieved without numbering the units, by measuring them off.

sample average is an unbiased estimator of $\bar{Y}$ only if N is an integral multiple of n^0 and so of n. In section 3 we shall give methods of sampling and estimation which partly or wholly overcome the disadvantages of the above method. In further sections, to simplify the presentation, we shall assume that $N = nk$, unless mentioned otherwise.

Another systematic sampling plan, which yields samples of size about ln (with l some number satisfying $1 < l < k$), is to draw a simple random sample (without replacement) of l natural numbers $x_1, \ldots, x_l$ between 1 and k. If $x_1 = t_1, \ldots, x_l = t_l$, the sample consists of the union of $\mathscr{s}_{t_1}, \mathscr{s}_{t_2}, \ldots, \mathscr{s}_{t_l}$. For example if $N = 120$ and $l = 2$, and if the desired total sample size is 10,

$$\mathscr{s}_1 = \{u_1, u_{25}, u_{49}, u_{73}, u_{97}\}, \ldots, \mathscr{s}_{24} = \{u_{24}, u_{48}, u_{72}, u_{96}, u_{120}\}.$$

This is called "systematic sampling *with l random starts*".

Systematic sampling can also be carried out in two (or more) dimensions. Since this is not quite as frequent an occurrence in sample surveys, we shall not discuss this matter here.†

Systematic sampling can, of course, be used together with stratified and cluster sampling; e.g., it can be used in subsampling of clusters, or in sampling from strata.

Systematic sampling of every kth can be regarded both as a modified form of stratified sampling and as a form of cluster sampling. "Every kth" sampling is *stratified* sampling in which the first stratum consists of $u_1, \ldots, u_k$; the second of $u_{k+1}, \ldots, u_{2k}$; $\ldots$; the $(n - 1)$th of $u_{(n-1)k+1}, \ldots, u_{nk}$ and the remaining u's; and we sample one from each stratum. However, the sampling does *not* proceed independently in each stratum, since in *each* stratum the xth unit is chosen. Similarly, in systematic sampling with l random starts as described above, we have n strata and sample l in each stratum, but not independently.

Systematic sampling with l ($l \geq 1$) random starts is a sample of l *clusters* (without subsampling), in the sense that we choose at random l of the possible clusters $\mathscr{s}_t$ ($t = 1, \ldots, k$).

2. Advantages and disadvantages

Systematic sampling of a population or subpopulation is often substantially simpler or more convenient to carry out than simple random sampling of $\mathscr{P}$, and usually the arrangement of the field work is easier. If it is desired to include a given percentage of the population in the sample, or if N is known in advance, the sample can be drawn without first listing or labelling the ele-

† See Vos, Rev. Intern. Statist. Inst., **32** (1964) 226–241, Matérn, Medd. fr. Statens Skogsforsknings Inst., **49** (1960) Nr. 5 (144 pp.), and references quoted there.

ments. Moreover, as we shall see below, in certain cases it may lead to substantially reduced variances.

The main disadvantage lies in the possibility of such periodicities in $\mathscr{P}$ as are not foreseen or guarded against in the design and which may substantially increase the variance of estimators.† Another disadvantage concerns the difficulties of estimating the variance of estimators; these will be explained in section 8.

3. Estimator for $\bar{Y}$

It is convenient to also use a double subscript notation in which we denote the jth member of $\mathscr{s}_t$ by u_{tj};‡ thus $\bar{Y}_{t.}$ is the average of the observations in $\mathscr{s}_t$. We wish to estimate

$$\bar{Y} = (y_1 + \ldots + y_N)/N$$

($= \bar{\bar{Y}}$ in the notation of cluster sampling). For $N = nk$, we also have

$$\bar{Y} = (1/k) \sum_{t=1}^{k} \bar{Y}_{t.}.$$

Let $a_{tx_g} = 1$ when $x_g = t$ and 0 otherwise, then we may write the sample averages corresponding to the different random starts as

$$\bar{y}_{sy}(g) = \sum_{t=1}^{k} a_{tx_g} \bar{Y}_{t.} = \bar{Y}_{x_g.} \quad (g = 1, \ldots, l).$$

3.1. N EQUAL TO nk

If $N = nk$, $\bar{y}_{sy}(g)$ is an unbiased estimator of $\bar{Y}$, since then $\mathscr{E} a_{tx_g} = 1/k$, and since $\bar{Y}$ is the unweighted average of $\bar{Y}_{1.}, \ldots, \bar{Y}_{k.}$. When $l = 1$ this is our estimator of $\bar{Y}$; when $l > 1$ we form the average of the l estimators:

$$\bar{y}_{sy} = \frac{1}{l} \sum_{g=1}^{l} \bar{y}_{sy}(g).$$

† As an extreme case consider a period k and denote the first k $\mathscr{Y}$-values by $a_1, \ldots, a_k$ with average $\bar{a}$. In the notation of the next section

$$(N-1)S^2 = \Sigma\Sigma(y_{tj} - \bar{Y})^2 = \sum_t \{\sum_j (y_{tj} - \bar{Y}_{t.})^2 + n(\bar{Y}_{t.} - \bar{Y})^2\}$$

equals $0 + n\Sigma(a_t - \bar{a})^2$, so that the variance of $\bar{y}$ in simple random sampling is

$$\frac{nk-n}{nk} \frac{1}{n} \frac{1}{N-1} n\Sigma(a_t - \bar{a})^2 = \frac{k-1}{k} \frac{N}{N-1} \frac{1}{n} \Sigma(a_t - \bar{a})^2/k,$$

whereas the variance of the sample average in a systematic sample of n is, by section 4, $\Sigma(a_t - \bar{a})^2/k$, about $\{(k-1)/k\}n$ times as large.

‡ Properly speaking, there should be a comma between the two subscripts so as to distinguish the double subscript notation from the single subscript one, in which tj denotes t times j; but usually the meaning is evident even without the comma.

3.2. N NOT EQUAL TO nk

A method to overcome wholly or partly the biasedness of the sample mean if $N \neq nk$, as well as the difference between n^0 and n if this is not small, is the following. Multiply N/n^0 by a power, p, of 10:† $(N/n^0)10^p$, and round to an integer K; p should be chosen such that the relative amount of rounding is very small. Select a random digit x between 1 and K, compute the smallest integers containing, respectively,

$$x10^{-p}, \quad (x + K)10^{-p}, \ldots, (x + (n^0 - 1)K)10^{-p}, \quad (x + n^0K)10^{-p}, \ldots.$$

Then select the u's with the resulting subscripts insofar as they exist. Each unit will then have probability $10^p/K$ of being included.

EXAMPLE. $N = 160$, $n^0 = 60$,

$$K = \frac{160}{60} \times 10 = 26.7,$$

which we may round to 27 or 26. Thus in the latter case

$$\mathscr{s}_1 = \{u_1, u_3, u_6, u_8, u_{11}, \ldots, u_{159}\},$$
$$\mathscr{s}_2 = \{u_1, u_3, u_6, u_8, u_{11}, \ldots, u_{159}\},$$
$$\mathscr{s}_3 = \{u_1, u_3, u_6, u_9, u_{11}, \ldots, u_{159}\},$$

etc. and the sample size is 62 or 61, while

$$\mathscr{E}\bar{y}_{sy} = \frac{1}{26}\left(\frac{Y_{1.}}{62} + \ldots + \frac{Y_{14.}}{62} + \frac{Y_{15.}}{61} + \ldots + \frac{Y_{26.}}{61}\right)$$
$$= \bar{Y} + \left\{\left(\frac{1}{1\,586} - \frac{1}{1\,600}\right)(Y_{15.} + \ldots + Y_{26.}) - \left(\frac{1}{1\,600} - \frac{1}{1\,612}\right)(Y_{1.} + \ldots + Y_{14.})\right\}.$$

It would be preferable to take $K = \frac{160}{60} \times 3 = 8$ exactly, and so any sample will be of size 60.

A method which entirely overcomes the above-mentioned disadvantages when $N \neq nk$ is the following. Select a random number x not between 1 and k, but between 1 and N. If x takes on the value t, the units selected will have the indices $t, t + k, \ldots, t + (n^0 - 1)k$, where those indices which exceed N are replaced by their excess over N (hence the name *circular* sampling). For example, if $N = 14$ and $n^0 = 5$, and we take $k = 2$ (according to its definition

† Some power of ten is suggested just for the sake of ease of application. Any integer will do; in particular n^0, or a factor of n^0 that does not divide N, in which case the sample mean will be exactly unbiased.

in section 1), we get for $t = 10$, the 5 indices 10, 12, 14, 2, 4. However, it is not necessary to define k as in section 1; thus if we take $k = 3$, we get the 5 indices 10, 13, 2, 5, 8. Under this system each unit will have exactly probability n^0/N of being included.

If more than a single random start is used, both of the above methods may lead to multiple inclusion of some units.

If we do not wish to modify the method of sampling described in section 1, we can keep the unbiasedness property by *modifying our estimator*. The sample mean for a single start is

$$\sum_{t=1}^{k} a_{tx} Y_{t.}/n_t,$$

where n_t, the sample size, is sometimes n and sometimes $n + 1$. If we want an unbiased estimator of $\bar{Y}$, we may use in its place

$$\sum_{t=1}^{k} a_{tx} Y_{t.}/(N/k).$$

Its expected value equals

$$\frac{1}{k} \sum_{b=1}^{k} \{ \sum_{t=1}^{k} a_{tb} Y_{t.}/(N/k) \},$$

but since

$$\sum_{b=1}^{k} a_{tb} = 1,$$

this equals

$$(1/N) \sum_{t=1}^{k} Y_{t.} = \bar{Y}.$$

4. The variance of $\bar{y}_{sy}$

When $l = 1$,

$$V(\bar{y}_{sy}(g)) = \frac{1}{k} \sum_{t=1}^{k} (\bar{Y}_{t.} - \bar{Y})^2.$$

This is immediate from the fact that $\bar{y}_{sy} = \bar{Y}_{x_g.}$ and the definition of a variance, but can, if desired, also be derived formally as follows:

$$V(\sum_{t=1}^{k} a_{tx_g} \bar{Y}_{t.}) = \sum_{l=1}^{k} V(a_{tx_g}) \bar{Y}^2_{t.} + \sum_{t \neq t'}\sum \operatorname{Cov}(a_{tx_g}, a_{t'x_g}) \bar{Y}_{t.} \bar{Y}_{t'.}$$

$$= \frac{1}{k}\left(1 - \frac{1}{k}\right) \Sigma \bar{Y}^2_{t.} - \frac{1}{k^2} \sum_{t \neq t'}\sum \bar{Y}_{t.} \bar{Y}_{t'.} = \frac{1}{k} \sum_{t=1}^{k} (\bar{Y}_{t.} - \bar{Y})^2.$$

For the estimator given at the end of the last section we get similarly

$$\frac{1}{k}\sum\left(\frac{Y_{t.}}{N/k}-\bar{Y}\right)^2=\frac{k}{N^2}\sum\left(Y_{t.}-\frac{Y}{k}\right)^2.$$

It is also customary to write

$$S^2(\bar{y}_{sy}(g))=\frac{k}{k-1}V(\bar{y}_{sy}(g))=\frac{1}{k-1}\sum_{t=1}^{k}(\bar{Y}_{t.}-\bar{Y})^2.$$

For more than one random start, $\bar{y}_{sy}$ is the arithmetic mean of $\bar{Y}_{x_1.}, \ldots, \bar{Y}_{x_l.}$, where $\{x_1, \ldots, x_l\}$ is a simple random sample drawn without replacement from $\{1, \ldots, k\}$, so that

$$V(\bar{y}_{sy})=\frac{k-l}{k}\frac{1}{l}S^2(\bar{y}_{sy}(g))=\frac{k-l}{k-1}\frac{1}{l}V(\bar{y}_{sy}(g)).$$

It is helpful to express the variance of $\bar{y}_{sy}$ in terms of the symbols used in Chapter V. Here we have to replace $\bar{M}$ by n, N by k and $\bar{\bar{Y}}$ by $\bar{Y}$, so, for $l = 1$,

$$S_w^2(t)=\frac{1}{n-1}\sum_{j=1}^{n}(y_{tj}-\bar{Y}_{t.})^2,$$

$$S_w^2=\frac{1}{k}\sum_{t=1}^{k}S_w^2(t),$$

$$S_l^2=\frac{1}{k-1}\sum_{t=1}^{k}(\bar{Y}_{t.}-\bar{Y})^2=S^2(\bar{y}_{sy}(g)),$$

$$S^2=\frac{1}{kn-1}\{k(n-1)S_w^2+(k-1)nS_l^2\}$$

$$=\frac{1}{kn-1}\{k(n-1)S_w^2+knV(\bar{y}_{sy}(g))\}.$$

Therefore,

$$V(\bar{y}_{sy}(g))=\frac{kn-1}{kn}S^2-\frac{k(n-1)}{kn}S_w^2$$

$$=\frac{N-1}{N}S^2-\frac{n-1}{n}S_w^2=\sigma^2-\sigma_w^2=\sigma_l^2.$$

In terms of the intracluster correlation coefficient – the correlation between pairs of units corresponding to the same potential start – we have

$$S^2(\bar{y}_{sy}(g))=\frac{nk-1}{(k-1)n^2}S^2\{1+(n-1)r\},$$

so

$$V(\bar{y}_{sy}(g)) = \frac{N-1}{N}\,\frac{S^2}{n}\{1+(n-1)r\}$$

$$= \frac{N-n}{N}\,\frac{S^2}{n} + \frac{n-1}{N}\,\frac{S^2}{n}\{1+(N-1)r\},$$

and for l starts

$$V(\bar{y}_{sy}) = \frac{N-nl}{N}\,\frac{S^2}{nl} + \frac{(n-1)(k-l)}{N(k-1)}\,\frac{S^2}{nl}\,\{1+(N-1)r\}.$$

So the variance is smaller than for simple random sampling with replacement if $r < 0$, without replacement if $r < -(1/(N-1))$. We already saw that r is small if the clusters are heterogeneous with respect to the $\mathscr{Y}$-values of its units. Thus it may be advantageous, if possible, to *arrange* the units, prior to sampling, in approximately increasing or decreasing order of their $\mathscr{Y}$-values; we then achieve a large heterogenity of clusters. (However, for large k or extreme values of $y_{\min}$ and $y_{\max}$, the range of possible values of the sample mean may become substantial.) Actually, as we shall see in section 6.2, stratified sampling may give still smaller variance in case of monotone order. In most naturally occurring populations and for most $\mathscr{Y}$, adjacent units are positively correlated, and then, in the absence of periodicities and if the skipping interval is not very small, r will be negative. Then, even if $|r|$ is small, $1 + (N-1)\,r$ may be much less than 1.

5. Example

Let us take again the population

$$\mathscr{P} = \{u_1, u_2, u_3, u_4\}$$

and the characteristic $\mathscr{Y}$

$$y_1 = 3, \quad y_2 = 4, \quad y_3 = 3, \quad y_4 = 5.$$

Let us sample systematically with $k = 2$, $n = 2$ and $l = 1$.

(a) If the order in the population is

$\mathscr{P}$	u_1	u_2	u_3	u_4
$\mathscr{Y}$	3	4	3	5

then the possible samples are

Sample	$\mathscr{Y}$	$\bar{y}_{sy}$	$\bar{y}_{sy} - \mathscr{E}\bar{y}_{sy}$
$\{u_1, u_3\}$	(3, 3)	3	$-\frac{3}{4}$
$\{u_2, u_4\}$	(4, 5)	$4\frac{1}{2}$	$\frac{3}{4}$
		$\mathscr{E}\bar{y}_{sy} = 3\frac{3}{4}$	0

and so

$$V(\bar{y}_{sy}) = \tfrac{1}{2}[(\tfrac{3}{4})^2 + (\tfrac{3}{4})^2] = \tfrac{9}{16}.$$

(b) If the order in the population is

$\mathscr{P}$	u_1	u_3	u_2	u_4
$\mathscr{Y}$	3	3	4	5

then the possible samples are

Sample	$\mathscr{Y}$	$\bar{y}_{sy}$	$\bar{y}_{sy} - \mathscr{E}\bar{y}_{sy}$
$\{u_1, u_2\}$	(3, 4)	$3\frac{1}{2}$	$-\frac{1}{4}$
$\{u_3, u_4\}$	(3, 5)	4	$\frac{1}{4}$
		$\mathscr{E}\bar{y}_{sy} = 3\frac{3}{4}$	0

therefore

$$V(\bar{y}_{sy}) = \tfrac{1}{2}[(\tfrac{1}{4})^2 + (\tfrac{1}{4})^2] = \tfrac{1}{16}.$$

We see (cf. the end of the previous section) that the variance of the sample mean is smaller when the variances within the possible samples are larger [in (a) the variances are 0 and $\frac{1}{4}$, in (b) they are $\frac{1}{4}$ and 1].

6. Comparisons of simple random sampling, systematic sampling and stratified sampling for various types of populations

6.1. RANDOMLY ORDERED POPULATIONS

We first examine the case in which the ordering in the population may be considered essentially *random*. In that case we would expect that formation of strata or clusters do not have a pronounced effect on the variance if the sample is not small.

A way to examine the situation is to suppose that $u_1, \ldots, u_N$ have themselves been drawn at random from a superpopulation (see Chapter VIII) such that

$$\mathscr{E} y_i = \mu, \quad \mathscr{E}(y_i - \mu)(y_{i'} - \mu) = 0 \quad (i \neq i' \in \{1, \ldots, N\}).$$

If we draw from $\mathscr{P}$ a simple random sample of n, the variance of its mean is

$$V_{\text{ran}} = \frac{N-n}{N}\frac{1}{n}\sum_{i=1}^{N}(y_i - \bar{Y})^2/(N-1).$$

Now

$$\sum_{i=1}^{N}(y_i - \bar{Y})^2 = \sum_{i=1}^{N}(y_i - \mu)^2 - N(\bar{Y} - \mu)^2.$$

Writing σ_i^2 for $\mathscr{E}(y_i - \mu)^2$, the mean of

$$\sum_{i=1}^{N}(y_i - \mu)^2$$

is

$$\sum_{i=1}^{N}\sigma_i^2,$$

and the mean of $(\bar{Y} - \mu)^2$ is

$$(1/N^2)\sum_{i=1}^{N}\sigma_i^2,$$

so

$$\mathscr{E} V_{\text{ran}} = \frac{N-n}{N}\frac{1}{n}\sum_{i=1}^{N}\sigma_i^2/N.$$

If we draw from $\mathscr{P}$ a stratified sample with as strata $\{u_{1j}, \ldots, u_{kj}\}$ $(j = 1, \ldots, n)$, drawing 1 observation from each stratum, independently from stratum to stratum, we have for the variance of the sample mean, since $N_j = k$ and $n_j = 1$,

$$V_{\text{st}} = \sum_{j=1}^{n}\frac{k}{N^2}\frac{k-1}{1}\sum_{t=1}^{k}(y_{tj} - \bar{Y}_{.j})^2/(k-1).$$

But

$$\begin{aligned}\mathscr{E}\sum_{j=1}^{n}\sum_{t=1}^{k}(y_{tj} - \bar{Y}_{.j})^2 &= \mathscr{E}\sum_{j=1}^{n}\sum_{t=1}^{k}(y_{tj} - \mu)^2 - k\mathscr{E}\sum_{j=1}^{n}(\bar{Y}_{.j} - \mu)^2 \\ &= \left(1 - \frac{1}{k}\right)\sum_{i=1}^{N}\sigma_i^2,\end{aligned}$$

since

$$\mathscr{E}(y_{tj} - \mu)^2 = \sigma_{tj}^2 \quad \text{and} \quad \mathscr{E}(\bar{Y}_{.j} - \mu)^2 = \sum_{t=1}^{k} \sigma_{tj}^2/k^2.$$

So

$$\mathscr{E} V_{\text{st}} = \frac{k-1}{N^2} \sum_{i=1}^{N} \sigma_i^2 = \frac{N-n}{N} \frac{1}{n} \sum_{i=1}^{N} \sigma_i^2/N.$$

If we draw from $\mathscr{P}$ a systematic sample,

$$V_{\text{sy}} = \frac{1}{k} \sum_{t=1}^{k} (\bar{Y}_{t.} - \bar{Y})^2.$$

Now

$$\mathscr{E}(\bar{Y} - \mu)^2 = \sum_{i=1}^{N} \sigma_i^2/N^2 \quad \text{and} \quad \mathscr{E}(\bar{Y}_{t.} - \mu)^2 = \sum_{j=1}^{n} \sigma_{tj}^2/n^2,$$

so

$$\mathscr{E} \sum_{t=1}^{k} (\bar{Y}_{t.} - \mu)^2 - k\mathscr{E}(\bar{Y} - \mu)^2 = \sum_{i=1}^{N} \sigma_i^2 \left(\frac{1}{n^2} - \frac{k}{N^2}\right)$$

and

$$\mathscr{E} V_{\text{sy}} = \frac{1}{k} \left(\frac{1}{n^2} - \frac{k}{N^2}\right) \sum_{i=1}^{N} \sigma_i^2 = \frac{1}{N}\left(\frac{1}{n} - \frac{1}{N}\right) \sum_{i=1}^{N} \sigma_i^2 = \frac{N-n}{N} \frac{1}{n} \sum_{i=1}^{N} \sigma_i^2/N.$$

Therefore

$$\mathscr{E} V_{\text{ran}} = \mathscr{E} V_{\text{st}} = \mathscr{E} V_{\text{sy}}$$

in this case. Of course, for any particular one population $\mathscr{P}$ drawn from such a superpopulation there may well be substantial differences between V_{ran}, V_{st} and V_{sy}.

6.2. NONAUTOCORRELATED POPULATIONS IN GENERAL AND WITH LINEAR TREND

More generally, if $\mathscr{E} y_i = \mu_i$ instead of μ and, for $i \neq j$,

$\mathscr{E}(y_i - \mu_i)(y_j - \mu_j) = 0,$

$$\mathscr{E} \sum_{i=1}^{N} (y_i - \bar{Y})^2 = \sum_{i=1}^{N} \mathscr{E}(y_i - \mu_i)^2 - 2\mathscr{E} \bar{Y} \sum_{i=1}^{N} (y_i - \mu_i)$$

$$+ 2 \sum_{i=1}^{N} \mu_i(\mathscr{E} y_i - \mu_i) + \sum_{i=1}^{N} \mathscr{E}(\bar{Y} - \mu_i)^2.$$

The first term on the right-hand side equals

$$\sum_{i=1}^{N} \sigma_i^2,$$

the second

$$-2\left\{\mathscr{E}\frac{1}{N}\sum_{i'=1}^{N}(y_{i'}-\mu_{i'})\sum_{i=1}^{N}(y_i-\mu_i)+\sum_{i'=1}^{N}\frac{1}{N}\mu_{i'}\sum_{i=1}^{N}\mathscr{E}(y_i-\mu_i)\right\}$$
$$= -\frac{2}{N}\sum_{i=1}^{N}\sigma_i^2,$$

the third 0, and the fourth

$$N\mathscr{E}(\bar{Y}-\bar{\mu})^2-2\mathscr{E}(\bar{Y}-\bar{\mu})\sum_{i=1}^{N}(\mu_i-\bar{\mu})+\sum_{i=1}^{N}(\mu_i-\bar{\mu})^2$$
$$=\frac{1}{N}\sum_{i=1}^{N}\sigma_i^2+\sum_{i=1}^{N}(\mu_i-\bar{\mu})^2.$$

Therefore

$$\mathscr{E}V_{\text{ran}}=\frac{N-n}{N}\frac{1}{n}\sum_{i=1}^{N}\sigma_i^2/N+\frac{N-n}{N}\frac{1}{n}\sum_{i=1}^{N}(\mu_i-\bar{\mu})^2/(N-1).$$

Also

$$\mathscr{E}\sum_{j=1}^{n}\sum_{t=1}^{k}(y_{tj}-\bar{Y}_{.j})^2=\sum_{j=1}^{n}\sum_{t=1}^{k}\mathscr{E}(y_{tj}-\mu_{tj})^2-2\mathscr{E}\sum_{j=1}^{n}\bar{Y}_{.j}\sum_{t=1}^{k}(y_{tj}-\mu_{tj})$$
$$+2\sum_{j=1}^{n}\sum_{t=1}^{k}\mu_{tj}\mathscr{E}(y_{tj}-\mu_{tj})+\sum_{j=1}^{n}\sum_{t=1}^{k}\mathscr{E}(\bar{Y}_{.j}-\mu_{tj})^2.$$

The first term on the right-hand side is

$$\sum_{i=1}^{N}\sigma_i^2,$$

the second equals

$$-\frac{2}{k}\sum_{j=1}^{n}\mathscr{E}\sum_{t'=1}^{k}y_{t'j}\sum_{t=1}^{k}(y_{tj}-\mu_{tj})=-\frac{2}{k}\sum_{j=1}^{n}\sum_{t=1}^{k}\sigma_{tj}^2=-\frac{2}{k}\sum_{i=1}^{N}\sigma_i^2,$$

the third equals 0, and the fourth equals

$$k\sum_{j=1}^{n}\mathscr{E}(\bar{Y}_{.j}-\bar{\mu}_{.j})^2-2\sum_{j=1}^{n}\mathscr{E}(\bar{Y}_{.j}-\bar{\mu}_{.j})\sum_{t=1}^{k}(\mu_{tj}-\bar{\mu}_{.j})+\sum_{j=1}^{n}\sum_{t=1}^{k}(\mu_{tj}-\bar{\mu}_{.j})^2$$
$$=k\sum_{j=1}^{n}\sum_{t=1}^{k}\sigma_{tj}^2/k^2+\sum_{j=1}^{n}\sum_{t=1}^{k}(\mu_{tj}-\bar{\mu}_{.j})^2=\frac{1}{k}\sum_{i=1}^{N}\sigma_i^2+\sum_{j=1}^{n}\sum_{t=1}^{k}(\mu_{tj}-\bar{\mu}_{.j})^2.$$

So

$$\mathscr{E}V_{\text{st}} = \frac{N-n}{N}\frac{1}{n}\sum_{i=1}^{N}\sigma_i^2/N + \frac{1}{n}\sum_{j=1}^{n}\sum_{t=1}^{k}(\mu_{tj} - \bar{\mu}_{.j})^2/N.$$

Finally

$$\mathscr{E}\sum_{t=1}^{k}(\bar{Y}_{t.} - \bar{Y})^2 = \mathscr{E}\sum_{t=1}^{k}(\bar{Y}_{t.} - \bar{\mu}_{t.})^2$$

$$- 2\mathscr{E}\sum_{t=1}^{k}(\bar{Y} - \bar{\mu}_{t.})(\bar{Y}_{t.} - \bar{\mu}_{t.}) + \mathscr{E}\sum_{t=1}^{k}(\bar{Y} - \bar{\mu}_{t.})^2.$$

The first term on the right-hand side is

$$\sum_{t=1}^{k}\left(\sum_{j=1}^{n}\sigma_{tj}^2/n^2\right) = \frac{1}{n^2}\sum_{i=1}^{N}\sigma_i^2,$$

the second

$$-\frac{2}{k}\sum_{t=1}^{k}\left(\sum_{j=1}^{n}\sigma_{tj}^2/n^2\right) = -\frac{2}{kn^2}\sum_{i=1}^{N}\sigma_i^2,$$

and the third

$$k\mathscr{E}(\bar{Y} - \bar{\mu})^2 - 2\sum_{t=1}^{k}(\bar{\mu}_{t.} - \bar{\mu})\mathscr{E}(\bar{Y} - \bar{\mu}) + \sum_{t=1}^{k}(\bar{\mu}_{t.} - \bar{\mu})^2$$

$$= \frac{1}{n^2k}\sum_{i=1}^{N}\sigma_i^2 + \sum_{t=1}^{k}(\bar{\mu}_{t.} - \bar{\mu})^2.$$

So

$$\mathscr{E}V_{\text{sy}} = \frac{k-1}{k^2n^2}\sum_{i=1}^{N}\sigma_i^2 + \frac{1}{k}\sum_{t=1}^{k}(\bar{\mu}_{t.} - \bar{\mu})^2$$

$$= \frac{N-n}{N}\frac{1}{n}\sum_{i=1}^{N}\sigma_i^2/N + \frac{1}{k}\sum_{t=1}^{k}(\bar{\mu}_{t.} - \bar{\mu})^2.$$

Suppose, e.g., that $\mu_i = \alpha + \beta i$ (model of *linear trend*); then

$$\sum_{i=1}^{N}(\mu_i - \bar{\mu})^2/(N-1) = \beta^2\frac{N(N+1)}{12}.$$

So for all j

$$\sum_{t=1}^{k}(\mu_{tj} - \bar{\mu}_{.j})^2/(k-1) = \beta^2\frac{k(k+1)}{12}.$$

If $\mu_i = \alpha + \beta i$, the mean for $\mathscr{Y}$ of the $(t + 1)$st potential cluster exceeds that of the tth by β, so

$$\sum_{t=1}^{k} (\bar{\mu}_{t.} - \bar{\mu})^2/(k - 1) = \beta^2 \frac{k(k + 1)}{12}.$$

Therefore, if we write

$$A = \frac{N - n}{N} \frac{1}{n} \sum_{i=1}^{N} \sigma_i^2/N,$$

we have

$$\mathscr{E} V_{\text{ran}} = A + \frac{N - n}{N} \frac{1}{n} \beta^2 \frac{N(N + 1)}{12} = A + \beta^2 \frac{(k - 1)(kn + 1)}{12},$$

$$\mathscr{E} V_{\text{st}} = A + \frac{1}{nN} \beta^2 \sum_{j=1}^{n} \frac{k(k - 1)(k + 1)}{12} = A + \beta^2 \frac{1}{n} \frac{(k - 1)(k + 1)}{12},$$

$$\mathscr{E} V_{\text{sy}} = A + \frac{1}{k} \beta^2 \frac{k(k - 1)(k + 1)}{12} = A + \beta^2 \frac{(k - 1)(k + 1)}{12}.$$

Hence, it is evident that for a linear trend

$$\mathscr{E} V_{\text{st}} \leq \mathscr{E} V_{\text{sy}} \leq \mathscr{E} V_{\text{ran}},$$

with the sharp inequalities holding only if $n > 1$.

The result is not surprising. Stratified and systematic sampling restricts the possibilities of the sample means taking extreme values: the former method forces the superpopulation values corresponding to the units observed in the successive strata to be not more than $2\beta k$ apart, the latter forces them to be exactly βk apart. Moreover, in contrast to the former method, the latter method does not allow any compensation in the position of these units due to n independent drawings; so that, under systematic sampling, there is a much larger probability that all of them (and so their average) fall at the left or the right ends of the ranges for the h strata than under stratified sampling.

NOTE. If we suppose that exist α and β such that, for u_i in $\mathscr{P}$, $y_i = \alpha + \beta i$, we evidently obtain the same results with

$$\sum_{i=1}^{N} \sigma_i^2 = 0.$$

However, such an assumption is rarely justified (see Chapter VIII, section 6). If more generally we suppose $y_i = \alpha + \beta i + w_i$ (and without loss of generality take

$$\sum_{i=1}^{N} w_i = 0),$$

we obtain

$$V_{\mathrm{ran}} = \frac{N-n}{N}\frac{1}{n}\sum_{i=1}^{N} w_i^2/(N-1) + 2\frac{N-n}{N}\frac{1}{n}\beta\sum_{i=1}^{N} iw_i/(N-1) + \beta^2\frac{(k-1)(kn+1)}{12},$$

$$V_{\mathrm{st}} = \frac{1}{n}\sum_{j=1}^{n}\sum_{t=1}^{k}(w_{tj}-\bar{w}_{.j})^2/N + \frac{2}{n}\beta\sum_{t=1}^{k} t\bar{w}_{t.}/k + \beta^2\frac{1}{n}\frac{(k-1)(k+1)}{12},$$

$$V_{\mathrm{sy}} = \sum_{t=1}^{k}\bar{w}_{t.}^2/k + 2\beta\sum_{t=1}^{k} t\bar{w}_{t.}/k + \beta^2\frac{(k-1)(k+1)}{12}.$$

However, it would be difficult to formulate reasonable assumptions concerning the behaviour in $\mathscr{P}$ of the terms involving the w_i such as would allow a comparison between V_{ran}, V_{st} and V_{sy}. Therefore, it may be worthwhile to examine a model of superpopulations such as the above.

6.3. AUTOCORRELATED POPULATIONS

Often we find that in linearly ordered populations $\mathscr{P}$ the values of a characteristic $\mathscr{Y}$ of members which are located close together are positively correlated, but that the correlations tend to decrease with the distance between members. A simple interpolatory formula for a superpopulation from which $\mathscr{P}$ may be considered to have been drawn states that the correlation coefficient between values of $\mathscr{Y}$ of two members v units apart is a function, ρ_v, of v only, and that ρ_v is a nonnegative, nonincreasing function of v. (In most cases it would be unreasonable to state this for $\mathscr{P}$ itself in any exact sense; but we can consider using an interpolatory formula here, leading to a model of sampling from a superpopulation.) COCHRAN, Ann. Math. Statist., **17** (1946) 164–177 has shown that in this case, if† $\mathscr{E}y_i = \mu$, and if $\mathscr{E}(y_i - \mu)^2 = \sigma^2$, then

$$\mathscr{E}V_{\mathrm{st}} \leq \mathscr{E}V_{\mathrm{ran}},$$

with the ratio increasing with increasing (total) sample size; and if moreover the second differences

$$\Delta^2\rho_v = \rho_{v+2} - 2\rho_{v+1} + \rho_v$$

are nonnegative then

$$\mathscr{E}V_{\mathrm{sy}} \leq \mathscr{E}V_{\mathrm{st}},$$

with equality only in case the second differences are 0. The condition about the second differences means that correlation coefficients rapidly fall off from

† For a relaxation of these conditions see QUENOUILLE, Ann. Math. Statist., **20** (1949) 355–375.

$\rho_0 = 1$ until they reach relatively small values. Such correlation functions appear to fit quite a few practical situations, especially if ρ_v can be considered approximately continuous in v. In this case the advantages of sampling by a method different from simple random sampling can be large. E.g., if

$$\rho_v = (L - v)/L$$

for $v \leq L$ and 0 for $v > L$, it can be shown that, for $L \geq N - 1$,

$$\mathscr{E}V_{\text{ran}}/\mathscr{E}V_{\text{sy}} = \mathscr{E}V_{\text{ran}}/\mathscr{E}V_{\text{st}} = \frac{kn+1}{k+1}.$$

Some empirical data are given in COCHRAN, Sampling Techniques (Wiley, 2nd ed., 1963) and YATES, Sampling Methods for Censuses and Surveys (Griffin, 3rd ed., 1960).

7. Effect of the number of starts

We saw that with l random starts $\bar{y}_{\text{sy}}$ is the average of l systematic estimators of $\bar{Y}$, and so we get the variance of $\bar{y}_{\text{sy}}$ by multiplying the variance of $\bar{y}_{\text{sy}}(g)$ by $\{(k - l)/(k - 1)\}l^{-1}$. This may be compared with drawing l elements without replacement from each of the n strata independently; in that case we obtain, as variance of the estimator of $\bar{Y}$, $\{(k - l)/(k - 1)\}l^{-1}$ times the variance for the case $l = 1$. It may also be compared with drawing a simple random sample of size nl, leading to a variance of the estimator of $\bar{Y}$ equal to

$$\{(N - nl)/(N - 1)\}\sigma^2/(nl)$$

rather than

$$\{(N - n)/(N - 1)\}\sigma^2/n,$$

the variance for the estimator based on a simple random sample of n observations; the ratio is again $\{(k - l)/(k - 1)\}l^{-1}$.

GAUTSCHI, Ann. Math. Statist., **28** (1957) 385–394, compares systematic random sampling with l random starts and a spacing of k (total sample size nl) with systematic sampling of n' with a single start but with spacing k' (every k'th), when $k' = k/l$, assumed an integer, and $n' = nl$ (so that the total sample size is the same for the two procedures).† Let us denote the variances of the estimators of $\bar{Y}$ by V_l and V_1, respectively. By the above, $\mathscr{E}V_l$ may be obtained by multiplying the $\{(k - l)/(k - 1)\}l^{-1}$ the expected value of the variance under systematic sampling of every kth (single start, total sample size n):

† Gautschi denotes our k' by l, and our l by s.

(a) Populations in random order:

$$\mathscr{E}V_1 = \frac{N - n'}{N}\frac{1}{n'}\sum_{i=1}^{N}\sigma_i^2/N = \frac{k'n' - n'}{k'n'}\frac{1}{n'}\sum_{i=1}^{N}\sigma_i^2/N = \frac{k' - 1}{N}\sum_{i=1}^{N}\sigma_i^2/N,$$

$$\mathscr{E}V_l = \frac{k - l}{k - 1}\frac{1}{l}\left(\frac{k - 1}{k}\frac{1}{n}\sum_{i=1}^{N}\sigma_i^2/N\right) = \frac{k' - 1}{N}\sum_{i=1}^{N}\sigma_i^2/N,$$

so that

$$\mathscr{E}V_1 = \mathscr{E}V_l.$$

(b) Populations with linear trend:

$$\mathscr{E}V_1 = \frac{k' - 1}{N}\sum_{i=1}^{N}\sigma_i^2/N + \beta^2\frac{(k' - 1)(k' + 1)}{12},$$

$$\mathscr{E}V_l = \frac{k - l}{k - 1}\frac{1}{l}\left\{\frac{k - 1}{k}\frac{1}{n}\sum_{i=1}^{N}\sigma_i^2/N + \beta^2\frac{(k - 1)(k + 1)}{12}\right\}$$

$$= \frac{k' - 1}{N}\sum_{i=1}^{N}\sigma_i^2/N + \beta^2\frac{(k' - 1)(k + 1)}{12},$$

so that for $l > 1$,

$$\mathscr{E}V_1 < \mathscr{E}V_l.$$

NOTE. $\mathscr{E}V_l$ may be compared with the expected value of the variance under random sampling with the same sample size, which, according to the previous section differs from $\mathscr{E}V_l$ in that the coefficient of $\frac{1}{12}\beta^2$ is $(k' - 1)(k'n' + 1)$ instead of $(k' - 1)(k'l + 1)$; so the variance advantage of systematic over random sampling is diminished on the average as l approaches $n' = ln$ (that is, as n gets small).

(c) Autocorrelated populations: Gautschi shows that if all the assumptions under section 6.3 are fulfilled and $l > 1$,

$$\mathscr{E}V_1 \leq \mathscr{E}V_l$$

if the second differences are nonnegative, with equality only if all the ρ_v are equal.

8. Estimation of the variance in systematic sampling

Since, with one random start, systematic sampling amounts to drawing a single cluster, no reasonable estimator of the variance of $\bar{y}_{sy}(g)$ is available in this case, unless further information is available.† Thus if the successive u_t have $\mathscr{Y}$-values

$$a, b, c, d, a, b, c, d, a, b, c, d,$$

a systematic sample of every fourth will always consist of 3 *equal* numbers! (Of course, if we know that the population is approximately periodic and also know the period to be about 4, we would not choose $k = 4$.) Under the assumption that the population is in random order (section 6.1),‡

$$\frac{N-n}{N}\frac{1}{n}\sum_{j=1}^{n}(y_{tj} - \bar{Y}_{t.})^2/(n-1)$$

is an unbiased estimator of the variance of $\bar{Y}_{x.}$, but in the sense that its $\mathscr{E}$-mean equals $\mathscr{E}V_{sy}$. For certain other models it is also possible to devise reasonable estimators, but such estimators are generally quite sensitive to errors in specification of the model.

With more than one start, the variance of $\bar{y}_{sy}$ can be estimated from the $\bar{y}_{sy}(g)$ by§

$$\frac{k-l}{k}\frac{1}{l}\sum_{g=1}^{l}\{\bar{y}_{sy}(g) - \bar{y}_{sy}\}^2/(l-1).$$

† However, as we saw in Chapter VII, sections 5.1 and 7, if we use cluster sampling with subsampling, variances can be reasonably estimated without estimates of the within cluster subsampling variance components being available, if a small fraction of the primary units is included in the sample or the sample of such units is drawn with replacement. Therefore, no difficulties arise in the case of systematic subsampling of clusters if l/k is small.

‡ The use of systematic sampling with a single start together with that of the variance estimator here quoted are extremely widespread. It would be desirable that such populations as are sampled be submitted to tests for randomness of the order of the $\mathscr{Y}$-values at least occasionally. Such tests cannot, of course, be based on systematic samples with a single start, but must be performed on some specially drawn samples or on preliminary samples. A number of such tests are described in KENDALL and STUART, The Advanced Theory of Statistics (Griffin, 1966) Vol. 3.

§ Note that in the notation of section 1 of Chapter V this expression may be written

$$\frac{k-l}{k}\frac{1}{ln}\mathrm{SS}_l/\mathrm{DF}_l.$$

If one is reasonably confident of the assumption that the population is in random order, and if l exceeds 1 but is small, one might like to use for $V(\bar{y}_{sy})$ the estimator

$$\frac{k-l}{k}\frac{1}{ln}\Sigma\Sigma(y_{x_g j} - \bar{y}_{sy})^2/(ln-1) = \frac{k-l}{k}\frac{1}{ln}\mathrm{SS}_t/\mathrm{DF}_t,$$

but perhaps not until conducting some formal test of the hypothesis of random order and finding that the hypothesis is not rejected by that test. Thus to guard against a possibility of approximate periodicity of period k, the test may consist in rejecting the hypothesis if $\mathrm{MS}_l/\mathrm{MS}_w$ exceeds the upper α-point of the F-distribution with $l-1$ and $l(n-1)$ degrees of freedom (cf. section 15.4 of Chapter II) provided F_N is not highly asymmetric and n not very small. For references to other tests, see the previous footnote.

If l is small, the precision of this estimator is low, since it is based on only $l - 1$ degrees of freedom. The discussion of section 7 should, however, warn against lightly deciding to choose $l > 1$ (and, if nonetheless we decide to take $l > 1$, against lightly choosing it much larger than 1). Note that, if we do systematic sampling within many strata, the precision of the estimators for the variance of an overall mean or total may be satisfactory even for small $l > 1$.

With a single start, the method of collapsed strata may be used.

9. Centered systematic samples

If $N = nk$ and we take as sample $u_c, u_{c+k}, \ldots, u_{c+(n-1)k}$, where $c = \frac{1}{2}(k + 1)$ if k is odd and $\frac{1}{2}k$ or $\frac{1}{2}(k + 2)$ if k is even, we speak about centered systematic sampling. It should be noted that there is no randomness in this sample.

The mean square error for the sample average $\bar{Y}_{c.}$ of such a sample equals $(\bar{Y}_{c.} - \bar{Y})^2$, whereas the variance of $\bar{y}_{sy} = \bar{y}_{x.}$ equals

$$(1/k) \sum_{t=1}^{k} (\bar{Y}_{t.} - \bar{Y})^2.$$

LEMMA. If $v_1 \leq \ldots \leq v_k$,

$$(v_c - \bar{v})^2 \leq (1/k) \sum_{t=1}^{k} (v_t - \bar{v})^2.$$

Proof. We first transform the numbers by subtracting v_c from each, which does not affect the desired inequality. Moreover, it suffices to show the inequality for the case in which all negative or zero v_t are replaced by 0 and all positive v_t by a, their average. For these latter transformations will increase the left-hand side (or leave it unchanged) and decrease the right-hand side (or leave it unchanged). If the number of vanishing v_t is denoted by h, the left-hand side equals $\{(k - h)a/k\}^2$ and the right-hand side $(1/k)(k - h)a^2 - \{(k - h)a/k\}^2$, and the inequality is satisfied since $2h \geq 2l \geq k$ and $h \leq k$.

It follows that if $\bar{Y}_{1.} \leq \ldots \leq \bar{Y}_{k.}$, the mean square error for $\bar{Y}_{c.}$ is less than or equal to the variance of $\bar{y}_{sy}$. (A similar result can be shown for the case in which ρ_v decreases monotonically.) Since the bias of $\bar{Y}_{c.}$ can be large, however, this fact is itself no argument for using $\bar{Y}_{c.}$ as an estimator for $\bar{Y}$.

10. Systematic sampling with unequal probabilities

In the chapter on unequal probability sampling we saw that such sampling may be somewhat complicated, especially when we desire to sample n without replacement. If the sampling is systematic there are, however, few difficulties in doing this. We consider sampling proportional to a positive integral-valued characteristic $\mathscr{Z}$. Let $k = Z/n$.

Consider sampling with a single random start. We draw again a number x at random from among $\{1, \ldots, k\}$. If it takes on the value t, we include u_i if, for some $j = 0, 1, \ldots, n-1$,

$$\sum_{h<i} Z_h \leq t + jk < \sum_{h \leq i} Z_h,$$

where $Z_i \leq k$ for each i. Since Z_i of the possible values of x lead to inclusion of u_i, the probability of including u_i is Z_i/k. (If some $Z_i > k$, some units may be included more than once and should then be included in estimators as often as they appear. Also with several random starts we may, without excluding the units chosen at the previous step from the calculation of k and the cumulative totals, get some of the same units in steps following the first one.) For the case of randomly ordered units, an asymptotic theory of this method is given by HARTLEY and RAO, Ann. Math. Statist., **33** (1962) 350–374 and an exact theory by CONNOR, J. Amer. Statist. Ass., **61** (1966) 384–390. The former paper gives as an approximation to the variance of the estimator of Y when each $np_i \leq 1$

$$\frac{1}{n}\sum\left(\frac{y_i}{p_i} - Y\right)^2 p_i k_i,$$

where $k_i = 1 - (n-1)p_i$ (in ordinary sampling with replacement with probabilities p_i, we get 1 instead of k_i).

If k is not an integer, we round the number Z/n to an integer. This, however, leads to a nonconstant sample size. To avoid this, we may carry out the sampling circularly and thus choose x between 1 and Z.

11. Deming's methods of cluster sampling with unequal probabilities (simple replicated designs)

In this section we present the main aspects of a set of very simple sampling plans which DEMING has described fully in his book Sample Design in Business Research (Wiley, 1960). Our presentation is given under the heading of systematic sampling, because the main idea of Deming's plans seems to stem from systematic sampling with several random starts, and so is perhaps more easily understood in connection with the previous discussion. Also Deming, in fact, often uses systematic sampling in his plans.

11.1. PLANS IN WHICH THE INCLUSION PROBABILITIES FOR CLUSTERS ARE PROPORTIONATE TO THEIR SIZES

What we have called strata in this chapter are referred to as *zones* by Deming. The important aspect of zones that he exploits is the fact that each of his n zones consists of a *constant* number of (small) units. We shall refer to this constant by k (Deming uses Z). For simplicity assume again that $N = nk$. We sample l in each zone. However, we may sample systematically, with l random starts, from the N units, without regard to zones; or we may draw the l samples independently in each zone (systematically or otherwise). The units of the population are assumed to be grouped, naturally or otherwise, into c clusters, containing $M_1, \ldots, M_c$ units, respectively. Therefore, the method leads to l sets of n drawings of clusters with replacement, each cluster having probabilities proportional to the size of that cluster.

The units included in the sample may consist of those indicated by the nl different serial numbers obtained as previously indicated. This need not, however, be the case; it may be more convenient to select, in each selected cluster independently, Σf_g units without replacement, where f_g is the number of times the cluster in question is drawn into sample g. Let us illustrate both cases with an example from Deming's book (the reader is advised to examine Chapter 6B, and Chapter 10 of that book for further details).

EXAMPLE 1. We wish to take $l = 10$ samples of $n = 10$ employees each from among those employed in $c = 12$ factories, which employ a total of $N = 9\,600$ employees (in this example Deming's method of sampling is systematic sampling of every kth with l random starts).

Since $k = N/n = 960$, if we have a serially numbered list of employees arranged by factories, the first 960 on the list will constitute the first zone, the second 960 the second zone, etc.

Actually, we need a list of employees only for those factories that will be included in the sample; for the others we need only a knowledge of the numbers of employees. Moreover, there is no harm if we overestimate these numbers: if our sampling procedure leads to a serial number which does not represent an employee, there will simply be no employee included in the sample corresponding to that number.

Deming gives a device whereby accurate counts are unnecessary for drawing the clusters. Suppose, e.g., that for factory 1 we have a rough estimate, 800, of the number of employees, but that there are in fact 822. In that case we may define a sampling unit to consist of one or two employees; e.g., sampling unit 1 consists of employees 1 and 801 on the list, . . ., sampling unit 22 of employees 22 and 822, sampling unit 23 of employee 23, . . ., sampling unit 800 of employee 800. Every employee has a chance of 1 in 960 to be included in the sample, since every sampling unit has that chance, and so every employee in every sampling unit has that chance. (But the probability of both employees 1 and 801 to fall in the sample is not the same as the probability of both employees 2 and 801 falling in the sample.)

REMARKS. It is clear that there are situations in which this device may lead to misleading results. Deming no doubt chose the particular allocation of employees to units mentioned here in the belief that characteristics of employees far apart on the list are less correlated than of those close together on the list.

Note that the units are (sub)clusters and that per employee averages will have to be estimated as ratios.

EXAMPLE 2. Here the sampling units are parts ("segments") of city blocks, which are intended to contain on the average 3 dwelling units each. In advance of sampling we need to fix the *number* of segments (r_i for the ith block) into which each block is to be subdivided, but not the boundaries of the segments. In the absence of previous information, the approximate number and location of dwelling units may be obtained by means of a quick preliminary survey.

The aim is a 25% sample of the dwelling units contained in a certain area in which we distinguish 79 such sampling units. For convenience we shall make $N = 80$ by adding a "blank". Take $l = 2$, then n is half of 25% of 80; i.e., $n = 10$, and $k = 80/10 = 8$.

Deming selects 2 random numbers without replacement between 1 and 8 for each of the $n = 10$ zones independently, and adds them to the serial number of the last segment in the preceding zones. The sums determine, for each of the $l = 2$ samples, which blocks are included.† Then one random number is selected between 1 and r_i in block i if this block is included in one sample only, and two numbers between 1 and r_i (without replacement) if it is included in both samples.

The interviewers go to the selected blocks and define accurately the segments. These should be defined primarily with a view to unequivocal identification, and thus their size may differ very much from 3 dwelling units and differ even on the average from 3 dwelling units.

This will be so particularly if the "rough survey" proves to have been rough indeed, or if the previous information lost validity. In that case Deming again reverts to the device employed in Example 1: a sampling unit may consist of more than one segment. This he achieves by choosing, for the selected blocks, besides random numbers between 1 and r_i, also random numbers between $r_i + 1$ and $2r_i$, and perhaps also between $2r_i + 1$ and $3r_i$ if some blocks ultimately contain more than twice the number of segments originally expected.‡

The simplicity of Deming's methods leads to very simple *estimators* as well. Consider the ratio estimator $(y/x)X$ of Y, which reduces to

$$(y/ln)kn = (k/l)y$$

† In Deming's example it so happened that, as a result of the drawing of the random numbers, no block was included more than once in either sample, so $f_1 = 1$ for each block included in sample 1 and f_2 for each block included in sample 2.

‡ Accordingly, in the descriptions of Table 5 for columns 3, 4, 7 and 8, the words "segments" should be replaced by "sampling units" – For cases in which one runs into many more segments than originally expected, one would be wise to create a separate stratum.

if $\mathscr{X}$ is 1 for sampling units in the sample and 0 for those not in the sample. Denote the n zone means by $\bar{Y}_{.1}, \ldots, \bar{Y}_{.n}$, and the jth of the adjusted zone variances

$$\sum_{t=1}^{k} (y_{tj} - \bar{Y}_{.j})^2/(k-1)$$

by S_j^2. Assuming that the random numbers were assigned independently in each zone, we have

$$V(\bar{y}_{.j}) = \frac{k-l}{k}\frac{1}{l}S_j^2,$$

$$V(y) = V(\sum_{j=1}^{n} l\bar{y}_{.j}) = \frac{k-l}{k} l \sum_{j=1}^{n} S_j^2,$$

$$V\left(\frac{k}{l}y\right) = k(k-l)\frac{1}{l}\sum_{j=1}^{n} S_j^2,$$

$$V\left(\frac{1}{nl}y\right) = \frac{k-l}{k}\frac{1}{ln}\bar{S}^2,$$

where

$$\bar{S}^2 = (1/n)\sum_{j=1}^{n} S_j^2;$$

we may estimate S_j^2 by s_j^2, where $(l-1)s_j^2$, is the sum of squares of deviations of the $\mathscr{Y}$-values of units from zone j in the sample around their average $\bar{y}_{.j}$.

The extension to ratio estimators is immediate.

REMARK. Deming also gives a short-cut estimator based on the average $\bar{w}$ of the ranges $w_1, \ldots, w_n$ of the sample values in the n zones. It may be shown that, for l between 3 and 10, for many F_N, $\{(k-l)/k\}^{\frac{1}{2}} w_j$ is a good estimator of $\{V(l\bar{y}_{.j})\}^{\frac{1}{2}}$; replacing w_j by $\bar{w}$ if the w_j do not vary much (he suggests: if

$$\max_j w_j - \min_j w_j \leq \bar{w}),$$

we may estimate $\{V(y)\}^{\frac{1}{2}}$ by $\{(k-l)k^{-1}n\}^{\frac{1}{2}}\bar{w}$.

These formulas for the variance of the estimator and for its estimator are not correct for systematic sampling (at least, not without some further assumption on the population). For that case we must use the following estimator of the variance of $(nl)^{-1}y$:

$$\frac{k-l}{k}\frac{1}{l}\sum_{g=1}^{l}\left\{\bar{y}(g) - \frac{y}{nl}\right\}^2 \Big/ (l-1),$$

where $\bar{y}(g)$ is the average for the gth sample. (The same formula is valid if $\bar{y}(g)$ is replaced by a ratio estimator.) This estimator is also valid for the case in which sampling is done independently from zone to zone, but it is based on $l - 1$ instead of $n(l - 1)$ degrees of freedom. The advantage of using this estimator when zones were sampled independently is that the $\bar{y}(g)$ have a more nearly normal distribution than the original observations (see Chapter III, section 14.2). One way nonnormality may sometimes be partially overcome is by combining, in the tabulations, the zones into a smaller number of "thick" zones. Moreover, in thick zones rare characteristics may have a better chance to appear with a frequency sufficient for providing good estimators of variance than in the original zones.†

11.2. OTHER PLANS

In the preceding plans the probability of including a cluster is exactly proportionate to the number of sampling units it contains (and in the examples may be only approximately proportionate to the number of workers or segments or dwelling units it contains). These probabilities do not explicitly appear in our formulae. If we wish the probabilities to be roughly proportional not to the number of workers or dwelling units but to, say, the square of these, we can achieve that by defining the sampling units such that the number of sampling units in a cluster is roughly proportional to the square of the number of workers or dwelling units.

Deming's plans make use of whatever natural stratification there is in the arrangement of the sampling frame and is preserved in the zones, and uses proportional allocation (since all the zones contain the same number of sampling units). Often it is advantageous, however, to introduce additional stratification and/or an allocation that is not proportional. If allocation is proportional, the method of estimation is not affected by further stratification, but, if, for confidence intervals, we should use Student's distribution, we cannot simply add degrees of freedom. For nonproportional allocation, the strata are dealt with separately for building up the l subsample estimates of totals, averages or ratios. One may estimate the variance of such estimators from these subsample estimates, as given above, or estimate them for each stratum separately (by using the subsamples or by the usual methods) and combine the results appropriately.

† When considering how to combine their zones into thick ones, one ought to keep in mind the fact that the precision of our estimator of $V\{y/nl\}$ is unfavorably affected by large variation among the S_j. To avoid such variation, Deming advises systematic or random selection of thin zones to form thick zones.

CHAPTER X

NONSAMPLING ERRORS

1. Introduction

In this chapter we examine some errors which are not due to sampling fluctuations, and thus also arise in complete enumerations. If a survey is a large scale one, it is not always possible to use expensive measurement instruments; to engage staff with high expertise; nor to maintain uniform interpretations of the instructions and questions, and uniform attitudes towards interviewees. Furthermore, for such surveys, data usually need to be coded, which also may introduce errors. From this it follows that a complete census may well have much larger errors than a sample survey.†

We shall discuss a number of classes of errors:

(a) During the period of the survey we may not succeed in locating all the units which according to our plans are to form part of the sample. But even if all units are located, there may remain a discrepancy between the sampled population and the target population (see Chapter I, section 9).

(b) For some of the units located we may not be able to secure all the desired measurements.

Errors under (a) and (b) will be referred to together as "nonresponse errors".

(c) The measurements on the units included may be subject to error. In measurement of physical quantities the personnel and devices we have to use may not give as precise measurements as the best available; in fact, even the best available may not conform to the concept desired. If so, there is a difference between the true value μ_i, of $\mathscr{Y}$, and ν_i (as defined in Chapter VIII, section 6). Such discrepancies arise even to a larger extent if the people conducting the survey cannot measure the properties directly, but can only obtain

† In modern practice small surveys (*post-enumeration surveys and quality checks* and *reinterview surveys*) are often conducted to check on various aspects of a census or large survey. Thus large errors have been revealed (such as underenumeration of some 6 million people in the 1960 population census for the U.S.A.), and have led to remedial measures in later censuses and surveys. For references and recent applied research on nonsampling errors see ZARKOVIC, Quality of Statistical Data, United Nations Food and Agriculture Organization, 1966.

the measurements through the cooperation of owners of the units, etc. (e.g., measurement of output via records of producers or their memories). This introduces correlation between the measurements and the owners' attitudes towards them. Subject to still larger error are measurements of human recollections, attitudes or intentions. Here there is a particularly marked possibility of interaction between these measurements and the people collecting the data.

(d) Errors may arise in the processing of the data: classifying coding, etc.

In planning surveys, attention should be given to the allocation of sufficient resources for the prevention of errors of this kind,† as against reduction of sampling error. Some of the former kind of errors can be assessed from previous surveys, and we have to discuss ways of doing this. Finally, we shall see how conclusions from a survey may be modified in the presence of nonsampling errors.

2. Nonresponse errors

2.1. REPLACEMENT OF UNITS

A method sometimes used to overcome nonresponse errors is to replace each unit for which a response is not obtained by some other unit. For example, we may not succeed in obtaining an interview with a person who is rarely at home during interviewing hours. If these are the evening hours, there is likely to be a high correlation between his use of electricity and his inclusion in or exclusion from the sample. This may seriously bias an estimate of use of electricity. In most cases the correlation would not be so evident, but there would not usually be a way of making sure the correlation coefficient is negligible. For further comments see section 2.5 below. Similar problems arise if an interviewee fails to respond to part of the questionnaire.

2.2. MODELS OF NONRESPONSE; EFFECT OF NONRESPONSE ON CONFIDENCE INTERVALS

Therefore a model of nonresponse should distinguish the different subpopulations according to response factors. One subpopulation may consist of units that are readily located and readily and fully respond, another of units that are readily located but give incomplete responses, another of units that are possible to locate only after two (or more) attempts but give full response,

† Discussion of techniques for locating such errors falls outside the scope of this book. We merely mention the possibility of spotting errors in the detailed data using methods of statistical quality control and (univariate and multivariate) outlier techniques. In addition, one can make tests of consistency of the overall estimates arrived at, both tests with outside data and internal tests.

and so on. Such a model may be used to evaluate the various contributions to error and to properly allocate the total effort between such subpopulations. The probability for any unit to belong to one of these subpopulations will not usually be known, but may be estimated from the sample.

As a simple case, consider two subpopulations, called "response" and "nonresponse", containing N_1 and N_2 units, and with average $\mathscr{Y}$-values $\bar{Y}_1$ and $\bar{Y}_2$, respectively. Take a simple random sample of n with replacement. Let $\mathscr{s}_1$ be the subsample of those which fall in the response subpopulation; let n_1 be their number, which we shall assume to be positive, and let $\bar{y}_1$ be their arithmetic mean [we assume that n_1 comes out positive *and* that the probability of it being 0 is negligible]. The bias of $\bar{y}_1$ is

$$\mathscr{E}\bar{y}_1 - \bar{Y} = \mathscr{E}\mathscr{E}\{\bar{y}_1|\mathscr{s}_1\} - \bar{Y} = \bar{Y}_1 - \bar{Y}$$

$$= \bar{Y}_1 - \left(\frac{N_1}{N}\bar{Y}_1 + \frac{N_2}{N}\bar{Y}_2\right) = \frac{N_2}{N}(\bar{Y}_1 - \bar{Y}_2).$$

For large n_1, and when $\mathscr{Y}$ is 1 or 0, we would use for $\bar{Y}_1$ a 95% confidence interval

$$[a - 2w, a + 2w],$$

when $\bar{y}_1$ is at least 0.2 or at most 0.8 (say); here†

$$a = (y' + 2)/(n_1 + 4), \quad w = \{y'(1 - \bar{y}_1) + 1\}^{\frac{1}{2}}/(n_1 + 4),$$

if we may ignore the nonresponse. For $\bar{y}_1 = 0.20$, $n_1 = 225$, and $n - n_1 = 25$, this would give for $\bar{Y}_1$ and $\bar{Y}$ the confidence interval [0.152, 0.258]. We may consider using, instead of this, the confidence interval for $\bar{Y}$:

$$\left[\frac{n_1}{n}(a - 2w), \frac{n_1}{n}(a + 2w) + \frac{n - n_1}{n}\right],$$

which for $\bar{y}_1 = 0.20$, $n_1 = 225$, $n - n_1 = 25$ gives [0.137, 0.343]. This interval is obtained by taking for the average value in the nonresponse subpopulation, $\bar{Y}_2$, the extreme values 0 and 1, respectively. It is possible to obtain slightly narrower intervals without making any assumptions about $\bar{Y}_2$ (see COCHRAN, MOSTELLER and TUKEY, Statistical Problems of the Kinsey Report (Amer. Statist. Assoc., 1954) pp. 280–281). One can also ask what n has to be in order to have a confidence interval for $\bar{Y}$ of the same length (here 0.106) as in the case of absence of nonresponse. We find in our case that, with a nonresponse of 5% and $\bar{y}_1 = 0.20$, n has to be more than 4 times as large to give a confidence interval of the same length; for a nonresponse of 7%, more than 11 times. (Evidently, the situation can be much worse with other types of $\mathscr{Y}$!) This gives some idea as to the worth of efforts to keep nonresponse down. Unfortunately, many investigators do not succeed in keeping the nonresponse

† See section 4.3 of Chapter IIA.

rate to certain questions below 20 or 30%, or even more. Official statisticians in Australia have generally been more successful than those in the U.S. in keeping nonresponse rates low on comparable subjects – perhaps they have devoted relatively more resources to this.

A less pessimistic way of looking at nonresponse bias is to indicate the effect of certain values of $\delta = \bar{Y}_2 - \bar{Y}_1$. Thus from section 15.8 of Chapter II we find for our example that, when

$$\delta = \pm 0.1,$$

so that

$$(n - n_1)\delta/(nw) = \pm 0.37,$$

the probability that $[a - 2w, a + 2w]$ covers $\bar{Y}$ is about 0.934, and when $\delta = 0.2$ about 0.885. For $\delta = \pm 0.1$, the proper confidence interval is $[a - 2.10w, a + 2.10w]$, for $\delta = \pm 0.2$: $[a - 2.40w, a + 2.40w]$. In our case these give [0.150, 0.261] and [0.142, 0.269], respectively. In practice this type of information should be presented, together with some discussion as to how big δ might be, on the basis of whatever further information is available or may be obtained.

2.3. CALL-BACK PLANS

A scheme of call-backs is often usefully employed in sample surveys. Thus a plan can be devised which subsamples the nonresponses. The following is a simple plan (for a generalization, see EL BADRY, J. Amer. Statist. Ass., **51** (1956) 209–227):

First take a simple random sample of n, of which n_2 do not respond, and then sample fn_2 out of the n_2. Each time the sampling is without replacement. We shall assume that a response is obtained for all of these fn_2 units.

Let the total cost be $c_0 n + c_1 n_1 + c_2 fn_2$, so the expected cost is

$$\{c_0 + c_1 N_1/N + c_2 fN_2/N\}n.$$

If y_1 and y_2 are the totals for the observed units in the two strata, we estimate $\bar{Y}$ by $(y_1 + f^{-1}y_2)/n$, or in terms of the sample means $\bar{y}_1$ and $\bar{y}_2$ (which are unbiased estimators of $\bar{Y}_1$ and $\bar{Y}_2$):

$$\bar{y}_d = \frac{n_1}{n}\bar{y}_1 + \frac{n_2}{n}\bar{y}_2.$$

In order to compute the variance conveniently, we shall also introduce the symbol $\bar{y}^*$ for the average $\mathscr{Y}$-value of the $(1 - f)n_2$ units in the sample which are not observed, and write

$$\bar{y}_d = \frac{1}{n}\{n_1\bar{y}_1 + fn_2\bar{y}_2 + (1 - f)n_2\bar{y}^*\} + \frac{n_2}{n}(1 - f)(\bar{y}_2 - \bar{y}^*).$$

The first term on the right equals the sample mean of $\mathscr{Y}$ if all n units of the sample could have been observed, and therefore has variance $\{(N-n)/N\}n^{-1} \times S^2$. It has correlation zero with $(\bar{y}_2 - \bar{y}^*)$, as the latter has zero mean over the set of all possible subsamples from the fixed sample of size n_2, and the former does not vary on this set, given also the sample of first respondees. The fixed sample of size n_2 has average

$$\bar{y}(n_2) \equiv f\bar{y}_2 + (1-f)\bar{y}^*,$$

and so, as $\bar{y}_2$ is the average for a subsample of fn_2 from n_2, the conditional variance of

$$\frac{n_2}{n}(1-f)(\bar{y}_2 - \bar{y}^*) = \frac{n_2}{n}\{\bar{y}_2 - \bar{y}(n_2)\}$$

given this fixed sample is

$$\left(\frac{n_2}{n}\right)^2 \frac{1-f}{fn_2} S_2^2,$$

where S_2^2 is the adjusted variance of $\mathscr{Y}$ for those that do not respond on the first call. The mean of this is V_{s}/n with

$$V_{\mathrm{s}} = \frac{N_2}{N}\frac{1-f}{f} S_2^2.$$

Since $\bar{y}_{\mathrm{d}}$ is unbiased, we have

$$V(\bar{y}_{\mathrm{d}}) = \frac{1}{n}(S^2 + V_{\mathrm{s}}) - \frac{1}{N}S^2,$$

where the term V_{s}/n is due to subsampling.

For n and f to minimize the expected cost subject to $V(\bar{y}_{\mathrm{d}})$ not exceeding V_0, f has to minimize the product of

$$\left\{c_0 + c_1\frac{N_1}{N} + c_2 f\frac{N_2}{N}\right\}$$

and

$$\left\{S^2 + \frac{N_2}{N}S_2^2\left(\frac{1}{f} - 1\right)\right\},$$

which is of the form $(a/f) + bf + c$ with a and b positive (assuming $S^2 > N_2S_2^2/N$). The minimum for nonnegative f is achieved† for f satisfying $f^2 = a/b$; that is for

$$f^2 = \frac{c_0 + c_1N_1/N}{c_2}\,\frac{S_2^2}{S^2 - N_2S_2^2/N},$$

† $f = (a/b)^{\frac{1}{2}}$, the function equals $2(ab)^{\frac{1}{2}} + c$, and in general it differs from this value by $\{(a/f)^{\frac{1}{2}} - (bf)^{\frac{1}{2}}\}^2$, which is positive except at $f = a/b$.

which, if $S_2 = S$, becomes

$$f^2 = \frac{c_0 + c_1 N_1/N}{c_2 N_1/N}.$$

Moreover, since $(S^2 + V_s)/n$ equals $V_0 + S^2/N$, we have

$$n = \frac{S^2}{V_0 + S^2/N}\left\{1 + \frac{V_s}{S^2}\right\}.$$

With stratification a reasonable approximation to optimum values of n_h and f_h for stratum h is obtained as follows. The f_h are obtained separately from the formula given above in terms of the information for each stratum. The n_h are obtained by first deriving their optimum values in the absence of nonresponse, and then multiplying them, respectively, by the factors $1 + V_s/S^2$ for the corresponding strata.

If only a safe upper bound, N_2', is known for N_2, first draw a sample of size

$$\frac{S^2}{V_0 + S^2/N}\left\{1 + \frac{V_s'}{S^2}\right\},$$

where V_s' is the expression previously given for V_s with N_2 replaced by N_2', and is the square root of

$$\frac{c_0 + c_1(N - N_2')/N}{c_2} \quad \frac{S_2^2}{S^2 - N_2' S_2^2/N}.$$

Then, if n_2 do not answer, subsample a fraction f from the n_2, where f is determined from

$$\left(1 - \frac{n}{N}\right) S^2 + \left(\frac{1}{f} - 1\right)\frac{n_2}{n} S_2^2 = nV_0.$$

In practice it is frequently economical to plan more than one call-back. DEMING, J. Amer. Statist. Ass., **48** (1953) 743–772 has studied call-back policy and has found that as many as 5 call-backs may often be worthwhile. Poststratification may be used to weight the responses from the different response subpopulations; BARTHOLOMEW, Appl. Statist., **10** (1961) 52–59 claims that if information on stay-at-home habits of absent respondees can be obtained at the first call and used to time the second, this will often lead to nearly unbiased estimators based on a single call-back only.

2.4. POLITZ-SIMMONS ADJUSTMENTS

POLITZ and SIMMONS, J. Amer. Statist. Ass., **44** (1949) 9–31 and **45** (1950) 136–137 discuss a plan which adjusts – without call-backs – for the biases due to the fact that the sample is not distributed proportionally over the response subpopulation. Suppose all interviews are conducted during the evenings, excluding Saturday night and Sunday night. The respondent who is found at home is asked on how many, $(t - 1)$, of the 4 preceeding nights (excluding Saturday and Sunday) he was at home at the time of the interview. Each $\mathscr{Y}$-value is given a weight inversely proportional to t, since $\frac{1}{5}t$ is taken to estimate his probability of being included in the sample. The method reduces the bias due to people not being at home, but it increases the variance of the estimator of $\bar{Y}$, because unequal and estimated† weights are used. The method was originally proposed to avoid call-backs altogether; but, when circumstances permit call-backs, it will usually be advisable to use them (unless time does not permit this). Thus DURBIN and STUART, J. Roy. Statist. Soc., **A117** (1954) 387–428 found that, for the cases they examined, the weighted results were much more like the results of the first call than like the combined results for the different calls. SIMMONS, J. Marketing, **11** (1954) 42–53, has given a plan combining weighting with call-backs.

2.5. MORE ON REPLACEMENT OF UNITS

Substitution of a nonresponse unit by another may be used to achieve a sample size fixed in advance;‡ but it *does not reduce biases*, as it replaces $\mathscr{Y}$-values of units belonging to one subpopulation (a subpopulation with low response probability) by values for units of another subpopulation (with high response probability). KISH and HESS, Amer. Statistician, **13** (1959) 4, 17–19, have used units not found at home in earlier surveys as replacements for not-at-homes in a current survey. This plan can only be used in special circumstances; thus it requires that few of the people have moved meanwhile or changed their habits of going out. Even when it is suitable for diminishing the bias from not being at home, it may not be suitable for overcoming refusal to answer, since the latter probability will often differ much more from survey to survey than the probability of not being at home at the time of the visit.

† The bias may not be much reduced by the adjustments discussed if errors in t are highly correlated with errors in $\mathscr{Y}$. Deming found that if this correlation is not high, a Politz-Simmons plan with k periods has about the same bias as a call-back plan with $k - 1$ call-backs.

‡ In the absence of further information, some form of replacement may become necessary in all sampling plans other than simple random ones, if there are differences among response rates in the groups that are included in the sample with different probabilities. The simplest form is to inflate any subsample total for such a group to what it would have been if all members of the group that were planned to be included in the sample would have responded exactly like those that did respond. Sometimes, however, there is a clear indication of trends in the observed characteristics for the successive response classes in some of the groups, and one may be able to use these to get improved estimators. Thus one often finds that the first response strata contain too small a proportion of shift workers or of single- or two-person families.

In Chapter III, section 12.4 reference has been made to quota-sampling. Clearly in quota sampling substitutions may be freely made, and therefore the survey is not delayed by call-backs. SUDMAN, Reducing the Cost of Surveys (Aldine Publishing Co., 1967) reports that in recent years a new quota procedure has been widely used in which interviewees are selected according to a preassigned probability scheme, but substitutes are used for those not available at first call. The resulting biases are believed to be kept low by additional stratification into strata within which one may expect at-home probabilities to vary little. Precise descriptions of such procedures, which Sudman refers to as *probability sampling with quotas*, do not appear to be available in published form. Sudman reports that their *only* advantage lies in speeding up the survey results, their costs being about the same as that of ordinary probability methods which include call-backs. Apart from the remaining biases, the applicable theory is that of inverse hypergeometric sampling, referred to in section 14 of Chapter II.

3. Measurement of response errors, and processing errors

Let us develop further the model of Chapter VIII, section 6.2. There we showed that the variance of $\bar{y}$ is the sum of two components,† the *sampling variance component*

$$\{(N-n)/Nn\}S_v^2,$$

and the *response variance component*

$$(1/n)\sigma_e^2(1+(n-1)r_e).$$

Here σ_e^2 is referred to as the average variance of the measurement errors or the *simple response variance*, and r_e as the average *correlation* of the response deviations.

3.1. REPETITION OF THE MEASUREMENTS

Consider now the case in which $\mathscr{Y}$ is measured twice on an identical sample. We then have two sets of measurements on $\mathscr{Y}$. Denote everything relating to the second set by a prime ('). $\bar{y}-\bar{y}'$ is called the *net difference rate*. If the set of responses is obtained under identical conditions, the mean of $\frac{1}{2}(\bar{y}-\bar{y}')^2$ is equal to

$$\begin{aligned}\tfrac{1}{2}\mathscr{E}(\bar{e}-\bar{e}')^2 = \mathscr{E}(\bar{e})^2 &= \frac{1}{n^2}\mathscr{E}\Big(\sum_{i=1}^{N} a_i e_i\Big)^2\\ &= \frac{1}{n^2}\mathscr{E}\mathscr{E}\Big\{\Big(\sum_{i=1}^{N} a_i e_i\Big)^2 \Big| a_1, \ldots, a_N\Big\}\\ &= \frac{1}{n^2}\mathscr{E}\sum_{i=1}^{N}\sum_{j=1}^{N} a_i a_j \sigma_{ij}\\ &= \frac{\sigma_e^2}{n}(1+(n-1)r_e),\end{aligned}$$

† The components may be derived directly by writing

$$\mathscr{E}(\bar{y}-\bar{v})^2 = \mathscr{E}(n^{-1}\Sigma a_i v_i - \bar{v})^2 + \mathscr{E}\bar{e}^2.$$

the response variance. A slight difference between σ_e^2 and $\sigma_{e'}^2$ and between r_e and $r_{e'}$ can be neglected, but a difference between $\bar{v}$ and $\bar{v}'$ increases $\frac{1}{2}\mathscr{E}(\bar{y}-\bar{y}')^2$ by $\frac{1}{2}(\bar{v}-\bar{v}')^2$, which may not be negligible.† So generally $\frac{1}{2}(\bar{y}-\bar{y}')^2$ will be somewhat of an overestimate of the response variance. If there is a positive correlation between the response errors on the two occasions, a further negative contribution to $\frac{1}{2}\mathscr{E}(\bar{y}-\bar{y}')^2$ can make for a more serious underestimate. (If

$$\mathscr{E}\{e_i e_i' | a_i\} = \rho_e \sigma_e^2,$$

and, for $i \neq j$,

$$\mathscr{E}\{e_i e_j' | a_i, a_j\} = 0,$$

this contribution is $-\rho_e\sigma_e^2/n$.) Such correlations may arise from the memory of the respondee or from attempts of the surveyor to "reconcile" responses obtained on two occasions.

Hansen, Hurwitz and Pritzker‡ define

$$g = \frac{1}{n}\sum_{i=1}^{N} a_i(y_i - y_i')^2$$

and call it the *gross difference rate*.

Half its mean is, if $\rho_e = 0$, equal to

$$\tfrac{1}{2}\mathscr{E}g = \tfrac{1}{2}\mathscr{E}\frac{1}{n}\sum_{i=1}^{N} a_i(v_i - v_i')^2 + \tfrac{1}{2}\mathscr{E}\frac{1}{n}\sum_{i=1}^{N} a_i(e_i - e_i')^2$$

$$= \tfrac{1}{2}\sum_{i=1}^{N}(v_i - v_i')^2/N + \sigma_e^2;$$

and, if $\rho_e > 0$, smaller by $\rho_e\sigma_e^2$. If $\bar{v} = \bar{v}'$, the first term on the right-hand side is zero,§ but it will often be outweighed by a term due to positive correlation. *Therefore:* under favorable circumstances we can estimate σ_e^2 and r_e^2 from the gross and net difference rates, and if $\bar{v} = \bar{v}'$ we can compute an approximate correction for response error to the estimator $v(\bar{y})$ of $V(\bar{y})$.§§ We illustrate the procedure numerically:

† We also neglect $\frac{1}{2}\{(N-n)/Nn\}\Sigma\{(v_i-\bar{v})-(v_i'-\bar{v}')\}^2/(N-1)$ which equals $\frac{1}{2}\{(N-n)/(N-1)\}n^{-1}D$ or $-\frac{1}{2}\{(N-n)/(N-1)\}n^{-1}(N-1)C$, where D is the variance between the $v - v'$ values of identical units, and C the covariance between the $v - v'$ values of the different units.

‡ In: Contributions to Statistics presented to Professor P. C. Mahalanobis (Statistical Publishing Society, Calcutta 1964) 111–136.

§ This holds provided we also neglect $\frac{1}{2}D$, with D defined in the penultimate footnote; for $\frac{1}{2}\mathscr{E}g$ equals $\frac{1}{2}(\bar{v}-\bar{v}')^2 + \frac{1}{2}D + \sigma_e^2$ when $\rho_e = 0$.

§§ $\frac{1}{2}\mathscr{E}\{(\bar{y}-\bar{y}')^2 - g/n\} = \{\sigma_e^2 r_e + \frac{1}{2}(\bar{v}-\bar{v}')^2 + \frac{1}{2}C\}(n-1)/n$ (even if $\rho_e \neq 0$). As seen in Chapter 8, section 6, the difference between $V(\bar{y})$ and $\mathscr{E}v(\bar{y})$ is $(\sigma_e^2/N) + N^{-1}(N-1)r_e\sigma_e^2$.

EXAMPLE. In a survey 10 units were interviewed in a certain stratum containing 200 units and reinterviewed later. The following measurements were obtained on a characteristic $\mathscr{Y}$:

	Unit number										
	1	2	3	4	5	6	7	8	9	10	Totals
First interview	205	196	196	200	211	100	200	194	199	198	1 899
Reinterview	200	195	192	200	211	104	195	191	195	196	1 879
Differences	5	1	4	0	0	−4	5	3	4	2	20

It is desired to estimate $V(\bar{y})$ and its components as well as possible, assuming that the conditions under which the reinterview was conducted were identical with those of the original interview and that the correlation between the errors on the two occasions is not more than 0.1.

Solution:

$$\begin{aligned} s^2 &= \tfrac{1}{18}\{\Sigma(y_j - \bar{y})^2 + \Sigma(y'_j - \bar{y}')^2\} \\ &= \tfrac{1}{18}\{\Sigma y_j^2 + \Sigma y_j'^2 - \tfrac{1}{10}(y^2 + y'^2)\} \\ &= \tfrac{1}{18}[9\,199 + 8\,153 - \tfrac{1}{10}\{1^2 + 21^2\}] \\ &= 17\,308/18 = 961.6. \end{aligned}$$

(We subtracted 190 from all the numbers; this does not affect s^2.) $\bar{y} - \bar{y}' = 2$ and so $\frac{1}{2}(\bar{y} - \bar{y}')^2 = 2$ estimates $\frac{1}{10}\sigma_e^2(a + 9r_e)$, where $0.9 \le a \le 1.0$.

$$g = \tfrac{1}{10}[25 + 1 + 16 + 0 + 0 + 16 + 25 + 9 + 16 + 4] = 11.2,$$

so

$$\tfrac{1}{2}\{(\bar{y} - \bar{y}')^2 - (g/n)\} = 1.44$$

estimates $\{(n-1)/n\}r_e\sigma_e^2$. So $r_e\sigma_e^2$ is estimated by 1.60, and we would estimate σ_e^2 to be between 5.6 and 6.2, and the response variance component will be estimated to be between 2.00 and 2.06. $s^2 = 961.6$ estimates

$$\mathscr{E}s^2 = S_y^2 + \sigma_e^2(1 - r_e)$$

and $\sigma_e^2(1 - r_e)$ is estimated as lying between $5.60 - 1.60$ and $6.22 - 1.60$, i.e., between 4.00 and 4.62; so S_y^2 as lying between $961.6 - 4.62$ and $961.6 - 4.00$, i.e., as 957. We may then estimate

$$V(\bar{y}) = \left(\frac{1}{n} - \frac{1}{N}\right) S_y^2 + \frac{1}{n}\sigma_e^2\{1 + (n-1)r_e\}$$

as lying between $(\frac{1}{10} - \frac{1}{200})957 + 2.00$ and $(\frac{1}{10} - \frac{1}{200})957 + 2.06$, i.e., as 93.

Hansen, Hurwitz and Pritzker study rough ways of bounding ρ_e for certain problems when the second survey is conducted under improved conditions (post-enumeration survey). In general they confine their attention to characteristics $\mathscr{Y}$ which can take on only values 0 or 1. In that case, response errors are classification errors and

$$\begin{aligned} ng &= \Sigma a_i(y_i - y'_i)^2 \\ &= \Sigma a_i y_i + \Sigma a_i y'_i - 2\Sigma a_i y_i y'_i \\ &= \Sigma a_i y_i(1 - y'_i) + \Sigma a_i y'_i(1 - y_i), \end{aligned}$$

which equals the total number of *switches* in classification. In the case of such $\mathscr{Y}$ we also have

$$\sigma_e^2 = \frac{1}{N}\Sigma\sigma_i^2 = \frac{1}{N}\Sigma\nu_i(1-\nu_i) = \bar{\nu}(1-\bar{\nu}) - \frac{1}{N}\Sigma(\nu_i - \bar{\nu})^2 = \bar{\nu}(1-\bar{\nu}) - \sigma_\nu^2,$$

which leads Hansen, Hurwitz and Pritzker to define

$$I_e = \frac{\sigma_e^2}{\bar{\nu}(1-\bar{\nu})} = \frac{\sigma_e^2}{\sigma_e^2 + \sigma_\nu^2}$$

as the *index of unreliability, or inconsistency, of classification.*† Note that, for fixed $\bar{\nu}$, the larger is the simple response variance, the smaller must be the sampling variance, and vice versa.

We saw before that in certain circumstances $\frac{1}{2}g$ may be a good estimator of σ_e^2; and a good estimator of $\bar{\nu}$ is $\bar{y}$ or $\frac{1}{2}(\bar{y} + \bar{y}')$. For most items of the 1960 U.S. census, the estimated values of I_e ranged from 0.02 for sex, to 0.26 for the average index for educational attainment classes of members of the U.S. population over 24 years of age. For age (in 5-year age classes) the average value of I_e was estimated at 0.05.‡ The estimates of I_e are biased downwards, since they are based on procedures which tend to lead to positive ρ_e. But their relative orders of magnitude are not surprising.

3.2. INTERPENETRATING SUBSAMPLES

We saw above that, as a method of measuring variances and their components, repetition of a survey has certain limitations. Some of these may be overcome by using interpenetrating subsamples.

Consider first the following situation. A simple random sample of size $n = \bar{n}k$ is selected from $\mathscr{P}$ and partitioned at random into k subsamples of size $\bar{n}$, each allocated to a different enumerator (or interviewer).§ Supervision and processing are supposed to be arranged in such a way that the response deviations for any two different subsamples are as much as possible uncorrelated. As shown below under section 3.3, it is possible to modify the sampling design itself in such a way as to assure this lack of correlation without further special arrangements or assumptions. We also assume that the

† In sampling with replacement (or, approximately, if n/N small) with $r_e = 0$, the variance of $\bar{y}$ is $(\sigma_e^2 + \sigma_\nu^2)/n$. The formula for sampling without replacement is found in section 6.2. of Chapter VIII.

‡ The first named quantity was not measured in 1950; for education I_e was estimated as 0.39, and for age as 0.07 in 1950. Generally I_e was smaller for 1960 than for 1950, indicating improved methods of measurement in the later census.

§ The method is also used as a way of keeping the survey results free from various sources of error, namely by comparing, as they come in, the results obtained by different interviewers who have been assigned very similar units.

response fluctuations of those units visited by one enumerator are not affected by sample fluctuations among those units visited by another enumerator.

Let r_w be the correlation of response deviations between units allocated to the same interviewer, and r_b between units allocated to different interviewers.

If $\bar{y}(h)$ is the average $\mathscr{Y}$-value for the hth interviewer, then $\bar{n}$ times the sum of squares of deviations of $\bar{y}(h)$ from $\bar{y}$ is the between interviewers sum of squares, based on $k - 1$ degrees of freedom. Denote its mean square by B; and denote the within interviewers mean square (deviations from $\bar{y}(h)$), based on $k(\bar{n} - 1)$ degrees of freedom, by w. Then, since $\bar{n}\mathscr{E}\{n\bar{e}(h)\, n\bar{e}(h')|h, h'\}$ equals $\bar{n}$ times the sum of the covariances of the response deviations for all pairs of units enumerated by two different interviewers,

$$\begin{aligned}\mathscr{E}B &= S_v^2 + \sigma_e^2\{1 + (\bar{n} - 1)r_w\} - \bar{n}\sigma_e^2 r_b \\ &= S_v^2 + \bar{n}\sigma_e^2(r_w - r_b) + \sigma_e^2(1 - r_w).\end{aligned}$$

We derived in Chapter VIII, section 6 that

$$\mathscr{E}\left\{\Sigma e_i^2 - \frac{1}{N}(\Sigma e_i)^2\right\} = \sigma_e^2(1 - r_e);$$

in a similar way we obtain that

$$\mathscr{E}W = S_v^2 + \sigma_e^2(1 - r_w).$$

It then follows that

$$\mathscr{E}(B - W)/\bar{n} = \sigma_e^2(r_w - r_b).$$

We also readily obtain for the variance of the sample mean

$$\frac{N - \bar{n}k}{N}\,\frac{1}{\bar{n}k}\,S_v^2 + \frac{1}{\bar{n}k}\,\sigma_e^2\{1 + (\bar{n} - 1)r_w + \bar{n}(k - 1)r_b\},$$

since, if h indicates the interviewer and j one of the $\bar{n}$ persons interviewed by him, and if δ is 1 if its two subscripts coincide and zero otherwise, then

$$\mathscr{E}e_{hj}e_{h'j'} = \sigma_e^2\{\delta_{hh'}\delta_{jj'} + \delta_{hh'}(1 - \delta_{jj'})r_w + (1 - \delta_{hh'})r_b\}$$

so

$$\mathscr{E}\bar{e}^2 = \mathscr{E}e_{hj}\bar{e} = \sigma_e^2\{1 + (\bar{n} - 1)r_w + \bar{n}(k - 1)r_b\}/(k\bar{n}).$$

Due to the assumption about supervision and processing mentioned above, the r_b will usually be a small fraction of the r_w; then $(B - W)/\bar{n}$ gives a slight underestimate of $r_w\sigma_e^2$, and $\{B + (\bar{n} - 1)W\}/\bar{n}$ a slight underestimate of $S_v^2 + \sigma_e^2$, which except for the finite multiplier is $k\bar{n}$ times the variance of the sample mean we would have obtained if r_w were zero (and r_b zero). Note that, if r_b can be neglected, $B/(\bar{n}k)$ overestimates the variance of the sample mean by S_v^2/N.

EXAMPLE. Assume that the data from the previous example represents the result of the work of 2 interviewers in an interpenetrating design in which each interviewed 10 units. When $N = 400$, estimate the variance of $\bar{y}$ and its components as well as possible, assuming that r_w and r_b are nonnegative, but that the latter is much smaller than r_w.

Solution:

$$W = 961.6,$$
$$B = \tfrac{1}{10}\{1^2 + 21^2\} - \tfrac{1}{20} \times 22^2 = 20.0,$$
$$\bar{n} = 10.$$

$(B - W)/\bar{n}$, which here comes out as -94.2, is an unbiased estimator of $(r_w - r_b)\,\sigma_e^2$; but because of our assumptions it is not reasonable to estimate it to be a negative quantity and so we shall estimate it to be close to 0, and also estimate $r_w\sigma_e^2$ and $r_b\sigma_e^2$ to be close to 0; because of the magnitude of $(B - W)/\bar{n}$, we would guess σ_e^2 to be several hundred. $S_v^2 + \sigma_e^2(1 - r_w)$ is estimated by 961.6, and S_v^2 well below this amount. The variance of $\bar{y}$ is

$$(\tfrac{1}{20} - \tfrac{1}{400})S_v^2 + \tfrac{1}{20}\sigma_e^2(1 + 9r_w - 10r_b)$$
$$= \tfrac{1}{20}\{S_v^2 + \sigma_e^2(1 - r_w)\} + \tfrac{1}{2}\sigma_e^2(r_w - r_b) - \tfrac{1}{400}S_v^2,$$

where the first term on the right-hand side is estimated at 48, the second at 0, and $\tfrac{1}{400}S_v^2$ is less than 2; so a reasonable estimate of $V(\bar{y})$ is 46. (As to a comparison with the previous example, recall that there $\bar{y}$ was defined as a quantity based on 10 rather than 20 observations.)

FELLEGI,† J. Amer. Statist. Ass., **59** (1964) 1016–1041, has pointed out that if $\bar{n}k/N$ is not small and if sampling is carried out such that the coefficient of correlation between the response fluctuations of the units visited by one enumerator and the sampling fluctuations among the different‡ units visited by the same enumerator are positively correlated, the above estimator of $r_w\sigma_e^2$ may well be seriously underestimated. As an example he mentions that enumerators whose assignments contain many unemployed may have become more alert in their questioning concerning employment status.

Fellegi studies primarily a combination of interpenetration and repetition of surveys, and thus is able to obtain estimators which, under certain assumptions, may improve upon those based on interpenetration only. His method involves pairing each of k random subsets in a random way with some different subset, such that the set of pairs embraces every one of the k subsets twice; and assigning the enumerators at random to the *pairs* of subsets instead of to the subsets. Each enumerator then surveys both parts of a pair of subsets.

REMARK. If we write the model as

$$y_i = v_i + d_i + g_i,$$

† The reference is to page 1033, where he calls his coefficient α.

‡ On our assumptions, the corresponding correlation for identical units is zero.

where the g_i and d_i are distributed independently of each other with 0 mean (the g_i are thought of as effects ascribed to the interviewers), then

$$\mathscr{E}B = S_v^2 + \sigma_d^2 + \bar{n}\sigma_g^2,$$
$$\mathscr{E}W = S_v^2 + \sigma_d^2.$$

Here σ_d^2 corresponds to $\sigma_e^2(1 - r_w)$ and σ_g^2 to $\sigma_e^2(r_w - r_b)$.

Note that analysis of variance tables are based on the assumption of k strata within which $\bar{n}$ units are chosen out of $\bar{N} = N/k$, so that $\mathscr{E}B$ has a factor $(\bar{N} - \bar{n})/\bar{N}$, which does not appear here. If the k interviewers are drawn at random out of a pool of K, we also need a factor $(K - k)/K$ to adjust for the finiteness of the pool of available interviewers.

3.3. A MORE COMPLEX SETUP

Consider the following design.† Let us divide the population into k equal strata (k an even number). Each stratum is subdivided into l areas, each under the supervision of a crew leader. Each of these areas is subdivided into m interviewing assignment areas, containing a constant number of individuals, of which $\bar{\bar{n}} = \bar{n}/(lm)$ are interviewed. There are k data processors, which are assigned at random to the k strata. Again the response variance component is

$$\mathscr{E}\bar{e}^2 = \frac{1}{k\bar{n}}\sigma_e^2\{1 + r_w(\bar{n} - 1) + r_b\bar{n}(k - 1)\},$$

but here the average correlations r_w and r_b refer to response deviations of units within the strata, and between units in different strata, respectively. (r_b need not be zero, because there may be influences which cut across strata; but it is not influenced by crew leaders and processors.)

For simplicity let $l = 2$. In order to be able to estimate the components of variance, let us allocate the $2m$ interviewers in each stratum at random to the 2 crew leaders. Let strata 1 and 2 be contiguous, and similarly 3 and 4, etc. Every interviewer assignment area in stratum 1 is paired with a contiguous interviewer assignment area in stratum 2, and similarly for strata 3 and 4, etc. Divide the $2\bar{\bar{n}}$ units in each such pair of original interviewer assignment areas into 2 equal random groups. Consider one such pair of random groups and the corresponding 2 interviewers. By the flip of a coin assign the first interviewer to one of the random groups of units, the other to the other group of the pair. The work load of no interviewer has increased, but the work is more widespread geographically.

In place of the above definition of B, we now have $\bar{n}$ times the sum of the squares of deviations of the strata means from the overall mean divided by

† HANSEN, HURWITZ and BERSHAD, Bull. Intern. Statist. Inst., **38**, pt 2 (1961) 359–374.

$k - 1$; if $k = 2$, this equals $\bar{n}$ times $\frac{1}{2}(\bar{y}[1] - \bar{y}[2])^2$, where $\bar{y}[1]$ and $\bar{y}[2]$ are the two strata means. W is still defined as the within interviewer deviations mean square; this is now computed as the sum of klm squared deviations divided by $klm\,(\bar{n} - 1)$, the number of degrees of freedom.

3.4. FURTHER EXTENSIONS

The methods above extend immediately to the case in which $\mathscr{P}$ is one of several strata or clusters, provided all errors are uncorrelated from stratum to stratum or from cluster to cluster. This holds exactly only if in each stratum (or cluster, respectively) there are different enumerators and processors. It is also possible to set up an analysis for the case in which this is not the case (e.g., the interviewer may be differently affected in his work by differences in the social or age strata to which the interviewees belong).

4. Correction of estimates based on biased measurements with the aid of a small sample of unbiased measurements

We refer again to the model of section 6.2 of Chapter VIII. If we are able to measure μ_i for a small *subsample* of m observations, we may estimate the population average of the μ_i by

$$\bar{y}_{\mathrm{D}} = \bar{\mu}_m + \bar{y} - \bar{y}_m,$$

where $\bar{\mu}_m$ and $\bar{y}_m$ are the averages of the μ_i and the y_i from the subsample. From section 13.2 of Chapter IIA it follows that

$$V(\bar{y}_{\mathrm{D}}) = \frac{N-n}{N}\frac{1}{n}S_\mu^2 + \frac{n-m}{n}\frac{1}{m}S^2(\mathscr{D}),$$

where $d_i = y_i - \mu_i$. With α_i and e_i uncorrelated,

$$S^2(\mathscr{D}) = S_\alpha^2 + \sigma_e^2(1 - r_e).$$

We can estimate $V(\bar{y}_{\mathrm{D}})$ by replacing S_μ^2 and $S^2(\mathscr{D})$ by their usual estimators s_μ^2 and $s^2(\hat{\mathscr{D}})$ based on the subsample. If the μ_i and the α_i are uncorrelated, another unbiased estimator is

$$\frac{N-n}{N}\frac{1}{n}s^2(\mathscr{Y}) + \left(\frac{1}{m} - \frac{2}{n} + \frac{1}{N}\right)s^2(\hat{\mathscr{D}}),$$

since the expected value of the first term is

$$\frac{N-n}{N}\frac{1}{n}\{S_v^2 + \sigma_e^2(1 - r_e)\}$$

and of $s^2(\hat{\mathscr{D}})$ is

$$S_\alpha^2 + \sigma_e^2(1 - r_e).$$

If the μ_i and the α_i are positively correlated, it leads to overestimation. For m much smaller than n and a very limited correlation between the μ_i and the α_i, the second estimator may be preferable, $s^2(\mathscr{Y})$ being based on a much larger number of observations than s_μ^2.

It may be that the unbiased measurements would also not be constant if repeated, but subject to fluctuations f_i (with average variances σ_f^2 and average covariances $r_f\sigma_f^2$). If we denote these unbiased measurements by x_i, the estimator becomes

$$\bar{y}_D = \bar{x}_m + \bar{y} - \bar{y}_m$$

and in the variance formula S_μ^2 is replaced by $S^2(\mathscr{X})$ estimated by $s^2(\mathscr{X})$.

If the fluctuations e_i and f_i are chiefly caused by the method of measurement, e_i and f_i may be largely uncorrelated. In that case, since d_i is here $y_i - x_i$, $V(\bar{y}_D)$ can be written

$$\left(\frac{1}{n} - \frac{1}{N}\right)\{S_\mu^2 + \sigma_f^2(1 - r_f)\} + \left(\frac{1}{m} - \frac{1}{n}\right)\{S_\alpha^2 + \sigma_e^2(1 - r_e) + \sigma_f^2(1 - r_f)\}$$

$$= \left(\frac{1}{n} - \frac{1}{N}\right) S_\mu^2 + \left(\frac{1}{m} - \frac{1}{n}\right)\{S_\alpha^2 + \sigma_e^2(1 - r_e)\} + \left(\frac{1}{m} - \frac{1}{N}\right)\sigma_f^2(1 - r_f).$$

On the other hand the mean of

$$\left(\frac{1}{n} - \frac{1}{N}\right) s^2(\mathscr{Y}) + \left(\frac{1}{m} - \frac{1}{N}\right) s^2(\hat{\mathscr{D}})$$

is

$$\left(\frac{1}{n} - \frac{1}{N}\right)\{S_\mu^2 + 2S_{\mu\alpha}\}$$

$$+ \left(\frac{1}{m} + \frac{1}{n} - \frac{2}{N}\right)\{S_\alpha^2 + \sigma_e^2(1 - r_e)\} + \left(\frac{1}{m} - \frac{1}{N}\right)\sigma_f^2(1 - r_f),$$

which under positive correlation of μ and α exceeds $V(\bar{y}_D)$. Therefore, if the f_i are not very much smaller than the e_i, m is much smaller than n, and the correlation between the μ_i and the α_i is very limited in extent, we may prefer to use the above alternative estimator of $V(\bar{y}_D)$.

Another case is one in which the e_i and f_i are primarily caused by the measured objects themselves (e.g., the respondees) and thus may be strongly correlated. For the extreme case that $f_i = ke_i$, we have

$$d_i = y_i - x_i = \alpha_i + (1 - k)e_i,$$

so $V(\bar{y}_D)$ is

$$\left(\frac{1}{n}-\frac{1}{N}\right)\{S_\mu^2+k^2\sigma_e^2(1-r_e)\}+\left(\frac{1}{m}-\frac{1}{n}\right)\{S_\alpha^2+(1-k)^2\sigma_e^2(1-r_e)\}$$

$$=\left(\frac{1}{n}-\frac{1}{N}\right)S_\mu^2+\left(\frac{1}{m}-\frac{1}{n}\right)S_\alpha^2$$

$$+\left\{\left(\frac{1}{n}-\frac{1}{N}\right)k^2+\left(\frac{1}{m}-\frac{1}{n}\right)(1-k)^2\right\}\sigma_e^2(1-r_e).$$

The expectation of

$$\left(\frac{1}{n}-\frac{1}{N}\right)s^2(\mathscr{Y})+\left(\frac{1}{m}-\frac{2}{n}+\frac{1}{N}\right)s^2(\hat{\mathscr{D}})$$

is

$$\left(\frac{1}{n}-\frac{1}{N}\right)\{S_\mu^2+2S_{\mu\alpha}\}+\left(\frac{1}{m}-\frac{1}{n}\right)S_\alpha^2$$

$$+\left\{\frac{1}{n}-\frac{1}{N}+\left(\frac{1}{m}-\frac{2}{n}+\frac{1}{N}\right)(1-k)^2\right\}\sigma_e^2(1-r_e),$$

which exceeds $V(\bar{y}_D)$ by

$$2\left(\frac{1}{n}-\frac{1}{N}\right)S_{\mu\alpha}+2k(1-k)\left(\frac{1}{n}-\frac{1}{N}\right)\sigma_e^2(1-r_e).$$

Therefore, if m is much smaller than n, and there is little correlation between the μ_i and the α_i, the above variance estimator may be preferred.

In case the m observations form an *independent* sample,

$$V(\bar{y}_D)=\frac{N-n}{N}\,\frac{1}{n}\,S^2(\mathscr{Y})+\frac{N-m}{N}\,\frac{1}{m}\,S^2(\mathscr{D}),$$

which may be estimated by substituting $s^2(\mathscr{Y})$ for $S^2(\mathscr{Y})$ and $s^2(\hat{\mathscr{D}})$ for $S^2(\mathscr{D})$.

CHAPTER XI

REPEATED SURVEYS

1. Some special aspects of sequences of surveys

We are often interested in certain characteristics of a population as they change over time. This interest is reflected most strikingly in the periodic censuses of population, housing, manufacturing, etc. that are conducted in many countries.

Over time some units may fall out from the population and new units may be added. In fact, the main purpose of periodic censuses may be simply to obtain information on the size of the population at regular intervals. For this, however, a census may not be necessary. Thus it may be that data are obtained which allow us to keep the total up to date. E.g., in some countries a card is kept on each inhabitant and an attempt is made to record all the relevant changes. More usually, the total may be kept up to date statistically; in demography one uses statistics of births, deaths and migration. Theoretically, these methods should give accurate results (provided, of course, the original census is accurate); but (like, to some extent, the census) the registration of changes is often quite incomplete and subject to errors. In specific fields of application, methods have been developed for adjusting the results with the help of as much information on the population and comparable populations as is obtainable. Some of these methods – an extensive discussion is found in the literature on demography and animal populations – are of a general nature.

Another approach to the updating of a census is to conduct a survey. If efficient, the design of such a survey and the analysis of its results incorporate (in the frame, the stratification, the use of auxiliary variables and unequal probabilities) data from the censuses and all the other information referred to above. This need not be discussed here again. There are, however, some novel features which come up in connection with the design of a *sequence* of surveys, intercensal or otherwise, to some of which we shall give attention in this chapter.

As mentioned previously, estimation of the total number of units in the population falls outside the scope of the present work. However, all the above remarks apply as well to characteristics other than the total size.† That is, even in the case of a fixed population, at different dates any one unit may fall in different categories. Therefore there may be changes in the identities or sizes of subpopulations of interest, or in the total or average values for certain of their characteristics. Mathematically there is no essential difference between these two cases, but in practical terms they may be much different. In the latter case it is usually possible to use the original listing of the units in the strata or the clusters. In the former case and in the case of populations which are not fixed this may not be possible; but sometimes it is feasible and advantageous to retain the original subpopulation frame and adjoin to it a new frame of additions to the subpopulation, treating the two frames as two strata (or, if the first has L_1 strata and the second L_2 strata, as $L_1 + L_2$ strata).

When it is intended to conduct a sequence of surveys, it is usually worthwhile to plan each with a view to the others to follow. Thus, the fact that the cost of a frame may be shared by several surveys may greatly affect the design of these surveys (and even their feasibility).‡ It may be worthwhile to create a more or less permanent regional field staff if surveys are going to be made regularly.

Repeated surveys are of special importance from an *analytic* point of view, in particular for the study of changes in certain characteristics or for relating these changes to changes in other characteristics. Sometimes we are in a position to formulate a suitable *model* for this prior to conducting these surveys. In the sections that follow we assume that this is not the case.

In the sequel we shall consider a characteristic $\mathscr{Y}$ and denote it by $\mathscr{Y}_h$ when we consider its value an occasion h.

2. Sampling on two occasions from a fixed population – a simple case

Consider the following plan: on the first occasion we draw, without replacement, a simple random sample n' units, denoting the average of the $\mathscr{Y}_1$ values of this sample by $\bar{y}_1$. On the second occasion a random subset of $\alpha n'$ of these units are dropped (with average $\mathscr{Y}_1$-value $\bar{y}_{1\text{o}}$ – o stands for omitted), and the remaining $(1 - \alpha)n'$ retained (with average $\mathscr{Y}_1$-value $\bar{y}_{1\text{r}}$ and average $\mathscr{Y}_2$-value $\bar{y}_{2\text{r}}$ – r stands for retained); $\alpha n'$ fresh units replace the dropped units (with average $\mathscr{Y}_2$-value $\bar{y}_{2\text{f}}$ – f stands for fresh). In the following example $n' = 7$ and $\alpha n' = 4$:

† Incompleteness of and inaccuracies in registration of changes are usually relatively much larger for subpopulations than for the entire population.

‡ This remark is, of course, also relevant to the conduct of more or less simultaneous surveys on different topics.

Drawn on and evaluated as of	Sample units			Sample averages			
Occasion 1	× × × ×	× × ×		$\bar{y}_{1o}$	$\bar{y}_{1r}$	
Occasion 2		× × ×	× × × ×		$\bar{y}_{2r}$	$\bar{y}_{2f}$

Let us estimate $\bar{Y}_2$ on the second occasion. Consider first doing so on the basis of the $(1 - \alpha)n'$ retained units only, i.e., when no new units are obtained. This is the situation considered in section 13.2 of Chapter IIA (generalized difference estimation with double sampling) with $\mathscr{X} = \mathscr{Y}_1$, and $\mathscr{Y} = \mathscr{Y}_2$, subsampling $(1 - \alpha)n'$ from n'. The difference estimator for this is

$$\hat{\bar{Y}}_{2D} = \bar{y}_{2r} + k(\bar{y}_1 - \bar{y}_{1r}) = \bar{y}_{2r} + k\alpha(\bar{y}_{1o} - \bar{y}_{1r})$$

with variance

$$\left(\frac{1}{(1-\alpha)n'} - \frac{1}{N}\right)S^2(\mathscr{Y}_2) - \left(\frac{1}{(1-\alpha)n'} - \frac{1}{n'}\right)(2kS(\mathscr{Y}_1\mathscr{Y}_2) - k^2S^2(\mathscr{Y}_1))$$
$$= S^2\left\{\frac{1}{n'}\,\frac{1-\alpha k(2\rho - k)}{1-\alpha} - \frac{1}{N}\right\},$$

if, as we shall assume, $S^2(\mathscr{Y}_1) = S^2(\mathscr{Y}_2) = S^2$ (say), and if ρ is the correlation coefficient between $\mathscr{Y}_1$ and $\mathscr{Y}_2$.

Let us now try to form a weighted average of this unbiased estimator of $\bar{Y}_2$ and $\bar{y}_{2f}$, the average $\mathscr{Y}_2$-value of the freshly added units, in such a way that the variance is as small as possible. This turns out to be rather complicated, so we shall just consider the case in which n'/N is negligibly small. Then the correlation between, on the one hand, the $\mathscr{Y}_h$-value of a unit in the part of the sample on occasion h that is retained from the previous occasion or held over to the next occasion with, on the other hand, the $\mathscr{Y}_h$-value of a unit in the other part of that sample may be shown to be negligible, and so we can apply the following, widely used lemma.

LEMMA. If the random variables $\mathscr{X}_1, \ldots, \mathscr{X}_l$ have the same means, positive variances $\sigma_1^2, \ldots, \sigma_l^2$, and zero covariances, then, among all linear combinations

$$\beta_1\mathscr{X}_1 + \ldots + \beta_l\mathscr{X}_l$$

(with nonrandom coefficients $\beta_1, \ldots, \beta_l$) which have the same mean as the common mean of the $\mathscr{X}$'s, regardless of what be the value of this mean,

$$\frac{\sigma_1^{-2}}{\Sigma\sigma_h^{-2}}\mathscr{X}_1 + \ldots + \frac{\sigma_l^{-2}}{\Sigma\sigma_h^{-2}}\mathscr{X}_l$$

has the smallest variance, namely $1/\Sigma\sigma_h^{-2}$.

Proof. Let the common mean be ξ; then

$$\mathscr{E}(\beta_1\mathscr{X}_1 + \ldots + \beta_l\mathscr{X}_l) = \xi\,\Sigma\beta_h$$

equals ξ, regardless of what the value of ξ is, only when $\Sigma\beta_h = 1$.

The Schwarz-Cauchy inequality (Chapter III, section 8), for $a_h = \beta_h\sigma_h$ and $b_h = \sigma_h^{-1}$, gives

$$\Sigma\beta_h^2\sigma_h^2\Sigma\sigma_h^{-2} \geq (\Sigma\beta_h)^2,$$

and the right-hand side equals 1. So $\Sigma\beta_h^2\sigma_h^2$, the variance of $\Sigma\beta_h\mathscr{X}_h$, is at least equal to $1/\Sigma\sigma_h^{-2}$; and is equal to it for (and only for) $\beta_h\sigma_h^2 = \lambda$, a constant, i.e., for $\beta_h = \lambda\sigma_h^{-2}$, with (as $\Sigma\beta_h = 1$) $\lambda = 1/\Sigma\sigma_h^{-2}$. Another proof may be given using Lagrange multipliers.

Thus the minimum variance combination of $\bar{y}_{2f}$ and $\hat{\bar{Y}}_{2D}$ has weights about proportional to α and $(1 - \alpha)/\{1 - \alpha k(2\rho - k)\}$, respectively.† The variance of this combination is about

$$\frac{S^2}{n'}\frac{1 - \alpha k(2\rho - k)}{1 - \alpha^2 k(2\rho - k)}.$$

Clearly, when $k > 2\rho$, the variance of $\bar{y}_2$ for an entirely fresh sample of n' is smaller than this.

Assume now that $k < 2\rho$ and consider the value of α which minimizes the above expression. Differentiating with respect to α and setting the result equal to 0 gives a quadratic equation with two roots, one of which,

$$\alpha_0 = [1 + \{1 - k(2\rho - k)\}^{\frac{1}{2}}]^{-1},$$

is real since $k(2\rho - k)$ has a maximum at $k = \rho$ (the other root exceeds 1). Also, at α_0 the second derivative is positive, so that for $\alpha = \alpha_0$ we obtain the smallest variance for the combination, namely

$$\frac{S^2}{n'}\frac{1}{2}[1 + \{1 - k(2\rho - k)\}^{\frac{1}{2}}].$$

† On occasion 2 we can prepare a revised estimate of $\bar{Y}_1$, which is obtained from the estimator of $\bar{Y}_2$ by reversing the arrow of time, i.e., it is a weighted average of

$$\bar{y}_{1o} \quad \text{and} \quad \bar{y}_{1r} + k(\bar{y}_2 - \bar{y}_{2r});$$

the weights then come out to be the same. The difference between $\bar{Y}_2$ and $\bar{Y}_1$ is then estimated as the similarly weighted average of

$$\bar{y}_{2f} - \bar{y}_{1o} \quad \text{and} \quad (1 + k)(\bar{y}_{2r} - \bar{y}_{1r}) + k(\bar{y}_1 - \bar{y}_2)$$

and the sum of $\bar{Y}_1$ and $\bar{Y}_2$ as the similarly weighted average of

$$\bar{y}_{2f} + \bar{y}_{1o} \quad \text{and} \quad (1 - k)(\bar{y}_{2r} + \bar{y}_{1r}) + k(\bar{y}_1 + \bar{y}_2).$$

The variance of the above estimator of the difference can be substantially smaller than that of $\bar{y}_2 - \bar{y}_1$, especially if ρ is at all large. On the other hand, calculation shows that the variance of the above estimator of the sum is little smaller (relatively speaking) than of $\bar{y}_1 + \bar{y}_2$ unless ρ is very close to 1. For $k = \rho$, the variance of the estimator of the difference is about $2(S^2/n')(1 - \rho)/(1 - \alpha\rho)$, that of the estimator of the sum about $2(S^2/n')(1 + \rho)/(1 + \alpha\rho)$.

This expression is decreasing in ρ, from its largest value S^2/n' (attained when $\rho = \frac{1}{2}k$) to a value equal to at least $S^2/(2n')$. The smallest value is attained by setting k equal to ρ; in that case, and also in the case in which we set k equal to 1, the value $S^2/(2n')$ is actually attained at $\rho = 1$. Should we know that ρ is only a few per cent below 1, we would probably use $k = 1$ and α equal to 1 or close to 1. For smaller ρ, numerical calculation shows that the variance is quite insensitive† to the value chosen for α, and also that taking α less than $\frac{1}{2}$ can never diminish the variance much more than taking it to be $\frac{1}{2}$. This is fortunate, since, on the basis of the two samples we may obtain only a poor estimate‡ of ρ; and since, moreover, in one survey one often estimates a number of characteristics which may have quite different ρ values.

COCHRAN, Sampling Techniques (Wiley, 2nd ed., 1963) (where only the case $k = \rho$ is considered), recommends taking for α, when ρ is not very large, $\frac{2}{3}$ or $\frac{3}{4}$ as a good compromise value. If, as is often the case, field costs decrease with a decrease in α, or if, in addition to current estimates, it is also desired to estimate the difference§ $\bar{Y}_2 - \bar{Y}_1$, the optimum α will be lower than α_0. The

† The relative decrease in variance due to retaining in the sample the units from the first occasion also appear to depend very little on the choice of k, especially in cases in which the advantage is not small.

‡ If we use for ρ its estimator from the two samples (so that our estimators become regression estimators – compare Chapter IIA, section 13.3) and if n' is not very small, the variance of the estimator of $\bar{Y}_2$ is usually only a little larger than the one using the true value of ρ, but, of course, the estimator of $\bar{Y}_2$ may then be subject to not inconsiderable bias.

§ Note that reference is made to an estimator of *net* change. Often we are also interested in changes in the mean $\mathscr{Y}$-values of units with given $\mathscr{Y}_1$-values. As an example consider the following population frequencies:

On the first occasion	On the second occasion			Total
	For	Against	Undecided	
For	10	25	5	40
Against	26	20	4	50
Undecided	4	5	1	10
Total	40	50	10	100

For each of the three categories the net change is 0, but 69 of the 100 units changed their position. Designs which minimize the variance of the estimator of $\bar{Y}_2 - \bar{Y}_1$ may not be efficient for *analytic* purposes like these. Without special assumptions, the latter clearly would require $\alpha = 0$, unless it would be possible to obtain, at low cost, a sample of additional units from among those in the population which on the first occasion had the different specified $\mathscr{Y}_1$-values separately for each such $\mathscr{Y}_1$-value (in the example that means that it would be inexpensive to sample separately from among those in the population who on the first occasion were for, those who were against, and those who were undecided).

former is evident, and also it is clear that $\bar{Y}_2 - \bar{Y}_1$ is estimated with smallest variance if we use the same units on both occasions, as long as $\rho > 0$: $V\{y_2(u) - y_1(u')\}$ equals $\sigma^2(\mathscr{Y}_2) + \sigma^2(\mathscr{Y}_1) - 2\rho\sigma(\mathscr{Y}_1)\sigma(\mathscr{Y}_2)$ for $u = u'$, whereas for $u \neq u'$ and n'/N small, it equals about $\sigma^2(\mathscr{Y}_2) + \sigma^2(\mathscr{Y}_1)$. On the other hand, if we are also interested in estimating $\frac{1}{2}(\bar{Y}_1 + \bar{Y}_2)$, this should lead us to increase α, since $V\{y_1(u) + y_2(u')\}$ equals $\sigma^2(\mathscr{Y}_1) + \sigma^2(\mathscr{Y}_2) + 2\rho\sigma(\mathscr{Y}_1)\sigma(\mathscr{Y}_2)$ for $u = u'$.

Often there are difficulties connected with including a unit on more than one occasion (refusal; or unit has changed address) and sometimes the fact of earlier inclusion changes the $\mathscr{Y}_2$-value. Occasionally we may, instead of using a unit on more than one occasion, use units matched to it which may be expected to have $\mathscr{Y}$-values strongly correlated with the $\mathscr{Y}$-value of the unit for which they substitute. Optimum choice of α when a cost function is given is considered in much detail by KULLDORF, Rev. Intern. Statist. Inst., **31** (1963) 24–57.

3. Sampling on two occasions – some generalizations

The case in which $S(\mathscr{Y}_1) \neq S(\mathscr{Y}_2)$ and in which the number of fresh units is different from the number dropped has also been considered; see PATTERSON, J. Roy. Statist. Soc., **B12** (1950) 241–255.

If on the first occasion we sample n' with probabilities p_i with replacement, and on the second occasion subsample $(1 - \alpha)n'$ of these with equal probabilities without replacement, adding $\alpha n'$ to these by sampling $\alpha n'$ with probabilities p_i with replacement, we can still use the above results with the following modifications: replace $\bar{y}_1$, $\bar{y}_{hr}$, $\bar{y}_{2f}$ respectively, by

$$\frac{1}{n'}\sum_{i=1}^{N}\frac{t_{1i}y_{1i}}{Np_i}, \qquad \frac{1}{(1-\alpha)n'}\sum_{i=1}^{N}\frac{t_{1i}a_i y_{hi}}{Np_i}, \qquad \frac{1}{\alpha n'}\sum_{i=1}^{N}\frac{t_{2i}y_{2i}}{Np_i},$$

where t_{1i} is the number of times u_i is taken in the sample on the first occasion, t_{2i} the number of times it is taken in the fresh sample drawn on the second occasion, and $a_i = 1$ if u_i is retained and 0 if not. Also $S^2(\mathscr{Y}_h)$ and ρ are replaced by the "variance" of the $(Np_i)^{-1}y_{hi}$ around $\bar{Y}_h$:

$$\Sigma p_i\{(Np_i)^{-1}y_{hi} - \bar{Y}_h\}^2$$

and the "correlation" between the $(Np_i)^{-1}y_{1i}$ and the $(Np_i)^{-1}y_{2i}$, respectively.

For cluster sampling with a given cost function, optimum choice of α has been considered by SAITO, Rep. Stat. Appl. Res., JUSE, **4** (1957) 125–131. Some further results on cluster sampling have been obtained by TIKKIWAL, J. Ind. Statist. Ass., **3** (1965) 125–133. One may consider replacing also some of the subunits belonging to clusters which are retained.

Instead of unbiased estimators one may also consider biased ones, such as ratio estimators; the theory given above may then be used as an approximation.

We may also use some auxiliary characteristic observed on the first occasion in order to define strata or new probabilities for the selection of units on the second occasion,† or for improving the estimators. For double sampling for stratification (with or without clustering) on successive occasions see SINGH and SINGH, J. Amer. Statist. Ass., **59** (1965) 784–792.

4. Sampling on three or more occasions from a fixed population

As in section 2, we shall confine our attention to simple random sampling in which as many ($\alpha n'$) new units are added each time as are dropped, and in which $S^2(\mathscr{Y}_h)$ is equal to the same number, S^2, on each occasion. Moreover, we shall assume that the correlation of $\mathscr{Y}$-values of a unit on occasion h and occasion $h+k$ is ρ^k. More general results are found in PATTERSON, J. Roy. Statist. Soc., **B12** (1950) 241–255. A different setup is studied by RAO and GRAHAM, J. Amer. Statist. Ass. **59** (1964) 492–509.

The notation now has to be generalized. While $\bar{y}_{hr}$ and $\bar{y}_{hf}$ will still denote the average $\mathscr{Y}_h$-value for units retained on occasion h and units freshly added on occasion h, respectively, we need some new notation for the average $\mathscr{Y}_{h-1}$-value of the units which will be retained on occasion h; we shall call it $\bar{x}_{h-1,r}$. This is illustrated in the following example for $n' = 7$, $\alpha n' = 4$ and $h = 1, 2, 3, 4$ (see display on the top of the next page):

For $h = 1$, we shall continue to use $\hat{Y}_1 = \bar{y}_1$; and for $\hat{Y}_2$ the estimator of smallest variance of the form (found in section 2):

$$w_h\bar{y}_{hf} + (1 - w_h)\{\bar{y}_{hr} + k_h(\hat{Y}_{h-1} - \bar{x}_{h-1,r})\}.$$

Obtain estimators of this form for all $h > 1$ with coefficients that minimize their variances, using the following proposition proved in section 7.

PROPOSITION. We denote the observations on $\mathscr{Y}_k$ by $y_{k1}, y_{k2}, \ldots$, and by $\mathscr{D}$ the class of linear functions of these observations (the hth one only of the observations $y_{k1}, y_{k2}, \ldots$ for $k \leq h$) which are unbiased estimators for the $\bar{Y}_h$. A necessary and sufficient condition for the set of Q_h, functions in $\mathscr{D}$, to be of minimum variance in this class is that for all h and $k \leq h$

$$\text{Cov}\,(y_{kl}, Q_h)$$

† In this connection see KEYFITZ, J. Amer. Statist. Ass., **46** (1951) 105–109, KISH, Proc. Soc. Statist. Section, Amer. Statist. Assoc. (1963) 124–131, and FELLEGI, J. Amer. Statist. Ass., **58** (1963) 183–201. Kish also discusses changing strata.

Drawn on and evaluated as of — Sample units and notation of averages computed from them

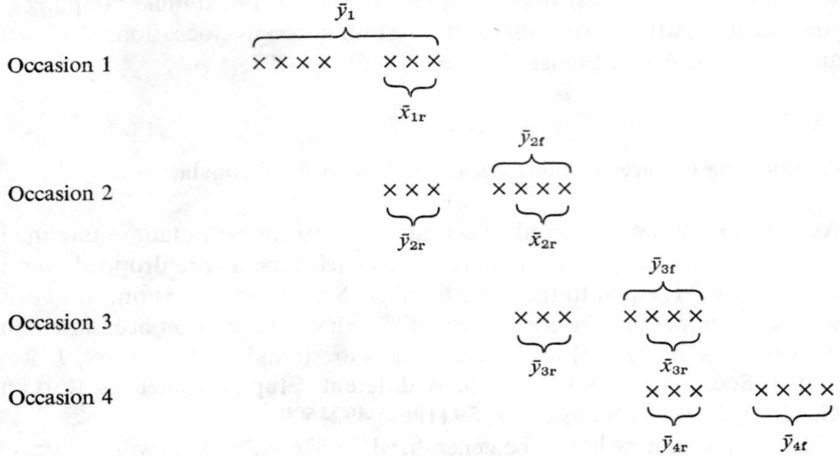

is the same for all l. Consequently, if the Q_h are of minimum variance in $\mathscr{D}$,

$$\operatorname{Cov}(Q_k, Q_h) = \operatorname{Cov}(y_{kl}, Q_h).$$

Moreover, if Q_h is of minimum variance in $\mathscr{D}$ with $\mathscr{E} Q_h = \bar{Y}_h$, no other function of $\mathscr{D}$ is.

In using the Proposition, if the $\mathscr{Y}_k$-value of some unit not included, some unit retained, or some unit freshly added on occasion h is observed, denote its $\mathscr{Y}_k$-value by y_{koh}, y_{krh} or y_{kfh}, respectively. Note that all the $\mathscr{Y}$-values as of occasion k ($\leq h$) of all units observed up to occasion h, inclusive, are of one of these three kinds. Of course, y_{kfh} is defined only for $k = h$.

For our purpose we also assume sampling with replacement (i.e., neglect n'/N). Then

$$\operatorname{Cov}(y_{h-1,r,h}, \hat{\bar{Y}}_h) = (1 - w_h)\{\operatorname{Cov}(y_{h-1,r,h}, \bar{y}_{hr}) + k_h \operatorname{Cov}(y_{h-1,r,h}, \hat{\bar{Y}}_{h-1}) - k_h \operatorname{Cov}(y_{h-1,r,h}, \bar{x}_{h-1,r})\}$$

and

$$\operatorname{Cov}(y_{h-1,o,h}, \hat{\bar{Y}}_h) = (1 - w_h) k_h \operatorname{Cov}(y_{h-1,o,h}, \hat{\bar{Y}}_{h-1}).$$

If the variances of the unbiased estimators are to be minimal, we must have by the above proposition that

$$\operatorname{Cov}(y_{h-1,\mathrm{r},h}, \hat{Y}_h) = \operatorname{Cov}(y_{h-1,\mathrm{o},h}, \hat{Y}_h)$$

and

$$\operatorname{Cov}(y_{h-1,\mathrm{r},h}, \hat{Y}_{h-1}) = \operatorname{Cov}(y_{h-1,\mathrm{o},h}, \hat{Y}_{h-1}),$$

and so by substituting these we get

$$(1 - w_h)\{\operatorname{Cov}(y_{h-1,\mathrm{r},h}, \bar{y}_{h\mathrm{r}}) - k_h \operatorname{Cov}(y_{h-1,\mathrm{r},h}, \bar{x}_{h-1,\mathrm{r}})\} = 0.$$

But the left-hand side equals

$$(1 - w_h)(\rho - k_h)\frac{S^2}{(1-\alpha)n'},$$

so that, whenever any units are retained at all, the value of the k_h for which the estimators of the $\bar{Y}_h$ are of minimum variance is ρ also for $h > 2$.

Now we use the last part of the above proposition: if $\hat{Y}_{h-1}$ and $\hat{Y}_h$ are linear unbiased minimum variance estimators,

$$V(\hat{Y}_{h-1}) = \operatorname{Cov}(y_{h-1,\mathrm{r},h}, \hat{Y}_{h-1})$$

and

$$V(\hat{Y}_h) = \operatorname{Cov}(y_{h\mathrm{f}h}, \hat{Y}_h) = w_h \operatorname{Cov}(y_{h\mathrm{f}h}, \bar{y}_{h\mathrm{f}}),$$

with the latter equal to $w_h S^2/(\alpha n')$, and so the former equal to $w_{h-1}S^2/(\alpha n')$. Also $\hat{Y}_{h-1}$ is a linear combination of the $\mathscr{Y}$-values, on occasions before h, of units observed before occasion h and

$$\operatorname{Cov}(y_{h\mathrm{r}h}, y_{h-k,\mathrm{r},h}) = \rho^k \operatorname{Cov}(y_{h-k,\mathrm{r},h}, y_{h-k,\mathrm{r},h}),$$
$$\operatorname{Cov}(y_{h\mathrm{r}h}, y_{h-k,\mathrm{o},h}) = 0,$$

so

$$\operatorname{Cov}(y_{h\mathrm{r}h}, \hat{Y}_{h-1}) = \rho \operatorname{Cov}(y_{h-1,\mathrm{r},h}, \hat{Y}_{h-1}),$$

which equals

$$\rho w_{h-1}(S^2/\alpha n').$$

Therefore,

$$\operatorname{Cov}(y_{h\mathrm{r}h}, \hat{Y}_h) = (1 - w_h)\{\operatorname{Cov}(y_{h\mathrm{r}h}, \bar{y}_{h\mathrm{r}}) - \rho \operatorname{Cov}(y_{h\mathrm{r}h}, \bar{x}_{h-1,\mathrm{r}})\} + (1 - w_h)\rho \operatorname{Cov}(y_{h\mathrm{r}h}, \hat{Y}_{h-1})$$

equals

$$(1 - w_h)(1 - \rho^2)\frac{S^2}{(1-\alpha)n'} + (1 - w_h)\rho^2 w_{h-1}\frac{S^2}{\alpha n'}.$$

Equating this to Cov $(y_{hfh}, \hat{Y}_h)$ yields

$$(1 - w_h)\left\{\frac{1-\rho^2}{1-\alpha} + \frac{\rho^2 w_{h-1}}{\alpha}\right\} = \frac{w_h}{\alpha}$$

or

$$(1 - w_h)\frac{1 - \alpha\rho^2 + \rho^2(1-\alpha)w_{h-1}}{\alpha(1-\alpha)} = \frac{1}{\alpha},$$

where we may write the left-hand side in terms of $\bar{w}_h = 1 - w_h$, thus obtaining

$$\bar{w}_h = \frac{1-\alpha}{1 - \rho^2\{2\alpha - 1 + \bar{w}_{h-1}(1-\alpha)\}}, \quad V(\hat{Y}_h) = w_h \frac{S^2}{\alpha n'}.$$

Moreover, in section 2 we obtained (substituting $k = \rho$)

$$\bar{w}_2 = \frac{1-\alpha}{1-\alpha^2\rho^2}.$$

(Note that, if we put $\bar{w}_1 = 1 - \alpha$, the recursion formula also yields this $\bar{w}_2$; this means interpreting the so far undefined symbol $\bar{y}_{1r}$ as $\bar{x}_{1r}$ – the average of the $\mathscr{Y}_1$-values of units retained on the second occasion – and the symbol $\bar{y}_{1f}$, which has not previously been defined, as $\bar{y}_{1o}$ – the average of the $\mathscr{Y}_1$-values of units omitted on the second occasion; the expression following k_1 in the formula for $\hat{Y}_h$ with $h = 1$ is defined as 0.)

We now check whether all conditions for minimality of the variances are fulfilled. The condition

$$\text{Cov}\,(y_k', \hat{Y}_h) = \text{Cov}\,(y_k'', \hat{Y}_h)$$

where y_k' and y_k'' represent $\mathscr{Y}_k$-values of units both belonging to the same class (among the three considered) is evidently fulfilled. Just like we showed

$$\text{Cov}\,(y_{hrh}, \hat{Y}_{h-1}) = \rho\,\text{Cov}\,(y_{h-1,r,h}, \hat{Y}_{h-1}),$$

we show that for $j = 0, 1, 2, \ldots$

$$\text{Cov}\,(y_{h-j,r,h}, \hat{Y}_h) = \rho^{|j-1|}\,\text{Cov}\,(y_{h-1,r,h}, \hat{Y}_h),$$

and for $j = 1, 2, \ldots$

$$\text{Cov}\,(y_{h-j,o,h}, \hat{Y}_h) = \rho^{j-1}\,\text{Cov}\,(y_{h-1,o,h}, \hat{Y}_h),$$

to the extent that the left-hand sides are defined. We used and so fulfilled

$$\text{Cov}\,(y_{h-j,r,h}, \hat{Y}_h) = \text{Cov}\,(y_{h-j,o,h}, \hat{Y}_h) \qquad (*)$$

for $j = 1$; the previous two equations show that (*) also holds for $j = 0$ and for $j = 2, \ldots$. We have now examined the covariance of *all* the units observed with $\hat{Y}_h$ for a fixed but arbitrary occasion h, and so for all occasions.

Consequently the estimators we derived are of minimum variance, among all linear unbiased estimators which for estimating $\bar{Y}_h$ depend only on the observations up to and including occasion h.

By similar methods Patterson finds the minimum variance unbiased linear estimator of $\bar{Y}_h - \bar{Y}_{h-k}$. For $k = 1$, this is, not surprisingly, of the form

$$\hat{\bar{Y}}_h - \hat{\bar{Y}}_{h-1} + v_h(\hat{\bar{Y}}_h - \bar{y}_{ht});$$

here $v_h = \rho w_{h-1}$. The variance of this estimator is therefore

$$\frac{S^2}{\alpha n'}[w_h(1 + \rho w_{h-1})^2 + w_{h-1}\{1 - \rho(2 + \rho w_{h-1})\}]$$

and the variance of the corresponding estimator of $\hat{\bar{Y}}_{h-1} + \hat{\bar{Y}}_h$

$$\frac{S^2}{\alpha n'}[w_h(1 - \rho w_{h-1})^2 + w_{h-1}\{1 + \rho(2 - \rho w_{h-1})\}].$$

Patterson also investigates the case $k > 2$; he finds that the variance of the minimum variance linear unbiased estimator of $\bar{Y}_h - \bar{Y}_{h-1}$ computed from data available on occasion $h + k$ is very close to that of its minimum variance linear unbiased estimator computed on occasion h, so that it is *not worthwhile* to keep revising the estimate of changes.

One may also obtain the weights which will minimize the variance of the estimator of $\bar{Y}_h$ if the k's are preassigned and different from ρ. For that purpose one cannot, of course, use the Proposition, but must calculate k for the third occasion, then for the next, etc. This is somewhat complex, but simplifies greatly if $\alpha = \frac{1}{2}$. In that case we may use a simple notation (illustrated for $n' = 6$):

Drawn on and evaluated as of	Sample units and notation for averages computed from them			
Occasion 1	× × × $\bar{y}_1(1)$	× × × $\bar{y}_1(2)$		
Occasion 2		× × × $\bar{y}_2(2)$	× × × $\bar{y}_2(3)$	
Occasion 3			× × × $\bar{y}_3(3)$	× × × $\bar{y}_3(4)$
Occasion 4				× × × $\bar{y}_4(4)$	× × × $\bar{y}_4(5)$

$$\hat{Y}_1 = \tfrac{1}{2}\{\bar{y}_1(1) + \bar{y}_1(2)\},$$
$$\hat{Y}_2 = w_2\bar{y}_2(3) + \bar{w}_2[\bar{y}_2(2) + k_2\{\hat{Y}_1 - \bar{y}_1(2)\}]$$
$$= w_2\bar{y}_2(3) + \bar{w}_2\{\bar{y}_2(2) + \tfrac{1}{2}k_2(\bar{y}_1(1) - \bar{y}_1(2))\},$$
$$\hat{Y}_3 = w_3\bar{y}_3(4) + \bar{w}_3[\bar{y}_3(3) + k_3\bar{w}_2\{\bar{y}_2(2) - \bar{y}_2(3) + \tfrac{1}{2}k_2(\bar{y}_1(1) - \bar{y}_1(2))\}],$$

etc., where variances are relatively easy to compute as $\bar{y}_i(j)$ and $\bar{y}_{i'}(j')$ are uncorrelated if $j' \neq j$, and

$$V\{\bar{y}_i(j)\} = 2S^2/n',$$
$$\text{Cov}\,\{\bar{y}_{i-1}(i), \bar{y}_i(i)\} = 2\rho S^2/n'.$$

Note that no assumption is necessary on the correlation of $\mathscr{Y}$-values more than one occasion apart.

5. Two-level rotation sampling from a fixed population

Another plan, which is sometimes practicable, is to obtain *a fresh sample of n* on each occasion and to record its values of $\mathscr{Y}$ on the current occasion as well as on the previous one. This is called two-level rotation. The basic assumption is that it is possible to obtain the retroactive data on a basis which is *comparable* to the data on a current basis.

We shall confine our attention to simple random sampling, with $S^2(\mathscr{Y}_h) = S^2$ and a constant correlation ρ between successive $\mathscr{Y}$-values of a unit. For the case in which n varies from occasion to occasion see ECKLER, Ann. Math. Statist., **26** (1955) 664–685. We shall denote by $\bar{y}_{hh}$ the average $\mathscr{Y}_h$-values of the units drawn on occasion h, by $\bar{y}_{h-1,h}$ the average of their $\mathscr{Y}_{h-1}$ values. The following illustrates the matter for $n = 3$.

	Sample units drawn on occasion				Averages from samples drawn on occasion			
	1	2	3	4	1	2	3	4
Value on occasion 0	× × ×				$\bar{y}_0$			
Value on occasion 1	× × ×	× × ×			$\bar{y}_{11}$	$\bar{y}_{12}$		
Value on occasion 2		× × ×	× × ×			$\bar{y}_{22}$	$\bar{y}_{23}$	
Value on occasion 3			× × ×	× × ×			$\bar{y}_{33}$	$\bar{y}_{34}$
Value on occasion 4				× × ×				$\bar{y}_{44}$

It is natural to consider estimators of the form

$$\hat{\bar{Y}}_h = \bar{y}_{hh} + c_h\{\hat{\bar{Y}}_{h-1} - \bar{y}_{h-1,h}\}$$

with $c_1 = 0$. (It is clear that, without prior restriction on the $\bar{Y}_h$, $\bar{y}_0$ does not help in the estimation of $\bar{Y}_1$.)

We apply again the Proposition of the previous section, which implies that if the $\hat{\bar{Y}}_h$ are to be linear unbiased of minimum variance we must have the equality of

$$\begin{aligned}\operatorname{Cov}(\bar{y}_{h-1,h}, \hat{\bar{Y}}_h) &= \operatorname{Cov}(\bar{y}_{h-1,h}, \bar{y}_{hh}) - c_h V(\bar{y}_{h-1,h}) + c_h \operatorname{Cov}(\bar{y}_{h-1,h}, \hat{\bar{Y}}_{h-1})\\ &= n^{-1}S^2(\rho - c_h) + 0\end{aligned}$$

and

$$\begin{aligned}\operatorname{Cov}(\bar{y}_{h-1,h-1}, \hat{\bar{Y}}_h)&\\ &= \operatorname{Cov}(\bar{y}_{h-1,h-1}, \bar{y}_{hh}) - c_h \operatorname{Cov}(\bar{y}_{h-1,h-1}, \bar{y}_{h-1,h})\\ &\quad + c_h \operatorname{Cov}(\bar{y}_{h-1,h-1}, \hat{\bar{Y}}_{h-1})\\ &= 0 - 0 + c_h \operatorname{Cov}\{\bar{y}_{h-1,h-1}, \bar{y}_{h-1,h-1} + c_{h-1}(\hat{\bar{Y}}_{h-2} - \bar{y}_{h-2,h-1})\}\\ &= c_h(n^{-1}S^2 - c_{h-1}\rho n^{-1}S^2),\end{aligned}$$

for $h > 1$, which gives

$$c_h = \frac{\rho}{2 - c_{h-1}\rho}$$

for $h > 1$ ($c_1 = 0$); and by the Proposition

$$\begin{aligned}V(\hat{\bar{Y}}_h) &= \operatorname{Cov}(\bar{y}_{hh}, \hat{\bar{Y}}_h)\\ &= \operatorname{Cov}(\bar{y}_{hh}, \bar{y}_{hh} - c_h\bar{y}_{h-1,h})\\ &= n^{-1}(1 - \rho c_h)S^2.\end{aligned}$$

Now check whether all other conditions for minimality of the variances are fulfilled, namely, the equality for $j = 1, \ldots$ of

$$\operatorname{Cov}(\bar{y}_{h-j-1,h-j-1}, \hat{\bar{Y}}_h)$$

and

$$\operatorname{Cov}(\bar{y}_{h-j-1,h-j}, \hat{\bar{Y}}_h).$$

By repeated substitution, the former is

$$\begin{aligned}\operatorname{Cov}\{\bar{y}_{h-j-1,h-j-1}, c_hc_{h-1}\ldots c_{h-j}(\bar{y}_{h-j-1,h-j-1} - c_{h-j-1}\bar{y}_{h-j-2,h-j-1})\}&\\ = c_hc_{h-1}\ldots c_{h-j}(1 - c_{h-j-1}\rho)S^2/n,&\end{aligned}$$

and the latter

$$\begin{aligned}\operatorname{Cov}\{\bar{y}_{h-j-1,h-j}, c_h\ldots c_{h-j+1}(\bar{y}_{h-j,h-j} - c_{h-j}\bar{y}_{h-j-1,h-j})\}&\\ = c_h\ldots c_{h-j+1}(\rho - c_{h-j})S^2/n.&\end{aligned}$$

But the relations derived previously imply that the two right-hand sides are equal. So $\hat{\bar{Y}}_h$ is, indeed, the required minimum variance unbiased linear estimator of $\bar{Y}_h$.

Eckler notes that there is a formal connection between the estimator for the two-level rotation and that for the one-level rotation with $\alpha = \frac{1}{2}$ and $n = \frac{1}{2}n'$. In the latter case we have, on occasion h, an extra n observations, and so the estimator for one-level rotation is a weighted average of the two-level estimator and $\bar{y}_{hf}$. From simple direct considerations Eckler finds the weight of the two-level estimator to be $(2 - c_i\rho)^{-1}$.

Eckler also discusses estimation of differences. The minimum variance linear unbiased estimator of $\bar{Y}_{h-k}$ on occasion h turns out to differ from $\hat{\bar{Y}}_{h-k}$ by a multiple of $\hat{\bar{Y}}_k$, the multiple being the ratio of the variance of $\hat{\bar{Y}}_{h-k}$ to the sum of the variances of $\hat{\bar{Y}}_{h-k}$ and $\hat{\bar{Y}}_k$. (Evidently this result does not hold for $h - k \leq k$. Thus by direct calculation $\frac{1}{2}(\bar{y}_{11} + \bar{y}_{12})$ is found to be the unbiased linear estimator for $\bar{Y}_1$ on occasion 2 of minimum variance.) This gives us the minimum variance unbiased linear estimator of $\bar{Y}_h - \bar{Y}_{h-k}$:

$$\hat{\bar{Y}}_h - \hat{\bar{Y}}_{h-k} + \frac{1 - c_{h-k}\rho}{2 - c_{h-k}\rho - c_k\rho}(\hat{\bar{Y}}_{h-k} - \hat{\bar{Y}}_k) \quad (k < \tfrac{1}{2}h).$$

Moreover, the unbiased linear estimator on occasion h of $\bar{Y}_{h-k}$ of minimum variance with two-level rotation sampling equals the unbiased linear estimator on occasion $h - k$ of $\bar{Y}_{h-k}$ of minimum variance with one-level rotation sampling when $n = \frac{1}{2}n'$ $(k \geq 1)$.

Finally Eckler discusses higher level rotation sampling; the formulas rapidly get more complicated. One may also consider a combination of partial replacement of units and two more levels.

In the monthly Retail Trade Survey,† conducted by the U.S. Census, $\bar{Y}_h$ has been estimated by a weighted average of $\bar{y}_{hh}$ and $(\bar{y}_{hh}/\bar{y}_{h-1,h})\hat{\hat{\bar{Y}}}_{h-1}$:

$$\hat{\hat{\bar{Y}}}_h = (1 - c_h)\bar{y}_{hh} + c_h(\bar{y}_{hh}/\bar{y}_{h-1,h})\hat{\hat{\bar{Y}}}_{h-1}.$$

(Actually in that Survey instead of c_h a constant c is used which approximately minimizes the variance of the estimator of $\bar{Y}_h$ for large h.)

† In that survey an important concern is that a number of stores are established and go out of business within a very short period. Annual censuses or surveys do not therefore give a positive probability of inclusion to some of these stores. In addition, there is the advantage of dividing the work of surveying over the year by observing each month only a twelfth of the population to be included in the sample; and the advantage of getting monthly figures and (by returning to the same stores each 12 months) getting a valid estimator of annual trend, free from seasonals. Rotation sampling also offers a way of handling unusually large observations that may reduce variances considerably. A discussion of the procedures of the Retail Trade Survey is found in the book by Hansen, Hurwitz and Madow, and in Woodruff, J. Amer. Statist. Ass., **58** (1963) 454–467.

By writing $\hat{Y}_h$ in the form

$$\hat{Y}_h = (1 - c_h)\bar{y}_{hh} + c_h\{\bar{y}_{hh} - \bar{y}_{h-1,h} + \hat{Y}_{h-1}\},$$

we see that the c_h found above furnish reasonably good weights: if $\bar{Y}_h$ and $\bar{Y}_{h-1}$ differ little, and the biases of the $\hat{\hat{Y}}_h$ are small, then it follows by the approximation methods of Chapter IIA that both estimators have about the same variance. A similar approach does not yield a reasonable estimator of $\bar{Y}_h - \bar{Y}_{h-1}$. Instead one may revise the estimator for $\bar{Y}_{h-1}$ on occasion h to

$$k_h \hat{\hat{Y}}_{h-1} + (1 - k_h)\bar{y}_{h-1,h},$$

with k_h chosen to minimize the (approximate) variance.

Note that for $c_h = 1$ for all h, we have the chained index

$$\hat{\hat{Y}}_h = \bar{y}_{hh} \prod_{j<h} (\bar{y}_{jj}/\bar{y}_{j,j+1})$$

(for identical units it would be $\bar{y}_{hh}$). As is known, a chained index tends to drift from the correct level as h increases due to accumulation of biases and to increase in variance with h. By taking $c_h < 1$, this drift is partially prevented.

6. Estimating characteristics of a population that is not fixed

In this section we make some brief remarks on estimation of a characteristic $\mathscr{Y}$ (other than size) of a population that is not fixed, as of one or more points of time, by a model-free approach.

First suppose that we have a listing or frame of all the N_1 units of the population at the base date (1) and know Y_1, the total value at that date. We wish to estimate Y_2 or $\bar{Y}_2 = Y_2/N_2$, the total or average value at date 2, and perhaps compare total or average values at the two dates. Denote the units common to both populations by $\mathscr{P}_c$, the other units of the population at the second date by $\mathscr{P}_b$, the other units of the population at the first date by $\mathscr{P}_d$ (b and d stand for born and died, but the dates need not be in chronological order, and there may be changes of all sorts, such of migration of inhabitants; promotion, resignation, engagement or transfer of personnel; change in tax-exempt status of taxpayers; etc.). Then $\bar{Y}_{1c}$ will, e.g., denote the average $\mathscr{Y}_1$-value for $\mathscr{P}_c$, and $\bar{y}_{1c}$ the corresponding sample value.

One way to estimate $\bar{Y}_2$ may be as follows. We take a simple random sample of units from the list, and find from it $\bar{y}_{1c}$, $\bar{y}_{2c}$ and $\bar{y}_{1d}$. $\bar{Y}_{1c}$ may be estimated by $\bar{y}_{1c}$, but if N_d is relatively small, a more efficient estimator may be

$$\hat{Y}_{1c} = (Y_1 - N_d\bar{y}_{1d})/(N_1 - N_d),$$

where N_d may be known or estimated (e.g. as N_1 times the fraction in the sample of those belonging to $\mathscr{P}_d$). Then $\bar{Y}_{2c}$ may be estimated by $\hat{Y}_{1c}\bar{y}_{2c}/\bar{y}_{1c}$.

(The variance of this expression differs little from $\bar{Y}_{1c}^2 V(\bar{y}_{2c}/\bar{y}_{1c})$ if N_d is relatively small; in general it is the variance of a product-cum-ratio estimator.) If $\bar{Y}_{2b}$ is estimated on the basis of a separate sample by $\hat{\bar{Y}}_{2b}$, the following estimator results for $\bar{Y}_2$:

$$\hat{\bar{Y}}_2 = (N_b \hat{\bar{Y}}_{2b} + \hat{\bar{Y}}_{1c} \bar{y}_{2c}/\bar{y}_{1c})/(N_1 - N_d + N_b).$$

If there are relatively few in $\mathscr{P}_d$ and $\mathscr{P}_b$, this may be an efficient estimator of $\bar{Y}_2$. Otherwise, and especially if $\hat{\bar{Y}}_{2b}$ must be based on a sample from the entire population at a later date, it may be more efficient to use that sample to estimate $\bar{Y}_2$ by $\bar{y}_2$, or by the above formula with $\hat{\bar{Y}}_{1c}$ based on a separate sample from $\mathscr{P}_d$ or the entire list. Especially if N_d is not relatively small, it may, however, be more efficient to estimate the Y_{2c} component on the basis of a combination of samples from the population at both dates in the manner of section 2, but taking into account the necessity of estimating as well the Y_{2b} component and the greater variety of cost factors involved. It should also be considered in this that besides $\bar{Y}_2$ one may be interested in estimating as well $Y_2 - Y_1$, $\bar{Y}_{2c}/\bar{Y}_{1c}$ (index number), etc. Moreover, sampling need not be simple random sampling; and $\bar{y}_{1c}$, $\bar{y}_{2b}$, etc. may be replaced by other estimators of their expected values.

When $\bar{Y}_1$ is not known, we may estimate the Y_{1c} and Y_{2c} components like in section 2 with appropriate modifications as mentioned above. One can also work out methods for several periods.

7. Appendix: Some results on minimum variance unbiased estimators

7.1. A GENERAL RESULT

Let $\mathscr{C}$ be a class of statistics on the space of samples from a population with a real parameter θ, and let $\mathscr{C}_\theta$ be the subclass of $\mathscr{C}$ for which, if $\mathscr{T}$ belongs to $\mathscr{C}_\theta$, $\mathscr{T}$ is unbiased for θ and of finite variance.† Let $\mathscr{N}$ be a class‡ of statistics (on the same sample space) with finite variances and expectation 0, such that,§ if $\mathscr{T}$ belongs to $\mathscr{C}$ and $\mathscr{Z}$ to $\mathscr{N}$, $\mathscr{T} + \lambda\mathscr{Z}$ belongs to $\mathscr{C}$ for all λ in some open interval containing 0.

Define $\mathscr{C}_\theta^0$ as the set of statistics $\mathscr{S}$ in $\mathscr{C}_\theta$ for which $V(\mathscr{S}) \leq V(\mathscr{T})$ for all $\mathscr{T}$ in $\mathscr{C}_\theta$.

† In an infinite population it is possible for a variance not to be finite.

‡ $\mathscr{N}$ for null, $\mathscr{Z}$ for zero.

§ The latter condition is automatically fulfilled if $\mathscr{C}$ is the class of *all* statistics on the sample space.

THEOREM.† $\mathscr{S}$ is in $\mathscr{C}_\theta^0$ if and only if $\mathscr{S}$ is in $\mathscr{C}_\theta$ and

$$\mathscr{E}(\mathscr{S}\mathscr{Z}) = 0$$

for all $\mathscr{Z}$ in $\mathscr{N}$. In particular, if $\mathscr{S}$ is in $\mathscr{C}_\theta^0$,

$$\operatorname{Cov}(\mathscr{S}, \mathscr{T}) = V(\mathscr{S})$$

for all $\mathscr{T}$ in $\mathscr{C}_\theta$. Also if $\mathscr{S}$ and $\mathscr{T}$ are in $\mathscr{C}_\theta^0$, $V(\mathscr{T} - \mathscr{S}) = 0$.

Proof of necessity. The hypothesis implies that the following expression is nonnegative for λ sufficiently small:

$$V(\mathscr{S} + \lambda\mathscr{Z}) - V(\mathscr{S}) = \lambda^2 V(\mathscr{S}) + 2\lambda\mathscr{E}(\mathscr{S}\mathscr{Z}).$$

It follows that

$$2\lambda\, \mathscr{E}(\mathscr{S}\mathscr{Z}) \geq -\lambda^2 V(\mathscr{Z}).$$

Then, for any sufficiently small positive λ,

$$\mathscr{E}(\mathscr{S}\mathscr{Z}) \geq -\tfrac{1}{2}\lambda V(\mathscr{Z}),$$

and so, for any $\mathscr{Z}$ in $\mathscr{N}$, $\mathscr{E}(\mathscr{S}\mathscr{Z})$ does not fall below any negative number (with arbitrarily small absolute value). Therefore $\mathscr{E}(\mathscr{S}\mathscr{Z}) \geq 0$ for any $\mathscr{Z}$ in $\mathscr{N}$. Consider now any sufficiently small $-\lambda > 0$, then $\mathscr{E}(\mathscr{S}\mathscr{Z}) \leq -\tfrac{1}{2}\lambda V(\mathscr{Z})$ and so $\mathscr{E}(\mathscr{S}\mathscr{Z})$ does not exceed any arbitrarily small positive number. So $\mathscr{E}(\mathscr{S}\mathscr{Z}) \leq 0$ for any $\mathscr{Z}$ in $\mathscr{N}$. Together the inequalities prove that $\mathscr{E}(\mathscr{S}\mathscr{Z}) = 0$ for any $\mathscr{Z}$ in $\mathscr{N}$.

Proof of sufficiency. By the hypothesis, the last term in the following identity for arbitrary $\mathscr{S}$ and $\mathscr{T}$ in $\mathscr{C}_\theta$ vanishes:

$$V(\mathscr{T}) = V(\mathscr{S}) + V(\mathscr{T} - \mathscr{S}) + 2\mathscr{E}\{\mathscr{S}(\mathscr{T} - \mathscr{S})\},$$

and so $V(\mathscr{T}) \geq V(\mathscr{S})$.

Proof of the last part. If $\mathscr{S}$ and $\mathscr{T}$ belong to $\mathscr{C}_\theta^0$, the last equation becomes

$$V(\mathscr{S}) = V(\mathscr{S}) + V(\mathscr{T} - \mathscr{S}),$$

as then $V(\mathscr{T}) = V(\mathscr{S})$, so $V(\mathscr{T} - \mathscr{S}) = 0$. Also

$$\mathscr{E}\{\mathscr{S}(\mathscr{T} - \mathscr{S})\} = \operatorname{Cov}(\mathscr{S}, \mathscr{T}) - V(\mathscr{S}),$$

so, if $\mathscr{E}\{\mathscr{S}(\mathscr{T} - \mathscr{S})\} = 0$,

$$\operatorname{Cov}(\mathscr{S}, \mathscr{T}) = V(\mathscr{S}).$$

† Adapted from LEHMANN and SCHEFFÉ, Sankhya, **10** (1950) 305–340. The ideas go back to FISHER, Proc. Cambridge Phil. Soc., **22** (1925) 700–725. Note that in general $\mathscr{C}_\theta^0$ may be empty even if $\mathscr{C}_\theta$ is not. Also, if $\mathscr{C}_\theta$ is not empty, $\mathscr{N}$ is not empty, since, if $\mathscr{T}$ is in $\mathscr{C}_\theta$, $\mathscr{T} - \mathscr{T}$ is in $\mathscr{N}$.

COROLLARY. Let $\mathscr{C}$ be a class of statistics such that, if $\mathscr{T}$ belongs to $\mathscr{C}_\theta$, and if $\mathscr{T}_1$ and $\mathscr{T}_2$, with $\mathscr{E}\mathscr{T}_1 = \mathscr{E}\mathscr{T}_2$ and with finite variances, belong to $\mathscr{C}$, then $\mathscr{T} + \lambda(\mathscr{T}_1 - \mathscr{T}_2)$ belongs to $\mathscr{C}$ for all λ in an open interval containing 0. Then $\mathscr{S}$ is in $\mathscr{C}_\theta^0$, if and only if $\mathscr{S}$ is in $\mathscr{C}_\theta$ and

$$\text{Cov}\,(\mathscr{S}, \mathscr{T}_1) = \text{Cov}\,(\mathscr{S}, \mathscr{T}_2)$$

for all $\mathscr{T}_1$ and $\mathscr{T}_2$ in $\mathscr{C}$ with the same expected value and finite variances. In particular, if $\mathscr{S}$ is in $\mathscr{C}_\theta^0$,

$$\text{Cov}\,(\mathscr{S}, \mathscr{T}) = \text{Var}\,(\mathscr{S})$$

for all $\mathscr{T}$ in $\mathscr{C}_\theta$. Also if $\mathscr{S}$ and $\mathscr{T}$ are in $\mathscr{C}_\theta^0$,

$$V(\mathscr{T} - \mathscr{S}) = 0.$$

7.2. APPLICATION TO CLASSES OF LINEAR ESTIMATORS

The class $\mathscr{C}$ of all linear estimators on the sample space satisfies the condition mentioned in the Corollary for all λ.

Note that if $\mathscr{C}_\theta^0$ is not empty, it has only one element, since here

$$V(\mathscr{T} - \mathscr{S}) = 0$$

implies $\mathscr{T} = \mathscr{S}$.

To derive the proposition of section 4, we apply the Corollary separately for the estimation of $\bar{Y}_1, \bar{Y}_2, \ldots$.†

7.3. REMARKS

The Proposition of section 4 and its application are due to PATTERSON, J. Roy. Statist. Soc., **B12** (1950) 241–255. He proves the necessary conditions for the linear case by differentiation along the following lines: If we seek estimators of the form $\Sigma\Sigma a_{il}(h)y_{il}$, unbiasedness implies that

$$\sum_l a_{il}(h) = \delta_{ih},$$

where $\delta_{hi} = 1$ if $i = h$, and 0 otherwise. Then setting the derivatives of the variance minus

$$2\sum_i \lambda_{ih} \sum_l a_{il}(h)$$

equal to zero (where the λ_{ih} are Lagrange multipliers), we get

$$\text{Cov}\,(y_{il}, Q_h) = \lambda_{ih},$$

† There are two ways to satisfy the requirement that for estimating $\bar{Y}_h$ we take into account only the observations up to and including occasion h. One is to restrict $\mathscr{C}$ to estimators based on such observations only – $\mathscr{C}$ then evidently satisfies the restrictions of the Corollary; another is to deal with an expanding sequence of sample spaces, such that, for $k > 0$, the sample space for estimating $\bar{Y}_{h+k}$ contains more elements than the sample space for estimating $\bar{Y}_h$.

which may be written as

$$\sum_{i'l'}\sum \text{Cov}\,(y_{il}, y_{i'l'})\, a_{i'l'}(h) = \lambda_{ih},$$

or in the matrix form

$$Fa(h) = \lambda(h),$$

where we write the transpose of the row

$$a_{11}(h), a_{12}(h), \ldots; a_{21}(h), \ldots$$

as $a(h)$, write the covariance matrix F conformably, and write the transpose of the row

$$\lambda_{1h}, \lambda_{1h}, \ldots; \lambda_{2h}, \lambda_{2h}, \ldots; \ldots$$

as $\lambda(h)$. Then, if we write the elements of G, the inverse of F, conformably to the matrix of the Cov $(y_{il}, y_{i'l'})$, we have

$$a_{il}(h) = \sum_s\sum_t g_{i,l;s,t}\lambda_{sh},$$

and the other conditions on the $a_{il}(h)$ become

$$\sum_s(\sum_l\sum_t g_{i,l;s,t})\lambda_{sh} = \delta_{ih}.$$

If the expression in parentheses has an inverse, these equations may be solved for the λ_{sh}, and, in fact, the matrix of the λ_{sh} equals this inverse. We can then substitute these solutions in the formulas for the $a_{il}(h)$.

Patterson implicitly assumes the inverse to exist. He remarks that the conditions given are also sufficient, since if also the

$$Q_h' = \Sigma\Sigma b_{il}(h) y_{il}$$

satisfy

$$\text{Cov}\,(y_{il}, Q_h') = \mu_{ih}$$

for some set of numbers μ_{ih}, we get

$$Fb(h) = \mu(h),$$

together with the unbiasedness conditions for the b's; and so (the equations for the μ's and b's being identical with those for the λ's and a's), we find the same solutions for $\mu(h)$ and $b(h)$ as for $\lambda(h)$ and $a(h)$.

Even given the existence of the inverse, when using this method one has to verify whether the solution obtained by differentiation is actually a (conditional) minimum.

In our case, we deal with the submatrix $F(h)$ instead of F, since the $a_{il}(h)$ are restricted to zero for $i > h$.

CHAPTER XII

SOME REMARKS ON ANALYTIC USES OF SURVEYS

1. Introduction

The majority of surveys have been *designed* as descriptive surveys, even when intended to be used analytically. The question of efficient design of an analytic survey has only very recently come under investigation, namely in Sedransk's doctoral dissertation of 1964, followed by a number of his publications (the latest to hand being in J. Amer. Statist. Ass., **62** (1967) 1121–1139).

We shall therefore confine ourselves to the study of analytic ways of *using* survey results. Even so, the scope of our treatment of this important subject will have to be severely limited, since the subject has received very little attention in the literature and since, moreover, it involves principles whose study is not needed for the other topics of this book. In particular, we shall have to confine ourselves to the discussion of *simple subtractive comparisons of two populations or subpopulations*. This may serve to bring out a few of the main new points that come up in analytic uses of surveys. More in general, many analytic uses of survey results are special cases of regression analysis of survey data, for which we may refer to Konijn, J. Amer. Statist. Ass., **57** (1962) 590–606, and Morgan and Sonquist, J. Amer. Statist. Ass., **58** (1963) 415–434.†‡

2. Standardization and models

In section 3 of Chapter I, Example 9, we discussed comparison of the average stature of men and that of women in a given population. Evidently, such a comparison must be affected by the age distribution of the men and of the women, since differences in stature are not the same at every age z. If we are interested in the average difference for the given population, we should weight the *age-specific* differences, $\bar{Y}_{mz} - \bar{Y}_{fz}$, by the relative sizes of the different

† See also the third footnote in section 8 of Chapter VIII.

‡ In their formula (7), $-N\bar{X}^2$ is omitted in the numerator.

age groups. However, these relative sizes, $w_{\mathrm{m}}(z)$ and $w_{\mathrm{f}}(z)$, are not usually the same for the males and for the females, so that a weighted difference $\bar{Y}^*_{\mathrm{m}} - \bar{Y}^*_{\mathrm{f}}$ of the form

$$\sum_z w(z)(\bar{Y}_{\mathrm{m}z} - \bar{Y}_{\mathrm{f}z}) = \sum_z w(z)\,\bar{Y}_{\mathrm{m}z} - \sum_z w(z)\,\bar{Y}_{\mathrm{f}z}$$

would not in general be equal to

$$\bar{Y}_{\mathrm{m}} - \bar{Y}_{\mathrm{f}} = \sum_z w_{\mathrm{m}}(z)\,\bar{Y}_{\mathrm{m}z} - \sum_z w_{\mathrm{f}}(z)\,\bar{Y}_{\mathrm{f}z}.$$

We say that the former formula *standardizes* the stature comparison of males and females with regard to their age distribution, and call the particular function w used, the standardizing distribution. The two components of the standardized sex difference in stature are called the standardized average male stature and the standardized average female stature, respectively.† One possible choice of w may be the age distribution of the entire population. Such a choice would, in a sense,‡ *eliminate* the effect of the difference between the two age distributions from the stature comparison. It is not, however, always the most appropriate choice of standardizing distribution for this purpose.

It may be, e.g., that the age distribution of males should be considered distinctly atypical, due to certain epidemics at an earlier time, so that we had rather use w_{f} or some *adjusted* w as standardizing distribution. Or it may be that the age distribution of this population as a whole is not representative of the population to which we wish to apply the result, due to postponement of marriages in a preceding war or depression, and that we may therefore need to *adjust* w to a more "normal" one.

We see, therefore, that the choice of standardizing distribution depends on the *use* we plan to make of the figure to be arrived at, and that, moreover, this may involve formulating a *model* which will imply a standardizing distribution. (In practice one encounters relatively infrequently a case in which such a

† Standardized averages can also be used to compare more than two populations; instead of differences, one then often uses ratios like $I(\mathrm{m}|*) = \bar{Y}^*_{\mathrm{m}}/\bar{Y}_*$, called *index numbers*. For the purpose of standardizing, instead of the $w(z) = w_*(z)$ – the relative number in the standard population for each age –, we may use the $\bar{Y}_*(z)$ – the average stature in the standard population for each age. Instead of $I(\mathrm{m}|*) = \Sigma w(z)\,\bar{Y}_{\mathrm{m}z}/\Sigma w(z)\,\bar{Y}_{*z} = \bar{Y}_{\mathrm{m}}{}^*/\bar{Y}_*$, this yields $J(\mathrm{m}|*) = \Sigma w_{\mathrm{m}}(z)\,\bar{Y}_{\mathrm{m}z}/\Sigma w_{\mathrm{m}}(z)\,\bar{Y}_{*z} = \bar{Y}_{\mathrm{m}}/\bar{Y}^{\mathrm{m}}_*$; and, instead of $\bar{Y}_{\mathrm{m}}{}^* = I(\mathrm{m}|*)\,\bar{Y}_*$, $J(\mathrm{m}|*)\,\bar{Y}_*$. Such standardization is called *indirect standardization*. Its main advantage is that it does not require knowledge of the *age-specific* mean male statures. (If estimates are to be found for them from a sample in which the age groups constitute domains, their variances will be much inflated for those (m, z)-cells which with an appreciable probability will only yield a small number of sample observations.)

‡ Note that it is implicitly assumed that the quantities $\bar{Y}_{\mathrm{m}z} - \bar{Y}_{\mathrm{f}z}$ would not be different in the hypothetical case, in which the age distributions for both sexes would be that of the standardizing distribution, from what they are in the given population.

model is formulated *explicitly*, and even more rarely a case in which the model is sufficiently detailed so as to imply a unique standardizing distribution.)

As another example, consider the case in which m and f refer not to sex but to two subpopulations living under different circumstances. The purpose of the inquiry may be to find out what, if any, would be a $\mathscr{Y}$-difference if conditions for the group living under less favorable circumstances would be brought up to the level of the more favored group; then we would standardize on the latter's age distribution. There are, however, some objections to this procedure: the attempt to bring the former group's conditions up to the level of the latter's may affect conditions of both groups; and a change in conditions may itself affect the age distribution. Only by specifying a model governing the phenomena under consideration (which would, among others, describe how age and circumstances of life are related) can we arrive at the "right" standardization. Nonetheless, a vague specification may give us enough of a clue to what standardization should be approximately appropriate. And in many cases standardizing distributions that are not *very* different give very nearly the same numerical results; but the *sensitivity* of the results to the specification of the standardizing distribution should in each case be investigated.†

In all these cases we wish to estimate expressions of the form

$$\Sigma w(z)(\bar{Y}_{1z} - \bar{Y}_{2z}),$$

where the function w may be known or may itself have to be estimated from the sample.

There are circumstances under which weighting should *not* be done *by means of a distribution*. An important instance would be one in which we feel justified in supposing that $\bar{Y}_{1z} - \bar{Y}_{2z}$ is about the same for all z, and we

† In many analytic surveys the two populations are those in which a factor is present or absent, respectively, and characteristics like $\mathscr{Y}$ are observed in search of an explanation of that factor. For example, the factor may be a certain coronary disease and $\mathscr{Y}$ may measure the number of cigarettes smoked per day by an individual. In such surveys one may well wish to take as the second, contrasting, population, not all those who do not have this disease, but only the individuals among the latter who are comparable with the members of the first population in some relevant sense. Thus one would exclude from the second population young children, or even people not old enough to have developed the disease if they had started smoking at an early age. In general one would do a certain amount of *matching* of individuals on a number of characteristics ($\mathscr{Z}$) in order to define the appropriate second population, and this diminishes the sensitivity of the results of standardized comparisons to the choice of w. This sensitivity may be further diminished by stratification into as many appropriate $\mathscr{Z}$-classes as possible, and avoided altogether if individuals from the samples in the two populations are *paired* according to their $\mathscr{Z}$-values. The latter procedure is, however, often inadvisable, or even not feasible, since to many of the $\mathscr{Z}$-values represented in the sample from the first population there may be few or no individuals in the second population.

wish to estimate that common quantity. In that case, $\Sigma w(z)(\bar{Y}_{1z} - \bar{Y}_{2z})$ will be about equal to that same quantity, no matter what be the function w satisfying $\Sigma w(z) = 1$. In a sample, however, $\Sigma w(z)(\bar{y}_{1z} - \bar{y}_{2z})$ may be expected to be dependent on the choice of w, and so the question arises as to what w is best to use. Usually one will choose that w which minimizes the variance of $\Sigma w(z)(\bar{y}_{1z} - \bar{y}_{2z})$. For example, if the $\mathscr{Z}$-groups are sampled independently, the best choice of w would be the one in which $w(z)$ is inversely proportional to $V(\bar{y}_{1z} - \bar{y}_{2z})$, provided these variances are known (see the lemma of section 2 of Chapter XI). The same result holds if the $\mathscr{Z}$-groups are domains in simple random sampling, since, as is shown in section 14 of Chapter IIA, $\bar{y}_{1z}$ and $\bar{y}_{2z}$ are uncorrelated with $\bar{y}_{1z'}$ and $\bar{y}_{2z'}$ for $z \neq z'$.

If the model for comparisons implies sampling from an *infinite* superpopulation, the finite population multiplier should, of course, not be used.

3. Standardized comparisons

Consider the estimation of $\Sigma w(z)(\bar{Y}_{1z} - \bar{Y}_{2z})$ for given $w(z)$. Estimators and their biases and variances, as well as estimators of the latter, are easily obtained from the results given in the previous chapters when $\bar{Y}_{1z}$ and $\bar{Y}_{2z}$ are estimated from independent samples. In the chapter on repeated sampling the case of partially overlapping samples has been considered. If $\bar{Y}_{1z}$ and $\bar{Y}_{2z}$ are estimated for two nonoverlapping domains (numbered 1 and 2) from one and the same sample, with independent simple random samples for each z, the results of section 14 of Chapter IIA apply to $\bar{Y}_{1z} - \bar{Y}_{2z}$. From these one obtains at once the relevant results for $w(z)(\bar{Y}_{1z} - \bar{Y}_{2z})$, and so for the sum of these expressions.

Sometimes $\bar{Y}_{1z}$ and $\bar{Y}_{2z}$ are estimated for all z from one single simple random sample, subdivided into two domains, and further into subdomains according to the $\mathscr{Z}$-values. As mentioned above

$$\text{Cov}\,(\bar{y}_{1z} - \bar{y}_{2z}, \bar{y}_{1z'} - \bar{y}_{2z'})$$

for $z \neq z'$ is zero, and so the variance of $\Sigma w(z)(\bar{y}_{1z} - \bar{y}_{2z})$ equals

$$\Sigma w^2(z) V(\bar{y}_{1z} - \bar{y}_{2z}),$$

where the latter is found in section 14 of Chapter IIA.

The estimators of $\bar{Y}_1$, $\bar{Y}_2$, $\bar{Y}_{1z}$ and $\bar{Y}_{2z}$ may themselves be based on stratified samples. Then one has to decide whether or not the effect of the difference (if any) in $w^{(1)}$ and $w^{(2)}$ is to be eliminated, where $w^{(i)}$ is the $\mathscr{Z}$-value distribution of the units of $\mathscr{D}_i$. If not, the relevant theory is that of section 16 of Chapter III. Otherwise, we have to estimate expressions of the form

$$\Sigma w(z)\{Y'(z)/X'(z) - Y''(z)/X''(z)\}.$$

Here

$$Y'(z) = \sum_h Y'_h(z),$$

where $Y'_h(z)$ is the total $\mathscr{Y}$-value for units both in $\mathscr{S}_h$ and $\mathscr{D}_{1z}$. If we estimate $Y'_h(z)$ by $N_h \bar{y}'_h(z)$ and $X'_h(z)$ by $N_h \bar{x}'_h(z)$, our estimator for $Y'(z)/X'(z)$ is of the form $\hat{Y}_{R1}$ of section 16 of Chapter III with 1 replaced by $(1, z)$. From that section we find approximations to the variance of one such ratio and to the covariance of the ratio for two domains such as $\mathscr{D}_{1z}$ and $\mathscr{D}_{2z}$, or $\mathscr{D}_{1z}$ and $\mathscr{D}_{1z'}$ $(z \neq z')$, and so can find approximations to the variance of our estimator and an estimator of this variance, provided the biases are not too large in the separate domains.

One may also consider cluster sampling (see section 14 of Chapter VII).

4. Avoiding unnecessary distinctions

We remarked in section 2 that sometimes $\bar{Y}_{1z} - \bar{Y}_{2z}$ is about the same for all z, and that this simplifies greatly the question of the choice of w. Since there may be many characteristics whose effects, if any, ought to be eliminated from the comparison under study, it may be very much worthwhile to sort out characteristics that have no appreciable effect. We shall therefore give some attention to ways of *testing* the (mathematically over-precise†) hypothesis that $\bar{Y}_{1z} - \bar{Y}_{2z}$ is the same for all z, or for a subset of z's.

In the example of average stature of male and female students, the students may have been classified by their age and by the area in which they were brought up (the latter because dietary habits of children vary much by area and appear to affect stature later in life). Some of the possible age-area cells may not be distinguished, because they would be represented by only a small number of observations and thus would greatly inflate the variance of the estimated comparison.‡ Others, even when observed in adequate numbers, may be combined, because for their constituents the differences $\bar{Y}_{1z} - \bar{Y}_{2z}$ are judged to be nearly the same.

A similar question arises as to whether we can make the stronger supposition that both $\bar{Y}_{1z}$ and $\bar{Y}_{2z}$ may be about the same for certain z.

† Compare Chapter VIII, section 3 sub (d).

‡ Estimated averages, totals or ratios for small subpopulations reported in statistical publications have often misled users interested in one of these subpopulations who were not aware of their substantial standard errors.

We shall here examine the first hypothesis only. In the case in which this concerns a subset $\{z', z''\}$, of 2 z's only, the hypothesis takes the form $\Delta = 0$, where

$$\Delta = \Delta(z', z'') = \bar{Y}_{1z'} - \bar{Y}_{2z'} - (\bar{Y}_{1z''} - \bar{Y}_{2z''}).$$

We may easily obtain a test of this hypothesis directly, or, as we shall do in the following, by using the confidence interval we gave for Δ. For a confidence interval for a parameter is defined to have the property that, no matter what be the value of that parameter, the probability of the interval covering this value is (at least) equal to the confidence coefficient β (which we generally choose close to 1). So the probability of the confidence interval for Δ not covering the value $\Delta = 0$ is (at most) $1 - \beta$, a small number. If, therefore, we adopt as a test of $\Delta = 0$ the procedure which rejects the hypothesis $\Delta = 0$ whenever the value 0 lies outside our confidence interval, we shall, if Δ is actually 0, erroneously reject our hypothesis with a probability of only $1 - \beta$ (at most). Thus, from the formula we employed for the confidence interval for Δ (with the use of the normal approximation), the resulting test of $\Delta = 0$ rejects this hypothesis if $|d|$ exceeds $t\{v(d)\}^{\frac{1}{2}}$, where

$$d = \bar{y}_{1z'} - \bar{y}_{2z'} - (\bar{y}_{1z''} - \bar{y}_{2z''}),$$

and where $t = 1.96$ for the case $\beta = 0.95$. Rejection means that the apparent deviation of the observed value of d from the hypothesis value 0 is *significant* in the sense that it is unlikely to have come about by random fluctuations only. However, even if we reject the hypothesis that $\Delta = 0$, the confidence interval obtained for Δ may indicate that Δ is of only minor importance.

Usually the number, k, of $\mathscr{Z}$-values to be compared *exceeds two*. Denote $\bar{Y}_{1z} - \bar{Y}_{2z}$ by A_z and $\bar{y}_{1z} - \bar{y}_{2z}$ by $\hat{A}_z$. We shall test the hypothesis that the A_z are equal for all z of the set or subset considered (to some unspecified number A). We shall only discuss the case in which, for any $z \neq z'$, $\hat{A}_z$ and $\hat{A}_{z'}$ are uncorrelated (which, as we saw, holds in the case of simple random sampling).

For known $V_z = V(\hat{A}_z)$, we know from the lemma of Chapter XI, section 2, that, among all unbiased estimators of A that are linear combinations of the $\hat{A}_z$,

$$\hat{A} = \Sigma V_z^{-1}\hat{A}_z/\Sigma V_z^{-1}$$

has smallest variance. We show in the appendix that if the $\hat{A}_z$ are independent normal variables with means A_z and variances V_z and if the A_z are equal

$$C^2 = \Sigma(\hat{A}_z - \hat{A})^2/V_z$$

is distributed like χ^2_{k-1} (chi-square with $k - 1$ degrees of freedom), that is like the sum of squares of $k - 1$ independent standard normal variables. There are widely available tables of this distribution, and in particular tables of $\chi_f^2(\alpha)$, the upper α-point of χ_f^2 – the value of a chi-square variable with f degrees of freedom which is exceeded with probability α exactly. For example, $\chi_1^2(0.05) = 3.84 = (1.96)^2$, since the absolute value of a standard normal variable exceeds 1.96 with probability 0.05. We shall assume that for each $\mathscr{Z}$-value the number of observations is sufficiently large for the normal approximation to be usable, also when we replace everywhere V_z by its estimator. Then a test at level of significance α of the hypothesis of equality of the A_z, based on the (modified) C^2, rejects this hypothesis if $C^2 > \chi_{k-1}^2(\alpha)$. Note that for the case $k = 2$

$$(V_{z'}^{-1} + V_{z''}^{-1})(\hat{A}_{z'} - \hat{A}) = (V_{z'}^{-1} + V_{z''}^{-1})\hat{A}_{z'} - (V_{z'}^{-1}\hat{A}_{z'} + V_{z''}^{-1}\hat{A}_{z''})$$
$$= V_{z''}^{-1}(\hat{A}_{z'} - \hat{A}_{z''}),$$
$$(V_{z'}^{-1} + V_{z''}^{-1})(\hat{A}_{z''} - \hat{A}) = V_{z'}^{-1}(\hat{A}_{z''} - \hat{A}_{z'}),$$

so that

$$C^2 = (\hat{A}_{z'} - \hat{A}_{z''})^2/(V_{z'} + V_{z''}),$$

which is equal to $d^2/V(d)$, in accordance with the test for $k = 2$ given above.

Denote by Δ_z the deviation of A_z from $\Sigma V_z^{-1}A_z/\Sigma V_z^{-1}$ (for $k = 2$, $V = V_{z'} + V_{z''}$: $\Delta V^{-1} = \Delta_{z'}V_{z'}^{-1} - \Delta_{z''}V_{z''}^{-1}$). We shall suppose that the smallest deviations considered of some importance are those for which $\lambda = \Sigma\Delta_z^2/V_z$ takes on at least some specified positive value λ_0. In the appendix we give a test of the hypothesis that λ does not exceed λ_0, for which the level of significance is approximately equal to some specified α'.

Sometimes our model describes how $\mathscr{Z}$ relates to $\mathscr{Y}$. In that case we may compute *adjustments* for $\bar{y}_{1z} - \bar{y}_{2z}$ such that the adjusted values estimate a quantity which is the same for all z (of course, the variance formulas for these quantities will be also affected). We may test the adequacy of the adjustments along the lines indicated above (as a rule the postulated relation of $\mathscr{Z}$ to $\mathscr{Y}$ involves parameters whose value the model does not specify, but must be estimated from the data). We may also combine standardizing or stratifying with respect to a crude classification into broad classes of $\mathscr{Z}$-values with adjustments of the abovementioned kind within these classes. The result would not be as sensitive to an erroneous specification of the model used as when the model would be applied without standardizing or stratifying.

Finally, we remark that, when the $\bar{Y}_{1z}$ and $\bar{Y}_{2z}$ are relative frequencies, the tests of dependence on $\mathscr{Z}$ are discussed in most textbooks under the heading of contingency table tests. In certain cases it is preferable to use the Poisson approximation (see Chapter II, section 15.1).

APPENDIX. The distribution of C^2 for normal $\hat{A}_z$ and its use for testing. If we write

$$x_z = (\hat{A}_z - A_z)/V_z^{\frac{1}{2}}, \quad A = \Sigma V_z^{-1} A_z / \Sigma V_z^{-1}, \quad \Delta_z = A_z - A, \quad \delta_z = \Delta_z / V_z^{\frac{1}{2}},$$

then

$$\begin{aligned} C^2 &= \Sigma V_z^{-1} \{(\hat{A}_z - A_z) - (\hat{A} - A) + (A_z - A)\}^2 \\ &= \Sigma x_z^2 - 2(\hat{A} - A)^2 \Sigma V_z^{-1} + (\hat{A} - A)^2 \Sigma V_z^{-1} + 2\Sigma x_z \delta_z + \Sigma \delta_z^2. \end{aligned}$$

If

$$c_{1z} = V_z^{-\frac{1}{2}} / (\Sigma V_z^{-1})^{\frac{1}{2}},$$

we consider an orthogonal transformation of the k independent standard normal variables x_z to k independent standard normal variables x_z^* by means of a matrix which has the c_{1z} as elements of the first row. Then

$$\Sigma x_z^2 = \Sigma x_z^{*2} \quad \text{and} \quad (\hat{A} - A)^2 / \Sigma V_z^{-1} = x_1^{*2},$$

so

$$C^2 = \Sigma x^{*2}_z - x_1^{*2} + 2\Sigma x_z \delta_z + \Sigma \delta_z^2.$$

It follows that if the Δ_z are all 0,

$$C^2 = \sum_{z=2}^{k} x_z^{*2},$$

and so has the χ^2_{k-1} distribution.

The distribution of the sum of squares of f independent normal variables, each having unit variance, and having means whose sum of squares is λ is called $\chi'^2_{f,\lambda}$ (noncentral chi-square with f degrees of freedom and noncentrality λ). If we transform the δ_z to δ_z^* by the same transformation as that applied to the x_z, we obtain that in general

$$C^2 = \sum_{z=2}^{k} (x_z^{*2} - 2x_z^* \delta_z^* + \delta_z^{*2}) = \sum_{z=2}^{k} (x_z^* - \delta_z^*)^2,$$

where we have used the fact that

$$\delta_1^* = \Sigma c_{1z} \delta_z = 0.$$

This shows that in general C^2 has the $\chi'^2_{k-1,\lambda}$ distribution. A test of the hypothesis $\lambda \leq \lambda_0$ (of approximate level of significance α') mentioned previously may be carried out as follows, using HAYNAM, GOVINDARAJULU and LEONE,

Tables of the Cumulative Non-Central Chi-Square Distribution (Case Institute of Technology, 1962). On different pages tables are given for different "alpha"; we look for that "alpha" for which the "power" equals the desired level of significance of our test (α'); then the α'-point of χ'^2_{k-1,λ_0} is χ^2_{k-1} (alpha). For example, if $\lambda_0 = 1.00$, $k = 2$ and $\alpha' = 0.05$, we find

"alpha"	"power"
0.001	0.011 0
0.005	0.035 4
0.010	0.057 7

So "alpha" is about 0.008, and from the (ordinary) χ^2_1 table we find that $\chi^2_1(0.008)$ is about 6.9. Therefore we reject the hypothesis if C^2 exceeds 6.9.

AUTHOR INDEX

SUBJECT INDEX